Ulrich J. Krull, Chemical & Physical Sciences
Professor of Analytical Chemistry and
Astra Zeneca Chair in Biotechnology
University of Toronto at Mississauga
South Building, Room 2035
3359 Mississauga Road North
Mississauga, Ontario
Canada L5L 1C6

TRANSPORT PHENOMENA IN MICROGRAVITY

ANNALS OF THE NEW YORK ACADEMY OF SCIENCES

Volume 1027

TRANSPORT PHENOMENA IN MICROGRAVITY

Edited by S.S. Sadhal

The New York Academy of Sciences
New York, New York
2004

Library of Congress Cataloging-in-Publication Data

Microgravity Transport Processes in Fluid, Thermal, Materials, and
   Biological Sciences Conference (2003 : Davos, Switzerland)
   Transport phenomena in microgravity / edited by S.S. Sadhal.
      p. ; cm. — (Annals of the New York Academy of Sciences ; v. 1027)
   "This volume is the result of the Microgravity Transport Processes
in Fluid, Thermal, Materials, and Biological Sciences Conference,
held September 14–18, 2003, in Davos, Switzerland"—Contents p.
   Includes bibliographical references and index.
   ISBN 1-57331-563-X (cloth : alk. paper) — ISBN 1-57331-564-8 (pbk. : alk. paper)
   1. Reduced gravity environments—Physiological effect—Congresses.
2. Transport theory—Congresses. 3. Biological transport—Congresses.
4. Crystal growth—Effect of reduced gravity on—Congresses.
I. Sadhal, S.S. II. Title. III. Series.
   [DNLM: 1. Biological Transport—physiology. 2. Membrane Transport
Proteins——physiology. 3. Weightlessness. W1 AN626YL
v.1027 2004 / QH 509 M626 2004]
Q11.N5 vol. 1027
[QP82.2.G7]
500 s—dc22
[620'.419]

                                                            2004015855
                                                                 CIP

K-M Research/CCP
Printed in the United States of America
ISBN 1-57331-563-X (cloth)
ISBN 1-57331-564-8 (paper)
ISSN 0077-8923

ANNALS OF THE NEW YORK ACADEMY OF SCIENCES

Volume 1027
November 2004

TRANSPORT PHENOMENA
IN MICROGRAVITY

Editor
S.S. SADHAL

Associate Editors
N. CHAYEN, V.K. DHIR, H. OHTA, AND R.W. SMITH

This volume is the result of a conference entitled **Microgravity Transport Processes in Fluid, Thermal, Materials, and Biological Sciences**, held September 14–18, 2003, in Davos, Switzerland.

CONTENTS

Part II: Biotransport Phenomena

Part III: Crystal Growth and Materials Technology

Part IV: Boiling Phenomena

Part VII: Acoustic, Electrostatic, and Electromagnetic Levitation

Part VIII: Space Systems

Financial assistance was received from:

- ENGINEERING CONFERENCES INTERNATIONAL (ECI)
- NATIONAL AERONAUTICS AND SPACE ADMINISTRATION (NASA)
  —Grant NNG04GA06G
- UNITED STATES NATIONAL SCIENCE FOUNDATION (NSF)
  —Grant 0322207

Attendees of the Microgravity Transport Processes Conference, Davos, Switzerland, September 2003.

Preface

In a continuing series of successful interdisciplinary conferences, we assembled once again as a diverse group of microgravity scientists, focused on the fundamental area of *transport phenomena*. The purpose of this series of conferences is to promote the exchange of technical information and new ideas in *microgravity fluid, thermal, biological, and materials sciences*. The theme consists in furthering the established interdisciplinary aspects of *microgravity science and technology*, and addressing the cross-cutting issues therein. The participants in this interdisciplinary conference gained considerable knowledge and insight into new areas, and by sharing expertise, attempts were made at forging new frontiers in this emerging science. This volume is the result of the latest conference in the series, held in Davos, Switzerland, in 2003.

As a result of our interaction, we hope to continue to provide direction for microgravity research in transport processes that form a common basis for many investigations. Fluid, thermal and mass-transport aspects have gained considerable importance in the biological and the materials sciences, and this has raised the necessity for the sharing and complementing of expertise. In addition, with long-term manned space missions in the near future, technical problems encompassing several of these disciplines have been envisioned. This volume provides a focus on the interdisciplinary aspects of transport phenomena, and for the inception of new ideas under this common theme. It is the hope of the organizers that this conference series and the volumes that follow will keep on stimulating interdisciplinary activity and lead to new research programs that cross over the traditional boundaries of scientific expertise.

The conference organizing committee thanks Professor Frank Schmidt who, as technical liaison for Engineering Conferences International, has been very supportive of this conference series and has provided the necessary help to acquire endorsement from the ECI. In addition, various members of the scientific committee gave valuable input in putting the conference together. Thanks are also due to Eugene Trinh and Bradley Carpenter (NASA) as well as Richard Smith (U.S. National Science Foundation) for financial support from the respective U.S. government agencies. On behalf of all the conference participants, we thank Barbara Hickernell and Kevin Korpics at the Brooklyn Office of ECI for a great deal of organizational work.

The participants are thanked for putting together their technical papers, making their way to Davos for the event, and contributing toward the peer-review process.

Finally, we thank the editorial staff of the New York Academy of Sciences for their role in publishing this collection of scientific papers in the *Annals of the New York Academy of Sciences*.

SATWINDAR SINGH SADHAL (USA), Chair & Scientific Secretary
NAOMI CHAYEN, Co-Chair (U.K.)
VIJAY K. DHIR, Co-Chair (USA)
HARUHIKO OHTA, Co-Chair (Japan)
REGINALD W. SMITH, Co-Chair (Canada)

Ann. N.Y. Acad. Sci. 1027: xi (2004). ©2004 New York Academy of Sciences.
doi: 10.1196/annals.1324.000

Effects of Buoyancy-Driven Convection on Nucleation and Growth of Protein Crystals

CHRISTO N. NANEV,[a] ANITA PENKOVA,[a] AND NAOMI CHAYEN[b]

[a]Institute of Physical Chemistry, Bulgarian Academy of Sciences, Sofia, Bulgaria

[b]Biological Structure and Function Section, Division of Biomedical Science, Sir Alexander Fleming Building, Imperial College, London, United Kingdom

ABSTRACT: Protein crystallization has been studied in presence or absence of buoyancy-driven convection. Gravity-driven flow was created, or suppressed, in protein solutions by means of vertically directed density gradients that were caused by generating suitable temperature gradients. The presence of enhanced mixing was demonstrated directly by experiments with crustacyanin, a blue-colored protein, and other materials. Combined with the vertical tube position the enhanced convection has two main effects. First, it reduces the number of nucleated hen-egg-white lysozyme (HEWL) crystals, as compared with those in a horizontal capillary. By enabling better nutrition from the protein in the solution, convection results in growth of fewer larger HEWL crystals. Second, we observe that due to convection, trypsin crystals grow faster. Suppression of convection, achieved by decreasing solution density upward in the capillary, can to some extent mimic conditions of growth in microgravity. Thus, impurity supply, which may have a detrimental effect on crystal quality, was avoided.

KEYWORDS: protein crystals; hen-egg-white lysozyme (HEWL); porcine pancreatic trypsin; crustacyanin; nucleation; growth; buoyancy-driven convection; mimicking microgravity

INTRODUCTION

X-ray structure determination is the most powerful tool for revealing the fundamental relation between molecular structure and biological function of proteins in biosolutions. However, obtaining high-quality crystals suitable for X-ray diffraction remains the major-bottleneck to structure determination. A profound knowledge of the details of the crystallization process is indispensable in order to achieve high-quality crystals.

Several different approaches can be used to grow relatively large and high-quality protein crystals. As well as control over crystal nucleation,[1,2] in the work reported in this paper, we use (in a batch method) buoyancy-driven convection in an attempt to replenish the exhausted solution in the crystal vicinity, thus ensuring faster growth and larger protein crystals. Conversely, by suppressing the buoyancy-driven convection, microgravity conditions were to some extent mimicked in ground experiments.

Christo N. Nanev, Institute of Physical Chemistry, Bulgarian Academy of Sciences, 1113 Sofia, Bulgaria. Voice: 359-2 873 40 67; fax: 359-2 9712688.
nanev@ipchp.ipc.bas.bg

Ann. N.Y. Acad. Sci. 1027: 1–9 (2004). ©2004 New York Academy of Sciences.
doi: 10.1196/annals.1324.001

No matter how complex a crystal growth process is two main consecutive stages occur, namely mass transport to the crystal surface and incorporation of the elementary building blocks on it. (Simultaneously, during the second stage, latent heat is dissipated to the ambient solution.)

Nutrition is usually the rate-determining step under pure molecular diffusion. Forced agitation, which enhances solute delivery, is used in the practical growth of large and relatively homogeneous inorganic (e.g., optic and segnetoelectric) crystals. Convection replenishment was used long ago in the so-called hydrothermal synthesis of quartz crystals.[3,4] It is more effective than diffusion, since the material is transported from (almost) the entire solution volume. Slow growth under pure diffusion control leads to increased exposure times of (broader) interstep terraces, and hence, to enhanced impurity incorporation.[5] The need of fine-tuning the agitation velocity is evident, since due to the agitation impurities are carried as well. The excessive mass brought by intensive flow cannot be utilized if the increased impurities result in worse crystals. Experimental skill is, therefore, required to adjust the flow rate in order to obtain an acceptable ratio of impurities versus crystal building blocks that impinge on the crystal surface. Usually a gentle flow, created by means of slow solution stirring, has proved to be optimal. Under microgravity conditions buoyancy driven convection is avoided but Marangoni convection is always present, leading to solution agitation.

Buoyancy driven convection was evoked or suppressed in our work by putting a glass capillary tube with the protein solution in a temperature gradient, which creates a density gradient, directing it in a desired position with respect to the gravity vector.

To enhance buoyancy driven convection the heaviest solution, at 4°C, is placed at the top of a capillary tube that holds the protein solution, and the lightest solution at, 20°C, is placed at the bottom of the capillary. The heavier liquid flows down in this case and slowly mixes the protein solution. A reversed temperature gradient was used in an attempt to suppress buoyancy driven convection.

Temperature gradients have been used for growing protein crystals and other crystals.[6–8] Luft et al.[8] describe a gradient established by running hot and cold water through channels machined into opposite ends of an aluminum plate. Micropipettes containing the protein solution equilibrate thermally with the plate by surface contact. In this way, the authors[8] performed a blind search for optimal crystallization conditions of proteins. Recently, Mao et al.[9] used a high-tech experimental approach to create a temperature gradient.

Temperature gradients were used in the present work for two purposes. First, to create (or suppress) buoyancy driven convection, and second, to create a supersaturated gradient with proteins whose solubility is temperature dependent. The generated supersaturated gradient enabled us to find optimal crystallization conditions for these proteins.

EXPERIMENTAL

Experimental Procedure

The principle of creating a temperature gradient is very simple—heat is extracted from the one end of a special tube that contains the glass capillary. The temperature gradient was created with the chosen orientation, vertical or horizontal.

Glass capillary tubes were filled with the desired solutions. The capillary tubes were closed tightly with Teflon caps, inserted in the temperature gradient, and fixed there.

Nucleation of protein crystals was investigated by means of the so-called double-thermal-pulse technique.[10] Quenching created the higher supersaturation that was necessary for nucleation. By cooling (the entire glass capillary containing the solution) we enforced crystal nucleation only in that part of the capillary tube where the supersaturation exceeded that corresponding to the metastable zone. After a lapse equal to the nucleation time, the glass capillary tube was removed from the temperature gradient and replaced at ambient temperature, corresponding to the metastable zone, in order to grow visible sized protein crystals. These crystals were counted and their number related to the corresponding nucleation time.[10]

Investigation Strategy

The essential idea in this project is as follows.

1. To evoke motion in the solution contained in the capillary tube, we used buoyancy driven convection. Hence, the coldest (4°C) tube-end was put on top, the bottom being maintained at 20°C. In this way we created conditions favoring buoyancy driven convection; we created an unstable configuration of the density gradient. (As is well known, water density reaches its maximum at 4°C.) Due to gravity, the difference in the solution density induces gravity driven flow. It is initiated by horizontally directed density inhomogeneity due to cooling from the outside. Another factor that favors buoyancy is water viscosity. This is almost twice as low on the capillary bottom, at 20°C, compared with that at 4°C.

However, water at 4°C is only about 0.2% heavier than at 20°C, protein solutions being heavier than pure water. Thus, the convection driving force is rather small. Therefore, an extremely simple test was performed in order to prove the action of buoyancy driven convection in the case under consideration. We compared the replacement of a colored boundary in two (identical) glass capillary tubes, each 10-cm long. Both glass capillary tubes were filled with pure water. On the top of each we put small (1 μl) colored droplets containing crustacyanin, a blue-colored protein (or in other experiments coomassie blue, or blue ink). Contact was then made between the colored droplet and the water. Putting one tube in the temperature gradient with 4°C on top versus 20°C on the bottom, and the second tube acting as a control at 20°C, we measured the propagation distances of the colored solutions. Indeed, they depend on the diameter of the glass tube. For comparison purposes glass capillary tubes with different inside diameters, 0.55 mm and 1.8 mm, were used in the present investigation.

2. In an attempt to eliminate the cause of the buoyancy driven convection (thus mimicking microgravity conditions on the ground) we put the coldest (4°C)

capillary tube end, with the heaviest solution, on the bottom. (It was supported by the air bubble that results from the tight closure, the cap.) Since the upper layer of solution is at a higher temperature, it is lighter. Of course, Marangoni convection remains under both microgravity and Earth conditions. However, the more intensive the flow, the more impurities impinge on the crystal surface.

3. We started our investigations with nucleation of protein crystals. Capillary tubes with a smaller inside diameter (0.55 mm) were used to limit protein consumption. The idea was to use the nucleation as a fine tool in exploring the effects of buoyancy driven convection on protein crystallization, without disturbing the system. Since the nuclei are extremely small[1,2,10] they do not appreciably decrease the solute concentration, and do not create any appreciable local density change. Thus, they could not create any buoyancy driven convection (or so-called natural convection). In contrast (due to the appreciable solute consumption), the macroscopic size growing crystals often evoke (buoyancy created) convection plumbs. Such phenomena have been repeatedly observed by interferometry.[11,12]

During the present investigation the gradient tube was put in both vertical positions, namely with the coldest tube end at the top and at the bottom. Using the double (thermal)-pulse technique,[10] thereby separating nucleation and growth processes, and comparing with the results for those for the horizontal tube position, we tested the effect of the buoyancy on protein crystal nucleation. Provided that protein solubility versus temperature dependence is known, quantitative data on nucleus number versus supersaturation can be obtained (compare also with Refs. 1, 2, and 10).

Protein Solutions

Hen-egg-white lysozyme (HEWL) and porcine pancreatic trypsin were chosen for the crystallization investigations. Reliable solubility data for HEWL are available in the literature for temperatures higher than 5°C.[13–15] (Moreover, a systematic study of the nucleation rate was performed recently under exactly the same conditions as in this paper but in horizontal tube position.[10]) SEIKAGAKU, 6× crystallized lysozyme was used at 30 mg/ml protein in 50 mM sodium acetate, pH = 4.5. The precipitant solution contained 3 (w/v)% NaCl. Porcine pancreatic trypsin, the solubility of which is also temperature dependent[16] was obtained from Sigma and used at 15 mg/ml in 33% $(NH_4)_2SO_4$, tris-HCl-buffer, pH = 8.2. Since pouring $(NH_4)_2SO_4$ cause the solution to become turbid it was added in small portions.

RESULTS AND DISCUSSION

Microhydrodynamics

We observed that the (blue) color spread differently in the two (measuring and control) capillary tubes during the same time span. Longer distances were measured in the temperature gradient, 4°C on top (for both capillary diameters), as compared with the distances at room temperature, 20°C, despite the lower diffusivity at the lower temperatures. The difference increased as more time passed. The results for capillary tubes with an inside diameter of 0.55 mm are given in TABLES 1 and 2 for crustacyanin and TABLE 3 for coomassie blue. Since we observed that after 10 h

TABLE 1. Data for α-crustacyanin, 5 mg/ml

Duration (hours)	Vertical Tube 4°C on top (mm)	Reference Sample at room temperature, 20°C (mm)
2	7.1 ± 0.5	4.0 ± 0.5
6	13.0 ± 0.5	6.0 ± 0.5
10	15.5 ± 0.5	7.5 ± 0.5

NOTE: The table shows the distances (in mm) to which the blue color spreads in the temperature gradient and at room temperature, depending on the time span.

TABLE 2. Data for β-crustacyanin, 5 mg/ml

Duration (hours)	Vertical Tube 4°C on top (mm)	Reference Sample at room temperature, 20°C (mm)
2	8.1 ± 0.5	5.0 ± 0.5
6	15.0 ± 0.5	7.0 ± 0.5
10	18.5 ± 0.5	9.5 ± 0.5

NOTE: The table shows the distances (in mm) to which the blue color spreads in the temperature gradient and at room temperature, depending on the time span.

TABLE 3. Data for Coomassie

Duration (hours)	Vertical Tube 4°C on top (mm)	Reference Sample at room temperature, 20°C (mm)
2	4.5 ± 1.0	2.0 ± 1.0
3	6.5 ± 1.5	3.0 ± 1.5
4	8.2 ± 1.5	4.0 ± 1.5
6	9.0 ± 1.5	4.3 ± 1.5

NOTE: The table shows the distances (in mm) to which the blue color spreads in the temperature gradient and at room temperature, depending on the time span.

the colored substance dilutes substantially in the temperature gradient, longer times were not used. For longer times the colored boundary bleaches and its determination becomes uncertain. Similar results were also obtained with blue ink.

The reason that modest differences were measured may be that not only pure molecular diffusion is acting at 20°C; some flow may be initiated because the colored liquid is heavier. Although not large, the differences unambiguously show the presence of an additional transportation driving force under the temperature gradient; that is, buoyancy driven convection is taking place.

Experiments with wider glass capillary tubes (1.8 mm inside diameter) were very instructive. They showed that the colored boundary is replaced about five times faster, both for the temperature gradient and at room temperature. Moreover, the shape of the colored boundary is conic.

Cooling the solution from room temperature to 4°C at the beginning of the experiment makes the solution heavier and it starts to flow down. A narrow trickle should be formed along the capillary axis, where the viscous resistance is smallest. The conic form of the blue colored boundary proves this. It was already noted that toward the bottom of the capillary, the stream is facilitated due to the two-fold lower water viscosity there. Simultaneously, an upcoming fluid flow should be created nearer the capillary walls, despite the larger resistance there. This has to be more intensive in the wider capillary tubes. One can imagine that (after a sufficiently long time) the result is slow solution circulation, although we did not observe it due to the limited sensitivity of our eye. The presence of circulation explains why Poiseuille's law for the flow rate, which depends on the fourth power of the capillary radius, is not obeyed in the case under consideration.

Most important for the present work is the fact that the upcoming flow should replenish the crystal nutrition zone, contributing to faster crystal growth. Presumably, protein crystals, which are big enough and suitable for X-ray structure determination, can be grown in this way. This method of additional "feeding" should work both with proteins of normal and reverse temperature dependent solubility.

Results and Discussion on Protein Crystallization

Due to the small capillary cross-section (diameter 0.55 mm) and the relatively low driving force, the buoyancy driven convection causes a gentle slow motion of the solution. The visibly measured rates (in a 0.55-mm capillary) are roughly about 0.5 μm per sec. Therefore, we applied relatively long nucleation times of 2, 1, and 0.5 hours. The result for the nucleation of HEWL crystals in both scenarios of the vertical tube (averaged over three to eight separate experiments) is presented in FIGURE 1.

Nucleation of HEWL crystals, which has already been studied in the intermediate case, in the horizontal tube position,[10] was used as a reference in FIGURE 1. (Since we changed HEWL sample the data presented are averaged from new control experiments, performed in the framework of the present study.)

It is evident that the tube position is important. First, the number of HEWL crystals nucleated in both vertical tube positions is decreased substantially. The explanation is that, before growing to visible sizes, some crystals fall down because they are loosely connected to the glass wall. However, the assumption that the crystals may be "homogeneously" nucleated has to be rejected for several reasons.

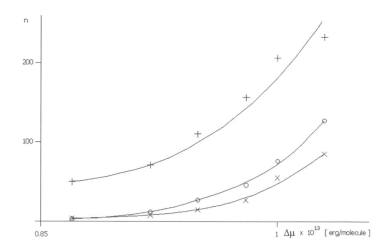

FIGURE 1. Dependence of the number of crystal nuclei n on supersaturation $\Delta\mu$ for two hours: +, in a horizontal tube position; $\bigcirc$, in a vertical tube with suppressed buoyancy; and $\times$, in a vertical tube with buoyancy convection (due to buoyancy, the $\Delta\mu$ value is changed and we are only able to show the nominal $\Delta\mu$ values).

1. In previous papers[1,2] we have described the effects of templates on crystal nucleation. A monomolecular layer of poly-L-lysine deposited on a glass substrate diminishes the number of nucleated hen-egg-white lysozyme (HEWL) crystals. In contrast, hydrophobization of glass surfaces by means of hexamethyl-disilazane stimulates nucleation. The fact that the templates were already exerting strong effects during the nucleation stage convinced us that heterogeneous nucleation was prevalent.

2. Most HEWL crystals stick strongly to glass or to the template,[1] the adhesion strength showing clear anisotropy.

3. True homogeneous nucleation is hardly possible. According to (classical) nucleation theory heterogeneous (crystal) nucleation is easier. Although "born" in the bulk solution some crystals may be nucleated heterogeneously on larger protein molecules that are present as impurities in the solution.

As is shown in FIGURE 1, when the coldest solution is at the top of the capillary the number of the HEWL crystals is reduced, as compared with their number in the reverse vertical tube position. The effect persists for a nucleation time of one hour, but seems to disappear for shorter times; for example, 30 min. We consider this fact as evidence that slow buoyancy-driven convection is acting. It appears that by trying to equalize supersaturation along the capillary, the convection reduces the highest supersaturation values.

Our study on convection effects exerted on the protein crystal growth was also performed using porcine pancreatic trypsin. The experiments show that trypsin crystals grew to visible sizes, by means of the thermal-double-pulse nucleation technique, for 24h (in some cases for 21h) only in the case when the coldest solution (4°C) was at the top of the capillary. No crystals were observed in the control sample, under exactly the same conditions but in the horizontal capillary tube. We found that under conditions without agitation, due to the buoyancy driven convection, trypsin crystals required substantially (many times) longer times for their growth to visible sizes. Evidently, buoyancy contributes to faster growth; *vice versa*, this fact is another, indirect proof for the presence of buoyancy driven convection.

CONCLUSIONS

Buoyancy effects on the nucleation (and growth) of protein crystals were investigated. This was done by monitoring the results of a series of experiments on protein crystallization in both vertical and in horizontal tube positions. We draw the conclusion that solution replenishment due to buoyancy driven convection can contribute to faster growth of protein crystals.

Bringing fresh solution, the buoyancy acts in two different ways. On one hand, the additional feeding results in fewer larger crystals. However, we suppose that convection brings impurities, which cause detrimental effects on crystal quality. Therefore, in the present work we are attempting to suppress (at least to some extent) buoyancy driven convection in order to grow purer protein crystals of high quality.

ACKNOWLEDGMENTS

We acknowledge the technical help of Engineer K. Goranov.

REFERENCES

1. TSEKOVA, D., S. DIMITROVA & C. NANEV. 1999. Heterogeneous nucleation (and adhesion) of lysozyme crystals. J. Cryst. Growth **196:** 226–233.
2. NANEV, C. & D. TSEKOVA. 2000. Heterogeneous nucleation of Hen-egg-white lysozyme—molecular approach. Cryst. Res. Technol. **35:** 189–195.
3. BALLMAN, A.A. & R.A. LAUDISE. 1963. The Art and Science of Growing Crystals. J.J. Gilman, Ed. New York.
4. LAUDISE, R.A. 1970. The Growth of Single Crystals, Prentice-Hall, Inc., Englewood Cliffs.
5. FRANK, F.C. 1958. Growth and Perfection of Crystals. R.H. Doremus, B.W. Roberts & D. Turnbul, Eds.: 411–418. J. Wiley & Sons, Inc. New York.
6. ATAKA, M. & S. TANAKA. 1979. Growth of large crystals. Chem. Abstr. **90:** 46940t.
7. ZEPPEZAUER, M. 1971. Enzyme purification and related techniques. *In* Methods in Enzymology, Vol. XXII. W.B. Jakoby, Ed.: 253–269. Academic Press, New York.
8. LUFT, J.R., D.M. RAK & G.T. DETITTA. 1999. Microbatch macromolecular crystallization on a thermal gradient. J. Cryst. Growth **196:** 447–449.
9. MAO, H., T. YANG & P.S. CREMER. 2002. A microfluidic device with a linear temperature gradient for parallel and combinatorial measurements. J. Am. Chem. Soc. **124:** 4432–4435.
10. PENKOVA, A., N. CHAYEN, E. SARIDAKIS & C. NANEV. 2002. Nucleation of protein crystals in a wide continuous supersaturation gradient. Acta Crystallographica D **D58:** 1606–1610.
11. ONUMA, K., K. TSUKAMOTO & I. SUNAGAWA. 1988. Role of buoyancy driven convection in aqueous solution growth; a case study of $Ba(NO_3)_2$ crystal. J. Cryst. Growth **89:** 177-188.

12. ONUMA, K., K. TSUKAMOTO & I. SUNAGAWA. 1989. Effect of buoyancy driven convection upon the surface microtopographs of $Ba(NO_3)_2$ and CdI_2 crystals. J. Cryst. Growth **98:** 384–390.

13. RIES-KAUTT, M. & A.F. DUCRUIX. 1992. Phase diagrams. *In* Crystallization of Nucleic Acids and Proteins: A Practical Approach. A.F. Ducruix & R. Giegé, Eds.: 195–218. Oxford University Press, Oxford.

14. ROSENBERGER, F. & S.B. HOWARD, J.W. SOWERS & T.A. NYCE. 1993. Temperature dependence of protein solubility—determination and application to crystallization in X-ray capillaries. J. Cryst. Growth **129:** 1–12.

15. FORSYTHE E.L. & M.L. PUSEY. 1996. The effects of the acetate buffer concentration on lysozyme solubility. J. Cryst. Growth. **168:** 112–117. (See also references therein.)

16. CHRISTOFER, G.K., A.G. PHIPPS & R.J. GRAY. 1998. Temperature-dependent solubility of selected proteins. J. Cryst. Growth **191:** 820–826.

Numerical Analysis of the Depletion Zone Formation Around a Growing Protein Crystal

HIROAKI TANAKA,[a] KOJI INAKA,[b] SHIGERU SUGIYAMA,[b]
SACHIKO TAKAHASHI,[a] SATOSHI SANO,[c] MASARU SATO,[c]
AND SUSUMU YOSHITOMI[c]

[a]Japan Space Utilization Promotion Center, Nishi-Waseda, Shinjyuku-ku, Tokyo, Japan

[b]Maruwa Food Industries, Inc., Japan

[c]Japan Aerospace Exploration Agency, Japan

ABSTRACT: It is expected that a protein depletion zone and an impurity depletion zone are formed around a crystal during protein crystal growth if the diffusion field around the crystal is not disturbed. The growth rate of the crystal
may be decreased and the impurity uptake may be suppressed to result in highly ordered crystals if these zones are not disturbed. It is well known that a
microgravity environment can reduce convective fluid motion, and this is
thought to disturb the depletion zones. Therefore, we expect that crystals
grown in space can attain better quality than those grown on the ground. In
this study, we estimate the depletion zone formation numerically and discuss
the results of crystallization in space experiments. In case of α-amylase, most
of the crystals form a cluster-like morphology on the ground using PEG 8000
as a precipitant. However, in space, we have obtained a single and high-quality
crystal grown from the same sample compositions. We have measured the viscosity of the solution, the diffusion coefficient, and the growth rate of protein
crystals on the ground. Applying numerical analysis to these values a significant depletion zone was expected to form mainly due to higher values of the
viscosity. This might be one of the main reasons for better quality single crystals grown in space, where the depletion zone is thought to remain undisturbed. For protein crystallization experiments, salts are widely used as a
precipitant. However, in that case, reduced concentration depletion zone
effects can be expected because of a low viscosity. Therefore, if it is possible to
increase the viscosity of the protein solution by means of an additive, the depletion zone formation effect would be enhanced to provide a technique that
would be especially effective in space.

KEYWORDS: protein crystal growth; protein depletion zone; impurity
depletion zone; viscosity; space experiment

INTRODUCTION

When a crystal grows in a supersaturated solution, significant concentration
gradients of materials occur that include, not only the protein molecule, but also

Address for correspondence: Hiroaki Tanaka, Japan Space Utilization Promotion Center,
3-30-16, Nishi-Waseda, Shinjyuku-ku, Tokyo, 169-8624, Japan. Voice: 81-3-5273-2442; fax:
81-3-5273-0705
    PXW01674@nifty.ne.jp

Ann. N.Y. Acad. Sci. 1027: 10–19 (2004). ©2004 New York Academy of Sciences.
doi: 10.1196/annals.1324.002

impurities, microcrystals, and small nuclei around the growing crystals.[1] These concentration gradients form depletion zones around the growing crystal that help to form high quality crystals with an impurity filtration effect, lowering the protein concentration to reduce the crystal growth rate to allow time for the molecules to become ordered in crystal.[2–4] However, on the ground, because of gravity, these depletion zones are disturbed by convective fluid motions and sedimentation.[5–8]

It is well known that a microgravity environment may maintain ideal depletion zones for protein crystal growth and may contribute to obtaining high-resolution diffracting crystals, with better internal order and fewer defects.[9,10] Using the applied protein crystallization facility (APCF) in space, a quasiperfect diffusive field was observed.[11]

Alpha-amylase (E.C.3.2.1.1) derived from *Aspergillus oryzae*, a glycoprotein that catalyzes the hydrolysis of the α-1,4-glycosidic linkage in starch, was chosen for the model protein because it is a well known and readily available protein that is useful as a model protein for many physiologically significant protein crystal growth processes. According to our previous experiments, a major problem of crystallization of α-amylase using PEG 8000 as a precipitant on the ground was formation of a cluster-like morphology. We performed α-amylase crystallization experiments during the Odissea mission kindly donated by ESA, in 2002, and obtained high quality crystals with different morphology from that of ground-grown crystals, without cluster-like formation.[12] We expected that the following points would be emphasized by the space experiments:

1. The protein depletion zone might decrease the density of protein molecules around the growing crystal, so that the crystal growth rate would be reduced and the protein molecules would be incorporated in order in the crystal and the quality of the crystal would be improved.

2. The impurity depletion zone might reduce the concentration of impurities around the growing crystal, so that impurity incorporation into the crystal would be decreased and the quality of the crystal improved.

3. The protein depletion zone might decrease the probability of the nuclei formation on the surface of the growing crystal, so that the possibility of the cluster-like morphology would be reduced.

In this report, we use numerical analysis of the depletion zone formation around a growing protein crystal to estimate the effects of depletion zones.

MATERIALS AND METHODS

Crystallization

Alpha-amylase in 2 mM $CaCl_2$ and 50 mM *tris*-HCl buffer pH 6.0 was crystallized using PEG 8000 as a precipitant in 2 mM $CaCl_2$ and 50 mM *tris*-HCl buffer pH 6.0 at 20°C by batch method.

Measurement of Solution Viscosity

The coefficient of viscosity of the PEG 8000 solution was measured by the capillary method. The ratio of the coefficient of viscosity between 18% and 22% PEG 8000 solution, approximated by measured values, was used to calculate the diffusion coefficient of α-amylase in 22% PEG 8000 solution.

Measurement of the Diffusion Coefficient

The thin layer diffusion pair method was used to measure the diffusion coefficients of α-amylase in an 18% PEG 8000 solution, which did not crystallize the protein.[12] The diffusion coefficient in 22% PEG 8000, at which concentration the crystallization experiment would be performed in space, was estimated from the kinetic viscosity ratio.

Estimation of Solubility

To estimate the approximate solubility of α-amylase in PEG 8000 solution, crystallization experiments of α-amylase using PEG 8000 as a precipitant were performed by a batch method. Several concentrations of the protein and the precipitant were prepared. After several days crystal growth was observed. The approximate solubility was estimated as a slightly lower concentration that that at which crystallization occurred.

Measurement of Growth Rate

Crystal growth in conditions of 30 mg/ml α-amylase and 22% PEG 8000 was observed by an optical microscope during a crystallization experiment performed by a batch method. Using the time required to grow to half size, β was calculated according to the following approximate estimation method.

Estimation of β

Since it is difficult to measure the correct value of β, we used the following model to approximate β easily. If the crystal is presumed spherical, the protein molecule weight (mg) in the crystal is

$$1000 \cdot n \cdot \frac{4}{3} \cdot \pi R^3(t), \tag{1}$$

where $R(t)$ is the radius of the crystal (cm) at time t and n is the relative density of the crystal.

Although several crystals with various sizes usually grew in one batch, we presumed that all the crystals grew to the same size at once, and that one of the crystals grew from the solution with spherical shape, radius L (cm). In this case, the amount of protein removed from the protein solution is

$$C(0) \cdot \frac{4}{3} \pi L^3 - C(t) \cdot \left(\frac{4}{3} \pi L^3 - \frac{4}{3} \pi R^3(t) \right), \tag{2}$$

where $C(0)$ and $C(t)$ are the concentrations of protein (mg/ml) at the beginning and at time t, respectively. Combining (1) with (2), gives the following equation:

$$C(t) = \frac{C(0)L^3 - 1000nR^3(t)}{L^3 - R^3(t)}. \tag{3}$$

When the crystallization is complete, L can be obtained from

$$L^3 = \frac{1000n - Ce}{C(0) - Ce} \cdot R^3(\infty), \tag{4}$$

where Ce is the protein solubility and $R(\infty)$ is the radius of the final crystal.

The following differential equation is related to the crystal growth rate

$$\frac{dR(t)}{dt} = \beta\omega(C'(t) - Ce'), \tag{5}$$

where $C'(t)$ and Ce' are the number of protein molecules in a unit volume $(1/cm^3)$ at time t after the crystallization occurred and the protein solubility concentration, respectively; ω can be expressed as $\omega = M/nN$ (cm^3), where M is the protein molecular weight and N is Avogadro's number.

The relationship between C (mg/ml) and C' $(1/cm^3)$ is

$$\omega C' = \frac{M}{nN} \times \frac{CN}{1000M} = \frac{C}{1000n}. \tag{6}$$

Substituting **(6)** in **(5)**,

$$\frac{dR(t)}{dt} = \frac{\beta(C(t) - Ce)}{1000n}. \tag{7}$$

By substituting **(3)** and **(4)** in **(7)** we obtain

$$\frac{dR(t)}{dt} = \frac{\beta}{1000n} \frac{R^3(\infty) - R^3(t)}{\dfrac{R^3(\infty)}{C(0) - Ce} - \dfrac{R^3(t)}{1000n - Ce}}. \tag{8}$$

We obtain the time-course of crystal growth from this differential equation, **(8)**. Approximate β values can be obtained by using various β and comparing the measured and calculated time required to grow to half size.

Estimation of Protein Depletion Zone Formation

The equation used to estimate protein depletion zone formation was[1]

$$C(R) = \frac{\dfrac{R\beta Ce}{D} + C(\infty)}{1 + \dfrac{R\beta}{D}}, \tag{9}$$

where $C(R)$, $C(\infty)$, and Ce (mg/ml) are the concentrations of the protein on the surface of the crystal, of the protein far away from the crystal, and of the protein solubility, respectively; β is the kinetic coefficient for protein trapping into the crystal; D is the diffusion coefficient of the protein molecules; and R is the radius of the crystal. From this equation, we defined the driving force ratio (*DFR*) for microgravity $(0g)$ versus Earth $(1g)$ as follows:

$$DFR = \frac{C(R) - Ce}{C(\infty) - Ce} = \frac{1}{1 + \dfrac{R\beta}{D}}. \tag{10}$$

Estimation of Impurity Depletion Filtration Effects

Impurity uptake into the crystal was driven by attachment to the crystal surface, which depended on the concentration of the impurity.[1] Here we ignore the dissociation of the impurity from the crystal because it was supposed to be considerably smaller than uptake.

The equation used to estimate impurity concentration on the crystal surface is

$$C_i(R) = \frac{C_i(\infty)}{1 + \dfrac{R\beta_i}{D_i}}. \tag{11}$$

The equation used to estimate the impurity uptake ratio versus the protein uptake (IUR_{0g}) is

$$IUR_{0g} = \frac{\beta_i C_i(R)}{\beta(C(R) - Ce)} = \frac{\beta_i}{\beta} \cdot \frac{1 + \dfrac{R\beta}{D}}{1 + \dfrac{R\beta_i}{D_i}} \cdot \frac{C_i(\infty)}{C(\infty) - Ce},$$

where R is the radius of the crystal, β_i is the kinetic coefficient for impurity trapping into the crystal, D_i is the diffusion coefficient of the impurity molecules, $C_i(R)$ and $C_i(\infty)$ are the concentrations of the impurity on the surface of the crystal and far away from the crystal, respectively.

If there is no formation of an impurity depletion zone, the impurity uptake ratio (IUR_{1g}) becomes

$$IUR_{1g} = \frac{\beta_i C_i(\infty)}{\beta(C(\infty) - Ce)}.$$

Therefore, the ratio of the impurity uptake for $0g$ versus $1g$ (IUF, impurity uptake filtration) because of the impurity depletion zone formation around the crystal is,

$$IUF = \frac{IUR_{0g}}{IUR_{1g}} = \frac{1 + \dfrac{R\beta}{D}}{1 + \dfrac{R\beta_i}{D_i}} = \frac{DFR_i}{DFR}, \tag{12}$$

where DFR_i is the driving force ratio of impurity depletion zone. For simplicity, substitute Equation (13) into (12)

$$A = \frac{\beta_i D}{\beta D_i} \tag{13}$$

to obtain,

$$IUF = \frac{1 + \dfrac{R\beta}{D}}{1 + A \cdot \dfrac{R\beta}{D}}. \tag{14}$$

This equation was used to estimate the filtering effect of the impurity by the depletion zone formation. If the crystal was small, $IUF \approx 1$, which means that there is little effect of the impurity depletion zone. If the crystal was large, $IUF \approx 1/A$, which means that the impurity depletion zone is effective.

RESULTS AND DISCUSSION

Estimation of Solubility

Occurrence crystallization was demonstrated by using a batch method for each experiment. In FIGURE 1 each $*$ indicates the concentration of α-amylase and PEG 8000 when crystallization occurred. Each ◆ indicates when crystallization did not occur. The solubility of α-amylase was roughly estimated to be 10mg/ml for PEG 8000 concentrations greater than 22%, a concentration at which crystallization was estimated to occur in a space experiment.

Coefficients Obtained from the Experimental Results

The values obtained from experiments performed to determine the coefficients are shown in TABLE 1. Using these results gave $\beta = 0.09\,\mathrm{mm\,hr^{-1}}$. This value is about four times larger than the β value for lysozyme.[1]

The diffusion coefficient of α-amylase in 22% PEG 8000 was estimated from the ratio of the kinetic viscosity in 18% and 22% PEG 8000 solutions. The diffusion coefficient of α-amylase in 18% PEG 8000 was divided by the ratio (1.67) to obtain the value in 22% PEG 8000.

Numerical Estimation of Protein Depletion Zone Formation Around a Crystal

As shown in Equation (**9**), the larger the crystal grew, and the smaller the diffusion coefficient, the lower the protein concentration on the crystal surface. In the case of α-amylase, if the values described above are substituted in Equation (**10**), the protein depletion zone formation is obtained, as shown in FIGURE 2. As the figure shows, for an α-amylase crystal whose size was suitable for X-ray diffraction data collection (about several dozen μm), DFR on the surface of the crystal decreased significantly. The larger the crystal, the weaker the driving force.

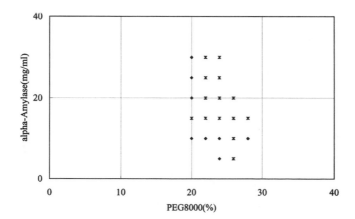

FIGURE 1. Crystallization of α-amylase in PEG 8000 solution by batch method: $*$, crystallization occurred; ◆, crystallization did not occur.

TABLE 1. Numerical data obtained from the experiments

Parameter		Numerical Data	Detail
Solubility, Ce		10 mg/ml	estimated in 22% PEG 8000
Terminal crystal size, $R(\infty)$		0.15 mm	measured if $C(0)$ was 30 mg/ml in 22% PEG 8000
Time required to grow half size		50 hours	measured if $C(0)$ was 30 mg/ml
Relative density		1.1	roughly estimated
Diffusion coefficient of α-amylase		2.97×10^{-11} m²/sec	measured in 18% PEG 8000
Diffusion coefficient of α-amylase		1.78×10^{-11} m²/sec	estimated from measurement in 22% PEG 8000
Coefficient of viscosity	18% PEG 8000	13.0 mPa	approximated from measured value
	22% PEG 8000	21.6 mPa	approximated from measured value

We obtained high quality α-amylase crystals, without cluster-like formation, grown in space during the Odissea mission, as compared to ground grown crystals. Numerical analysis of the depletion zone around the growing crystals as analyzed in this paper suggested that the reduction of DFR was significantly explained by: (1) reduced disorder of the crystal because of crystal growth in lower supersaturation conditions around the crystal surface, and (2) cluster-like formation was avoided because of suppression of nuclei formation on the surface of the crystal due to the lower supersaturation condition around the crystal surface.

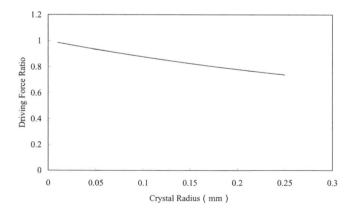

FIGURE 2. Relationship between crystal radius and driving force ratio for α-amylase.

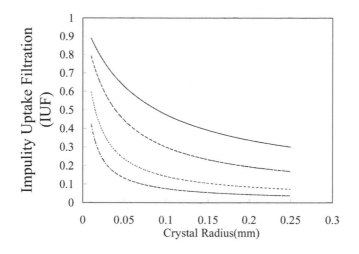

FIGURE 3. The ratio of impurity uptake filtration for α-amylase with PEG 8000: ——, $A = 10$; - - -, $A = 20$; ······, $A = 50$; - · - ·, $A = 100$.

Numerical Estimation of Impurity Depletion Zone Formation Around a Crystal

As shown in Equation **(14)**, if the kinetic coefficient for impurity trapping into the crystal (β_i) is large or the impurity diffusion coefficient (D_i) is small, the value A is large. If A is large, the impurity uptake filtration (IUF) is thought to be effective, as shown in FIGURE 3 for the case of α-amylase crystallization.

Since it was reported that the value of the impurity uptake coefficient was much larger than the value of the protein uptake coefficient,[13] the value of A was assumed

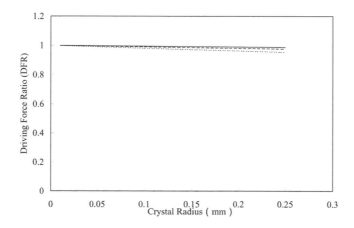

FIGURE 4. Relationship between crystal radius and driving force ratio for lysozyme: ——, standard; - - -, D half of standard; ······, D one fourth of standard.

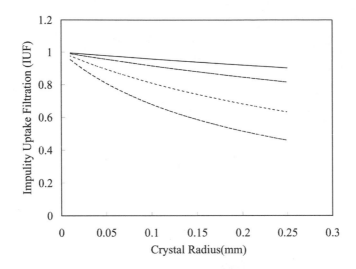

FIGURE 5. The ratio of impurity uptake filtration for lysozyme: ——, $A = 10$; ---, $A = 20$; ·····, $A = 50$; ---·, $A = 100$.

to be about 10–100, which seemed realistic. An impurity uptake suppression effect was expected, especially under conditions in which it was easy to form an impurity depletion zone. Therefore, from this numerical estimate, it was expected that impurity uptake could be decreased and a better quality crystal obtained that would be diffracted with higher resolution in the case of α-amylase crystallization in space.

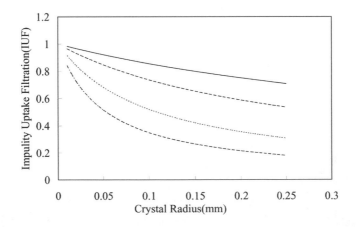

FIGURE 6. The ratio of impurity uptake filtration for lysozyme when the diffusion coefficient is one-fourth of the conditions in FIGURE 5: ——, $A = 10$; ---, $A = 20$; ·····, $A = 50$; ---·, $A = 100$.

Expected Effect of Depletion Zone Formation with Change in Diffusion Coefficient of the Solution

If the diffusion coefficient of lysozyme in a low viscosity precipitant solution, such as salt solution, is substituted in Equation (10), then the concentration of protein solution on the surface of the lysozyme crystal does not decrease sufficiently, as shown in FIGURE 4. It does not change even though the diffusion coefficient decreased to one-half or one-fourth, both of which are realistic. It suggested that the low viscosity precipitant solution cannot easily provide a sufficient depletion zone effect in the space experiment.

As shown in FIGURE 5, the impurity uptake filtration (*IUF*) is not as effective as α-amylase crystallization. However, if the coefficient of viscosity could somehow be made four times higher (the diffusion coefficient is one-fourth), the *IUF* effect might be expected to be enhanced, as shown in FIGURE 6. In this case, impurity uptake filtration by the impurity depletion zone was more effective than the protein depletion zone effect.

Thus, if the diffusion coefficient of the protein can be reduced several times, the impurity depletion zone effect might be expected, even in the salt solutions, and high quality crystal obtained. The addition of highly viscous material to the solution could be considered.

REFERENCES

1. CHERNOV, A.A. 1998. Crystal growth and crystallography. Acta Cryst. **A54:** 859–872.
2. MCPHERSON, A. 1999. Crystallization of Biological Macromolecules. Cold Spring Harbor Laboratory Press, New York.
3. THOMAS, B.R., *et al.* 2000. Distribution coefficients of protein impurities in ferritin and lysozyme crystals self purification in microgravity. J. Cryst. Growth **211.** 149–156.
4. LIN, H., *et al.* 2001. Lower incorporation of impurities in ferritin crystals by suppression of convection: modeling results. Cryst. Growth Des. **1**(1): 73–79.
5. BETH, M., *et al.* 1988. Preliminary observations of the effect of solutal convection on crystal morphology. J. Cryst. Growth **90:** 130–135.
6. PUSEY, M.L. & K. GERNERT. 1988. A method for rapid liquid–solid phase solubility measurements using the protein lysozyme. J. Cryst. Growth **88:** 419–424.
7. BOISTELLE, R. & J.P. ASTIER. 1988. Crystallization mechanisms in solution. J. Cryst. Growth **90:** 14–30.
8. GARCIA-RUIZ, J.M., *et al.* 2001. Agarose as crystallization media for proteins I: transport processes. J. Cryst. Growth **232:** 165–172.
9. LITTKE, W. & C. JOHN. 1984. Protein single crystal growth under microgravity. Science **225:** 203–204.
10. KUNDROT, C.E., *et al.* 2001. Microgravity and macromolecular crystallography. Cryst. Growth Des. **1:** 87–99.
11. OTALORA, F., *et al.* 2001. Experimental evidence for the stability of the depletion zone around a growing protein crystal under microgravity. Acta Cryst. **D57:** 412–417.
12. INAKA, K., *et al.* 2003. Crystallization of α-amylase using the counter-diffusion method under microgravity. Presented at International Symposium on Diffraction Structural Biology 2003, Tsukuba, Ibaraki, Japan, May 28–31.
13. THOMAS, B.R. & A.A. CHERNOV. 2001. Acetylated lysozyme as impurity in lysozyme crystals: constant distribution coefficient. J. Cryst. Growth **232:** 237–243.

Size and Shape Determination of Proteins in Solution by a Noninvasive Depolarized Dynamic Light Scattering Instrument

NAOMI CHAYEN,[a] MATTHIAS DIECKMANN,[b]
KARSTEN DIERKS,[c] AND PETRA FROMME[d]

[a]*Division of Biomedical Sciences, Imperial College of Science, Technology and Medicine, London, United Kingdom*

[b]*ESA/ESTEC, Noordwijk, the Netherlands*

[c]*Dierks + Partner Systemtechnik, Hamburg, Germany*

[d]*Department of Chemistry and Biochemistry, Arizona State University, Tempe, Arizona, USA*

ABSTRACT: Dynamic light scattering (DLS) is a well-known noninvasive technique for investigating interactions of protein molecules in solution. Unfortunately, DLS is not very sensitive to small size changes because covariables, such as temperature, viscosity, and refractive index, are not precisely known, or they vary as functions of an experiment run, making it difficult to resolve subtle size changes of only a few Angström. It is usually not possible, if these covariables are not systematically measured and brought into the DLS analysis, to separate monomers from dimers when both are present in solution. We present here measurements with a variant of DLS that determines rotational diffusion as well as translation diffusion. This technique, called depolarized dynamic light scattering (DDLS) is, like DLS, also an old method, but it is rarely used due to enormous practical difficulties. However, we have found that a combination of DLS with DDLS is very promising, because it allows for a rough shape determination of the molecule under study and it is more sensitive to subtle size changes. We built an instrument that overcomes some of the difficulties, and report measurements made with this instrument. One of the samples was Photosystem-I, a membrane protein for photosynthesis. Its dimensions were determined to be 9.6 nm thick and 26 nm in diameter, values that are in good agreement with the dimensions obtained from X-ray diffraction analysis of single crystals.

KEYWORDS: dynamic light scattering; depolarized scattering; particle sizing; shape determination

INTRODUCTION

Dynamic light scattering (DLS) is a well established technique, often called photon correlation spectroscopy (PCS) or quasielastic light scattering (QELS). This

Address for correspondence: K. Dierks, Dierks+Partner Systemtechnik, Haferweg 46, 22769 Hamburg, Germany. Voice: 49 40 85330119; fax: 49 40 8502803.
kdierks@aol.com

Ann. N.Y. Acad. Sci. 1027: 20–27 (2004). ©2004 New York Academy of Sciences.
doi: 10.1196/annals.1324.003

method exploits the fact that particles contained in a liquid undergo Brownian motion and the average velocity of this motion is determined by the particle size, their thermal energy, and by the viscosity of the medium and the particle geometry. The dependence is described by the Stokes–Einstein equation, which was formulated in the beginning of the twentieth century. When a vertically polarized laser beam is directed through a volume containing the particles to be investigated, light is scattered by the particles as a consequence of the increments of refractive indices between particle, possible shell molecules, and the surrounding medium. Because of the coherent laser light, the scattered light forms an interference pattern. When the particles are in motion, this pattern changes permanently, so that particle movement is transferred to a changing light intensity distribution.

A detector collects a fraction of the scattered light at a certain angle, thus seeing permanent light intensity fluctuations. An analysis of the temporal behavior of these fluctuations yields a measure for the average velocity of the Brownian motion, and finally, via Stokes–Einstein, to the particle size. The analysis is performed by calculating the autocorrelation function (ACF) of the light intensity. Usually the applicability of this method is from about 1 nm particle size to about 2 μm The lower limit results from the strong decrease in light intensity with decreasing size, so that the signal vanishes in the noise; the upper limit is dictated by other velocity components in the sample volume, for example, convection currents, which deteriorate the measurement statistics leading to overflows that destroy a proper ACF.

The DLS method uses only the translational movement of molecules. It can be expected that an additional analysis of the rotational movement of the particles would reveal more information about the particle morphologies. Although DLS is insensitive to rotational movements, by exploiting the depolarized component of the scattered light we obtain information about particle rotation. This method is thus called depolarized dynamic light scattering (DDLS). The depolarized component is only produced by optically anisotropic particles, which is generally the case for non-spherical particles. Unfortunately, the intensity component scattered off the linear polarized laser plane is very weak, a fact leading to the need for highly technical efforts to exploit and acquire the corresponding signal.

The main advantage of DDLS is that it is now possible to determine the *shape* of the particles, whereas DLS measures the hydrodynamic radius of the particles alone. Knowledge of the shape is valuable when, for example, considering conformation changes of molecules, or aggregation phenomena, such as dimerization reactions. Furthermore, the measured quantities (decay time constants) are proportional to r^3, compared to a simple r dependence for DLS, leading to a higher sensitivity against small changes in the particle radius r.

THEORY

The formula applied here for DDLS treats the molecules as rotational ellipsoids. They have an axis of rotation (a) and two (equal) perpendicular axes (b). Two cases may be distinguished: the oblate ellipsoid, axis a is shorter than axis b (more disk like); the prolate ellipsoid, axis a is longer than b (more stick like). "Normal" DLS covers only the case when $a = b$; that is, for spherical particles.

For DLS, as indicated, the relation between the translational diffusion constant D_t, derived from the decay time constant of the measured ACF, is given by the Stokes–Einstein equation,

$$D_t = \frac{k_B T}{6\pi\eta r_0},$$

whereas the equivalent for rotational spectroscopy is Perrin's formula,

$$D_r = \frac{3}{16}\frac{K_B T}{16\pi\eta a^3}\frac{1-(b^2/a^2)^2}{(2-b^2/a^2)G(a,b)-1}.$$

Here K_b is the Boltzmann constant, T is the temperature, and η is the viscosity. The radius r of a particles is now replaced by the half-lengths a and b of above ellipsoid. Thus, the factor $G(a,b)$ describes the shape dependence of the diffusion constant,[1]

$$G(a, b) = \ln\left(\frac{1+\sqrt{1-b^2/a^2}}{b/a}\right)\times\frac{1}{\sqrt{1-b^2/a^2}}, \quad a>b \tag{1}$$

and

$$G(a, b) = \frac{\mathrm{atan}\sqrt{1-b^2/a^2}}{\sqrt{1-b^2/a^2}}, \quad a<b. \tag{2}$$

The dependence between diffusion constants and decay time constants was taken from Reference 2. A similar equation holds for D_t, it is not shown here.

EVALUATION

When the two diffusion constants D_r and D_t are measured independently, it is possible to derive the desired values for a and b by applying the above equation.

FIGURE 1 shows the relation between decay time constants and particle dimensions as described by these equations. Because it is difficult to solve the equations algebraically, an iterative procedure was used instead. This procedure is an inversion of the equations for given diffusion constants (i.e., taken from measurements), yielding a relation between the half axes, a and b (as shown, e.g., in FIG. 4). Because two equations are used, two curves result, with up to two intersections. One intersection belongs to the oblate ellipsoid, the other belongs to the prolate ellipsoid. In some cases it may be difficult to decide which of these solutions is correct. Other sources of information need to be used in such cases (e.g., the molecular weight for both cases may be quite different). An independent determination is therefore helpful in making a determination.

INSTRUMENT

FIGURE 2 shows an overview of the set up. Light from a laser source is focused into a cuvette containing the sample solution. The lasers used depend on the acceptance of the sample to certain wavelengths. For example, in case of the pronounced photosensitivity of Photosystem I, two lasers were used: a Nd–YAG laser with 532 nm wavelength and 20 mW optical power for molecules showing no

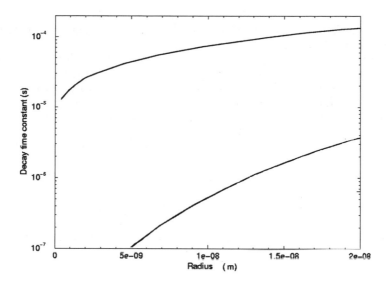

FIGURE 1. Decay constants for translational diffusion (*upper curve*) and rotational diffusion (*lower curve*) as a function of the particle radius for a given length of axis *a* (here 8 nm). One can see that for small particles the rotational decay times become very short (less than 100 nsec).

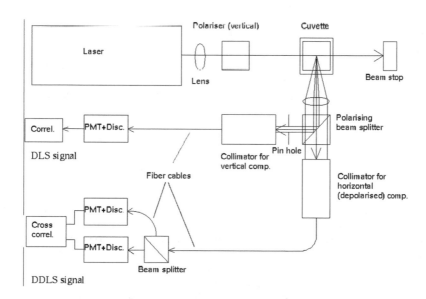

FIGURE 2. Arrangement of components for the DDLS instrument.

light sensitivity to this part of the spectrum, and a laser diode with 785 nm wavelength and a similar output power for Photosystem I. To maintain spectral stability of the NIR laser it was equipped with temperature stabilization and a Faraday isolator to avoid back reflection into the lasing cavity. The scattered light was collected by an objective and its horizontally (depolarized) and vertically polarized components were separated by a polarizing beam splitter. The vertically polarized signal represents a normal DLS signal, used to determine D_t. The other channel contains the DDLS signal for measurement of D_r. Thus, it is possible to measure both parameters simultaneously. For the DLS signal, a photomultiplier tube acts as detector, followed by a pulse shaper and an autocorrelator. The DDLS channel required a little more effort because many detectors show after-pulsing; that is, a certain probability of a distorted pulse response for short times (less than about 1 µsec). This after-pulsing is just in the region of interest for DDLS and therefore counter measures were necessary. A beam splitter separates the signal into two parts that were lead to two detectors. Cross-correlating the signals from these two detectors suppresses unwanted distortions. The sample time ranges from 800 nsec to several minutes for the DLS channel and from 50 nsec to several minutes for the DDLS channel.

MEASUREMENTS AND RESULTS

Photosystem-I (PS-I) is a membrane protein responsible in plants for photosynthesis (in combination with Photosystem-II). It is usually insoluble in water and the presence of detergent (e.g., glucosyl maltoside) is usually required to make it soluble. We performed the first successful DLS/DDLS measurements with this molecule. The measured auto/cross-correlation functions are shown in FIGURE 3. A wavelength of 785 nm was used for these measurements to avoid the spectral regions of strong absorption. It can be seen that the decay time constant of the DLS ACF with 100 µsec is much longer than for DDLS with 0.5 µsec. Furthermore, it can be seen that the correlation function of the DDLS signal also contains contributions from the translational diffusion with its long decay time. This is partly due to imperfect separation of the polarizing beam splitter and, according to theory, the DDLS signal indeed contains translational components as well. The advantage of the two-channel set up is now clear: it makes it possible to differentiate between the purely rotational signal and translational components.

From a measurement series with 10 single measurements the following values were obtained:

rotational decay time constant, 996 ± 100 nsec ($N = 9$);
translational decay time constant, 89.6 ± 5.5 µsec ($N = 7$).

Using these values, an iterative procedure was applied, leading to the curves shown in FIGURE 4. The desired values for a and b can now be taken from the position of the appropriate intersection. The molecule was determined to be disk-like, with a thickness of 9.6 nm and a diameter of 26 nm. When subtracting the thickness of the detergent surrounding the molecule, one obtains a diameter of 22 nm, which is in good agreement with the dimensions obtained from X-ray analysis. The lower intersection in FIGURE 4 was discarded, because it would belong to an almost spherical particle shape, producing no depolarized signal component.

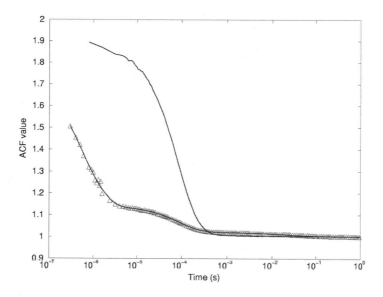

FIGURE 3. Example of correlation functions of the DLS (*upper curve*) and the DDLS signal ($\triangle$), obtained during the measurements with PS-I.

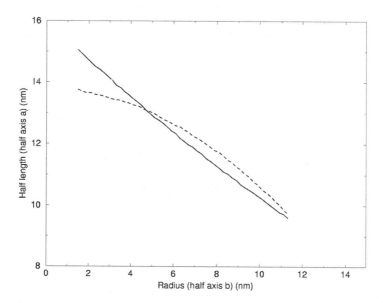

FIGURE 4. Half-axis b as a function of half-axis a for given D_r (*dotted line*) and D_t (*straight line*); similar to FIGURE 1, but here for the decay times measured with PS-1.

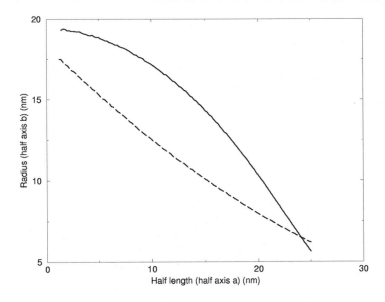

FIGURE 5. Half-axis b as a function of half-axis a for given D_r and D_t for tropomyosin. Here there is only one intersection, at $a = 24$ nm and $b = 7$ nm, the molecule is, therefore, considered a prolate ellipsoid.

Tropomyosin is a muscle protein with a rod-like shape, it was chosen as a model for a prolate ellipsoid with 2 nm diameter and about 40 nm length. A solution was prepared with a concentration of 10 g/l protein, an addition of 1.1 molar KCl and a 0.067 molar phosphate buffer. These concentrations and agents were chosen to give a monodisperse protein solution, but the first experiments indicated that this was not the case; that is, aggregation was observed. It is desirable to measure monomeric solutions, because aggregates may show a strong depolarization and thus deteriorate measurement of the molecule itself. Nevertheless a measurement series was started.

The peaks of the smallest component both in the DDLS and the DLS data were, therefore, rather weakly pronounced and the corresponding measurement errors rather large:

rotational decay time constant, 2.84 ± 1.84 μsec ($N = 14$);
translational decay time constant, 106 ± 26 μsec ($N = 19$).

According to FIGURE 5 these values yield molecular length of about 48 nm and a thickness (two times half-axis a, i.e., $2a$) of 12 nm. Aggregation can, therefore, be assumed; about 50 monomers in a roughly parallel arrangement would be necessary to form a bundle with this diameter.

CONCLUSIONS

DDLS was found to deliver valuable information beyond that already available from DLS. It was successfully used for measurements of large proteins (PS-1 has a molecular weight of about one million). Such proteins scatter the light rather strongly, so that measurements are relatively easy to perform. Nevertheless, a higher signal strength is desirable to reduce the measurement time needed and/or to improve the statistical accuracy. Additional experiments are necessary to find out how the instrument behaves when measuring smaller proteins.

ACKNOWLEDGMENTS

We thank Guido Laubender (Technical University of Berlin) for supplying tropomyosin and the ATP synthase.

REFERENCES

1. DUBIN, S.B., *et al.* 1971. Measurement of the rotational diffusion coefficient of lysozyme by depolarised light scattering: configuration of lysozyme in solution. J. Chem. Phys. **54**(12): 5158.
2. ZERO, K. & R. PECORA. 1985. Dynamic depolarized light scattering. *In*: Dynamic Light Scattering: Applications of Photon Correlation Spectroscopy. R. Pecora, Ed.: 59. Plenum Press, New York.

Scientific Approach to the Optimization of Protein Crystallization Conditions for Microgravity Experiments

IZUMI YOSHIZAKI,[a] HIROHIKO NAKAMURA,[b] SEIJIRO FUKUYAMA,[c] HIROSHI KOMATSU,[a,d] AND SHINICHI YODA[a]

[a]*National Space Development Agency of Japan, Tsukuba, Japan*

[b]*Mitsubishi Research Institute, Chiyoda-ku, Japan*

[c]*Advanced Engineering Services Co., Ltd., Tsukuba, Japan*

[d]*Iwate Prefectural University, Takizawa-mura, Japan*

ABSTRACT: The National Space Development Agency of Japan (NASDA) developed a practical protocol to optimize protein crystallization conditions for microgravity experiments. This protocol focuses on the vapor diffusion method using high density protein crystal growth (HDPCG)—hardware developed by the University of Alabama, Birmingham—that flew on the STS-107 mission. The objective of this development was to increase the success rate of microgravity experiments by setting crystallization conditions based on knowledge of crystal growth and fluid dynamics. The protocol consists of four steps: (1) phase diagram preparation, (2) estimation of condensation rate in the vapor diffusion method, (3) fluid dynamic property measurement, and (4) fluid dynamic simulation. First, a phase diagram was constructed. Crystallization characteristics were investigated by a microbatch method. The data were recalculated based on classical nucleation theory and the crystallization boundary was determined as a function of time. The second step was to develop a practical model to estimate the condensation rate of the crystallizing solution, including protein and precipitant, as a function of the precipitant concentration and solution volume. By considering the crystallization map and the vapor diffusion condensation model we were able to optimize the crystallization conditions that generate crystals in the desired time. This was particularly important in a shuttle mission whose mission duration is limited. The third step was fluid dynamic property measurement necessary for fluid dynamics simulation and crystal growth study. The last step was to estimate the mass transport in space on the basis of the fluid dynamics simulation transport model. It turned out that neither the vapor phase nor the solution phase was seriously affected by gravity until nucleation provided the hardware was set in a normal direction. Therefore, we concluded that the optimized crystallization conditions could be directly applied to microgravity experiments. By completing the approach, we were able to control the time for nucleation in the vapor diffusion method.

KEYWORDS: protein; crystal growth; microgravity; vapor diffusion; nucleation

Address for correspondence: Izumi Yoshizaki, Japan Aerospace Exploration Agency, ISS Science Project Office, 2-1-1 Sengen, Tsukuba, Ibaraki, 305-8505 Japan. Voice: 81-29-868-3654; fax: 81-29-868-3956.
yoshizaki.izumi@jaxa.jp

Ann. N.Y. Acad. Sci. 1027: 28–47 (2004). ©2004 New York Academy of Sciences.
doi: 10.1196/annals.1324.004

INTRODUCTION

Thousands of protein crystal growth experiments have been carried out under microgravity in the past. However, only 20–30% of this experiments are reported to have succeeded. One of the reasons for this may be because the experiment conditions, such as protein and precipitant concentrations and vapor diffusion droplet versus reservoir precipitant concentration ratio, were set based on experience. It is necessary to be aware of the fact that the best conditions in Earth experiments are not always the best conditions for space shuttle experiments. This is because the mass transport characteristics differ and also the experiment time and experiment setup are limited. Therefore, growth condition optimization based on fluid dynamics and crystallization kinetics is necessary.

NASDA developed a practical *optimization protocol* that focuses on STS-107 protein crystal growth experiments. This protocol was developed to obtain crystals efficiently by studying the phase diagram, to obtain sufficient size crystals during 16 days, and to customize the crystallization conditions for space experiments. Here we report the details of this protocol.

EXPERIMENTAL APPARATUS

The vapor diffusion method protein crystal growth apparatus used for STS-107 was high density protein crystal growth (HDPCG), developed by the University of Alabama, Birmingham. The apparatus is available for commercial use from SPACE-HAB Inc. FIGURE 1 shows a schematic drawing of the HDPCG equipment. The apparatus is in position A at launch. At this time the protein solution (containing a low concentration of precipitant) and the precipitant solution do not meet, thus vapor diffusion does not start. The vapor starts diffusing when the barrel is turned to position B by the crew. As the vapor diffuses toward the precipitant solution, the protein and precipitant concentration increases in the protein droplet, which drives nucleation and crystal growth (FIG. 1C). At the end of the experiment the barrel is turned back to the original position (FIG. 1D) to avoid the protein droplet falling off when the experiment returns to earth.

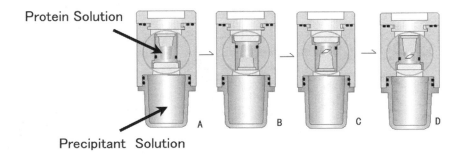

Protein Solution

Precipitant Solution

FIGURE 1. Schematic drawing of the HDPCG.

BASIC IDEA OF OPTIMIZATION

The most important point is to know the condensation profile in the protein droplet. This profile should be plotted on a phase diagram. FIGURE 2 shows the schematic image of the phase diagram and the condensation profile in the protein droplet. As the vapor diffuses, the concentration increases, and the concentration profile goes over the crystallization boundary. After nucleation, the protein concentration decreases due to protein consumption by the crystal. The important point is that the condensation profile should go over the crystallization boundary within a certain time interval to allow the crystal to grow to a sufficient size during the rest of the mission.

For this purpose, first the phase diagram with information about the crystallization boundary was constructed. Next, the condensation rate estimation model was built. Using this model, we could estimate the vapor diffusion rate with time at various concentration combinations by using just one experiment. Thus, we could optimize the protein droplet concentration and the precipitant concentration quickly, to initiate nucleation within a certain time interval.

The third task was to measure the fluid dynamic properties, such as diffusion constant, viscosity, and density. Efforts were made to measure these properties with the least amount of protein sample. This data was used in the fluid dynamic simulation to evaluate the difference in mass transport with and without gravity.

Protocol (1)—Phase Diagram

First, the phase diagram was constructed.[1] It is well known that the driving force necessary for nucleation is larger than that for crystal growth. Thus, there is a region in the phase diagram where crystals can grow but cannot nucleate. This region is

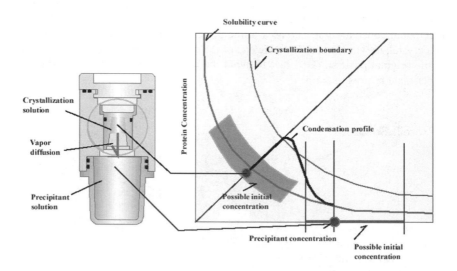

FIGURE 2. Basic idea of condition optimization.

called the metastable zone. The boundary between the metastable zone and the crystallization zone (crystallization boundary) varies with time. We developed a model to estimate the change in the crystallization boundary with time as a result of a simple experiment.

We used lysozyme as a model protein and investigated the crystallization time by a microbatch method (50μl droplet). Crystallization solutions with various protein and precipitant concentrations were prepared, and the time at which the first crystal was observed was recorded (see FIGURES 3 and 4). FIGURE 4 shows one of the results at 63 hours after the start of the experiment. With the same NaCl concentration, the crystal appeared sooner in a high lysozyme concentration solution, as expected. This result agreed well with previous results.[2]

Here we consider only homogeneous nucleation. According to classical nucleation theory, the nucleation rate I is represented as follows:[3–5]

$$I = \frac{\text{const}}{\eta} \times C \times \exp\left(-\frac{\Delta G^*}{kT}\right), \tag{1}$$

where

$$\Delta G^* = \frac{16}{3} \times \frac{\pi \gamma^3}{(\overline{\Delta \mu})^2} = \text{const} \times \frac{\gamma^3}{(\overline{\Delta \mu})^2}, \tag{2}$$

η is the solution viscosity, C is the initial protein concentration, γ is the surface tension, and $\overline{\Delta \mu} = \Delta \mu / v$ is the difference in chemical potential between the crystal and the solution. The following relation must be satisfied in order to have at least one crystal in the solution:

$$I \times V \times t = 1, \tag{3}$$

where V is the solution volume and t is the elapsed time. In this experiment V is constant. The solution viscosity η is also constant in this experiment (lysozyme and NaCl solution). Thus, Equation (3) can be simplified to

$$Ct = \exp\left(\frac{\Delta G^*}{kT}\right) \times \text{const.} \tag{4}$$

By writing Equation (4) in logarithmic form,

$$\ln(Ct) = \frac{\Delta G^*}{kT} \times \text{const}, \tag{5}$$

and then substituting for ΔG^* from Equations (2)–(5),

$$\ln(Ct) = \frac{1}{kT} \times \frac{\gamma^3}{(\overline{\Delta \mu})^2} \times \text{const}, \tag{6}$$

the driving force for crystallization $\Delta \mu$ can be expressed as follows:

$$\Delta \mu = kT \ln\left(\frac{C}{C_e}\right),$$

where C_e is the solute solubility. In this experiment, the temperature T was kept constant. The surface constant γ is likely to change with the solution concentration. However, this value cannot be measured easily, consequently, here we assume γ to be constant. Thus, the time before crystallization, solute concentration, and the solubility can be expressed as follows:

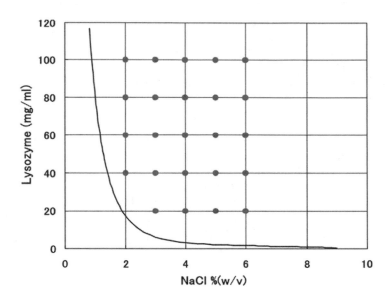

FIGURE 3. Crystallization matrix.

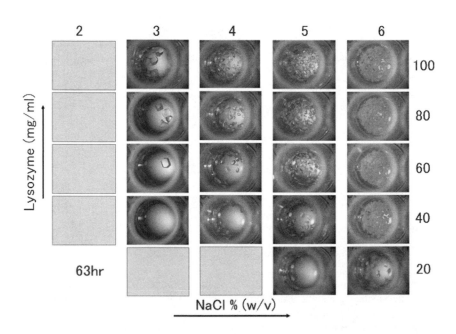

FIGURE 4. Crystallization results at 63 hours.

$$\left[\ln\left(\frac{C}{C_e}\right)\right]^{-2} = A\ln(Ct) + B, \tag{7}$$

where A and B are constant values.

Using Equation (7), we sorted the experiment results. The solubility was known in this experiment.[6] The time at which the first crystal was observed at each solution concentration was obtained from the results. Using our results, $\ln(Ct)$ was plotted as abscissa and $[\ln(C/C_e)]^{-2}$ was plotted as ordinate (see FIGURE 5). The data obtained are approximated by the following equation,

$$y = 0.0393x - 0.0443. \tag{8}$$

The inclination indicates A and the intercept indicates B of (7). Equation (7) is rewritten as follows:

$$[\ln(C) - \ln(C_e)]^2[A\{\ln(C) + \ln(t)\} + B] = 1.$$

By making the following replacements,

$$\ln(t) = D,$$

$$\ln(C_e) = E, \text{ and}$$

$$\ln(C) = X,$$

this equation can be written

$$X^3 + \left(D - 2E + \frac{B}{A}\right)X^2 + \left(E - 2D - 2\frac{B}{A}\right)EX + \left(D + \frac{B}{A}\right)E^2 - \frac{1}{A} = 0. \tag{9}$$

The values of A, B, D, and E are already known, so that (9) can be solved for the value of X for each crystallization condition. The crystallization boundary change with time was recalculated and plotted (dashed lines), together with the actual crystallization results (closed circles, see FIGURE 6).

Although we used a simplified model, the recalculated crystallization boundary change with time in FIGURE 6 shows fairly good agreement with the experimental results. By using this model, we can estimate the crystallization time in a simple microbatch experiment. The phase diagram (crystallization map) provides the basis

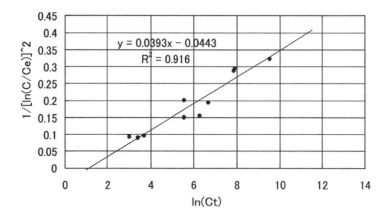

FIGURE 5. $\ln(Ct)$ versus $[\ln(C/C_e)]^{-2}$ plot.

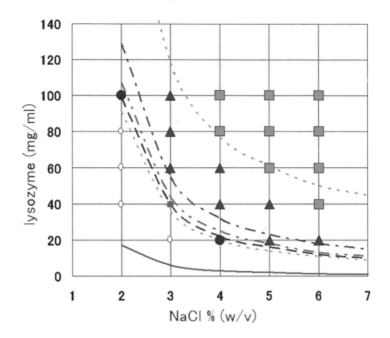

FIGURE 6. Crystallization boundary change with time: ———, C_e (mg/ml); ■, 0.5 h; - - -, 0.5 hc; ▲, 13 h; — ·, 13 hc; ■, 63 h; — ·, 63 hc; ●, 135 h; — —, 135 hc; - - -, 352 c.

to optimize crystallization conditions. This model can be modified for viscous precipitants, such as PEG, by taking the change in η value into account.

Protocol (2)—Estimation of Condensation Rate in the Vapor Diffusion Method

The second step was to develop a practical model for estimating the condensation rate of the crystallizing solution, including protein and precipitant, as functions of the precipitant concentrations and solution volume.[7] By considering the crystallization map and the vapor diffusion condensation model we were able to optimize the crystallization condition that generates crystals in the desired time.

Vapor Diffusion Model for HDPCG

The following five assumptions were applied.

(1) Local Equilibrium at Interfaces. The water vapor concentration in air, C_V, that contacts the solution was assumed to be identical to the vapor concentration of the solution.

(2) Vapor Concentration Dependence on Precipitant Concentration. The equilibrium vapor concentration was assumed to be a linear function of only the precipitant concentration in the crystallizing solution, C, as follows:

$$C_V(C) = C_{V_0} + k_{vp}C, \tag{10}$$

where, C_{V_0} means the equilibrium water vapor concentration in air over water without the precipitant and k_{vp} is the coefficient of the water vapor concentration relative to the coexisting precipitant concentration.

Although a more accurate treatment was applied by Fowlis *et al.*,[8] our simplification is sufficient as a first order approximation and it enables us to simplify the final form of the vapor diffusion model.

(3) Homogeneous Precipitant Concentration in the Solution. Precipitant concentration in the crystallizing solution was assumed to be homogeneous. Thus, the concentration can be represented as follows:

$$C = C_0 \frac{V_0}{V}, \tag{11}$$

where C_0, V_0, and V are the initial precipitant concentration, the initial solution volume, and the solution volume of the crystallizing solution, respectively.

(4) Constant Precipitant Concentration in the Reservoir. Since the volume of the reservoir solution was much larger than that of the crystallizing solution, the precipitant concentration in the reservoir, C_P, was assumed to be constant during the vapor diffusion process.

(5) Diffusion Dominant Vapor Transport. No convective effect was considered for vapor diffusion. This assumption was adequate since the water vapor–diffusion coefficient was relatively high and the vapor diffusion apparatus was relatively small.

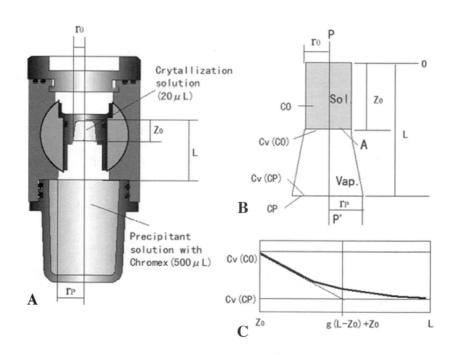

FIGURE 7. Schematic drawings of HDPCG: (**A**) actual configuration, (**B**) geometric model, (**C**) diffusion boundary layer image applied to the HDPCG.

Next, the model was considered. The characteristics of the HDPCG from a geometric point of view are (1) nearly cylindrical solution vessel, and (2) nearly flat solution free face. FIGURE 7 represents the geometry of the HDPCG model. It is clear that V is represented by the height of the solution surface from the bottom of the vessel, Z, and the surface area of the free surface, A, as $V = AZ$. The non-dimensional volume is, therefore, represented as follows:

$$\bar{V} \equiv \frac{V}{V_0} = \frac{Z}{Z_0} \equiv \bar{Z}, \tag{12}$$

where Z_0 is the initial solution height. In addition, the vapor concentration can be approximated as a monotonously decreasing function from the surface of the crystallizing solution to the surface of the reservoir solution, as shown in FIGURE 7C.

Therefore, the diffusion boundary layer thickness, δ, can be approximated by introducing the compensation coefficient, g, as $\delta = g(L - Z)$. Here, L denotes the length from the bottom of the crystallizing solution to the surface of the reservoir solution. Using these conditions, we obtain the following evaporation rate equation:

$$\frac{dV}{dt} = V_0 \frac{d\bar{V}}{dt} = AZ_0 \frac{d\bar{Z}}{dt} = -\frac{AV_wD_Vk_{VP}C_0\left(\frac{1 - C_R\bar{Z}}{\bar{Z}}\right)}{Z_0g\left(\frac{L}{Z_0} - \bar{Z}\right)}, \tag{13}$$

where D_V and V_w represent the diffusion coefficient of water vapor in air and the molar volume of water, and $C_R \equiv C_P/C_0$ is the precipitant concentration ratio—an important control parameter of the evaporation.

The evaporation rate in the initial stage of the evaporation is

$$\left.\frac{dV}{dt}\right|_{V = V_0} = AZ_0\left.\frac{d\bar{Z}}{dt}\right|_{Z = 1} = \frac{AV_wD_Vk_{VP}C_0(1 - C_R)}{Z_0g\left(\frac{L}{Z_0} - 1\right)}. \tag{14}$$

It is obvious that the evaporation rate tends to zero as $\bar{Z}$ approaches $1/C_R$.

Integrating Equation (13) after separating the variables, we obtain

$$\left[(1 - C_RL_F)\{C_Rx + \ln(-1 + C_Rx)\} + \frac{(C_Rx)^2}{2}\right]_1^{\bar{Z}} = \left(-\frac{C_R^3V_wD_Vk_{VP}C_0}{Z_0^2g}\right)t, \tag{15}$$

where $L_F \equiv L/Z_0$. The characteristic time scale for HDPCG is defined by

$$\tau_{HD} \equiv \frac{Z_0^2g}{C_R^3V_wD_Vk_{VP}C_0}. \tag{16}$$

Using this time scale, we rewrite the time in non-dimensional form, $\bar{t} \equiv t/\tau_{HD}$.

The relationship between non-dimensional time and non-dimensional solution height is, therefore,

$$F_{HD}(\bar{Z}) - F_{HD}(1) = \bar{t}, \tag{17}$$

where

$$F_{HD}(x) = (1 - C_RL_F)\{C_Rx + \ln(-1 + C_Rx)\} + \frac{(C_Rx)^2}{2}. \tag{18}$$

Next, consider the compensation factor g. For simplicity, we treat the compensation factor as a constant, although it is actually a function of the distance between the crystallization solution surface and the precipitant solution surface. Employing this treatment, we are able to calculate the factor analytically.

Consider the initial stage of the evaporation and ideal planes that are perpendicular to P–P' in the atmosphere, as represented in FIGURE 7B. We examine the mass balance through the planes. If steady state mass transport is achieved, the following relation must be satisfied:

$$D_V \frac{\partial C_V(Z_0)}{\partial Z} \pi r_0^2 = D_V \frac{\partial C_V(Z_0)}{\partial Z} \pi [r_0 + (Z - Z_0)\tan\theta]^2, \tag{19}$$

where $\tan\theta = (r_P - r_0)/(L - Z_0)$. We obtain the following integration form after the variable separation procedure,

$$\int_{C_V(Z_0)}^{C_V(Z)} dX = \frac{\partial C_V(Z_0)}{\partial Z} \int_{Z_0}^{Z} \left(1 + \frac{\tan\theta}{r_0}(Y - Z_0)\right)^{-2} dY. \tag{20}$$

The solution is

$$C_V(Z) - C_V(Z_0) = \frac{\partial C_V(Z_0)}{\partial Z} \left(\frac{Z - Z_0}{1 + \frac{\tan\theta}{r_0}(Z - Z_0)} \right). \tag{21}$$

Since $C_V(L) = C_V(C_P)$ at $Z = L$ and $C_V(Z_0) = C_V(C_0)$ at $Z = Z_0$, we can represent the concentration gradient at the surface of the crystallizing solution by

$$\frac{\partial C_V(Z_0)}{\partial Z} = \frac{\{C_V(C_P) - C_V(C_0)\} r_P}{r_0(L - Z_0)}. \tag{22}$$

Therefore, the initial value of the diffusion boundary layer thickness, δ_0, can be calculated by using the relation

$$C_V(C_P) = C_V(C_0) + \frac{\{C_V(C_P) - C_V(C_0)\} r_P}{r_0(L - Z_0)} \delta_0. \tag{23}$$

The resulting equation for δ_0 is as follows:

$$\delta_0 = \frac{r_0}{r_P}(L - Z_0), \tag{24}$$

so that $g \cong r_0/r_P$.

The measured values were $r_0 = 1.67 \times 10^{-3}$ m for a 20 μl cell and 1.8×10^{-3} m for a 40 μl cell, and $r_P = 3.47 \times 10^{-3}$ m, $g = 0.4805$ for the former case and $g = 0.5187$ for the latter case. We used this initial value for g during the entire process.

Experiments

We conducted a series of experiments to check the validity of the model. We selected NaCl solution with 50 mM sodium acetate (NaAc) buffer as the sample solution, since NaCl is used as a precipitant in real protein experiments. In addition, the equilibrium water vapor pressure dependence on the coexisting NaCl solution was reported to enable us to calculate the characteristic time scale, τ_{HD}.

Three solutions with different NaCl concentrations (2%, 3%, and 6%w/v) were prepared as the model crystallizing solutions, and NaCl solutions with 4% and 6%w/v concentrations were prepared as model reservoir solutions. The resulting experimental conditions were (2/4), (2/6), (3/6), and (6/6), representing C_0/C_P. Under these conditions, the (6/6) experiment was conducted to check the standard error during an experiment in which no vapor transport should be observed if the experiments were carried out ideally.

Initial solution volumes of the crystallizing solutions were fixed at 20μl except for one experiment with (2/6) concentration to evaluate the effect of the initial solution volume on vaporization rate. For this experiment, 40μl of initial solution volume was used. The reservoir solution volume was fixed at about 570μl (full volume of the reservoir) for HDPCG. The experimental conditions are summarized in TABLE 1.

Experiments were preformed according to the following procedure.

1. The desired number of cells was prepared for one experiment. Twelve cells were prepared for each experiment in this case.
2. Both the crystallizing solutions and the reservoir solutions with desired concentrations were loaded in the cells using a micropipette.
3. The cells were then set into the isothermal incubator. The temperature of the incubator was controlled at $20 \pm 0.1°C$.
4. After the desired time interval, we weighed the crystallizing solution in the desired cell.
5. The rest of the cells were kept in the incubator until the next measurement.
6. The operation was repeated until all the cells were expended.
7. After these measurements, we converted the weight data of the crystallizing solution to volume data and plotted it as a function of preservation time.

Next we measured the solution weight. An electric balance (Metler Tledo AT201, accuracy 0.01mg) was used to weigh the solutions. Small pieces of KIM wipe paper were weighed and prepared in advance. The crystallizing solution was then absorbed into the paper. The total weight of the paper with the absorbed solution was measured. The solution weight was obtained as the weight difference of the paper before and after the absorption. This measurement procedure was completed in five minutes. During the measurements, the room temperature was kept at $23 \pm 1°C$.

Prior to the actual experiments, we checked the applicability of this procedure using HDPCG cells. A model crystallization solution was prepared and 20μl of each

TABLE 1. Experimental conditions

ID	C_0 (% w/v)	C_P (% w/v)	C_R	V_0 (μl)
1	2	4	2	20
2	2	6	3	20
3	3	6	2	20
4	6	6	1	20
5	2	6	3	40

solution was loaded into the cells. After loading the solutions were weighed immediately based on the above procedure. We found that the weight error was less than
± 0.5mg EMBED through six independent measurements.

Quantitative Comparison Between Theory and Experiment

To calculate the characteristic time scale, we need to know D_V, V_W, and k_{VP}. Data
for D_V and V_W at 20°C have been reported as $2.42 \times 10^{-5}\,m^2/sec$ and $1.8048 \times 10^{-5}\,m^3/mol$, respectively.[9,10] k_{VP} at 20°C was calculated using saturated vapor
pressure data for NaCl solution as a function of NaCl concentration at 18°C,[11] satu-
rated vapor pressure data for pure water as a function of temperature,[12] and the vapor
pressure–vapor concentration relationship.

The saturated vapor concentration is well approximated by a linear function of
NaCl concentration, $C_V = 9.60 \times 10^{-1} - 4.68 \times 10^{-3}C$, in the NaCl concentration
range of 2% to 10%w/v, where the units for C_V and C EMBED are mol/m^3 and %w/v,
respectively. Therefore,

$$k_{VP} = -4.68 \times 10^{-3}\,\frac{mol/m^3}{\%w/v}.$$

Using these parameters and Equation **(14)**, we calculated t–V relations for each
experiment. FIGURE 8 shows the results. In this figure, marks indicate the experimental values and lines indicate the calculated values. Experimental data and calculated
data agreed well, even at various initial precipitant concentrations, initial solution
volumes, and initial concentration ratios. Based on these comparisons, we conclude
that the present simplified model reproduced the actual evaporation sequence well.

Practical Method to Evaluate k_{VP} Values

Unfortunately, we cannot directly calculate the evaporation sequence for all kinds
of commonly used precipitants, since there are no data from which to evaluate the

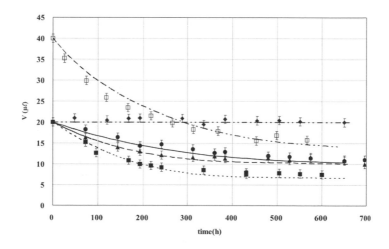

FIGURE 8. Volume change with time: •, 2-4Exp; ——, 2-4Th; ■, 2-6Exp; ····,
2-6Th; ▲, 3-6Exp; – –, 3-6Th; ◆, 6-6Exp; –··, 6-6Th; □, 2-6(40)Exp; –···, 2-6(40)Th.

k_{VP} values. Here, we propose a method to estimate k_{VP} values by using one evaporation experiment data set and the present model. The method is as follows.

1. An evaporation experiment for one experimental condition was performed following the procedure described in the EXPERIMENTS section.
2. The measured solution volume was then transformed to a non-dimensional volume by using Equation (12).
3. Non-dimensional time was calculated from the experimentally obtained non-dimensional volume, using Equations (17) and (18).
4. The non-dimensional time data were then plotted against the dimensional time data recorded during the experiment run.
5. The non-dimensional time versus dimensional time plot must be approximated as a straight line, which crosses the origin of the coordinate system. It is clear that the inclination of the line represents τ_{HD}.
6. From the experimentally obtained τ_{HD} value and Equation (16), we estimate the k_{VP} value, since this is the only unknown parameter.

FIGURE 9 shows the non-dimensional time versus dimensional time plot for the (3,6) evaporation experiment discussed previously. From the figure, we estimated the k_{VP} value as $4.55 \times 10^{-2}\,mol/m^3/\%w/v$, which is nearly equal to the tabulated data. The figure also revealed that our method was useful for estimating the k_{VP} value.

We developed a simple vapor transport model for a microgravity vapor diffusion apparatus, HDPCG, for practical use by experimenters. Evaporation experiments were performed to check the validity of the model. The calculated values reproduced the experimental evaporation sequence well. Moreover, a method to estimate the unknown parameter in the model, k_{VP}, was proposed. After checking the validity of the method, we estimated k_{VP} values for some precipitant systems (data not shown). The values are almost identical for the inorganic precipitant, if the mol/l unit were

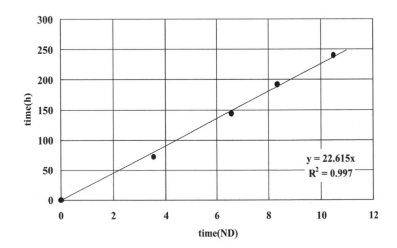

FIGURE 9. Non-dimensional time versus dimensional time plot.

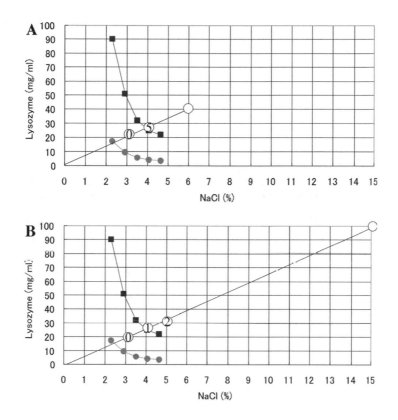

FIGURE 10. Vapor diffusion with 20 mg/ml lysozyme/3% NaCl droplet and (**A**) 6% NaCl reservoir or (**B**) 15% NaCl reservoir.

used for the precipitant concentration. We also found that the value for the PEG precipitant was much less than those for inorganic precipitants and the value may become larger as the degree of polymerization increases (data not shown).

By using Protocols (1) and (2), we were able to estimate the appropriate precipitant concentrations to obtain crystals in a certain time without running many experiments. FIGURE 10 is a phase diagram with a 24-hour crystallization boundary (squares indicate the 24-hour crystallization boundary and circles represent solubility). FIGURE 10A shows the calculation result of the condensation profile when a 6% NaCl reservoir was used. It takes five days for the protein droplet to cross the 24-hour crystallization boundary. On the other hand, a 15% NaCl reservoir was used in FIGURE 10B. In this case, it takes only one day to cross the boundary. The actual crystallization results agreed fairly well with the calculated results.

Protocol (3)—Fluid Dynamic Property Measurement

The third step was fluid dynamic property measurement, necessary for fluid dynamics simulation and the crystal growth study. The diffusion constant, viscosity, density of various samples were measured. Efforts were made to measure these properties with the least amount of protein sample. Here we describe a new method to measure the diffusion coefficient. It has four advantages:

1. it requires small amount of sample,
2. the experimental treatment is relatively simple,
3. there is no need to obtain reflective index data, although we use interferometry,
4. data treatment is simple.

Moreover, it is possible to measure the diffusion coefficient of protein in the precipitant solution. This data is necessary to estimate the actual crystallization kinctics in the crystallization solution. However, this method also enables us to measure both protein and precipitants (inorganic materials) separately. Using light scattering, it was impossible to measure the diffusion coefficient of inorganic materials.

Measurement Method

The measurement system consists of a diffusion pair and a Mach–Zhender interferometer with a 488 nm wavelength Ar laser. The diffusion pair is a general setup to measure the diffusion coefficient by interferometry. Generally, two kinds of solutions with different concentrations are carefully put in a cylindrical capillary vessel and the non-stationary one-dimensional diffusion process at the solution interface is calculated numerically. Here we used a rectangular thin and small vessel with dimensions, 15 mm height, 10 mm width, and 1.5 mm thickness as a diffusion pair (see FIGURE 11). This was useful to reduce both the sample volume and natural convection due to the density difference. The vessel consisted of two slide-glasses and

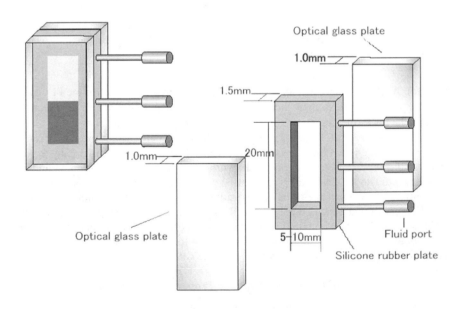

FIGURE 11. Diffusion pair.

a silicon rubber sheet 1.5 mm thick. Slide glasses were used in order to let the laser light pass through.

First, a high concentration of protein solution was poured into the vessel and then the low concentration protein solution was poured carefully on top of the high concentration solution. Then the protein molecules diffused from the high concentration solution to the low concentration solution. Only the protein molecules moved in the vessel, because these solutions were exactly the same except for the protein concentration. The optical lengths in the vessel altered with time due to protein molecule diffusion. The change of optical lengths was visualized as the interferometer fringes. The diffusion coefficient value can be extracted from the fringe information. Density driven convection was suppressed since we put the more dense solution at the lower half of the vessel, and also because we used a thin vessel. The interferometer image actually showed that material transportation was proceeding only by diffusion, but not by convection. An example of the Mach–Zhender interferometer image is shown in FIGURE 12.

It is necessary to know the concentration distribution along with time in order to calculate the diffusion coefficient. In a Mach–Zhender interferometer image, the region with a uniform concentration is shown as a straight line and the region with a concentration gradient is shown as a curved line (FIG. 12). If all information, such as the thickness of the vessel, the reflection index versus protein concentration, and the absolute value of the protein concentration at a certain point, is available, we can calculate the concentration gradient in the vessel by conventional means. However, there are no reflection index data versus protein concentrations for most proteins, nor is it easy to measure them. Therefore, we developed a novel analytical method to calculate the diffusion coefficient without these values.

The solute concentration at a point x at time t can be derived from the diffusion equation with a diffusion pair configuration. It is analytically shown as follows:

$$C(x, t) = \frac{C_2 - C_1}{2}\left[1 - erf\left(\frac{x}{2\sqrt{Dt}}\right)\right] + C_1, \qquad (25)$$

where C_1 and C_2 are the initial solute concentrations in the $x > 0$ and $x < 0$ regions, respectively. The error function is

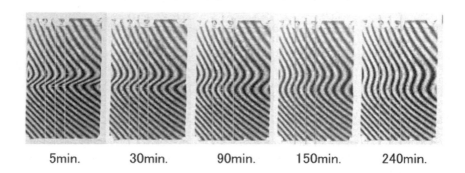

| 5min. | 30min. | 90min. | 150min. | 240min. |

FIGURE 12. Interferometry image.

$$erf(x) = \frac{2}{\sqrt{\pi}} \int_0^x \exp(-\eta^2) d\eta.$$

The concentration gradient in the system is

$$\frac{\partial C}{\partial x} = -\frac{C_2 - C_1}{2\sqrt{\pi D t}} \exp\left(-\frac{x^2}{4Dt}\right). \tag{26}$$

The extreme point of this concentration gradient is where the following equation is maximum:

$$\frac{\partial^2 C}{\partial x^2} = \frac{(C_2 - C_1)}{2\sqrt{\pi D t}} \frac{x}{Dt} \exp\left(-\frac{x^2}{4Dt}\right). \tag{27}$$

The value of this maximum point is equal to the point where the third derivative is 0. In other words, the slope of the fringe changes is maximum at

$$x = \sqrt{2Dt}.$$

We define this length, $L(t)$, as a characteristic length.

The maximum curved point of the fringe can be read from the Mach–Zhender interferometer image directly. Therefore, we can decide the characteristic length $L(t)$. The time change of this length enables us to calculate the diffusion coefficient. During data processing it is important to minimize the interfacial error when the diffusion pair was made and the time lag before starting the observation.

The data obtained from the experiment are the time t from the start of the measurement and the corresponding characteristic length $L(t)$. These data include the error Δx_i of the initial diffusion pair formation and the time lag t_0. Thus, assuming the interfacial concentration gradient in the initial stage of the experiment is the diffusion dominant concentration distribution, the value of Δx_i can be modified by changing the zero time point

$$\Delta t_i = \frac{\Delta x_i^2}{2D}.$$

In other words,

$$L(t) = \sqrt{2D(t + t_0 + \Delta t_i)} \tag{28}$$

is satisfied. Consequently, the square of both sides of this equation is

$$L(t)^2 = 2D(t + t_0 + \Delta t_i) = 2Dt + \text{const.} \tag{29}$$

From this equation, it is possible to approximate a straight line by plotting the progress time and characteristic length $L(t)$ as t–$L(t)^2$. The slope of this line is twice the diffusion coefficient. By using this analytical method, the time lag and the error when the diffusion pair was made is included in the segment. Therefore, these errors do not affect the diffusion coefficient.

Measurement

We used lysozyme as a model protein. This was to check the validity of our measurement and analysis system, because the diffusion coefficient of lysoyme was already known.[13]

Lysozyme, 25mg/ml, 2% NaCl, 100mM acetate buffer (pH4.5) (solution A) and 2% NaCl, 100mM acetate buffer (pH4.5) (solution B) were prepared; 120μl of solution A was poured into the diffusion pair vessel; and then 120 μl of solution B was added on top of solution A. The vessel was observed by using a Mach–Zhender interferometer. The fringe images were recorded on videotape. Later several images were captured from videotape and those images were analyzed on a computer. The characteristic length $L(t)$ versus time was defined from these images. For example, four images were used during the first hour (every 15 minutes) and after the first hour, images were captured every 30 minutes. Measurements continued until the $L(t)$ reached to half the vertical height of the vessel. In the case of lysozyme, the measurement took four hours. The $L(t)$ versus time plot was curved due to the interfacial effect of the container. The reliable data is from the initial time period, consequently the initial two hours (7,200 sec) were used for calculation.

The $t–L(t)^2/2$ plot for Lysozyme is shown in FIGURE 13. The slope corresponds to the diffusion coefficient value and the actual value was calculated to be $1.25 \times 10^{-10} m^2/sec$. Kim *et al.* reported that the lysozyme diffusion coefficient value varied within the range $1.2–1.4 \times 10^{-10} m^2/sec$.[13] Although our method requires small amounts of protein solution and the analysis is very simple, the acquired data corresponds relatively well with the data of Kim *et al.*

Diffusion constants for various proteins and precipitants were measured (data not shown). These values were used for fluid dynamics simulation and to estimate the degree of transport limited crystal growth.

Protocol (4)—Fluid Dynamic Simulation

The last step was to estimate the mass transport in space on the bases of the fluid dynamics simulation transport model. We developed a simulation code for HDPCG. The protein droplet and the vapor phase fluid dynamics were simulated using the diffusion constant, viscosity, and so forth measured in Protocol (3). Some of the simulation results are shown in FIGURE 14. It turned out that neither the vapor phase

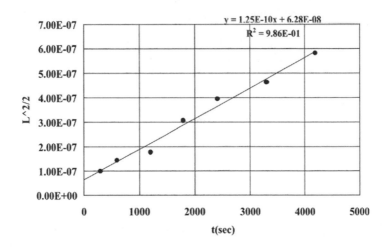

FIGURE 13. $t–L(t)^2/2$ plot for lysozyme.

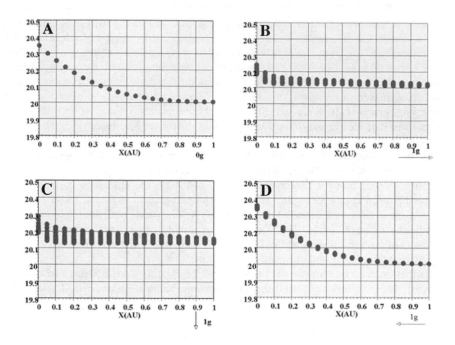

FIGURE 14. The lysozyme concentration in the protein droplet.

nor the solution phase was seriously affected by gravity until nucleation provided the hardware was set in a normal direction. In FIGURE 14, the horizontal axis indicates the protein insert depth. The left side of the horizontal axis is the insert opening and the right side is the insert bottom. The vertical axis is the lysozyme concentration. FIGURE 14 A shows the simulation result at $0g$ and FIGURE 14 B is when the hardware was set upsidedown. FIGURE 14 C is when the hardware was laid sideways and FIGURE 14 D is when the hardware was set in the right direction. It is clear that FIGURES 14 A and D are almost the same, which indicates that this hardware is not sensitive to gravity. (This simulation only deals with fluid dynamics before nucleation, after nucleation the fluid dynamics changes with/without gravity.) Therefore, we conclude that optimized crystallization conditions could be directly applied to microgravity experiments.

CONCLUSIONS

NASDA has developed a practical protocol to optimize the protein crystallization condition for microgravity experiments. By completing the approach, we were able to control the time for nucleation suitable for microgravity experiments. This approach is also useful for ground experiments. By making a phase diagram, it is much more efficient for deciding the vapor diffusion conditions.

ACKNOWLEDGMENTS

The authors thank Professors Takamitsu Kozuma of Ibaraki University, Takashi Yamane of Nagoya University, Junichi Oda of Fukui Pref. University, and Tanokura of Tokyo University for offering their precious protein solution. We also thank Ms. Ari Yamanaka for coordination.

REFERENCES

1. YOSHIZAKI, I., *et al.* 2002. Estimation of crystallization boundary of a model protein as a function of solute concentration and experiment time. Japn. Soc. Micrograv. Appl. **19**(1): 30–33. (Japanese)
2. ATAKA, M. & S. TANAKA. 1986. The growth of large single crystals of lysozyme. Biopolymers **25:** 337–350.
3. VOLMER, M. & A. WEBER. 1926. Nuclei formation in supersaturated states. Z. Phys. Chem. **119:** 277–301.
4. BECKER, R. & W. DORING. 1935. The kinetic treatment of nuclear formation in supersaturated vapors. Ann. Phys. **24:** 719–752.
5. TURNBULL, D. & J.C. FISHER. 1949. Rate of nucleation in condensed systems. J. Chem. Phys. **17:** 71–73.
6. CACIOPPO, E. & M.L. PUSEY. 1992. The solubility of the tetragonal form of hen egg white lysozyme from pH 4.0 to 5.4. J. Cryst. Growth **114:** 286–292.
7. NAKAMURA, H., *et al.* 2003. Theory of vapor diffusion in the high-density protein crystal growth device and its application to the measurement of the vapor pressures of aqueous solutions. J. Cryst. Growth **259**(1–2): 149–159.
8. FOWLIS, W.W., *et al.* 1988. Experimental and theoretical analysis of the rate of solvent equilibration in the hanging drop method of protein crystal growth. J. Cryst. Growth **90:** 117–129.
9. LIDE, D.R., Ed. 1997–1998. CRC Handbook of Chemistry and Physics, 78th edit. CRC Press. 6-205.
10. LIDE, D.R., Ed. 1997–1998. CRC Handbook of Chemistry and Physics, 78th edit. CRC Press. 6-4/6-5.
11. THE CHEMICAL SOCIETY OF JAPAN. 1993. Kagaku-Binran Kiso-hen. Maruzen. 2-139.
12. WEXLER, A. 1976. Vapor pressure formulation for water in range 0 to 100°C. A Revision. J. Res. Nat. Bur. Stand. **80A:** 775–785.
13. KIM, Y.-C. & A.S. MYERSON. 1994. Diffusivity of lysozyme in undersaturated, saturated and supersaturated solutions. J. Cryst. Growth **143:** 79–85.

High Resolution Imaging as a Characterization Tool for Biological Crystals

VIVIAN STOJANOFF,[a] B. CAPPELLE,[b] Y. EPELBOIN,[b] J. HARTWIG,[c]
A.B. MORADELA,[d] AND F. OTALORA[d]

[a]National Synchrotron Light Source, Brookhaven National Laboratories,
Upton, New York, USA

[b]LMCP, UMR 7590 CNRS, Université P.M. Curie, Paris, France

[c]ESRF, BP220, 38043 Grenoble, France

[d]LEC (IACT), Campus Fuentenueva, Granada, Spain

ABSTRACT: Biomolecular crystals consist of large unit cells that form a rather
flexible medium that is able to accommodate a certain degree of lattice distortion, leading to several interesting issues ranging from structural to physical
properties. Several techniques, from X-ray diffraction to microscopy, have
been adapted to study the structural and physical properties of biomolecular
crystals systematically. The use of synchrotron-based monochromatic X-ray
diffraction topography, with triple axis diffractometry and rocking curve measurements, to characterize biomolecular crystals is reviewed. Recent X-ray diffraction images from gel and solution grown lysozyme crystals are presented.
Defect structures in these crystals are discussed, together with reciprocal space
mapping, and compared with results obtained from crystals grown in a low
gravity environment.

KEYWORDS: high resolution X-ray diffraction; biomolecular crystals

INTRODUCTION

X-ray imaging is a widespread technique used in material characterization. A
review of the theoretical and experimental aspects of these methods was presented
recently by Bowen and Tanner[1] and by Authier.[2] With the advent of synchrotron
radiation sources, X-ray imaging methods have played a key role in the improvement
of crystal growth. Since the late 1990s, X-ray diffraction topography[3] and X-ray triple axis diffraction[4,5] have been used to better understand biomolecular crystal
growth.[6–10] The main goal behind the first studies was to find a relation between biomolecular crystal quality and defect structure and the three-dimensional molecular
structure determination of biomolecules.

Biomolecular crystal growth differs from that of other inorganic and organic
materials in that it is a multiparametric growth process that involves many more
parameters than the growth of crystals from inorganic materials. In this process we

Address for correspondence: Vivian Stojanoff, National Synchrotron Light Source,
Brookhaven National Laboratories, Upton NY 11973, USA. Voice: 631-344-8375; fax: 631-
344-3238.
    stojanof@bnl.gov

Ann. N.Y. Acad. Sci. **1027:** 48–55 (2004). ©2004 New York Academy of Sciences.
doi: 10.1196/annals.1324.005

need not only to worry about the physicochemical parameters, but also about biochemical, biophysical, and biological parameters, such as molecular sequence changes, quality of source materials, bacterial contamination, and aging factors.[11] Another very important aspect is the availability of the molecule. For most biomolecules only a few micrograms can be purified to the required level at one time. Typical crystals are of the order of 100 microns, and the tendency is to grow even smaller crystals. Recent developments at third generation synchrotron allow structural studies to be performed on even smaller crystals, 30 to 10 µm in size.

Among the three-dimensional molecular structures of biological molecules known today, 84.7% were determined by X-ray diffraction techniques <http://www.rcbs.org/pdb>. In this method a crystal is rotated through consecutive angular steps and a diffraction image collected at the end of each step, FIGURE 1. Structural biologists frequently describe the quality of their crystals in terms of the diffraction range (resolution) and an average mosaicity value determined from data reduction

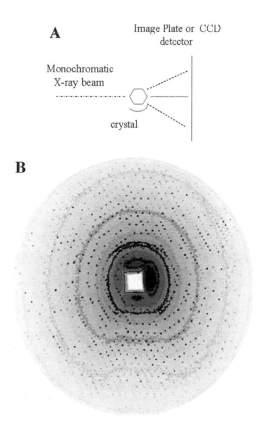

FIGURE 1. The rotation method.[12] **(A)** The crystal is rotated in the monochromatic beam in small angular steps. **(B)** The diffraction pattern is recorded on the X-ray detector at the end of each angular step. The size of the angular steps depends on the crystal lattice and the quality of the crystal. The total angular space covered depends on the crystal lattice.

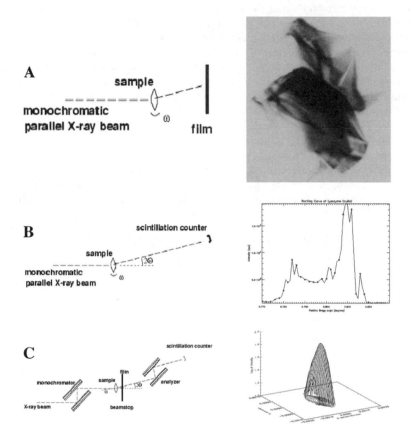

FIGURE 2. High resolution diffraction imaging techniques. **(A)** A sample is bathed in a monochromatic parallel X-ray beam. With X-ray diffraction topography imaging the spatial map of the diffracted intensity as a function of the position in the crystal is recorded on a film or nuclear imaging plate; the sample is kept stationary while a reflex is recorded on a film. The topograph shown is from a hen egg white lysozyme crystal grown by the batch method in gel media at room temperature (pH = 4.5, 5%w/v NaCl). The high contrast observed in this image is typical of crystals grown on the ground. **(B)** The simplest diffractometric measurement is the rocking curve; a proportional counter measures the diffracting power as a function of the incident angle as the sample is moved through the Bragg condition. The curve shown is the rocking curve of reflection (0 14 0) of a turkey egg white lysozyme grown by the vapor diffusion method in the sitting drop configuration. In reciprocal space mapping, the precise measurement of the incident and diffracted X-ray beam makes it possible to distinguish between the mosaic spread determined in **(A)** and the lattice strain; the determination of these directions is only possible with the help of a mono-chromator/analyzer crystal pair and is referred to as triple axis diffractometry. The typical reciprocal space map shown is of a HEWL crystal grown by the dialysis method (pH = 4.5, 5%w/v NaCl) at room temperature. The figure shows the three-dimensional map and the two-dimensional projection, $q_{perpendicular}$ and $q_{parallel}$ are, respectively, proportional to the mosaic spread and strain in the sample.[9]

programs. Although these values, and ultimately the electron density map, permit access to the quality of the crystal, the values are rather indirect and do not allow the study of defects in these structures.

High-resolution diffraction imaging methods provide direct information on the quality of a crystal (see FIGURE 2). These methods, which were essential to the development of growth techniques of nearly perfect silicon crystals, were recently employed by several groups to study the perfection and improve the growth parameters of biomolecular crystals. The fundamental idea is to employ monochromatic coherent synchrotron radiation and combine the rotation method with imaging techniques. The experimental procedure consists of recording a series of diffraction patterns and choosing a single reflection in the diffraction pattern. This reflection is then captured on film (FIG. 2 A). The spatial map of the diffracted intensity versus the position in the crystal is known as diffraction topograph. This image can provide information on the nature of defects and strain and stresses in the crystal. Additional information can be obtained if the film is replaced by a simple scintillation counter set at the Bragg position of the particular reflection under study. As the sample is

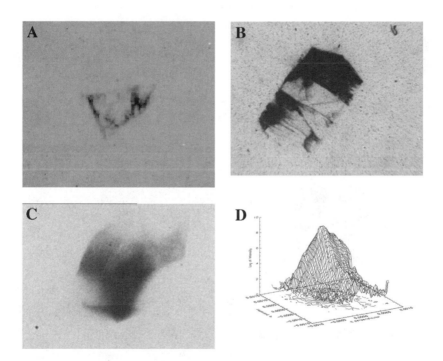

FIGURE 3. Effect of growth methods and growth parameters on the quality of crystals. (**A**) Topographic image of a tetragonal HEWL crystal grown by the dialysis method at room temperature, pH=4.5, 5%w/v NaCl in a 50 mM acetate buffer. (**B**) Image of reflection (16 16 0) of a tetragonal HEWL crystal grown by the batch method in gel media in low salt saturation conditions. (**C**) Topographic image and (**D**) reciprocal space map, at 3.0 Å resolution, of a HEWL crystal grown by the same method as the crystal in (**A**) except for the gravitational field.[9]

rotated through the Bragg reflection the diffracting power is measured as a function of the incident angle. The resulting rocking curve provides overall information on the mosaicity[12] of the sample (FIG. 2B). To distinguish between the mosaic spread and any lattice distortion it is necessary to measure precisely the incident and the diffracted beam direction. This requires the use of triple axis diffractometer and a monochromator analyzer pair (FIG. 2C).[4,5] The two-dimensional map that is obtained makes it possible to distinguish between the mosaic degree and strain in specific directions.[9] High resolution diffraction imaging methods are now routinely employed in the search for better quality biomolecular crystals. Various crystal growth methods were employed, from the traditional vapor diffusion to the batch method, in recently developed methods with oil[13,14] and gel acupuncture.[15,16] The effect of several physical and chemical parameters, such as purity, pH, temperature, protein concentration, and electrical, magnetic, and gravitational fields[9,17–20] have been studied. The main limitation of these methods is the current film–detector resolution limit. Therefore, studies are limited to crystals that are approximately 200 μm in size.

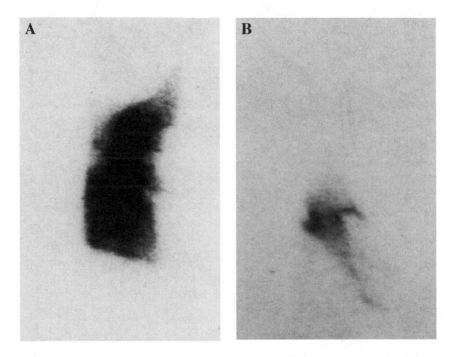

FIGURE 4. Effect of impurities on the quality of crystals. **(A)** Image and rocking curve of reflection (0 14 0) of a turkey egg white lysozyme crystal grown by the vapor diffusion method in the sitting drop configuration at room temperature.[25] The topographic image was taken at the peak of the curve. **(B)** image and rocking curve of reflection (12 8 14) of a turkey egg white lysozyme crystal grown by the vapor diffusion method in the sitting drop configuration at room temperature cocrystallized with 30% HEWL.[25]

As with inorganic materials, the goal is to employ these high resolution imaging techniques to control structural defects and obtain high quality biomolecular crystals. The effect of growth methods and growth parameters have been widely discussed in the literature. Examples are shown in FIGURES 3 and 4. The high contrast observed in the topographic images of tetragonal hen egg white lysozyme (HEWL) crystals grown by the dialysis method (FIG. 3A) and batch in the presence of gel media (FIG. 3B) are quite typical of crystals grown by different methods in a gravitational field. The uniform contrast observed in the image shown in FIGURE 3C from a HEWL crystal grown by the dialysis method in the absence of a gravitational field is also typical.

The defect formation in these biological crystals has been widely discussed. The two images shown in FIGURE 5 are simulations of dislocations in silicon (FIG. 5A) and in HEWL (FIG. 5B). The image shown in FIGURE 5B was obtained using lattice parameters and structure factors from HEWL and silicon elastic constants divided by 10. As one can see, the dislocation image is on a nearly empty background. The low background is due to the narrow Darwin width (0.01 arcsec) observed for protein crystals (the structure factor/cell volume is so small) that leads to a negligible integrated reflectivity. Recently it was shown that it is possible to image individual dislocations far from the Bragg reflection conditions by using weak beam conditions.[21]

The images shown in FIGURES 3 and 4 were obtained at the Bragg peak, when most of the crystal was in reflection. In this condition only a continuous dark contrast can be noticed. The width of the image is proportional to the Burgers vector, **b**, which means that the images are very large and individual contrasts overlap, not allowing the identification of individual dislocations in these images. Far from the

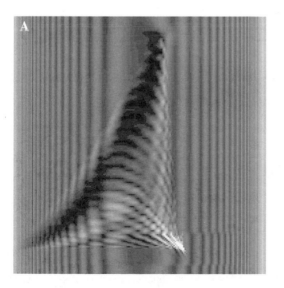

FIGURE 5. Simulation of dislocation diffraction power, **(A)** silicon crystal and **(B)** hen egg white lysozyme. The image in **(B)** was obtained using lattice parameters and structure factors from HEWL and silicon elastic constants divided by 10.[26]

Bragg condition the width of each contrast should become narrow enough to allow the imaging of individual dislocations.

To date, the identification of defects in biological crystals has relied on surface techniques, such as atomic force microscopy (AFM). Growth spirals have been reported by several authors and identified as screw dislocations.[22,23] Dislocations may also create spirals on the surface, as was explained by Strunk.[24] As discussed by Capelle and coworkers,[21] the AFM image for a the screw dislocation is represented by a circular spiral whereas in the case of a curved dislocation the image is elongated in the direction of the inclination of the line. This can in fact be observed in several AFM images presented in the literature; for example, in some of the images presented by Malkin et al.[22] and Yoshizake et al.[23] the circular spirals are elongated to the side.

High resolution diffraction imaging methods are a powerful suite of techniques that allow the characterization of growth defects not only in inorganic but also in biomolecular crystals. Systematic studies of different parameters will allow growth conditions to be optimized and high quality diffraction crystals to be obtained.

ACKNOWLEDGMENTS

The authors thank Dr. D.P. Siddons, for the simulations presented in FIGURE 5. They are indebted to Dr M.C. Robert who kindly helped with the analyses of the topographic images and identification of the various features of their contrast. The assistance of Dr. A. Mazuelas with the ID19 was greatly appreciated. The authors would also thank the ESRF and NSLS staff for their assistance during several beam line sessions that lead to the results shown here.

REFERENCES

1. BOWEN, D.K. & B.K. TANNER. 1998. High Resolution X-Ray Diffractometry and Topography. Taylor & Francis, London.
2. AUTHIER, A. 2002. Dynamical Theory of X-Ray Diffraction. Oxford University Press.
3. HART, M. 1975. Synchroton radiation—its application to high-speed, high-resolution X-ray diffraction topography. J. Appl. Cryst. **8:** 436–444.
4. HOLY, V. & P. MIKULIK. 1996. Theoretical description of multiple crystal arrangements. In X-Ray and Neutron Dynamical Diffraction: Theory and Applications. A. Authier, S. Lagormasino & B.K. Tanner, Eds: 259–268. Plenum, New York.
5. FEWSTER, P.F. 1996. Reciprocal space mapping. In X-Ray and Neutron Dynamical Diffraction: Theory and Applications. A. Authier, S. Lagormasino & B.K. Tanner, Eds: 269–287. Plenum. New York.
6. FOURME, R., A. DUCRUIX, M. RIES-KAUT & B. CAPELLE. 1995. The perfection of protein crystals probed by direct recording of Bragg reflection profiles with a quasi-planar X-ray wave. J. Synchrotron Rad. **2:** 136–142.
7. STOJANOF, V. & D.P. SIDDONS. 1996. X-ray topography of a lysozyme crystal. Acta Cryst. **A52:** 498–499.
8. IZUMI, K., S. SAWAMURA & M. ATAKA. 1996. X-ray topography of lysozyme crystals. J. Cryst. Growth **168:** 106–111.
9. BOGGON, T.J., J.R. HELLIWELL, R.A. JUDGE, et al. 2000. Synchrotron X-ray reciprocal-space mapping, topography and diffraction resolution studies of macromolecular crystal quality. Acta Cryst. **D56:** 868–880.

10. Volz, H.M. & R.J. Matyi. 2000. Triple-axis X-ray diffraction analyses of lysozyme crystals. Acta Cryst. **D56:** 881–889.
11. Ducruix, A. & R. Giege. 1992. Crystallization of Nucleic Acids and Proteins. Oxford University Press.
12. Cullity, B.D. 1978. Elements of X-Ray Diffraction. Addison Wesley Publishing Company, Inc. Reading.
13. Chayen, N.E., P.D. Shaw Stewart & D.M. Blow. 1992. Microbatch crystallization under oil—a new technique allowing many small-volume crystallization trials. J. Appl. Cryst. **23:** 297–302.
14. García-Ruiz, J.M. & A. Moreno. 1994. Investigations on protein crystal growth by the gel acupuncture method. Acta Cryst. **D50:** 484–490.
15. Moreno, A., E. Saridakis & N.E. Chayen. 2002. Combination of oils and gels for enhancing the growth of protein crystals. J. App. Cryst. **35:** 140–142.
16. Ng, J.D., J.A. Gavira & J.M. García-Ruiz. 2003. Protein crystallization by capillary counterdiffusion for applied crystallographic structure determination. J. Struct. Biol. **142:** 218–231. <http://dx.doi.org/10.1016/S1047-8477(03)00052-2" \t "_blank>
17. Stojanoff, V., D.P. Siddons, L.A. Monaco, *et al.* 1997. X-ray topography of tetragonal lysozyme grown by the temperature-controlled technique. Acta Cryst. **D53:** 588–595.
18. Dobrianov, I., C. Caylor, L.G. Lemay, *et al.* 1999. X-ray diffraction studies of protein crystal disorder. J. Cryst. Growth **196:** 511–523.
19. Otálora, F., J.M. García-Ruiz, J.A. Gavira & B. Capelle. 1999. Topography and high resolution diffraction studies in tetragonal lysozyme. J. Cryst. Growth **196:** 546–558. <http://lec.ugr.es/abstracts/abstracts/a1999_12.htm" \t "_blank>
20. Izumi, K., K. Taguchi, K. Kobayashi, *et al.* 1999. Screw dislocation lines in lysozyme crystals observed by Laue topography using synchrotron radiation. J. Cryst. Growth **206:** 155–158.
21. Capelle, B., Y. Epelboin, J. Härtwig, *et al.* 2004. Characterization of dislocations in protein crystals by means of synchrotron double crystal topography. J. Appl. Cryst. **37:** 67–71.
22. Malkin, A.J., Y.G. Kuznestov & A. McPherson. 1999. *In situ* atomic force microscopy studies of surface morphology, growth kinetics, defect structure, and dissolution in macromolecular crystallization. J. Cryst. Growth **196:** 471–488.
23. Yoshizaki, I., T. Sato, N. Igarashi, *et al.* 2001. Systematic analysis of supersaturation and lysozyme crystal quality. Acta Cryst. **D57:** 1621–1629.
24. Strunk, H.P. 1996. Edge dislocations may cause growth spirals. J. Cryst. Growth **160:** 184–185.
25. Hirschler, J., V. Biou, A. Thompson, *et al.* 1998. Presented at the 7th International Conference on the Crystallisation of Biological Macromolecules (ICCBM 7), Granada, Spain, May 3–8.
26. Siddons, D.P. 2000. Private communication.

Nucleation of Insulin Crystals in a Wide Continuous Supersaturation Gradient

ANITA PENKOVA, IVAYLO DIMITROV, AND CHRISTO NANEV

Institute of Physical Chemistry, Bulgarian Academy of Sciences, Sofia, Bulgaria

ABSTRACT: Modifying the classical double pulse technique, by using a supersaturation gradient along an insulin solution contained in a glass capillary tube, we found conditions appropriate for the direct measurement of nucleation parameters. The nucleation time lag has been measured. Data for the number of crystal nuclei versus the nucleation time were obtained for this hormone. Insulin was chosen as a model protein because of the availability of solubility data in the literature. A comparison with the results for hen-egg-white lysozyme, HEWL was performed.

KEYWORDS: protein crystal nucleation; direct measurement with insulin

INTRODUCTION

Nucleation is a key step in the crystallization process. Because it is the first stage, it predetermines to a great extent many features of the subsequent crystal growth. The nucleation rate, for instance, determines the number of crystals that may grow. Unfortunately, due to the inherent difficulties in monitoring the nuclei themselves, present knowledge about the nucleation stage in protein crystallization is insufficient.

Light scattering is widely used in the study of homogeneous protein crystal nucleation.[1,2] Although indirect, it is a powerful investigation technique. Unfortunately, "for data interpretation it is heavily dependent on assumptions about the interactions between the molecules."[3]

Direct determinations of nucleation rates, performed by the classical principle of separating nucleation and growth stages, the so called double pulse technique, have been recently reported.[3–5] The results of these studies were interpreted by applying the classical theory of crystal nucleation, developed originally for inorganic small-molecule crystals. Fundamental quantities, such as the work required for nucleus formation and the number of molecules in the critical nucleus, were determined from data for stationary nucleation rates.

However, the classical double pulse technique is time, labor, and moreover, material consuming. The last circumstance makes the classical double pulse technique almost inadequate, in particular when only limited protein amount is available. Therefore, in order to limit protein consumption we recently used a supersaturation gradient. During the nucleation stage, the supersaturation was varied continuously along a glass capillary tube filled with protein solution at crystallization conditions.

Address for correspondence: Anita Penkova, Bulgarian Academy of Sciences, Acad. G. Bonchev Str., bl. 11, Sofia 1113, Bulgaria. Voice: 3592-9792598; fax: 3592-9712688.
apenkova@ipchp.ipc.bas.bg

Ann. N.Y. Acad. Sci. 1027: 56–63 (2004). ©2004 New York Academy of Sciences.
doi: 10.1196/annals.1324.006

The continuous variation of supersaturation was achieved in this case by applying a continuous temperature gradient, for example, along the capillary.[6] After a given nucleation time, supersaturation throughout the capillary was brought to a lower value (corresponding to growth without additional nucleation) by placing the capillary at constant temperature. Using this method, quantitative studies on nucleation were performed with hen-egg-white lysozyme, HEWL,[7] a model protein with temperature dependent solubility. The same technique can be used with any protein, provided that the temperature dependence of its solubility[8] under given conditions (protein and precipitating agent concentration, pH, etc.) is known. To show this we performed experiments with insulin, since data on the temperature dependence of its solubility are available.[9]

Insulin is a hormone with a protein nature that is secreted from pancreas endocrine cells. It comprises two polypeptide chains of 51 amino acids linked by two disulfide bridges formed after some modifications on insulin precursors. Interest in the details of insulin crystallization is mainly due to the fact that slower drug uptake should be attained by growing equal-sized crystals thus ensuring retarded kinetics of dissolution. This could be achieved by simultaneous crystal nucleation. In this work we measured the nucleation time lag, a fundamental quantity that characterizes nucleation kinetics.[10,11] A comparison with the results for hen-egg-white lysozyme, HEWL was performed. Also data for the number of crystal nuclei, n, versus nucleation time, t, were obtained for this hormone.

EXPERIMENTAL

Measurement Principle

Strict separation in time of the nucleation and growth stages lies at the root of the classical approach for determining nucleation rates at a given supersaturation. The principle is very simple (see FIGURE 1 A). During the first stage, of higher supersaturation, the crystals are (predominantly) nucleated. By keeping this period relatively short, the crystals do not have enough time to grow. If the period is too long Ostwald ripening may occur.[7] After the (expected) onset of nucleation, the supersaturation is rapidly lowered to below the threshold necessary for nucleation, into the so-called metastable zone, so that no further nuclei appear during that second (growth) stage. (The presence or absence of crystals was used to determine the limits of the metastable zone in preliminary experiments.) Only the existing nuclei grow into crystals during the second stage, which is made long enough for crystals to become visible. The crystals are then counted under an optical microscope. The stationary nucleation rate is obtained simply by plotting the number n of crystals grown to visible sizes versus the nucleation time, t. The technique is usually called *double pulse technique*. It enables the measurement of rates of nucleation experimentally, without ever actually seeing the nuclei themselves.

However, in order to count the crystals unambiguously they have to be well formed and clearly separated. Furthermore, their number has to be counted over the entire interval of parameter values under investigation. Sometimes this is very hard and time-consuming task.

In this work we again used the modification of the double pulse technique,[7] the principle of which is depicted schematically in FIGURE 1 B. Instead of moving from one starting (nucleation) value for supersaturation to an end (growth) value, we produced a continuum of starting (nucleation) supersaturation levels, using a temperature gradient. Thus, several measurements of nucleation rates were performed at once, for various values of the supersaturation. The supersaturation gradient is applied along a (cylinder of) protein solution contained in a glass capillary tube (the *nucleation* area in FIG. 1 B). At each point of the capillary tube the supersaturation is kept constant throughout the entire nucleation time period, *t*, and in all experiments of a given series. Temperatures were fixed, at 5°C and 23°C, at the two ends of the capillary and, thus, a nearly linear temperature gradient was established along the tube. Good thermal insulation ensured that the temperature at each point on the capillary remained nearly constant during each stage of the experiment; that is, each "slice" of crystallization solution was moved from a single nucleation temperature, depending on its position along the gradient, to the unique growth temperature.

Although a complication (due to the temperature change), the use of a supersaturation gradient in this study offers practical advantages: it accelerates the measurement procedure, makes it easier, and perhaps more important, it uses limited amounts of protein. Another advantage of this technique is that a relatively wide

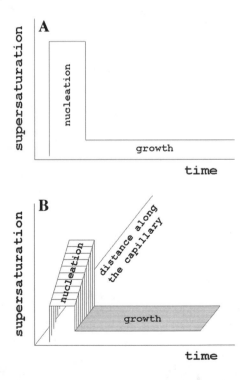

FIGURE 1. A. Schematic depiction of the principle of the classical double-pulse technique. **B.** A supersaturation gradient is applied along the capillary tube (the third axis) during the nucleation stage.

range of supersaturation values is investigated in a single experiment, ensuring that differences in nucleation rates are not due to the inherently imperfect reproducibility of the nucleation process.

Yet another advantage is that the nucleation time lag is measured directly and readily. This is done by simple microscopic inspection: The capillary was scanned and the supersaturation below which nucleation did not take place during this particular nucleation time was recorded. Thus, the time lag was determined without using an *n*–*t*-curve, independently in each measurement. (Usually, the time lag is read from the intersection of the *n* vs. *t* curves with the abscissa.[7])

Experimental Procedure

A glass capillary tube with an internal diameter of 550mm was used. The capillary was graduated in 5-mm segments (18 segments in all). It was then filled with insulin crystallization solution and placed horizontally in the temperature gradient for various time intervals, *t*. Our first task was to find conditions that are appropriate for nucleation investigation.

At the end of the preset nucleation time, the capillary was exposed to a constant temperature of 23 ± 0.5°C, in a constant-temperature room. The supersaturation was thus substantially lowered and kept constant both in time and along the capillary during the growth stage (FIG. 1 B). In preliminary experiments this temperature had been found to be sufficient for crystal growth, but not for nucleation. Growth for approximately two days yielded visible crystals. The number of crystals in each segment

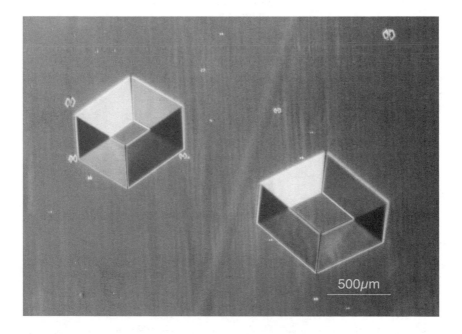

FIGURE 2. Rhombohedral insulin crystals. For growth conditions, see text.

was then counted and plotted as a function both of t and of the (initial, i.e., nucleation) supersaturation. This procedure was performed repeatedly for each t value in order to yield reliable statistics since the stochastic character of the nucleation process means a high inherent data scatter.

Insulin Solution Preparation

Our experiments were carried out with purified porcine insulin. The crystallization solution contained 2 mg/ml protein, 0.005 M $ZnCl_2$, 0.001 M HCl, 15%(v/v) acetone, and 50 mM citrate buffer with pH6.98. These crystallization conditions appeared to be appropriate for applying the thermal double-pulse technique. They produce well faceted rhombohedral insulin crystals, FIGURE 2.

The thermodynamic supersaturation is $\Delta\mu = k_B T\sigma$, where k_B is Boltzmann's constant, T the absolute temperature, and σ depends on the solute activities. Neglecting solution non-ideality, σ may be expressed as the logarithm of the ratio of the concentration, c, over the concentration at equilibrium, c_e, in mg/ml. Thus, in this case, $\sigma = \ln(2/c_e)$. Data for c_e as a function of temperature were taken from Vekilov.[8]

RESULTS AND DISCUSSION

We mentioned that an advantage of our method is the direct and ready determination of the nucleation time lag t. The rationale underlying our procedure is as follows: provided that a sufficiently long time passes, crystal nucleation only took place in the parts of the capillary where the supersaturation was greater than that corresponding to the upper limit of the metastable zone. Note that we do know this limit.

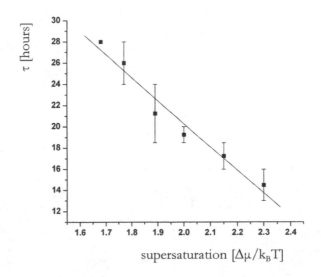

FIGURE 3. Plot of τ versus $\Delta\mu/k_B T$. The dependence is linear. (The point for $\Delta\mu/k_B T = 1.68$ is a result of a single measurement.)

It starts at about $\sigma = 1.29$, and was established in preliminary experiments, simply by presence or absence of insulin crystals.

However, we observed that the shorter the nucleation time we chose, the longer the part of the glass capillary without crystals, including segments where the supersaturation is above that of the upper limit of the metastable zone. This observation simply means that the time chosen was insufficient for nucleus formation. This was confirmed by the fact that, for longer times, we did observe crystals, also created under these lower supersaturations. In practice, we simply recorded the supersaturation segment where the last crystal was "born" (no crystals being present over the entire set of lower supersaturations) during the nucleation time t chosen by us. Consequently, we observed the time lag τ to be supersaturation-dependent. The higher the supersaturation, the shorter the time that appears to be sufficient for the creation of at least one cluster of critical size (i.e., a critical nucleus), see FIGURE 3. (Evidently, τ-data determined in this way corresponds exactly to the data read from the intersection of the n–t-curves with the abscissa.)

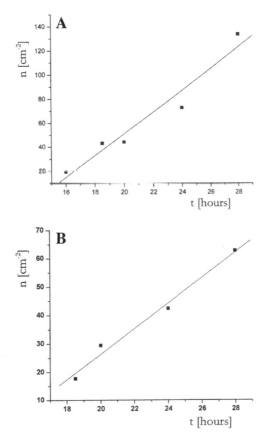

FIGURE 4. Number n of insulin crystals per cm^2 is plotted against t; the mean supersaturation $\Delta\mu/k_BT$ of the corresponding capillary segment, **(A)** $\sigma = 2.3$ and **(B)** $\sigma = 2.15$.

Comparison with the data for HEWL show (almost) the same linear dependence of the time lag τ on the supersaturation $\Delta\mu/k_BT$. Thus, we attributed τ to the corresponding supersaturation value $\Delta\mu/k_BT$. This means that the time needed to temper the glass capillary is only minutes, thus it can be neglected in the case under consideration. In agreement with general expectations, the number n of nucleated crystals increases exponentially with supersaturation, along the capillary. Conversely, the critical supersaturation threshold for nucleus formation depends on the nucleation time, ranging from $\Delta\mu/k_BT = 2.3$ to 1.68, as measured by us.

The physical explanation for the time lag (in the case of small molecule crystals) relies on the fact that steady state is established after a given time. The nucleation process cannot become stationary immediately after the conditions for steady state have been imposed. A certain time elapses before the establishment of the entire set of stationary populations of subcritical and supercritical clusters in the system, even under conditions ensuring steady state. One may have thought that this explanation also remains valid in the case under consideration, especially since large protein molecules move more slowly. An additional explanation lies in the fact that τ is very sensitive to the mechanism of monomer attachment to the nucleus; the difficulties in steric accommodation of the protein molecules into the clusters may, therefore, play an important role. The time required to reach the size of a critical nucleus should also be added (for a comprehensive discussion of the nucleation lag time, see Ref. 10, pp. 249–270).

We started (successfully) to plot n versus t, FIGURE 4. As can be seen, a linear dependence is obtained for the two highest supersaturations applied in our study. However, in order to calculate the work that is required for nucleus formation and the size of the critical cluster we need data for the steady state nucleation rate under other supersaturations. Unfortunately, the accuracy of the results for $\Delta\mu/k_BT = 1.77$, 1.89, and 2.0 is still insufficient, and they are not shown.

CONCLUSION

The applicability of the double pulse technique modification, which uses a supersaturation gradient, has been demonstrated also with insulin. The nucleation time lag depends linearly on the supersaturation; that is, it behaves in accord with the classical theory.

ACKNOWLEDGMENTS

We acknowledge the technical help of Engineer K. Goranov. This project was completed with the partial financial support of the Ministry of Education and Science of Bulgaria under contract YS-X-1305/03.

REFERENCES

1. KAM, Z. & H.B. SHORE & G. FEHER. 1978. On the crystallization of proteins. J. Mol. Biol. **123:** 539–555.

2. MALKIN, A.J. & A. MCPHERSON. 1994. Light-scattering investigations of nucleation processes and kinetics of crystallization in macromolecular systems. Acta Cryst. Sect. D **50:** 385–395.
3. GALKIN, O. & P. VEKILOV. 2000. Are nucleation kinetics of protein crystals similar to those of liquid droplets? J. Am. Chem. Soc. **122:** 156–163.
4. TSEKOVA, D., S. DIMITROVA & C.N. NANEV. 1999. Heterogeneous nucleation (and adhesion) of lysozyme crystals. J. Crystal Growth **196:** 226–233.
5. NANEV, C.N. & D. TSEKOVA. 2000. Heterogeneous nucleation of hen-egg-white lysozyme—molecular approach. Cryst. Res.Technol. **35:** 189–195.
6. LUFT, J.R., D.M. RAK & G.T. DETITTA. 1999. Microbatch macromolecular crystallization on a thermal gradient. J. Crystal Growth **196:** 447–449.
7. PENKOVA, A., N. CHAYEN, E. SARIDARIS & C.N. NANEV. 2002. Nucleation of protein crystals in a wide continuous supersaturation gradient. Acta Cryst. Sect. D **58:** 1606–1610.
8. CHRISTOFER, G.K., A.G. PHIPPS & R.J. GRAY. 1998. Temperature-dependent solubility of selected proteins. J. Crystal Growth **191:** 820–826.
9. BERGERON, L., L. FILOBELO, O. GALKIN & P.G. VEKILOV. 2003. Thermodynamics of the hydrophobicity in crystallization of insulin. Biophys. J. **85**(6): 3935–3942.
10. KASHCHIEV, D. 2000. Nucleation: Basic Theory and Application. Butterworth–Heinemann, Oxford.

Bone Cell Survival in Microgravity

Evidence that Modeled Microgravity Increases Osteoblast Sensitivity to Apoptogens

M.A. BUCARO,[a] J. FERTALA,[a] C.S. ADAMS,[a] M. STEINBECK,[a] P. AYYASWAMY,[b] K. MUKUNDAKRISHNAN,[b] I.M. SHAPIRO,[a] AND M.V. RISBUD[a]

[a]Department of Orthopaedic Surgery, Thomas Jefferson University, Philadelphia, Pennsylvania, USA

[b]School of Engineering and Applied Sciences, University of Pennsylvania, Philadelphia, Pennsylvania, USA

ABSTRACT: Studies were performed to evaluate the effects of modeled microgravity on the induction of osteoblast apoptosis. MC3T3-E1 osteoblast-like cells were cultured in alginate carriers in the NASA-approved high aspect ratio vessel (HARV). This system subjects the cells to a time-averaged gravitational field (vector-averaged gravity) to simulate low gravity conditions. Cells were cultured in the HARV for five days, and then examined for apoptosis. In simulated microgravity, the cells remained vital, although analysis of expressed genes indicated that there was loss of the mature osteoblast phenotype. Additionally, we noted that there was a loss of the mitochondrial membrane potential, a low level of the antiapoptotic protein Bcl-2, as well as Akt protein, and the redox status of the cells was disturbed. All of these parameters indicated that vector-averaged gravity disrupts mitochondrial function, thereby sensitizing osteoblasts to apoptosis. We then used a challenge assay to evaluate the apoptotic sensitivity of the cells subjected to vector-averaged gravity. When challenged with staurosporine, cells subjected to vector-averaged gravity evidenced elevated levels of cell death relative to control cell populations. Another objective of the study was to improve upon conventional carriers by using alginate encapsulation to support cells in the HARV. We have demonstrated that the alginate carrier system affords a more robust system than surface-seeded carriers. This new system has the advantage of shielding cells from mechanical damage and fluid shear stresses on cells in the HARV, permitting carefully controlled studies of the effects of vector-averaged gravity.

KEYWORDS: microgravity; osteoblast; apoptosis; HARV; alginate; encapsulation; vector-averaged gravity; bioreactor; clinostat; polymer scaffold; tissue engineering

Address for correspondence: I.M. Shapiro, Department of Orthopaedic Surgery, Thomas Jefferson University, 1015 Walnut St., 501 Curtis Bldg. Philadelphia, PA 19107, USA. Voice: 215-955-8754; fax: 215-955-9159.
  irving.shapiro@jefferson.edu

Ann. N.Y. Acad. Sci. 1027: 64–73 (2004). ©2004 New York Academy of Sciences.
doi: 10.1196/annals.1324.007

INTRODUCTION

Human space travel is presently limited by biological challenges that include atrophy of the muscular system, immune system dysfunction, fluid loss, electrolyte imbalance, and cardiovascular anomalies.[1–5] Perhaps the most severe is the reductions in bone mass, osteopenia, which accompanies extended spaceflight. We have proposed that the osteopenic effect of microgravity may be the result of an altered rate of osteoblast apoptosis. Within the past seven years, there has been intense study of the regulation of bone cell apoptosis, especially in terms of endocrine, growth factor, and pharmacologic control.[6–11] Although the percentage of cells undergoing apoptosis in bone at any one time is small,[12–15] most of the dying cells are found at sites of active bone resorption.[14,16] This study is directed at reexamining previous observations that vector-averaged gravity causes apoptosis, using a novel system to simulate microgravity in the high aspect ratio vessel (HARV).

To investigate the mechanism of bone loss during spaceflight, land based clinostat bioreactors have been used to simulate the weightlessness experienced in microgravity. The HARV is a popular clinostat used in which the cells are attached to small beads, or carriers, and cultured in the vessel chamber as it rotates. The cells spin on a horizontal axis, thereby causing the gravitational field vector (relative to the orientation of the cell) to be averaged over each revolution to near zero. Carriers with cells seeded on their outer surface have limited value because the cells are subjected to a potentially harsh and poorly controlled mechanical environment. This environment may include frequent collisions between carriers and between carriers and the walls of the HARV, as well as fluid shear stresses on the surface of the carriers, all of which can directly influence bone cell function. Moreover, the number of cells in surface-seeded carriers that can be cultured without a high frequency of collisions is limited and the three-dimensional nature of the adherent cell population is not maintained. To overcome the disadvantages inherent in conventional surface-seeded carriers, we have designed a novel carrier system in which osteoblasts are immobilized within a three-dimensional alginate polymer matrix, isolating cells from collisions and shear forces. By suspending these alginate polymer carriers in the HARV we can explore the impact of vector-averaged gravity on bone cell function, without many of the problems discussed above.

The major objective of the work described herein is to evaluate osteoblast function in vector-averaged gravity. We reasoned that one effect of microgravity is to lower the number of functioning osteoblasts. In this case, the loss of viable cells would be expected to markedly influence skeletal function. In this report, we evaluate cell death in osteoblasts cultured in the HARV to determine if vector-averaged gravity induces osteoblast apoptosis.

MATERIALS AND METHODS

Cell Culture in Alginate Carriers

MC3T3-E1 cells were cultured in Dulbecco's Modified Eagle's Medium (DMEM) supplemented with 10% fetal calf serum. For the initial study, cells at about 80% confluence were trypsinized and cultured in alginate gel beads as previously described. Briefly, cells in culture medium were suspended in an alginate (LVG, Pronova, Norway) stock solution so as to achieve 5×10^6 cells/ml of 1.2% (w/v) alginate. This alginate/cell suspension was forced through a 26-gauge tapered needle, using a droplet generator, into crosslinker bath (100 mM $CaCl_2$) and allowed to gel for 10 min with constant slow stirring. $CaCl_2$ solution was removed by aspiration and the resultant beads were washed twice with HBSS. The beads were then cultured in 15 ml culture medium for 24 h. A variety of parameters, including drop height, electrostatic potential, flow rate, and gelling bath $CaCl_2$ concentration, were optimized to consistently produce spherical carriers of the desired size and consistency. The HARVs were loaded with carriers (5×10^6 cells/ml of alginate) and 50 ml of 10% serum-containing medium and all air bubbles were removed. The cells were cultured in HARVs at 8 rpm for five days inside a cell culture incubator maintained at 37°C and 5% CO_2.

Osteoblast Apoptosis

To study the sensitivity of osteoblasts to apoptogens after exposure to vector-averaged gravity, cells were incubated with 1 μM staurosporine for 4 h in serum-free medium; for this experiment the results were compared to cells cultured in alginate carriers in static culture as our control. Following treatment, carriers were collected washed and dissolved using 25 mM sodium citrate and 10 mM EDTA solution in 150 mM NaCl. An equal number of cells (3×10^4) from each group were plated (in triplicate) in a 96-well plate. The cells were incubated with MTT (0.5 mg/ml in serum free DMEM) at 37°C for three hours. The supernatant was removed, and 200 μl of DMSO was added to each well. The optical density was read at 560 nm using an ELISA plate reader. The decrease in the absorbance is proportional to dehydrogenase activity and provided a direct measurement of the number of viable cells.

Western Blotting

Cells were washed with ice-cold PBS and harvested in 100 μl of mammalian protein extraction reagent buffer (MPER, Pierce, IL) containing NaF (5 mM) and Na_3VO_4 (200 μM). Lysates were centrifuged at 4°C for two minutes at 14,000 g and resolved on 12% SDS-polyacrylamide gels. Proteins were transferred by electroblotting to nitrocellulose membranes (Bio-Rad, CA). The membranes were blocked with 5% nonfat dry milk in TBS (50 mM Tris, pH 7.6, 150 mM NaCl, 0.1%) and incubated overnight at 4°C in 3% nonfat dry milk in TBS-tween with the indicated antibodies (anti-Akt, anti-phospho-Akt serine 473 and anti-actin). Immunolabeling was detected using enhanced chemiluminescence reagent (ECL, Amersham Biosciences, UK).

RT-PCR Analysis

Briefly, total RNA was isolated from cells cultured in alginate beads for five days using trizol reagent (Invitrogen, CA) following instructions from the the the manufacturer. Two micrograms of total RNA was reverse transcribed into cDNA using Superscript II RT enzyme (Invitrogen) and oligo (dT) primers. For PCR, 1 μl of cDNA template was used for each reaction and sequences were amplified using Taq DNA polymerase (Platinum Taq, Invitrogen). The PCR product was visualized on a Kodak 440 imaging station. The results were normalized to the expression of GAPDH.

Mitochondrial Function and Thiol Status

To evaluate cell viability, alginate carriers were rinsed twice and incubated in 2 μM Mitotracker Red, and/or 3 μM Celltracker Green (Molecular Probes, Eugene, Oregon) at 37°C for 20 min. After incubation, Celltracker Green and Mitotracker Red were excited at 488 nm and 567 nm using argon and krypton lasers, respectively, and the resulting fluorescence observed using an Olympus Fluoview Confocal microscope (Olympus, Japan) equipped with long working distance objectives. Optical scans were taken along the Z-axis and the layers were rendered to allow visualization of a 100 μm thick section of the carriers.

Detection of Apoptosis by Annexin-V and PI Dual-Color Flow Cytometry

Quantitation of live and dead osteoblasts that had been cultured for five days in HARV was performed using flow cytometric analysis. Cells were recovered from

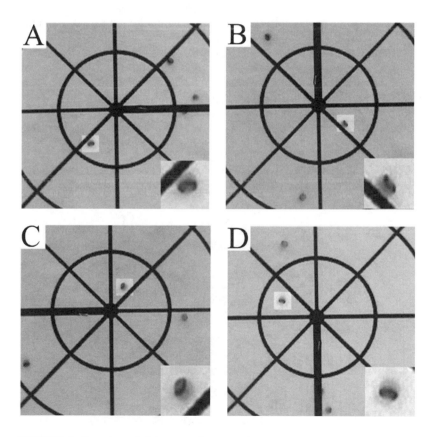

FIGURE 1. Rotation of alginate carriers over the course of one full rotation (counter clockwise) in the HARV with images taken each quarter turn (**A–D**). Alizarin Red and dextran blue were used to visualize their orientation. Note that the alginate carrier rotates with HARV rotation (**inset**).

carriers and then treated with RNase A (2 ng/ml) for 15 min at 37°C. Propidium iodide (PI) and fluorescein isothiocynate-conjugated annexin-V were added to the osteoblast-like cells. After five minutes, stained cells were detected using a Coulter flow cytometer. For each sample 10,000 events were recorded and analyzed. Cells were analyzed by forward and side light scatter into the four subpopulations, and the frequency distribution of live, early apoptotic, late apoptotic, and necrotic cells was determined.

RESULTS

The studies reported here were performed to evaluate the effects of vector-averaged gravity on osteoblast apoptosis in the alginate beads. To confirm that the cells in the alginate carriers experience vector-average gravity, the rotation of the carriers in the HARV was followed by video-microscopy. FIGURE 1 shows the movement of an alginate carrier over the course of one full rotation of the HARV. The insets show that the carrier rotates about itself on a horizontal axis at the same rate as the HARV rotation.

Next, we examined the phenotype of osteoblasts cultured in the alginate beads for a period of five days in static culture. FIGURE 2 shows that the MC3T3 cells maintained their phenotype and expressed a number of characteristic osteogenic genes. These genes included Runx 2, osteocalcin, and osteopontin. The impact of vector-averaged gravity on expression of osteogenic genes is shown in FIGURE 3. We observed profound decreases in the expression levels of osteocalcin, alkaline phosphatase, and collagen type I, as well as moderate reduction in Runx 2, in cells cultured in the HARV relative to static culture control cells.

Since the cells maintained their viability in vector-averaged gravity, we determined if the osteoblasts had become sensitized to apoptosis. For this study the osteoblasts

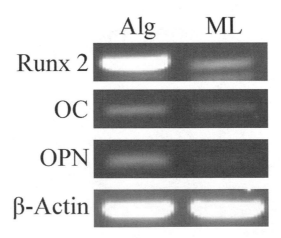

FIGURE 2. PCR analysis of osteoblast phenotypic markers following culture in alginate carriers (*Alg*) in comparison with monoloyer culture (*ML*). Note that alginate increases expression of all genes analyzed.

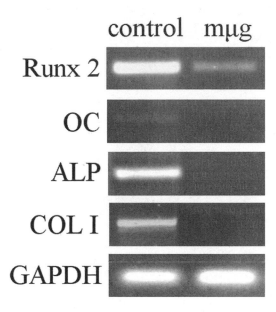

FIGURE 3. PCR analysis of the effects of vector-averaged gravity (mμg) on the osteoblast phenotype. Note that in vector-averaged gravity for five days, expression of Runx 2, osteocalcin (OC), alkaline phosphatase (ALP), and type I collagen (Col I) is dramatically reduced.

were challenged with staurosporine. FIGURE 4 A shows that the apoptogen had a minimal effect on the viability of control cells as measured by the MTT assay. However, cells maintained in the HARV were very sensitive to staurosporine challenge, showing approximately a 50% reduction in MTT signal. Four-quadrant flow cytometric analysis indicated that the cells were apoptotic. Osteoblasts that were exposed to vector-averaged gravity conditions for five days and then treated with the apoptogen exhibited a profound decrease in viability (low annexin fluorescence and increased propidium iodide staining (quadrants 2 and 4, compare FIG. 4B and C).

A similar experiment was performed to assess the effect of vector-averaged gravity on mitochondrial function. FIGURE 5 shows that when control cells and vector-averaged gravity cells were stained with Mitotracker Red, an indicator of the mitochondrial membrane potential, there was a well-defined fluorescence peak, characteristic of cells with functioning mitochondria. However, when the osteoblasts cultured in the HARVs were challenged with staurosporine, there was a loss of fluorescence indicating that these cells displayed a decreased mitochondrial membrane potential. A similar study was performed to evaluate the thiol status of the osteoblasts that had experienced vector-averaged gravity. In this case, the cells were stained with the thiol stain, Celltracker green. FIGURE 6 shows that exposure to vector-averaged gravity profoundly disturbs the oxidative status of the osteoblasts. Thus, there is a profound decrease in fluorescence indicating that there was a loss of reduced thiols.

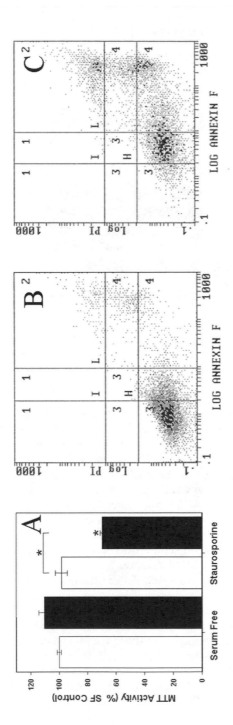

FIGURE 4. Sensitivity of osteoblasts to staurosporine in vector-averaged gravity. (**A**) Vitality of MC3T3-E1 cells maintained in vector-averaged gravity and challenged with 1 μM staurosporine for three hours, measured by MTT (values normalized to control cells treated with serum free media): ☐, control; ■, mμg. (**B**) Flow cytometric analysis of MC3T3-E1 cells maintained in static culture for five days and then treated for three hours with staurosporine and stained with Annexin V–propidium iodide (PI). (**C**) Flow cytometric analysis of MC3T3-E1 cells maintained in vector-averaged gravity for five days, treated for three hours with staurosporine and then stained with Annexin V–propidium iodide (PI). Annexin V and PI stained cells that had been vector-averaged gravity exposed osteoblasts stained with Annexin V. Note that vector-averaged gravity increases the number of apoptotic (*quadrant 4*) and necrotic (*quadrant 2*) cells. Live cells are shown in *quadrant 3*.

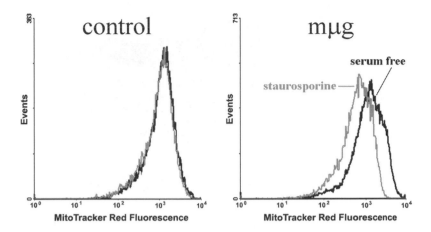

FIGURE 5. Effect of vector-averaged gravity on the response of osteoblast mitochondria to staurosporine. Osteoblasts were maintained in microgravity and then treated with staurosporine for three hours and then stained with Mitotracker Red. Flow cytometric analysis of the control cells showed no change in mitochondrial membrane potential staining. In contrast, vector-averaged gravity-treated osteoblasts showed a reduction in membrane potential following staurosporine treatment.

Finally, Western blot analysis was used to evaluate the status of two critical determinants of apoptosis, Bcl-2 and Akt. FIGURE 7 A shows that vector-averaged gravity depresses Blc-2 protein levels, thereby reducing the protection afforded by this protein against mitochondrial-dependent apoptosis. FIGURE 7 B indicates that the level of the survival factor, Akt (and pAkt) is reduced in the HARV.

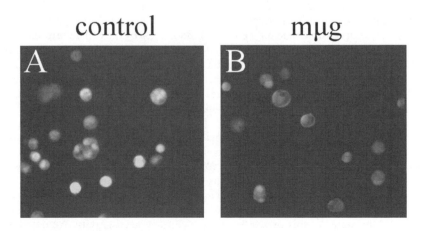

FIGURE 6. Effect of vector-averaged gravity (mμg) on the reduced state of osteoblasts. Following culture in static culture (**A**) and vector-averaged gravity (**B**) cells were stained with Mitotracker green and viewed by confocal microscopy. Note the profound fall in the thiol status of the modeled microgravity-treated cells.

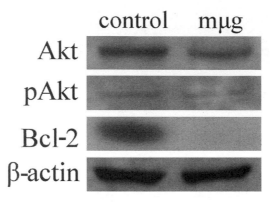

FIGURE 7. Effect of vector-averaged gravity (mμg) on the expression of Bcl-2 and Akt. Note that in microgravity, there is a profound decrease in Bcl-2 expression and a smaller decrease in Akt levels.

DISCUSSION

To overcome the inherent disadvantages of conventional surface-seeded carriers, we immobilized osteoblasts inside an alginate matrix. As FIGURE 1 indicates, these carriers rotate with the rotation of the HARV, vector-averaging the gravitational field over time. Use of these beads overcomes problems that are intrinsic to the use of conventional carriers and allows measurement of the clinorotation of the carriers.

Numerical analysis was used to predict the trajectory of carriers in the HARV.[17,18] These trajectory calculations are critical to the success of these experiments since they ensure that the carriers do not frequently collide with each other or with the boundary walls and elicit cell damage.[19,20] Numerical models allow analysis of the associated mass transfer of nutrients to the cells cultured in the HARV, permitting the calculation of the nutrient and oxygen consumption rates by the cell culture and helping to set the replenishment periods.[18] Aside from the previously mentioned benefits of the alginate carriers, they also allow the effects of vector-averaged gravity to be assessed on large numbers of cells and, therefore, allow the types of analysis used.

The results of the current study indicate that the effects of vector-averaged gravity on the induction of apoptosis are more subtle than previously recognized. Within an alginate matrix, osteoblasts exhibit many of the key indicators of osteogenic cells.[21] In vector-averaged gravity, the cells remain vital, although analysis of expressed genes indicate that the cells do not display a mature osteoblast phenotype. Nevertheless, there is little evidence of apoptosis. We conclude that although the cells remain viable, they are sensitized to staurosporine. Accordingly, these osteoblasts die at a higher rate than control cells when challenged with staurosporine.

Since mitochondria are known to participate in the induction of apoptosis we examined the status of this organelle under vector-averaged gravity. We noted that there was a loss of the mitochondrial membrane potential and a low level of the anti-apoptotic protein, Bcl-2. Correlated with this change, the thiol redox status of the cells was also disturbed and expression of the critical survival pathway enzyme Akt was also low. Since phosphorylated Akt also serves to regulate mitochondrial function, the change in the level of this protein would also serve to promote the apoptotic event in cells exposed to long-term microgravity.

In summary, this study describes a powerful new tool to study the influence of vector-averaged gravity on osteoblast function. We document that, although osteoblast-like cells survive in culture, there are underlying metabolic changes that predispose these cells to apoptosis. Evidence is presented that the cell displays changes at the level of the mitochondrion, possibly mediated by a disturbance in thiol redox, that sensitizes osteoblasts to local environmental challenge. These changes in osteoblast physiology may result in a loss of functioning osteoblasts and, in the whole tissue, development of an osteopenic state.

ACKNOWLEDGMENTS

This work was supported by NASA Grant NAG9-1400.

REFERENCES

1. ATKOV, O.Y. 1992. Adv. Space Res. **12:** 343–345.
2. LEACH, C.S. 1992. Micrograv. Quart. **2:** 69–75.
3. MANZEY, D. & B. LORENZ. 1999. Hum. Perform. Extreme Environ. **4:** 8–13.
4. STROLLO, F. 1999. Adv. Space Biol. Med. **7:** 99–129.
5. VERNIKOS, J. 1996. Bioessays **18:** 1029–1037.
6. DUNSTAN, C.R., R.A. EVANS, E. HILLS, *et al.* 1990. Calcified Tissue Intl. **47:** 270–275.
7. DUNSTAN, C.R., N.M. SOMERS & R.A. EVANS. 1993. Calcified Tissue Intl. **53**(Supl. 6):
8. FROST, H.M. 1960. J. Bone Joint Surg. **42A:** 138–143.
9. BENTOLILA, V., T.M. BOYCE, D.P. FYHRIE, *et al.* 1998. Bone **23:** 275–281.
10. TOMKINSON, A., J. REEVE, R.W. SHAW & B.S. NOBLE. 1997. J. Clin. Endocrinol. Metabol. **82:** 3128–3135.
11. TOMKINSON, A., E.F. GEVERS, J.M. WIT, *et al.* 1998. J. Bone Mineral Res. **13:** 1243–1250.
12. JILKA, R.L., R.S. WEINSTEIN, T. BELLIDO, *et al.* 1998. J. Bone Mineral Res. **13:** 793–802.
13. JILKA, R.L., R.S. WEINSTEIN, T. BELLIDO, *et al.* 1999. J. Clin. Invest. **104:** 439–446.
14. NOBLE, B.S., H. STEVENS, N. LOVERIDGE & J. REEVE. 1997. Bone **20:** 273–282.
15. WEINSTEIN, R.S., R.L. JILKA, A.M. PARFITT & S.C. MANOLAGAS. 1998. J. Clin. Invest. **102:** 274–282.
16. BRONCKERS, A.L.J.J., W. GOEI, G. LUO, *et al.* 1996. J. Bone Mineral Res. **11:** 1281–1291.
17. GAO, H., P.S. AYYASWAMY & P. DUCHEYNE. 1997. Micrograv. Sci. Technol. **10:** 154-165.
18. GAO, H. 2000. Numerical Studies of Microcarrier Particle Dynamics and Associated Mass Transfer in Rotating-Wall Vessels. Ph.D. Thesis, University of Pennsylvania.
19. DUCHEYNE, P., T. LIVINGSTON, I. SHAPIRO, *et al.* 1997. Tissue Engin. **3:** 219–229.
20. QIU, Q., P. DUCHEYNE, H. GAO & P. AYYASWAMY. 1998. Tissue Engin. **4:** 19–34.
21. MAJMUDAR, G., D. BOLE, S.A. GOLDSTEIN & J. BONADIO. 1991. J. Bone Mineral Res. **6:** 869–881.

The Importance of Being Asymmetric

The Physiology of Digesta Propulsion on Earth and in Space

C.P. ARUN

Institute of Urology, University College London, London, United Kingdom

ABSTRACT: In the embryo, the mammalian gastrointestinal tract (GIT) is a midline structure, but later becomes strikingly asymmetric. Such asymmetry is in contrast to other organ systems that are essentially bilaterally symmetric. Making a departure from the traditional straight tube model of the bowel, we offer a more realistic model—a kinked collapsible conduit disposed as a constrained kinematic chain. We examine evidence for the importance of its asymmetry to the unidirectional flow of digesta. A number of factors cooperate to ensure a unidirectional flow. The anatomical factors must include (1) the shape of the abdomen, the inverted truncated cone allowing several degrees of freedom of movement of the bowel allowing folds and twists; (2) the location of the liver and the stomach under the diaphragm, providing for efficient force transmission from the diaphragm (especially in the distended state in the case of the stomach); (3) cranio–caudal gradients in length of the small bowel mesentery and diameter of the bowel lumen. The physiological factors include (1) a deliberate conversion of ingested food into a non-Newtonian fluid with increasing viscosity, (2) nonlinearity of the tube law, (3) respiratory excursions of the diaphragm, and (4) a "Law of the Intestine". In microgravity, the bowel can be expected to float and exhibit loss of polarity of propulsion of digesta, but this can be compensated for by exercise (indirectly by increasing diaphragmatic movement). The asymmetry of the GIT is an ingenious device to ensure a unidirectional movement of digesta.

KEYWORDS: law of the intestine; peristalsis; digesta; low Reynolds number; collapsible tube; tube law

INTRODUCTION

At the time of writing, human travel to space is a reality and the first space tourist has safely returned to Earth. Thanks to the International Space Station (ISS), periods of stay in excess of a year have been achieved. The fact that humans have conquered such a harsh and forbidding environment as space is a testament not just to the strength of the human spirit but also to the endurance and adaptability of our organ systems. One such system whose value is often unappreciated is the gastrointestinal tract (GIT).

Address for correspondence: C.P. Arun, Department of Urology, Colchester General Hospital, Colchester CO4 5JL, England, U.K. Voice: 44-1206-742450; fax: 44-1206-742220.
arunpeter@yahoo.com

Ann. N.Y. Acad. Sci. 1027: 74–84 (2004). ©2004 New York Academy of Sciences.
doi: 10.1196/annals.1324.008

Mammals, like other animal life forms, need to extract nutrients from ingested food. This ingested food then undergoes digestion, involving mechanical and chemical processing to facilitate the later stage of nutrient absorption. As with other organs, the principle of form tailor-made to function is in evidence in the gastrointestinal tract. The type of food ingested and dietary habit involved determines the kind of movement required in its processing. For example, in ruminants, undigested food must be brought back into the mouth to be chewed with corresponding implications for design and function.

In humans, the aboral movement of digesta is important to nutrition (flow in the reverse direction can in certain conditions, be detrimental to health). This unidirectional movement of content appears to be achieved by a combination of factors involving the anatomy, physiology, soft tissue, and fluid mechanics of the GIT, its content, and neighboring structures, all aided no doubt by gravity. Given the circumstances in which it occurs, the predominantly aboral movement of digesta appears to be no small feat even in normal gravity. From classical physics, we know that a net movement of a particle requires an asymmetric net force. In most systems around us, we encounter asymmetry in the direction of the force imparted directly to the particle. The strategy of the gastrointestinal tract however, appears to involve asymmetry in its disposition, such that forces applied from various directions assist wall movement to produce a net aboral movement of content. An examination of how the asymmetry in the lie of the GIT is important to its function would require the application of skills from several fields of science: it is not surprising, therefore, that this issue has not been addressed by previous workers. A brief outline of the relevant aspects of anatomy of the GIT follows.

In humans, ingested food is transferred from the mouth to the esophagus. The esophagus is a conduit fixed at the upper end, the pharynx, and beyond, the cardiac sphincter, is continuous with the stomach. From the esophagus, boluses of food reach the stomach. Inside the abdomen, the presence or absence of a mesentery (a double fold of peritoneum that carries blood vessels and nerves) determines the degree of mobility of the viscus. The stomach is a strong muscular receptacle continuous with the esophagus above and the small intestine below. In turn, the small intestine (see FIGURE 1) is a 4–6-meter long tube separated for purposes of description into the fixed duodenum (except for the last inch or so) and the free jejunum and ileum. The duodenum is attached proximally to the stomach and is continuous with the jejunum and ileum. The jejunum and ileum have a 15-centimeter line of attachment by the mesentery to the posterior body wall. Beyond the ileo-cecal valve, the small intestine is continuous with the large intestine. The large intestine (see FIGURE 2) has, in succession, a fixed ascending part, called the ascending colon; a free part, the transverse colon with a mesentery (the transverse mesocolon); the fixed descending colon; and a free part, the sigmoid colon, that is in continuity with the fixed rectum and anal canal. The physiology of the processes involved in digestion is outlined next.

In the mouth, food is mixed with saliva, chewed, and digestion of carbohydrates is initiated. In the stomach, the majority of the water and all of the alcohol is absorbed, converting the ingested boluses into chyme. In the small intestine, digesta undergoes further chemical processing and more absorption of nutrient occurs. In the large intestine, water is recovered and the unwanted material is prepared for egestion.

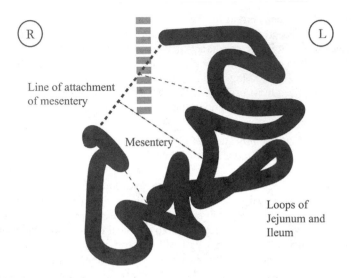

FIGURE 1. The lie of the small intestine. The small bowel is an asymmetrically arranged collapsible conduit with a small fixed part, the duodenum (held down to the posterior abdominal wall) and a long free part (invested by a mesentery that allows movement). Its small line of attachment and the considerably longer free margin allows it to kink (twist and fold) providing for *mobile* physiologic sphincters.

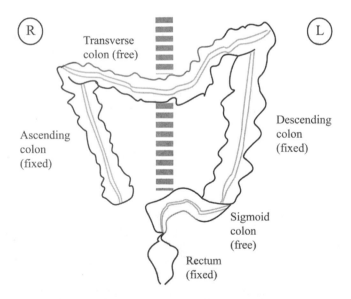

FIGURE 2. The lie of the large intestine. The large bowel is another asymmetrically arranged collapsible conduit with alternating fixed (held down to the posterior abdominal wall) and free parts (invested by a mesentery that allows movement). The junctions of the fixed and free parts must function as *fixed* physiologic sphincters.

Since there is redundancy in both the small and large bowel (much more so in the former) a much larger surface area is available for absorptive processes than a straight tube would allow. Also, since it is a collapsible tube, the folds produce a series of cascades with digesta undergoing "piecemeal digestion" in this "cascade reactor" as it moves along.[1]

Generally, the processes of digestion and absorption are slow and the digesta has to spend time in the lumen of the gut (termed residence time) in order to allow these processes to occur. Nature's solution to this need for long residence times appears to involve altering the consistency of ingested food to a fluid that is intrinsically resistant to quick movement. Common examples from every day life, of fluids that resist rapid motion, termed non-Newtonian fluids, include treacle and gum. In contrast, Newtonian fluids are obedient to Newton's laws of motion and are best exemplified by water (for the non-specialist, a useful *aide memoire*: Newtonian fluids fall with a splash, Non-Newtonian fluids fall with a splat). In the stomach, a nearly Newtonian fluid like milk undergoes processing involving precipitation of the protein and absorption of much of the water, and ends up a non-Newtonian fluid. It must be apparent from the above that digesta that leaves the stomach is essentially a complex (as opposed to simple Newtonian) fluid.

The world we inhabit is vastly different from the world of such complex fluid dynamics and to the majority of us, an unfamiliar one. Although the aboral flow of digesta is natural and normal it is by no means simple. The mechanics of fluid flow can be a trap for the unwary and the mere application of common sense can be unhelpful. Certain phenomena can be frankly counterintuitive (e.g., the Brazil nut phenomenon—after shaking a mixture of nuts and corn flakes, the heavier nuts are found at the top). An easy experiment to try at home is to compare the ease with which a candle flame can be blown out with efforts to try extinguishing the flame by sucking air in. A brief introduction to the relevant aspects of fluid dynamics of relevance to the propulsion of digesta is, therefore, in order here.

The behavior of fluids in flow or of bodies that are moving in fluid can be described using a dimensionless parameter termed the Reynolds number (Re). Named for the nineteenth century English engineer Osborne Reynolds, this number is given by $Re = \rho u d / \mu$, where ρ is density, u velocity, μ viscosity of the fluid, and d the dimension of the channel or body. When the Reynolds number is high, the flow is interpreted to be inertia driven; when low, to be dominated by viscous forces. To illustrate, from the above equation, the flight of a rocket would involve very high Reynolds numbers: the viscosity and density of the air are small, the body dimensions large, and the velocity extremely high. In contrast, digesta has a high viscosity and density, moves in a channel of small dimension, and has a very small velocity; a very low Reynolds number, therefore, governs such flow. Such very low Reynolds number regimes are not in the least uncommon; microorganisms (they far outnumber mammals and insects on earth) inhabit such an environment.

As is evident from the above, digesta in transit in the gut involve the strange world of what physicists would term low Reynolds number (LRN) flow. In such a regime, in the memorable words of Purcell,[2] "what you are doing at the moment is entirely determined by the forces that are exerted on you at that moment, and by nothing in the past." In everyday terms, this implies that there is no such thing as low Reynolds number football—a "kicked" ball would stop at the foot rather than move independently toward the goal. Therefore, for net movement to be successful under such conditions, the tactics employed have to be different. Attempts to move by a to-and-fro undulatory movement of the body employed by most fish or a to-and-fro tail waving laboratory based mechanical fish (see FIGURE 3) are unhelpful here. The pioneering work of Berg and Anderson[3] pointed out that bacteria swim by a peculiar helical action of their flagella. It was later shown that in fact, such asymmetric movements are a must in order to move in low Reynolds number regimes.[2] This is because, under such conditions, solutions of the fluid flow equations are unique and reversible.[4] Purcell[2] further proposed that for movement constrained to one plane, at the least a two-link mechanism would be necessary for locomotion at low Reynolds number. Following up on Purcell's work, more recently, Becker *et al.*[5] reiterated the importance of local anisotropy to swimming motions at low Reynolds number. Reversibility of fluid solutions in LRN regimes explains why a single-link (FIG. 3) mechanical fish can happily swim in water but cannot make progress in corn syrup.[4] Thus, mere to-and-fro movement, the mechanism employed by scallops swimming in water, is inadequate for treacle traversal. It is due to such considerations that attempts to explain unidirectional digesta propulsion by bowel wall movement alone come unstuck. Merely extrapolating from our understanding of Newtonian fluid behavior is inadequate to explain the mechanism of digesta propulsion in the gut.

Humans, like other mammals, have bilateral symmetry and the majority of body systems are more or less symmetric. For example, the left and right hand sides of the face, hands, and feet are normally almost perfectly symmetric. The one organ system that makes a gross departure from such symmetry happens to be the gastrointestinal tract (GIT). Since, in nature, form is firmly wedded to function, it is tempting to speculate that such asymmetry is indeed important to the function of the GIT. With an ingenious but curious layout, the biomechanics and kinematics of the bowel, coupled with LRN fluid dynamics constitute a problem of no mean magnitude for researchers. It is not surprising therefore that previous workers have dealt with very elementary models to try to explain gut propulsion. One such model termed the *law*

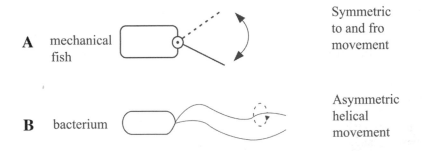

FIGURE 3. Fishy physics—the strange world of low Reynolds number flows. (**A**) The mechanical laboratory swimming device "fish" can move in water by a to-and-fro movement of its "tail". In non-Newtonian fluids like corn syrup or treacle, however, locomotion is not possible. (**B**) Microscopic beings like bacteria are able to achieve locomotion by using helical movements of appendages, such as flagella, because such movements are non-time-reversible.

of the intestine is a favorite with physiologists. In this paper, we examine evidence for GIT asymmetry being of importance to the unidirectional flow of digesta. Since this work is intended for a general scientific audience, various terms have been explained and excessive use of formulæ and equations avoided.

REVIEW OF LITERATURE

The movement of the intestines and the motion of digesta has been of interest to scientists for centuries. The model of Bayliss and Starling[6]—the *Law of the Intestine*—that has held sway for more than a century, presupposed a straight collapsible tube with waves of contraction that are directed distally. Interestingly, their experiments also included observations on solids placed in the bowel lumen—content that is quite unlike what the bowel is accustomed to conveying. Later work, both experimental and theoretical, has also employed a straight tube to model the bowel.[7–9] Such straight tube models are no doubt of relevance to propulsion in the esophagus (food bolus—semi-solid) and ureter (urine—Newtonian content); for the non-Newtonian fluid digesta traversing the loops of the small and large intestine, however, applicability of the law of the intestine is doubtful.

More recently, doubts have been expressed if the law of the intestine is valid—Hodgkiss,[10] working on the avian small intestine model, and Spencer *et al.*,[11] working on the guinea pig ileum, presented evidence against such a law. Work on human tissue is not available as this paper to goes to the press, but observations on bowel exposed at open operation reveals the bowel to be a writhing mass of tubes and no preferred direction of wall contraction is apparent. Clearly, a travelling wave of contraction that does not continuously and consistently occlude the bowel lumen cannot be credited with producing a unidirectional movement of digesta. Since waves of contraction in the bowel are only intermittent, some form of valve mechanism must exist to discourage a backward of flow of content between loops of bowel in the interval periods.

Other works of relevance include the paper by Amidon *et al.*,[12] which recognizes the importance of gravity for the aboral propulsion of digesta proposing that buoyancy, drag, and gravity are crucial factors in flow through the intestine. The monograph of Stevens[13] reviews the comparative anatomy and physiology of mammals and includes a study of the lengths of the various parts of the digestive tract among various species. The arrangement of the gut itself and its possible implications was not studied in these works.

The first model of the small bowel as a constrained kinematic chain was that of Arun[1] who pointed out that the length of the mesentery increases from the jejunum to the ileum and imposes a gradient that is antero-posterior in a supine position and directed inferiorly in the erect posture (i.e., cranio-caudally in the erect and supine positions). In the present work, we extend this model to help understand how bowel action occurs on Earth and how it continues to be possible in space.

MATERIAL AND METHODS

The literature on gastrointestinal physiology relating to propulsion of digesta was reviewed. The human GIT was examined from an anatomic, physiologic, and fluid mechanics standpoint to look for mechanisms that are likely to assist the net unidirectional propulsion of digesta.

RESULTS

By analyzing the GIT as a fluid propulsion system, we obtain interesting findings and a coherent theory of digesta propulsion is made possible. Unlike machines designed by humans, the functions of various biological organ systems are coupled. Such coupling is prominent in the structure and function of the GIT.

The alimentary tract is a system of collapsible conduits in continuity that have free and fixed portions. The distal small intestine is a kinematic chain held down at the duodenal flexure and the ileo-cecal junction, packed into a small conical cavity with play allowed by the length of its mesentery (see FIGURE 4) is kinked into forming loops. At the kinks, the bowel can not only be folded upon itself, but also, due to the significant anteroposterior diameter of the peritoneal cavity, it is capable of twisting (see FIGURE 5). The ends of loops must restrict the flow of content until such time

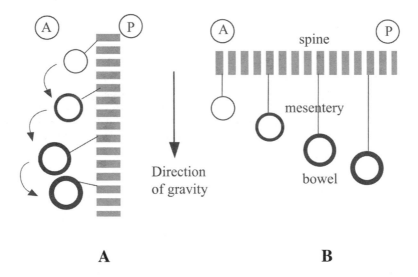

A   **B**

FIGURE 4. Gradient imposed by the small intestinal mesentery and cascading flow in the small intestine. Note the increasing length of the mesentery the cranio-caudal direction. The direction of the gravity gradient is marked at the center of the diagram. The anterior (A) and posterior (P) parts of the diagram are marked in circles. In the upright position (**A**) and the prone position (**B**), the gradient offered by the length of the length of the mesentery is retained. The direction of cascading flow of digesta is marked in (**A**). The increasing diameter and thickness of bowel wall is depicted as well.

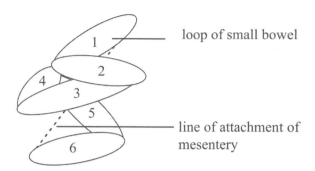

FIGURE 5. Digestion by numbers: the loopy dynamics of propulsion of digesta in the small intestine. In the schematic diagram, the digesta moves from the duodeno–jejunal junction at the start of loop 1, loop to loop, and at the end of loop 6; that is, beyond the ileo–cecal junction, joins the large intestine. The mesentery is depicted as a *dotted line*. Loops 1, 2, and 3 involve an anticlockwise spiral, whereas loops 4, 5, and 6 form a clockwise spiral.

when digestion and absorption is satisfactory. The alternation of fixed and free portions allows the junctions between the fixed and free parts to function as physiologic sphincters. The formation of loops is important to ensure an optimal residence time for digestive processes. When work on the portion of digesta is complete, a local reflex is initiated and wall movement provides a propulsive force for movement into the next bowel loop. Thus, due to the nonlinearity of the tube law for collapsible tubes,[14] the ends of bowel loops function as mobile valves.

The motive force for the movement of digesta is no doubt provided in large part by the contraction of the bowel wall. The mechanisms for implementing polarity though, are several. These may be separated into anatomic and physiologic features. Anatomic features of the GIT imposing polarity must include:

1. The shape of the abdomen a three-dimensional (3D) structure—an inverted truncated cone. The muscular abdomino-thoracic diaphragm forms the base; supporting the smaller lower end is the pelvic diaphragm. The mobility afforded by a 3D enclosure allows movement of the bowel with several degrees of freedom and content cascading from one loop to the next. This allows the junctions between free and fixed parts of the bowel to act as fixed physiologic "sphincters". As mentioned above, in the small intestine, kinks between successive bowel loops may involve not only a fold but also a twist, allowing for several such mobile physiologic sphincters (see FIG. 3).

2. The stomach is located directly under the diaphragm on the left side and correspondingly, the liver lies under the right hemi-diaphragm. Such an arrangement allows transmission of forces from the diaphragm to the intestines. The efficiency of such force transmission, of course, increases when, after a meal, the stomach is full of food.

3. The small intestinal mesentery increases in length from above downwards. This tends to cause smaller loops to form (and be constrained to remain) above (FIG. 5). This polarity persists, both in the erect as well as the prone position, but is lost in the supine position or in a microgravity environment. It is interesting to

note that researchers simulate a microgravity environment in terrestrial laboratories using the supine posture with a slight head down tilt.

4. The lumen of the bowel increases from above downwards. From the Young–Laplace relation ($P = T/R$, where P is the pressure, T the wall tension or hoop stress, and R the radius), and from force–tension relations for muscle, such an arrangement imposes a gradient in the strength of contraction possible.

The physiologic factors imposing polarity must include:

1. Gradients in viscosity of the content. The viscosity is an important determinant of the Reynolds number and the liability of a flow to reverse. The stomach appears to deliberately produce a non-Newtonian fluid from ingested food by precipitating protein and absorbing much of the water and alcohol content from ingested food. From the stomach to the colon, the viscosity increases disproportionate to the density and this inevitably imposes an increasing gradient.

2. The nonlinearity of the tube law for collapsible tubes that allows kinked portions (whether temporary or anatomical) and collapsed portions to function as sphincters.

3. Respiratory excursions of the diaphragm—the importance of the diaphragm to intra-abdominal physiology was recognized as early as the seventeenth century by William Harvey who called it "the engine of the abdomen".

4. Law of the intestine—aboral propagation of circular and longitudinal contractions.

DISCUSSION

The discovery of the circulation of blood by William Harvey (1578–1657) was a watershed event in physiology and the medical sciences. It placed the science of blood flow on a sound scientific basis and for coming up with such a radical idea, rewarded its discoverer, at least initially, with much notoriety. The heart and the circulation inspire much awe even in the mind of the layperson. Such interest is no doubt, justified: the flow of blood at the time of exsanguinations can be terrifying and if unchecked, the effect on life, drastic. Although the importance of the circulatory system cannot be disputed, the mechanism of blood flow is fairly simple when compared with that of the digesta.

In health, the propulsion of digesta is a languid affair and even if food is being enjoyed at the table, the mechanism of what is involved is not likely to be pleasant dinner time conversation. Given its complexity, it is not surprising that unravelling the mechanisms of digesta propulsion has taken much longer. Just as bacteria and spermatozoa employ asymmetric non-time-reversible movements in order to swim, the strategy of the gastrointestinal tract appears to be to employ asymmetry in its lie so as to ensure unidirectional movement of digesta. It is now possible to explain why, in the stage of the embryo, the gut that starts life as a midline structure eventually becomes starkly asymmetric. It appears that this asymmetry is a deliberate and ingenious device to ensure a unidirectional movement of digesta.

It appears that more than one feature is responsible for implementing polarity in digesta flow. An aboral gradient is implemented, not only by gravity, but also by the shape of the peritoneal cavity, its pressure–tension relations, and the influence of the diaphragm and other abdominal muscles. This arrangement emphasizes the coupling of one organ function to another that is a common theme in physiology. We formally recognize that propulsion in collapsible tubes in physiology depends not only on intrinsic wall contraction, termed peristalsis, but also by the self-explanatory neighboring organ assistance (covenant of NOA[15]) phenomenon. In the case of the bowel, NOA is provided by the movement of the diaphragm and the abdominal muscles. It appears that we are rediscovering the contribution of the diaphragm to bowel function that was originally recognized by William Harvey when he credited the diaphragm with being the engine of the abdomen.

The present model involves several subsystems working in collaboration to effect a unidirectional flow of digesta. It involves a multihierarchical multidimensional system. To validate this model, were it ever to be attempted, some very complicated experiments would have to be devised. Barium contrast studies of the small intestine in a microgravity setting would be interesting but are unavailable in the indexed literature. It is not unreasonable to assume that in microgravity the bowel floats freely inside the peritoneal cavity. This would to some extent upset the aboral gradient that the length of the mesentery would normally impose. Whereas the advice for healthy eating is to have a high roughage diet to avoid constipation, payload constraints dictate that astronauts consume a low residue diet. Since the bowel wall requires distension in order to propel content, a low residue can be expected to be passed onward with difficulty. Constipation is a well-known problem among astronauts and GI transit time is known to be reduced. Vigorous exercise that is required of astronauts in order to avoid osteoporosis of the bones can indirectly help the propulsion of digesta by increasing diaphragmatic movements and thereby bowel action. During periods of prolonged residence or flight in space, problems of ileus may occur and the management of such problems carries its own challenges. Robot technology and telepresence technology not withstanding, time lags between Earth and space may preclude safe surgery. The only safe option may be to have a couple of general surgeons on board for long duration flights in space.

The present work has brought home the realization that the gastrointestinal tract is a system with an awe-inspiring complexity of mechanism. No more can we explain gastrointestinal physiology in terms of itself (vitalism) or dismiss it as being simple—as the astute fictional detective Sherlock Holmes would probably have summed up for the perplexed doctor: "It is Alimentary, my dear Watson". The above analysis has demonstrated the vital importance of asymmetry to the unidirectional propulsion of digesta. Since several mechanisms in the alimentary tract, both anatomic and physiologic cooperate to impose polarity, we are afforded a more or less unidirectional propulsion of digesta even if one or more factors, such as gravity, are excluded. Without such asymmetry, let alone travel in space, human life on Earth as we know it would not be possible. This paper is a taster of more work on the propulsion of gastrointestinal and urinary tract content. Our future papers will involve numerical simulations including as many as possible of the several interacting systems from the above complex model.

ACKNOWLEDGMENT

I am grateful to Dr. Henry Oldwinkle, MBBS, House Officer at Colchester General Hospital for assessing the typescript for readability.

REFERENCES

1. ARUN, C.P. 2004. Small intestinal kinematics: a preliminary simulation study. Phil. Trans. Roy. Soc. A. In press.
2. PURCELL, E.M. 1977. Life at low Reynolds number. Am. J. Phys. **45:** 3–11.
3. BERG, H.C. & R.A. ANDERSON. 1973. Bacteria swim by rotating their flagellar filaments. Nature **245:** 380–382.
4. ACHESON, D.J. 1990. Very viscous flow. *In* Elementary Fluid Dynamics. 221–256. Clarendon Press, Oxford.
5. BECKER, L.E., S.A. KOEHLER & H.A. STONE. 2003. On self-propulsion of micromachines at low Reynolds number: Purcell's three-link swimmer. J. Fluid Mech. **490:** 15–35.
6. BAYLISS, W.M. & E.H. STARLING. 1899. The movements and innervation of the small intestine. J. Physiol. **24:** 99–143.
7. KAPUR, J.N. 1985. Models of flows for other biofluids. *In* Mathematical Models in Biology and Medicine. 390–395. East West Press, Pvt. Ltd., New Delhi.
8. BERTUZZI, A., R. MANCINELLI, M. PESCATORI & S. SALINARI. 1979. Biol. Cybernet. **35:** 205–212.
9. JAFFRIN, M.Y. & A.H. SHAPIRO. 1971. Peristaltic pumping. Ann. Rev. Fluid Mech. **3:** 13–35.
10. HODGKISS, J.P. 1986. Intrinsic reflexes underlying peristalsis in the small intestine of the domestic fowl. J. Physiol. **380:** 311–328.
11. SPENCER, N., M. WALSH & T.K. SMITH. 1999. Does the guinea-pig ileum obey the "law of the intestine"? J. Physiol. **517:** 889–898.
12. AMIDON, G.L., G.A. DEBRINCAT & N. NAJIB. 1991. Effects of gravity on gastric emptying, intestinal transit, and drug absorption. J. Clin. Pharmacol. **31:** 968–973.
13. STEVENS, C.E. 1988. The mammalian digestive tract. *In* Comparative Physiology of the Vertebrate Digestive System. 40–85. Cambridge University Press, New York.
14. SHAPIRO, A.H. 1977. Steady flow in collapsible tubes. J. Biomech. Eng. **99:** 126–147.
15. ARUN, C.P. 2003. The covenant of NOA: neighbouring organ assistance to soft tissue structures—a preliminary report. Proceedings of the IMECE, IMECE-42731, November 15–21, Washington D.C.

Modeling of Phosphate Ion Transfer to the Surface of Osteoblasts under Normal Gravity and Simulated Microgravity Conditions

KARTHIK MUKUNDAKRISHNAN,[a] PORTONOVO S. AYYASWAMY,[a] MAKARAND RISBUD,[b] HOWARD H. HU,[a] AND IRVING M. SHAPIRO[b]

[a]Department of Mechanical Engineering and Applied Mechanics, University of Pennsylvania, Philadelphia, Pennsylvania, USA

[b]Department of Orthopedic Surgery, Thomas Jefferson University, Philadelphia, Pennsylvania, USA

ABSTRACT: We have modeled the transport and accumulation of phosphate ions at the remodeling site of a trabecular bone consisting of osteoclasts and osteoblasts situated adjacent to each other in straining flows. Two such flows are considered; one corresponds to shear levels representative of trabecular bone conditions at normal gravity, the other corresponds to shear level that is representative of microgravity conditions. The latter is evaluated indirectly using a simulated microgravity environment prevailing in a rotating wall vessel bioreactor (RWV) designed by NASA. By solving the hydrodynamic equations governing the particle motion in a RWV using a direct numerical simulation (DNS) technique, the shear stress values on the surface of the microcarriers are found. In our present species transfer model, osteoclasts release phosphate ions (Pi) among other ions at bone resorption sites. Some of the ions so released are absorbed by the osteoblast, some accumulate at the osteoblast surface, and the remainder are advected away. The consumption of Pi by osteoblasts is assumed to follow Michaelis–Menten (MM) kinetics aided by a NaPi cotransporter system. MM kinetics views the NaPi cotransporter as a system for transporting extracellular Pi into the osteoblast. Our results show, for the conditions investigated here, the net accumulation of phosphate ions at the osteoblast surface under simulated microgravity conditions is higher by as much as a factor of three. Such increased accumulation may lead to enhanced apoptosis and may help explain the increased bone loss observed under microgravity conditions.

KEYWORDS: osteoblast; phosphate ion; rotating wall vessel; microgravity; direct numerical simulation; apoptosis

NOMENCLATURE:

C	concentration in molar units
D_{ij}	diffusion coefficient of species i through mixture j
f	constant flux od Pi from osteoclasts
$\mathbf{F}_i$	force imposed by the particle on the fluid
$\mathbf{g}$	body force on the fluid
$\mathbf{I}$	inertia moment matrix

Address for correspondence: P.S. Ayyaswamy, Department of Mechanical Engineering and Applied Mechanics, School of Engineering and Applied Science, University of Pennsylvania, Philadelphia, PA 19104-6315, USA. Voice: 215-898-8362; fax: 215-573-6334.
  ayya@seas.upenn.edu

Ann. N.Y. Acad. Sci. 1027: 85–98 (2004). ©2004 New York Academy of Sciences.
doi: 10.1196/annals.1324.009

K_m	Michaelis–Menten constant
m	particle mass
n	index for time step
$\hat{\mathbf{n}}$	unit normal vector
N	number of particles
p	pressure
$\mathbf{r}$	vector from the center of the particle to its surface
$\mathbf{R}$	vector from the center of HARV to the outer wall
t	time
T	moments imposed on the particle by the fluid
$\mathbf{u}$	flow velocity
$\mathbf{V}$	translational velocity of the particle
$\mathbf{x}$	position vector from the origin to any point in the particle
$\mathbf{X}$	position vector of the centroid of the particle
X	distance along the resorbing bone site
V_{max}	maximum rate of phosphate ion consumption
γ	osteoblast seeding density (cells/cm^2)
θ	orientation of the particle
μ	dynamic viscosity of the medium
ρ	density
δ	far-field boundary in the y-direction
ε	separation distance between osteoclasts and osteoblasts
$\boldsymbol{\sigma}$	stress tensor
ω	rotational velocity of the RWV
Ω	angular velocity of the particle
Γ	boundary of the incompressible fluid in the HARV
U	inlet fluid velocity
Subscripts	
f	fluid
i	particle
0	initial condition
∞	far away
m	mesh quantity

INTRODUCTION

Prolonged exposure of astronauts to microgravity leads to major bone loss. The precise mechanisms causing bone loss in microgravity are unknown at present. It may be noted that our knowledge of the adaptation of human bone cells to microgravity remains poor, despite long-term spaceflight experience and the availability of accurate techniques for bone mass measurements. Microgravity has been observed to modify the function of the bone cells and disturb metabolism.[1–4] Histomorphometric studies of osteopenic bone point to microgravity affecting the normal metabolic activities of both osteoclasts and osteoblasts. Cellular changes are reported to occur early and, if prolonged, result in osteopenia. Bone loss is thought to be due to a modest increase in bone resorption and a decrease in bone formation. Some studies suggest that reversibility of bone loss seems unlikely after six months of immobilization or weightlessness. For this reason, prolonged weightlessness of astronauts may present problems on return to Earth, with increased risk of fractures and premature osteoporosis. Thus, there is a great need to gain a fundamental understanding of the effects of microgravity on human bone cell function. Focusing on osteoblasts, we now ask: does microgravity modulate normal osteoblast development and induce cell death? What is the impact of microgravity on terminal activities of osteoblasts, namely the induction of the apoptotic response? In this context, recall that apoptosis is an evolutionary conserved physiologic strategy that ensures the elimination of unwanted or damaged cells from multicellular organisms.[5] The strategy is critical in embryogenesis, organogenesis, and morphogenesis; in the adult state it is required for the maintenance of tissue size, shape, and function. In a very precise study of osteoblasts in culture, Lynch *et al.*[6] clearly showed that, following differentiation, large numbers of osteoblasts became apoptotic. Studies of metaphyseal bone[7,8] indicate that many osteoblasts undergo apoptosis *in situ*. For this reason, it is probable that in areas of rapid bone formation, osteoblast terminal differentiation is short circuited, and instead of completing their maturation pathway, many osteoblasts die. When apoptosis is dysregulated, disease, malformations, and even organism death is evident. Thus, understanding osteoblast apoptosis under simulated microgravity conditions may shed light on the astronaut bone-loss problem.

Under unit gravity, the rate of cell turnover in bone is low. However, during the remodeling cycle, there is an elevated rate of both bone cell proliferation and death.[9] At remodeling sites, in tandem with resorption phase of the cycle, and the subsequent synthesis of new bone, only a few of the activated osteoblasts become terminally differentiated and subsumed into the bone matrix as osteocytes. Some become lining cells, whereas the remainder undergo apoptosis.[10] However, Silver *et al.*[11] have indicated that at resorption sites, there was as much as a ten-fold increase in the local calcium concentration. Although these workers did not measure the local phosphorus ion (Pi) concentration, it would not be unreasonable to assume that the increase in Ca^{2+} was paralleled by an elevation in the level of Pi. If this is the case, then the possibility exists that in localized resorptive environments, osteoblasts are exposed to levels of ions that could disturb normal function. Currently, little is known of agents that upregulate apoptosis and thereby trigger osteoblast death. In our previous studies of cell death in cartilage and bone, we observed that Pi can serve as a powerful apoptogen.[7,12–20] When exposed to 1–7 mM Pi, a dose-dependent increase in cell death was reported. At concentrations of Pi above 5 mM, almost 90% of cells were apoptotic. This effect was completely blocked by the Pi-transport inhibitor, phosphonoformic acid (PFA).[7,12–20] In discussing the physiologic significance of this activity, it was pointed out that at sites of bone remodeling and cartilage calcification, cells are exposed to high levels of Pi. From this point of view, the anion serves as a powerful physiologic cue for skeletal cell apoptosis. In a recent study of human osteoblasts[7,12–20] we have shown that a small increase in the medium Ca^{2+} level synergizes the apoptotic effects of Pi at unit gravity.

Based on these observations we now hypothesize that the local ion increase at the cell surface activates osteoblast apoptosis, and furthermore, the enhanced accumulation under microgravity conditions accelerates the apoptotic rate.

In this study, we numerically model the transport and accumulation of phosphate ions in a basic multicellular unit at a remodeling site of a trabecular bone consisting of osteoclasts and osteoblasts situated adjacent to each other in straining flows. Two such flows are considered; one corresponds to shear levels representative of trabecular bone conditions at normal gravity, the other corresponds to shear level that is representative of simulated microgravity conditions prevailing in a rotating wall

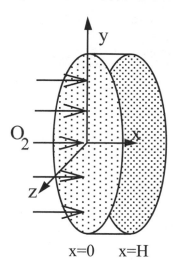

FIGURE 1. Schematic of high aspect ratio vessel (HARV).

bioreactor (RWV) culture designed by NASA. The latter is systematically evaluated in this study by solving the hydrodynamic equations governing particle motion in a RWV, employing a direct numerical simulation (DNS) technique.[21] In our present species transfer model, osteoclasts release phosphate ions among other ions at bone resorption sites. Some of the ions so released are absorbed by the osteoblast, some accumulate at the osteoblast surface, and the remainder are advected away. The consumption of Pi by osteoblasts is assumed to follow Michaelis–Menten (MM) kinetics aided by a NaPi cotransporter system. MM kinetics views the NaPi cotransporter as a system for transporting extracellular Pi into the osteoblast.

The RWV is a microcarrier culture system that simulates several aspects of microgravity on Earth.[22–31] In the RWV, suspension of the microcarrier is maintained by balancing sedimentation induced by gravity with centrifugation caused by the vessel rotation and fluid drag. In RWVs, cells and microcarriers, under carefully determined conditions, remain suspended without turbulence or large shear stresses while receiving adequate oxygenation and nutritional replenishment during the culturing process.

In RWVs, bone cells are usually cultured on microcarrier beads that serve as structural supports. The type of RWV we have studied is called a high-aspect ratio vessel (HARV) (see FIGURE 1). The microcarriers experience very low shear stress in a HARV and their trajectories are such that they eliminate cellular damage resulting from interactions among themselves, as well as with the outer wall of the HARV.

NUMERICAL MODELING

First we determine the trajectory of a typical spherical microcarrier for given rotation speed and the number of microcarriers. We choose a large number of microcarriers, each with a density less than that of the culture medium. This trajectory determination is important to make sure that the particles do not collide with each other or with the wall during their motion. We evaluate the average shear stress at microcarrier surface as a postprocessing calculation.

In the following, we describe the particle trajectory determination for solid particles using DNS.

Particle Trajectory Determination

Our modeling is based on a numerical procedure developed by Hu[21] for computing three-dimensional motions of large numbers of rigid particles. Consider an incompressible fluid in the HARV with the boundary denoted by $\Gamma(t)$. Let there be N particles of spherical shape and prescribed size freely moving in the fluid. We are interested in determining the motion of both fluid and individual particles. The fluid motion has to satisfy the conservation of mass

$$\nabla \cdot \mathbf{u} = 0 \tag{1}$$

and the conservation of momentum

$$\rho_f \left(\frac{\partial \mathbf{u}}{\partial t} + (\mathbf{u} \cdot \nabla)\mathbf{u} \right) = \rho_f \mathbf{f} + \nabla \cdot \boldsymbol{\sigma}, \tag{2}$$

where $\mathbf{u}$ is the fluid velocity vector, ρ_f is the fluid density, $\mathbf{f}$ is the body force, and $\boldsymbol{\sigma}$ is the stress tensor.

For a Newtonian fluid the stress tensor is given by the simple constitutive relation

$$\boldsymbol{\sigma} = -p\mathbf{I} + \eta(\nabla \mathbf{u} + \nabla \mathbf{u}^T). \tag{3}$$

The motion of a particle ($i = 1, 2, ..., N$) is represented by the following equations:

$$m_i \frac{d\mathbf{V}_i}{dt} - \mathbf{F}_i + \mathbf{G}_i, \quad \frac{d}{dt}(\mathbf{I}_i \Omega_i) - \mathbf{T}_i \tag{4}$$

$$\frac{d\mathbf{X}_i}{dt} = \mathbf{V}_i, \quad \frac{d\theta_i}{dt} = \Omega_i, \tag{5}$$

where m_i is the particle mass, $\mathbf{I}_i$ is the inertia moment matrix, $\mathbf{V}_i$ is the translational velocity and Ω_i is the angular velocity of the particle, $\mathbf{X}_i$ is the centroid position, and θ_i is the orientation of the particle. The force acting on the particle is expressed in two terms: $\mathbf{F}_i$ is the force imposed on the particle by the fluid and $\mathbf{G}_i$ is the body force exerted by external fields, such as gravity. The forces and the moments imposed on the particle by the fluid are given by

$$\mathbf{F}_i = \int \boldsymbol{\sigma} \cdot \hat{\mathbf{n}} d\Gamma, \text{ and} \tag{6}$$

$$\mathbf{T}_i = \int \mathbf{r} \times (\boldsymbol{\sigma} \cdot \hat{\mathbf{n}}) d\Gamma, \tag{7}$$

where the integration is performed over the surface of the particle, $\hat{\mathbf{n}}$ is the unit normal vector on the surface of the particle pointing outward, and $\mathbf{r} = \mathbf{x} - \mathbf{X}_i$ is a vector from the center of the particle to its surface.

The boundary conditions for the flow velocity are simply

$$\mathbf{u} = \omega \times \mathbf{R}_w, \text{ on all boundary sections of the vessel } \Gamma_0$$

$$\mathbf{u} = \mathbf{V}_i + \Omega_i \times \mathbf{r}, \text{ on surface of particle } \Gamma_i, i = 1, 2, ..., N, \tag{8}$$

where $|\omega|$ is the prescribed value of the speed of rotation of the vessel and $|R_w|$ is the radius of the HARV. Equation (8) represents the no-slip condition on the surface of the particle. Initially the flow field is assumed to be of uniform motion and the particles are distributed in the domain.

In the solution procedure, we employ a generalized Galerkin finite element formulation that incorporates both the fluid and particle equations of motion in a single variational equation. The spatial domain occupied by the fluid is discretized using the finite element scheme. In the temporal direction, a finite difference method is used. Furthermore, since the spatial domain occupied by the fluid changes with time, a moving finite element mesh is adopted to discretize the fluid domain. To handle the movement of the finite element mesh, an arbitrary Lagrange–Euler (ALE) technique is used. The nodes on the particle surface are assumed to move with the particle, and the grid is updated accordingly. The nodes in the interior of the fluid are computed using the Laplace equation. For a given finite element mesh, the ALE Galerkin finite element formulation can be discretized and be reduced to a non-linear system of algebraic equations, which is solved through a modified-Newton scheme. The corresponding linearized system is solved with an iterative solver using a preconditioned generalized minimal residual (GMRES) algorithm. In this manner, the trajectories and the velocities of the particles are accurately determined.

Initially, the fluid is flowing steadily around the particles. The motion of the combined fluid–particle system is simulated by the procedure developed by Hu.[21] The positions of the particles and the finite element mesh grid points in the fluid domain are updated explicitly, whereas the particle velocities and the fluid flow field are determined implicitly to avoid numerical instabilities. The details of the solution procedure are listed below.

1. Initialization
   $t_0 = 0$, $n = 0$, initialize particle information
   $\mathbf{X}_i(0)$, $\theta_i(0)$, $\mathbf{V}_i(0)$, $\Omega_i(0)$, for $i = 1, 2, ..., N$
   generate an initial mesh $\mathbf{x}_0$
   initialize flow field $\mathbf{u}(\mathbf{x}_0, 0)$, $p(\mathbf{x}_0, 0)$, $\mathbf{u}_m(\mathbf{x}_0, 0)$.

2. Explicit update
   select time step Δt_{n+1}: $t_{n+1} = t_n + \Delta t_{n+1}$, update particle position

$$
\begin{cases}
\mathbf{X}_i(t_{n+1}) = \mathbf{X}_i(t_n) + \Delta t_{n+1}\mathbf{V}_i(t_n) + (\Delta t_{n+1})^2\dot{\mathbf{V}}_i(t_n) \\
\theta_i(t_{n+1}) = \theta_i(t_n) + \Delta t_{n+1}\Omega_i(t_n) + (\Delta t_{n+1})^2\dot{\Omega}_i(t_n)
\end{cases}
$$

   update mesh nodes $\hat{\mathbf{x}}(t_{n+1}) = \mathbf{x}(t_n) + \Delta t_{n+1}\mathbf{u}_m(t_n)$.

3. Remeshing and projection (if the updated mesh $\hat{\mathbf{x}}(t_{n+1})$ is too distorted)
   generate a new mesh $\mathbf{x}(t_{n+1})$
   project the flow field from $\hat{\mathbf{x}}(t_{n+1})$ onto $\mathbf{x}(t_{n+1})$.

4. Flow solver: iteratively solve the new flow field

$$
\mathbf{u}(\mathbf{x}(t_{n+1}), t_{n+1}), \quad p(\mathbf{x}(t_{n+1}), t_{n+1}),
$$

$$
\mathbf{u}_m(\mathbf{x}(t_{n+1}), t_{n+1}), \quad \mathbf{V}_i(t_{n+1}), \quad \Omega_i(t_{n+1}).
$$

5. If the time t_{n+1} is less than a specified time go to step (2); otherwise stop.

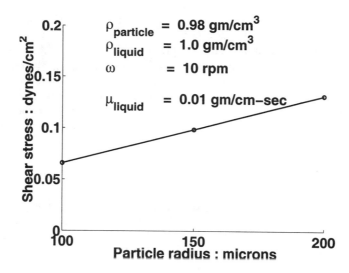

FIGURE 2. Shear stress variation as a function of particle size.

Using this procedure, the velocities of the particles and their trajectories in the HARV are calculated.

FIGURE 2 shows the variation in shear stress with particle radius for a solid microcarrier. As expected, shear stress increases with increasing particle size and the very small value of shear stress is noteworthy. FIGURE 3 shows the net acceleration

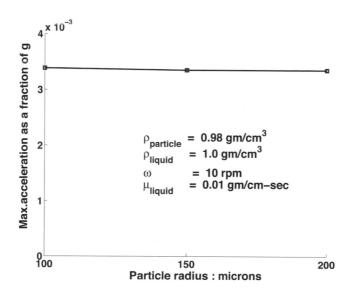

FIGURE 3. Net acceleration as a function of particle size.

experienced by a typical solid microcarrier[23,24] as a function of its size. It can be seen that the nearly neutrally buoyant microcarrier experiences simulated microgravity while suspended in the HARV during culture.

We observe that very low values of shear stress may be obtained by reducing the size of the microcarrier. After having determined the shear stress levels experienced by the osteoblasts on the microcarrier surface in a simulated microgravity environment, we now evaluate the mass transfer of phosphate ions to the surface of osteoblasts and its relation to apoptosis using a newly developed model.

MASS TRANSFER OF PHOSPHATE IONS

Currently, the exact role of Pi in apoptosis is unknown. However, in ground based studies at unit gravity, we have observed that exposure to high levels of phosphate ions induces apoptosis in osteoblasts. It is on this basis that we now hypothesize that Pi accumulation under simulated microgravity conditions enhances osteoblast apoptosis. However, to experimentally evaluate or numerically model this complex phenomenon in a direct way is difficult.

Fluid flow in the trabecular spaces (i.e., between trabeculæ) has been implicated in a number of important physiologic phenomena.[32] The pressure gradients in the marrow and interstitial fluid created by the mechanical loading of trabecular bone deform the trabecular matrix. The resulting fluid flow exerts shear stress on bone cells.[33–35] Recall that in remodeling sites of trabecular bone, osteoblasts and osteoclasts are lined up next to each other and osteoclasts resorb the hydroxyapatite mineral (calcified bone) and elevate the level of apatite ions. The apatite ions diffuse and convect in the prevalent shear field toward the osteoblasts. Depending on the rate of consumption of the ions, there will be a net accumulation at the osteoblast surface. If this accumulation exceeds a threshold value, it may lead to apoptosis. It is well known that in the normal resorbing bone at unit gravity, the level of Ca^{2+} concentration is in excess of $10\,mM$; although comparable figures on Pi are not available, they are probably about $4\,mM$; (Ca^{2+}/Pi apatite molar ratio is 1.67). With this information, we have now estimated the Pi accumulation at the surface of osteoblasts in shear fields with strengths representative of simulated microgravity and normal gravity.   We choose a basic multicellular unit consisting of osteoclasts adjacent to osteoblast.

A Model for the Study of Mass Transfer of Ions

The model consists of a line source of osteoclasts that release apatite ions at a constant rate, such that the serum level of Pi is elevated to about $4\,mM$ under normal gravity. Adjacent to this line source, is a line sink of osteoblasts. The source and the sink are taken to be one cell apart. The average size of the osteoblast is assumed to be $10\,microns$ and ten such osteoblasts are considered. In the prevailing shear field the ions released by the osteoclasts will both diffuse and advect toward the osteoblasts. At the osteoblast surface, the phosphate ion consumption is known to follow a saturable kinetics,[36] which can be described by Michaelis–Menten equations. Thus, the mass transport is given by

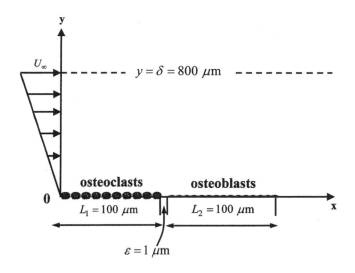

FIGURE 4. Schematic of the remodelling unit.

$$\frac{\partial C}{\partial t} + \mathbf{u}(\mathbf{x}) \cdot \nabla C(\mathbf{x}) = D_{ij}\nabla \cdot \nabla C(\mathbf{x}). \tag{9}$$

In this equation $C(\mathbf{x})$ is the concentration of Pi, $\mathbf{u}(\mathbf{x})$ is the prevailing velocity field (shear field), V_{max} is the maximum rate of phosphate ion consumption, and K_m is the Michaelis–Menten constant.[36] The initial condition is taken to be

$$C(\mathbf{x}, t = 0) = C_0 = 0.2\,\text{mM}, \tag{10}$$

where C_0 is the physiologic level of Pi. The geometry is shown in FIGURE 4. The boundary conditions are

Inlet: physiologic level of Pi concentration in the flow

$$x = 0, \quad 0 \le y \le \delta, \quad C = C_0. \tag{11}$$

Osteoclast surface: release of Pi

$$y = 0, \quad 0 < x \le L_1, \quad D_{ij}\frac{\partial C}{\partial y} = -f\,\frac{\mu\text{mole}}{\text{cm}^2\text{sec}}. \tag{12}$$

Intercellular Gap:

$$y = 0, \quad L_1 < x < L_1 + \varepsilon, \quad \frac{\partial C}{\partial y} = 0. \tag{13}$$

Osteoblast surface: consumption of Pi

$$y = 0, \quad L_1 + \varepsilon \le x \le L_1 + \varepsilon + L_2, \quad D_{ij}\frac{\partial C}{\partial y} = \frac{\gamma V_{max}C}{K_m + C}. \tag{14}$$

Far-field conditions:

$$y = 0, \quad L_1 + \varepsilon + L_2 < x < \infty, \quad \frac{\partial C}{\partial y} = 0 \tag{15}$$

$$0 \le y < \delta, \quad x \to \infty, \quad \frac{\partial C}{\partial x} = 0 \tag{16}$$

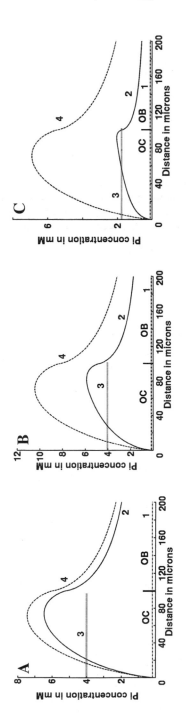

FIGURE 5. Pi accumulation on the remodeling surface at an imposed physiologic shear stress of **(A)** 0.01 dyne/cm^2, **(B)** 0.1 dyne/cm^2, and **(C)** 1.0 dyne/cm^2 and a microgravity shear stress of 0.001 dyne/cm^2.

$$0 < x < \infty, \quad y = \delta, \quad \frac{\partial C}{\partial y} = 0. \tag{17}$$

The above equations are solved for three possible levels of physical activity at normal gravity that correspond to three different stress values of 0.01, 0.1, and 1 dyne/cm^2 at both osteoclast and osteoblast surface. The formulation is also solved for one shear stress condition representative of simulated microgravity (0.001 dyne/cm^2). We note that in Equation (12), f denotes the flux of phosphate ions released by osteoclast at the resorption site. It is known that the average concentration of Pi in the vicinity of osteoclast at the resorption site is about 4mM. In our numerical calculations, we maintain this average concentration to be 4mM for the osteoclast region and, for a given shear stress level, determine the value of f. On this basis, for the osteoclast surface condition given by (12), f takes on values of 0.0085, 0.012, and 0.02 for imposed shear stress values of 0.01, 0.1, and 1 dyne/cm^2 in the flow field, respectively. The values of the other parameters used are $\gamma V_{max} = 0.0178$ nmol/(cm^{2}sec), $K_m = 448$ nmol/cm^3, $\gamma \approx 10^5$ osteoblasts/cm^2.[36]

RESULTS AND DISCUSSION

The transient formulation for the transport of ions given by Equation (12) subjected to initial and boundary conditions given by Equations (13)–(20) have been solved by finite element techniques. The solution tends to a finite steady state, as displayed in FIGURE 5. We have assumed that the flux of Pi release by the osteoclasts is constant, independent of the level of shear field. The value is such that the average concentration of Pi in the resorption site populated by the osteoclasts is approximately 4mM. In FIGURE 5, OC and OB denote osteoclasts and osteoblasts, characteristic 1 (dotted flat line) denotes the physiologic (serum) concentration of Pi in the extracellular fluid (0.2mM). Characteristic 2 describes the distribution of Pi at unit gravity and shear levels of 0.01, 0.1, and 1 dyne/cm^2 in FIGURE 5A, B, and C, respectively. The flat characteristic 3 denotes 4mM concentration level of Pi, which is an average level of Pi at the resorption site at normal gravity. Characteristic 4 describes the distribution of Pi at simulated microgravity and a shear level of 0.001 dyne/cm^2 that is representative of simulated microgravity conditions. As our results show, due to the weaker shear field prevailing under microgravity conditions compared to that under normal gravity, the accumulation of Pi at the osteoblast surface and hence the "dwell time" (which characterizes the duration of exposure of osteoblasts to elevated levels of Pi) are higher in all cases. These increased levels of concentration of Pi at the osteoblast surface may upregulate apoptosis and thereby trigger osteoblast death at an increased rate under simulated microgravity conditions compared to normal gravity.

CONCLUSIONS

Simulated microgravity conditions of the HARV impart a shear stress of 0.1 dyne/cm^2 at the surface of nearly neutrally buoyant 150-micron solid microcarrier. The magnitude of the net acceleration vector experienced is about 0.004g, where g is the acceleration due to gravity at the Earth surface.

Our calculations, with a model of basic multicellular unit that consists of osteo-clasts adjacent to osteoblasts in shear fields representative of microgravity and nor-mal gravity conditions, show that, for identical release of Pi by osteoclasts at the resorption site, the accumulation, and hence possibly the dwell time of Pi under sim-ulated microgravity conditions at the surface of osteoblasts, are very much higher than those encountered under normal gravity conditions. This supports the hypothe-sis that the local ion increase at the cell surface may activate osteoblast apoptosis, and furthermore, the enhanced accumulation under microgravity conditions may accelerate the apoptotic rate.

ACKNOWLEDGMENTS

The authors gratefully acknowledge support of this work by NASA under NAG 9-1400. We gratefully acknowledge the many insightful discussions with Pro-fessors Christopher S. Adams and Marla Steinbeck of Thomas Jefferson University, Philadelphia.

REFERENCES

1. FREED, L.E. & G. VUNJAK-NOVAKOVIC. 1995. Cultivation of cell-polymer tissue con-structs in simulated microgravity. Biotech. Bioeng. **46:** 306–313.
2. FREED, L.E., R. LANGER, I. MARTIN, *et al.* 1997. Tissue engineering of cartilage in space. Proc. Natl. Acad. Sci. USA **4:** 13885–13890.
3. Almeida, E. 2003. The effects of microgravity on cells. <http://generation.arc.nasa.gov/microgravity_ cells.pdf>.
4. SCHATTEN, H., M.L. LEWIS & A. CHAKRABARTI. 2001. Spaceflight and clinorotation cause cytoskeleton and mitochondria changes and increases in apoptosis in cultured cells. Acta Astronaut. **49:** 399–418.
5. ISHIZAKI, Y., L. CHENG, A.W. MUDGE & M.C. RAFF. 1995. Programmed cell death by default in embryonic cells, fibroblasts, and cancer cells. Mol. Biol. Cell **6:** 1443–1458.
6. LYNCH, M.P., C. CAPPARELLI, J.L. STEIN, *et al.* 1998. Apoptosis during bone-like tissue development *in vitro*. J. Cell. Biochem. **68:** 31–49.
7. HATORI, M., K.J. KLATTE, C.C. TEIXEIRA & I.M. SHAPIRO. 1995. End labeling studies of fragmented DNA in the avian growth plate: evidence of apoptosis in terminally differentiated chondrocytes. J. Bone Mineral Res. **10:** 1960–1968.
8. BRONCKERS, A.L., W. GOEI, G. LUO, *et al.* 1996. DNA fragmentation during bone for-mation in neonatal rodents assessed by transferase-mediated end labeling. J. Bone Mineral Res. **11:** 1281–1291.
9. NOBLE, B.S., H. STEVENS, J.R. MOSLEY, *et al.* 1997. Osteocyte apoptosis and func-tional strain in bone. J. Bone Mineral Res. **12:** O5–O5.
10. ZIMMERMAN, D.L., R. GLOBUS & C. DAMSKY. 1997. *In vivo* and *in vitro* models of altered integrin function in osteoblasts: effects on osteoblast maturation and bone remodeling. J. Bone Mineral Res. **12:** 211.
11. SILVER, I.A., R.J. MURRILLS & D.J. ETHERINGTON. 1998. Microelectrode studies on the acid microenvironment beneath adherent macrophages and osteoclasts. Exp. Cell Res. **175:** 266–276.
12. MELETI, Z., I.M. SHAPIRO & C.S. ADAMS. 2000. Inorganic phosphate induces apoptosis of osteoblast-like cells in culture. Bone **27:** 359–366.
13. MELETI, Z., I.M. SHAPIRO & C.S. ADAMS. 2000. Inorganic phosphate induces apoptosis in osteoblast-like cells. J. Dental Res. **79:** 170.

14. ADAMS, C.S., K. MANSFIELD, R.L. PERLOT & I.M. SHAPIRO. 2001. Matrix regulation of skeletal cell apoptosis—role of calcium and phosphate ions. J. Biol. Chem. **276:** 20316–20322.
15. PERLOT, R.L., I.M. SHAPIRO, K. MANSFIELD & C.S. ADAMS. 2002. Matrix regulation of skeletal cell apoptosis II: role of arg-gly-asp-containing peptides. J. Bone Mineral Res. **17:** 66–76.
16. SHAPIRO, I.M., K. MANSFIELD, C.M. TEIXEIRA, *et al.* 1999. Mechanism of induction of apoptosis in epiphyseal chondrocytes: role of phosphate ions and NO. J. Bone Mineral Res. **14:** S434.
17. ADAMS, C.S. & I.M. SHAPIRO. 2003. Mechanisms by which extracellular matrix components induce osteoblast apoptosis. Connect. Tissue Res. **44:** 230–239.
18. ADAMS, C.S., Z. MELETI, K. MANSFIELD & I.M. SHAPIRO. 1999. Osteoblast apoptosis is induced by inorganic calcium and phosphate ions. J. Bone Mineral Res. **14:** S346.
19. MANSFIELD, K., C.C. TEIXEIRA, C.S. ADAMS & I.M. SHAPIRO. 2001. Phosphate ions mediate chondrocyte apoptosis through a plasma membrane transporter mechanism. Bone **28:** 1–8.
20. TEIXEIRA, C.C., H. ISCHIROPOULOS, K. MANSFIELD & I.M. SHAPIRO. 2000. Pi-induced apoptosis of tibial chondrocytes is mediated by nitric oxide. J. Dental Res. **79:** 301.
21. HU, H.H., N.A. PATANKAR & M.Y. ZHU. 2001. Direct numerical simulations of fluid-solid systems using the arbitrary Lagrangian–Eulerian technique. J. Comput. Phys. **169:** 427–462.
22. AYYASWAMY, P.S. 1999. The culture of three-dimensional bone-like tissue under simulated microgravity conditions in NASA's rotating-wall vessels: experimental and numerical studies. *In* Microgravity Fluid Physics and Heat Transfer. V. Dhir, Ed.: 183–196. Begell House, Inc., New York.
23. GAO, H., P.S. AYYASWAMY & P. DUCHEYNE. 1997. Dynamics of a microcarrier particle in the simulated microgravity environment of a rotating-wall vessel. Micrograv. Sci. Technol. **X:** 154–165.
24. GAO, H. 2000. Numerical Studies of Microcarrier Particle Dynamics and Associated Mass Transfer In Rotating-Wall Vessels. Ph.D. Dissertation, Department of Mechanical Engineering and Applied Mechanics, University of Pennsylvania.
25. GAO, H., P.S. AYYASWAMY, P. DUCHEYNE & S. RADIN. 2001. Surface transformation of bioactive glass in bioreactors simulating microgravity conditions: Part II: numerical simulation. Biotech. Bioeng. **75:** 379–385.
26. GOODWIN, T.J., T.L. PREWETT, D.A. WOLF & G.F. SPAULDING. 1993. Reduced shear stress: a major component in the ability of mammalian tissues to form three-dimensional assemblies in simulated microgravity. J. Cell. Biochem. **51:** 301–311.
27. QIU, Q., P. DUCHEYNE, H. GAO & P.S. AYYASWAMY. 1998. Formation and differentiation of three-dimensional rat marrow stromal cell culture on microcarriers in a rotating-wall vessel. Tissue Engin. **4:** 19–34.
28. QIU, Q., P. DUCHEYNE & P.S. AYYASWAMY. 1999. Fabrication, characterization and evaluation of hollow bioceramic microspheres used as microcarriers for 3-D bone tissue formation in rotating bioreactors. Biomaterials **20:** 989–1001.
29. QIU, Q., P. DUCHEYNE & P.S. AYYASWAMY. 2000. New bioactive, degradable, composite microcarriers as tissue engineering substrates. J. Biomed. Mater. Res. **52:** 66–76.
30. QIU, Q., P. DUCHEYNE & P.S. AYYASWAMY. 2001. Three-dimensional bone tissue engineering with bioactive microspheres in simulated microgravity. *In vitro.* Cell. Develop. Biol.—Animal **37:** 165.
31. RADIN, S., P. DUCHEYNE, P.S. AYYASWAMY & H. GAO. 2001. Surface transformation of bioactive glass in bioreactors simulating microgravity conditions: Part I: experimental investigation. Biotech. Bioeng. **75:** 369–378.
32. OCHOA, J.A., A.P. SANDERS, D.A. HECK & B.M. HILLBERRY. 1991. Stiffening of the femoral-head due to intertrabecular fluid and intraosseous pressure. J. Biomech. Eng.—Trans. ASME **113:** 259–262.
33. NAUMAN, E.A., K.E. FONG & T.M. KEAVENY. 1999. Dependence of intertrabecular permeability on flow direction and anatomic site. Ann. Biomed. Eng. **27:** 517–524.
34. OCHOA, J.A., A.P. SANDERS, T.W. KIESLER, *et al.* 1997. *In vivo* observations of hydraulic stiffening in the canine femoral head. J. Biomech. Eng.—Trans. ASME **119:** 103–108.

35. DILLAMAN, R.M., R.D. ROER & D.M. GAY. 1991. Fluid movement in bone—theoretical and empirical. J. Biomech. **24:** 163–177.
36. CAVERZASIO, J., T. SELZ & J.P. BONJOUR. 1988. Characteristics of phosphate-transport in osteoblastlike cells. Calcified Tissue Intl. **43:** 83–87.

Magnetic Microspheres and Tissue Model Studies for Therapeutic Applications

NARAYANAN RAMACHANDRAN[a] AND KONSTANTIN MAZURUK[b]

[a]BAE SYSTEMS Analytical Solutions Inc., Huntsville, Alabama, USA

[b]University of Alabama in Huntsville, Alabama, USA

ABSTRACT: The use of magnetic fluids and magnetic particles in combinatorial hyperthermia therapy for cancer treatment is reviewed. The investigation approach adopted for producing thermoregulating particles and tissue model studies for studying particle retention and heating characteristics is discussed.

KEYWORDS: hyperthermia; cancer therapy; magnetic particles and microspheres; magnetic fluids; Curie temperature; thermoregulating particles

INTRODUCTION

Hyperthermia is a well-known cancer therapy and consists of heating a tumor region to elevated temperatures in the range of 40–46°C for an extended period of time (2–8 hours). This leads to thermal inactivation of cell regulatory and growth processes with resulting widespread necrosis, carbonization, and coagulation. Moreover, heat boosts the tumor response to other treatments, such as radiation, chemotherapy, or immunotherapy. Of particular importance is careful control of generated heat in the treated region and keeping it localized. Higher heating, to about 56°C can lead to tissue thermoablation. With accurate temperature control, hyperthermia has the advantage of having minimal side effects. Several heating techniques are employed for this purpose, such as whole body hyperthermia, radio-frequency (RF) hyperthermia, ultrasound technique, inductive microwave antenna hyperthermia, inductive needles (thermo-seeds), and magnetic fluid hyperthermia (MFH). MFH offers a unique targeting capability of magnetic particles in affected tissue through the use of magnetic fields. However, this technology still suffers from significant inefficiencies due to lack of thermal control.

This paper provides a review of the topic and outlines ongoing work in this area at NASA Marshall Space Flight Center (MSFC). The main emphasis is in devising ways to overcome the technical difficulty in hyperthermia therapy of achieving a uniform therapeutic temperature over the required region of the body and holding it steady for an extended period. The shortcomings stem from the non-uniform thermal properties of tissue and the point heating characteristics of present techniques without any thermal control. Our approach is to develop a novel class of magnetic fluids that have inherent thermoregulating properties. We have identified a few magnetic

Address for correspondence: Dr. N. Ramachandran, BAE SYSTEMS Analytical Services Inc., SD 46 NASA Marshall Space Flight Center, Huntsville, AL, USA. Voice: 256-544-8308; fax: 256-544-2102.

narayanan.ramachandran@msfc.nasa.gov

Ann. N.Y. Acad. Sci. 1027: 99–109 (2004). ©2004 New York Academy of Sciences.

doi: 10.1196/annals.1324.010

alloys that can serve as suitable nano- to micron-sized particle material. The objective of the investigation is to synthesize, characterize, and evaluate the efficacy of thermoregulating magnetic fluids (TRMF) for hyperthermia therapy. The development of a tissue model and testing the fluid dynamics of particle motion, settling, distribution in the tissue matrix, and heat generation are some of the other areas germane to this interdisciplinary project.

MAGNETIC FLUID HYPERTHERMIA RESEARCH

Hyperthermia therapy essentially consists of heating the affected regions of the body to temperatures between 40°C and 46°C that leads to the thermal inactivation of cell regulatory and growth processes. Furthermore, heat-treated cancer cells, in some cases, are recognized as foreign by the immune system and are subsequently inactivated. For some types of cancer, hyperthermia therapy combined with radiation- or chemotherapy has been found to be more efficacious. Several heating techniques are used for this purpose, such as whole body hyperthermia, RF hyperthermia, inductive microwave antenna hyperthermia, inductive needles, and MFH. The latter approach is part of the general field of magnetic fluid technology, which includes techniques, such as enhancement of magnetic resonance imaging, particle separation, blood cell separation, and protein purification.[1] Of particular importance is the careful control of generated heat in the treated region and keeping it localized. Higher heating, to about 56°C can lead to tissue thermoablation with attendant widespread necrosis, carbonization, and coagulation.[2]

The use of magnetic particles for hyperthermia in conjunction with alternating current (AC) magnetic fields dates back to the pioneering work of Gilchrist et al., in 1957.[3] MFH uses fine micrometer-size magnetic particles exposed to an alternating magnetic field. Typical field strengths are of the order of 5–30 kA/m with a frequency range of 100 to 500 kHz. The initial experiments resulted in a flurry of activity with the promise of realizing a powerful technique for cancer treatment. During the following decades, in vivo experiments with animals confirmed the general applicability of the technique to human patients. However, most of the studies were conducted with inadequate animal systems, inexact thermometry, and poor AC magnetic field parameters. These are the primary reasons why MFH cancer treatment has not achieved the state of clinical application until now. Renewed interest in this area stemmed from the work of Chan et al.[4] and Jordan et al.,[5] who showed that on a volume basis, nanometer size magnetic particles provide more heating power and more homogeneous heating than micron-size particles. Significantly, lower AC fields, more tolerable to humans, can then be used. Typically, AC fields with frequencies in the MHz range lead to muscle stimulation and high field strengths lead to dangerous point heating in tissue. Another factor in favor of the technique is the availability of new biological data on heat response of cells and tissues. The first prototype of a clinical MFH therapy system has just recently come on line.[6] Set up in the Clinic of Radiation Oncology in Berlin, the system operates at 100 kHz with the field strength from 0 to 15 kA/m and with a vertical aperture of 30–45 cm. The treatment of deep-seated tumors, especially in shielded areas, such as the pelvic region and the skull, still remains a challenge.

In hyperthermia therapy, magnetic particles can be introduced into the target regions inside the human body in various ways and once located, can be excited or activated to act as localized heat sources. The particles can be directly injected at the target site if possible, the most direct approach, or alternatively they can be introduced into an artery by inoculation and targeting can be achieved at the desired site using a static magnetic field. It is to be noted that the magnetic targeting technique is practically realizable only for larger particles, greater than $1.0 \mu m$. Recently, a new and promising technique has been demonstrated wherein 10nm size magnetic particles in dispersion (ferrofluids) have been shown to possess highly selective cellular adhesion to tumor cells. A tenfold particle uptake by tumor cells in comparison to normal cells was reported.[6] Hence, tumor cells can be selectively loaded with magnetic particles and subsequently killed with an AC field. The particular attraction of this method is its potential to reach body target sites, such as brain tumors, generally inaccessible by other techniques and the increased heat production capacity of the smaller particles. This fascinating idea may soon become an effective weapon against certain types of cancer.

There still remain serious problems with MFH that need to be understood and solved. Especially, the temperature distribution inside and outside the target region must be known precisely as a function of the AC magnetic field exposure time in order to provide optimum therapeutic temperature and to avoid overheating and damaging surrounding normal tissue. Here mathematical modeling is indispensable. Very few mathematical simulations of magnetic fluid hyperthermia have been performed to date. A one-dimensional, spherically symmetric, time dependent thermal model was proposed and solved analytically by Andra et al.[7] The results were compared with data from experiments studying RF heating of a cylindrical composite. Despite the obvious difference in geometry, close agreement was reported. Other published hyperthermia modeling attempts include ultrasound, capacitive radio frequency, and ferromagnetic implant techniques. All of these simulations involve simple mathematical modeling of heat generation and do not account for the effects of blood perfusion that are always present in the tumor area; Tomkins et al.[8]

Ongoing research is focused on specific issues relating to the optimization of the magnetic fluid properties. For example, for a given excitation frequency, a sharp particle size distribution is required in order to maximize the specific absorption rate (SAR) given in terms of Watt/gm. Further optimization may involve properties of the particle coating. For colloidal stabilization, a magnetic particle has to be coated with a surfactant layer. The interaction of this colloidal shell with the core of the particle in an AC magnetic field can be very complex. For example, the core can exhibit oscillations within the shell, or, alternatively, the whole particle can oscillate depending on elastic properties of the shell. Detailed study is required to describe particle oscillations in AC magnetic fields. In addition, the role of hydrodynamic and rheologic properties of the solvent is of importance. Selection of appropriate surfactants used for the particle shells is also of great importance since selective cellular adhesion can potentially be achieved. This is especially vital, since targeting of nanometer size particles by an exterior magnetic field is not very practical. Brownian motion keeps the particles suspended, even in strong magnetic fields, and in the presence of system flow the particles are transported with the flow.

Other approaches to cancer therapy using magnetic particles include reducing the blood circulation in the tumor area by overloading it with magnetic particles and thereby embolyzing it,[9] and overloading the tumor cell itself (uptake mechanism discussed earlier) leading to arrest of the cell division and to subsequent lyses.[10] Magnetic microspheres as drug carriers alone are another attractive research field overlapping hyperthermia. Clearly, the focus now is on the search for novel biocompatible ferrofluids with better SAR and precise cancer cell targeting selectivity.

An excellent overview of the applications of MMS and magnetic fluids in cancer therapy can be found in the book edited by Häfeli et al.[11]

REQUIREMENTS AND SPECIFICATIONS FOR MFH APPLICATIONS

The technical and medical requirements for hyperthermia application systems are rather advanced and difficult to fulfill. From a practical standpoint, the AC magnetic field homogeneity and its control, safety, accurate thermometry, and a precisely delineated treatment volume are the primary parameters that have to be met. This complex problem requires advanced thermal modeling complemented with experimental work. In order to make this presentation more complete, specific issues, such as particle size, toxicity, dissipation processes, and macroscopic heat and mass transport, are briefly discussed.

Particle Synthesis

In general, ferrofluids consist of three components: carrier, surfactant, and magnetic particles. The most important component responsible for its many unusual properties is obviously the dispersed nanoparticle phase. Varieties of techniques have been developed in the past to obtain these nanoparticles, which can be based on either chemical or physical processes. Physical methods have a higher yield in general, but chemical methods are more versatile and practical especially for a research laboratory environment. Examples of the methods include vacuum evaporation, electrolytic precipitation, chemical vapor deposition, spark erosion, wet chemical methods, and aerosol methods. Rosensweig[12] has written an excellent treatise on ferrofluids, their synthesis techniques and properties, as well as theoretical concepts dealing with their transport and behavior in magnetic fields. Many of the methods mentioned above suffer from a lack of control of the particle size distribution. Particle synthesis can be approached in a number of ways. We focus on a couple of tried and promising wet chemistry techniques, such as microemulsion-based approaches. The microemulsion approach is a reverse-micelle technique that gives a good narrow size distribution and is being successfully used in many laboratories for the synthesis of particles, for example Zeolites.[13] These water-in-oil microemulsions form a system of so called reverse micelles that act as micro-reactors. The size of these minute water pools influences the size of the synthesized micro-crystals to some extent. Size control is then accomplished by a simple variation of the molar ratio water/surfactant or water/oil in the system. Hori et al.[14] and Nonumara et al.[15] have recently prepared Pd/Ni alloy nanoparticles by wet chemistry.

Particle Size and Type

Particles stabilized by biocompatible substances must have appropriate shapes and sizes—less than a few micrometers—in order to pass through the capillary system without posing a threat of vessel embolism or physical irritation of the surrounding tissue. We focus on two categories, namely, magnetic fluids and magnetic microspheres, each with very specific chemical and physical characteristics.

Magnetic fluids are comprised of 1–100nm magnetic particles (magnetite, iron, nickel, etc.) freely dispersed in a carrier fluid. Due to their small dimensions, Brownian thermal molecular motion keeps them suspended and prevents sedimentation in gravitational or magnetic fields. A continuum fluid model has been shown to work well to describe these magnetic fluids.[12] The particles usually need to be coated with a surfactant (such as oleic acid or a biocompatible substance) or stabilizing polymer layer to prevent aggregation. This layer also provides the unique chemistry-driven surface adhesion properties of these particles to tissue. As a result, the particles in magnetic fluids interact with each other by means of magnetic dipoles, stearic or electrostatic repulsion, and hydrodynamic forces. The thickness of stabilizing layers is usually of the order of 2–3nm, so that the hydrodynamic size of the particles can be different from the magnetic core.

Magnetic microspheres (MMS) are magnetic particles in the 0.1–100μm size range, and have stronger dipole–dipole interactions than the thermal fluctuation energy. The particles can be used without coatings (such as iron particles with carbon) or specifically tailored for function by applying polymer, silan, or dextran type coatings. In strong magnetic fields, these particles form rigid spatial structures that can be used to selectively occlude blood vessels or to target them in a specific organ and retain them there for extended periods (days).

Toxicity

Magnetic fluid particles for *in vivo* applications have to be biocompatible, preferably biodegradable, and non-toxic. These criteria rule out many attractive magnetic materials from consideration. Toxicity studies of various types of commercially available microspheres. such as magnetic and non-magnetic polylactic acid microspheres, FeC, dextran-coated magnetite nanospheres, and magnetic polystyrene nanospheres, have shown no significant difference in viability, Häfeli *et al.*[16] The amount of material needed for magnetically targeted therapy is only a few milligrams. This is significantly lower than the lethal iron dosage of 200mg of iron per kg of human body. At lethal levels they can cause acidosis of the body or induce thrombosis of the vascular systems of the lung.[17] In summary, iron- and iron-oxide based particles show no adverse effects to humans or animals in small quantities.

Power Dissipation

The measure of AC power dissipation in ferrofluids is the specific absorption rate (in Watts per gram of fluid) or alternatively, the complex magnetic susceptibility of the ferrofluid. Only a few experimental reports on measurement of the magnetic susceptibility as a function of frequency are available in surveyed literature.[18] There is a significant body of work dealing with the theoretical development of the fundamentals of ferrofluids and their properties and these can be used to gain insight into microscopic processes responsible for the heating of ferrofluids when subjected to dynamic magnetic fields. Solving the equations governing the dynamics of the magnetic moment of a particle (Langevin approach) results in particle relaxation times. This, when used in conjunction with Debye theory, yields the required susceptibility.[19]

Heat Generation

Two rotational relaxation mechanisms may coexist in magnetic fluids when an AC magnetic field is imposed. In magnetic fluids that are composed of subdomain particles (nm size), the Neel relaxation,[20] when the magnetic moment moves with respect to the mechanical particle, and the Brownian relaxation corresponding to the rotation of the particle inside a fluid[21] are the main mechanisms of heat generation. The latter usually dominates the relaxation process in magnetic fluids. As the particles get larger, multidomain particles (micron size), hysteresis loss is the main mechanism of heat generation.[2] The SAR of magnetic fluids is proportional to H^2, where H is the magnetic field strength, and for larger particles, the heat production is proportional to H^3.[21]

Modeling heat production from magnetic particles from fundamentals is rather complex and involves the momentum conservation equation for the flow field, the magnetostatic equations, and a constitutive equation for the particle magnetization.[19] Our approach is to use the heat generation data from experiments in the numerical model to study its transport and diffusion when subjected to a flow field. This approach will provide a benchmarked predictive numerical model that can be used for the simulation of hyperthermia protocols for known system flows using known particle behavior and for dialed in values of the magnetic excitation field.

SYNTHESIS OF THERMOREGULATING PARTICLES

There are a couple of promising wet chemistry recipes for the development of magnetic particles, typically in the 1–10nm range. There also exist several candidate alloying metals that can give us the attributes required of such particles, namely, a fairly low Curie temperature (40 to 60°C), a large magnetic moment, and biocompatibility. Here we describe the reverse micelle or microemulsion technique for synthesizing particles. The microemulsion approach is a reverse-micelle technique that gives a good narrow size distribution and is being successfully used in many laboratories for the synthesis of nanoparticles. It consists of performing chemical synthesis in tiny droplets of water encapsulated in oil—the microreactors where nuclei germinate and grow into larger particles. This approach has been used for the synthesis of over 50 different materials[22] and has been shown to be a versatile way of producing inorganic nanoparticles. There is still considerable debate as to the exact mechanism of the formation of the particles. For a while, the radius of the droplets was thought to be the constraint on the resulting particle size[23] but experiments have yielded particles larger than the droplets of the microemulsion used.[24] Current thinking is a modification of the mechanism first proposed by LaMer[25] and holds that the particle size control is dictated by the kinetics of the elementary steps of particle formation.[24]

The microemulsions are mixtures of three components, water, oil, and surfactant. On the oil-rich side, a water–oil microemulsion consists of water droplets in the continuous oil phase, with the water and oil domains separated by a film of self-assembled surfactant molecules. They can be easily prepared by mixing water, oil (such as cyclohexane), and surfactant in a glass tube and then heating to an appropriate temperature. The microemulsion is formed spontaneously by gentle shaking. The radius of the micelles can be controlled by the water to oil ratio, or by the surfactant percentage, and can vary from a few nanometers to microns. As an example, consider the synthesis of palladium nanoparticles.[26] Palladium chloride microemulsion is formed using 75 wt% of cyclohexane (oil phase), 20 wt% of Marlipal O13/40 surfactant, and 5 wt% of aqueous solution of the reactants. A typical water droplet size of 1.7 nm is obtained. The two aqueous solutions (reactants) used in the process contain 0.2 mol/l $PdCl_2$.NaCl and 0.6 mol/l $NaH_2PO_2 \cdot H_2O$ of reducing agent, respectively. The resulting microemulsion is fed into a reactor and mixed. Fairly monodisperse palladium particles, 5.1 nm in diameter with a standard deviation of 0.5 nm are obtained. If just the reactant aqueous solutions (no emulsion) are mixed, palladium particles with a rather broad size distribution in the range of 2 to 15 nm diameter, are formed.

The phase diagram for nickel–palladium alloy is shown FIGURE 1. Also shown is the Curie temperature, which can be controlled by controlling the nickel composition in the alloy. Other possible alloys that have a tunable Curie temperature are Ni–Zn, Ni–W, Ni–Pt, Ni–Si, Ni–Sb, Co–Pd, and Co–Mn. Additional chemistry aspects include the addition of appropriate surfactants, such as oleic acid used in ferrofluid dispersions, and a carrier fluid (water or oil based). Obtaining a good dispersion of the synthesized particles is no trivial task and there is a wide choice of surfactants that can be tested and used. Furthermore, there is the issue of the biocompatibility of the dispersion agent. We plan to test, initially, surfactants that are in use for suspending commercially available MMS and that are used in therapy and clinical trials. The parameters to be tested for the synthesis protocol include the particle size, agglomeration characteristics, magnetic moment, and Curie temperature.

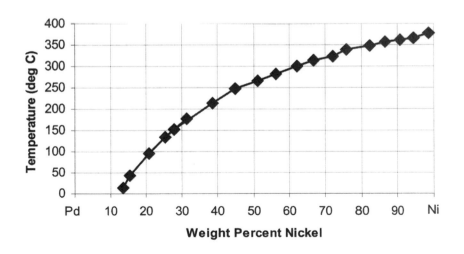

FIGURE 1. Curie temperature behavior of nickel–palladium alloy.

TABLE 1. Alloy combinations for thermoseed particle synthesis

Weight percent for target Curie temperature about 50°C
Pd (85%)–Ni (15%)
Pt (62%)–Ni (38%)
Ni (81%)–Zn (19%)

In addition to the Ni–Pd alloy we will attempt to synthesize two additional alloys from the candidates identified above. These are identified in TABLE 1 below. In addition, heavy earth based alloys, such as gadolinium alloys, as well as some semimagnetic semiconductors (compounds) also possess low Curie temperatures (near room temperature) and can be studied for potential applications.

MATERIALS PROCESSING AND TISSUE MODEL EXPERIMENTS

For targeted delivery studies using a static magnetic field, a fluid experiment model and associated diagnostics are required. Although experiments in test tubes and capillary tubes are easily performed and form a part of this investigation, they have a major shortcoming in that they do not simulate tissue even in the remotest sense. The length scales of tissue are in the 1–30 mm size (arterioles, venules, and capillaries) and more importantly, they are present as a diffuse network. Blood perfusion is an added complexity that cannot be ignored if one is to obtain meaningful and correlatable results. FIGURE 2 shows a schematic of the dimensions (diameter) of various types of blood vessels along with typical values of blood flow rates in the human body. Blood can be treated as a Newtonian fluid when flowing in arteries and veins exceeding 100 mm in diameter but a non-Newtonian model is more appropriate in smaller vessels, such as capillaries. The volumetric perfusion rate, usually given

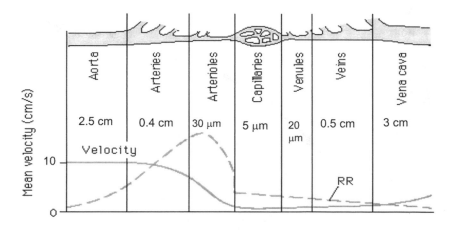

FIGURE 2. Typical dimensions and blood perfusion rates in human blood vessels. *RR* is the relative resistance, which is highest in the arterioles.

in units of ml (blood)/(cm^3 tissue·min) varies from 0.02 in skeletal muscle at rest to 2.0 during heavy exercise. It varies primarily in response to the oxygen demand of the body. The maximum flow Reynolds number in arteries is about 300 to 600.[27]

Surveyed literature has yielded no unique way of simulating tissue and blood flow. Investigators have used 40% glycerol in water to essentially get the working fluid viscosity to match the blood viscosity of 3.6cP at 37°C, but so far the perfusion framework has been limited to tubular and jagged tubular structures with a few bifurcations to simulate the hemodynamics in major blood vessels.[28–30] On the hyperthermia front, model experiments have essentially been conducted in gels[21] to measure heat production and dissipation. For this investigation, we propose an approach using the technique of refractive index matching for the simulation of a capillary bed, system perfusion, and hyperthermia studies. The technique is elaborated below.

MATCHED INDEX OF REFRACTION STUDIES

The basic premise of the technique is to match the refractive index n of the experiment model (cast acrylic, $n = 1.491$) and the working fluid. Specially formulated fluid mixtures and precise temperature control are used to match the refractive indices of the fluid and the model material, which essentially makes the internal geometry of the model transparent. Coupled with external flat surfaces, the experimenter can then use a laser doppler velocimeter (LDV), a particle image velocimeter (PIV), and a phase doppler particle analyzer (PDPA) to make non-intrusive particle and flow diagnostic measurements anywhere within the model. We have done some preliminary work in this area in order to identify a working fluid suitable for laboratory operations.[31] Some of the considerations in the study were to identify candidate fluids (or fluid blends) for proper refractive index matching of cast acrylic (typical model material). Requirements included good operational characteristic (reasonable temperature range), non-toxic (safety), chemical stability (inertness), cost, non-corrosive (does not consume the plastic model), low viscosity (good Reynolds number, Re, range from dynamic similitude standpoint), safe (combustion hazard), ease of operation (non-volatile—not requiring a sealed system), and being colorless and odorless. Three potential fluid combinations were identified as shown in TABLE 2. The choice of the appropriate fluid depends on compatibility with magnetic fluids and other factors identified above.

TABLE 2. Matched index of refraction fluids

1	73% Dow Corning 550 m silicone oil and 27% Union Carbide L42 fluid $n = 1.491$ at 22°C, $v = 188$ Cst, density 1.04 gm/cc
2	68.2% turpentine and 31.8% tetraline $n = 1.491$ at 25°C, $v = 1.63$ Cst, density 0.896 gm/cc
3	hydrocarbon liquid mixture, Cargille Laboratories, NJ (propriety composition) $n = 1.491$ at 25°C, $v = 27$ Cst, density 0.879 gm/cc

NOTES: Cast acrylic, $n = 1.491$; n, index of refraction; v, kinematic viscosity.

ACKNOWLEDGMENTS

This research is funded by the Department of Defense CDMRP Breast Cancer Research Program through an Idea grant. Use of research facilities provided by the Science Directorate at NASA MSFC for doing the work is acknowledged.

REFERENCES

1. ROATH, S. 1993. Biological and biomedical aspects of magnetic fluid technology. J. Magn. Magn. Mater. **122:** 329–334.
2. JORDAN, A., *et al.* 1999. Magnetic fluid hyperthermia: cancer treatment with AC magnetic field induced excitation of biocompatible superparamagnetic nanoparticles. J. Magn. Magn. Mater. **201:** 413–419.
3. GILCHRIST, R.K., *et al.* 1957. Selective inductive heating of lymph nodes. Ann. Surg. **146:** 596.
4. CHAN, D.C.F., *et al.* 1993. Synthesis and evaluation of colloidal magnetic iron oxides for the site-specific radifrequency-induced hyperthermia of cancer. J. Magn. Magn. Mater. **122:** 374.
5. JORDAN, A., *et al.* 1996. Cellular uptake of magnetic fluid particles and their effects on human adenocarcinoma cells exposed to AC magnetic fields *in vitro.* Int. J. Hypertherm. **12:** 705.
6. JORDAN, A., *et al.* 2001. Presentation of a new magnetic field therapy system for the treatment of human solid tumors with magnetic fluid hyperthermia. J. Magn. Magn. Mater. **225:** 118.
7. ANDRA, W., *et al.* 1999. Temperature distribution as function of time around a small spherical heat source of local magnetic hyperthermia. J. Magn. Magn. Mater. **194:** 197–203.
8. TOMPKINS, D.T., *et al.* 1994. Effect of interseed spacing, tissue perfusion, thermoseed temperatures and catheters in ferromagnetic hyperthermia: results from simulations using finite element models of thermoseeds and catheters. IEEE Trans. Biomed. Eng. **41:** 975–985.
9. LIU, J., *et al.* 2001. *In vitro* investigation of blood embolization in cancer treatment using magnetorheological fluids. J. Magn. Magn. Mater. **225:** 209–217.
10. HARUTYUNYAN, A.R., *et al.* 1999. Metal–organic magnetic materials based on cobalt phthalocyanine and possibilities of their application in medicine. J. Magn. Magn. Mater. **194:** 16.
11. HÄFELI, U., *et al.,* Eds. 1997. Scientific and Clinical Applications of Magnetic Microspheres. Plenum Press, New York.
12. ROSENSWEIG, R.E. 1985. Ferrohydrodynamics. Cambridge University Press.
13. DUTTA, P. 2001. Fundamental studies of crystal growth of microporous materials, NASA physical sciences research division program tasks—flight research. NAG8-1670.
14. HORI, H., *et al.* 2001. Magnetic properties of nano-particles of Au, Pd and Pd/Ni alloys. J. Magn. Magn. Mater. **226:** 1910–1911.
15. NONUMARA, N., *et al.* 1998. Magnetic properties of nanoparticles in Pd/Ni alloys. Phys. Letters A **249:** 524–530.
16. HÄFELI, U.O. & G.J. PAUER. 1999. *In vitro* and *in vivo* toxicity of magnetic microspheres. J. Magn. Magn. Mater. **194:** 76–82.
17. KUZNETSOV, O.A., *et al.* 1999. Correlation of the coagulation rates and toxicity of biocompatible ferromagnetic microparticles. J. Magn. Magn. Mater. **194:** 83–89.
18. HERGT, R., *et al.* 1998. Physical limits of hyperthermia using magnetite fine particles. IEEE Trans. Magn. **34:** 3745–3754.
19. SHLIOMIS, M.I. 1963. Magnetic fluids. Sov. Phys. Usp. **17:** 153–184.
20. NEEL, L. 1949. Influence of thermal fluctuations on the magnetization of ferromagnetic small particles. C.R. Acad. Sci. **228:** 664–668.
21. HIERGEIST, R., *et al.* 1999. Application of magnetite ferrofluids for hyperthermia. J. Magn. Magn. Mater. **201:** 420–422.

22. LADE, M., *et al.* 2000. On the nanoparticles synthesis in microemulsions: detailed characterization of an applied reaction mixture. Colloids Surfaces A **163:** 3–15.
23. TANORI, J. & M.P. PILENI. 1997. Control of the shape of copper metallic particles by using a colloidal system as template. Langmuir **13:** 639–646.
24. HIRAI, T., *et al.* 1993. Mechanism of formation of titanium dioxide ultrafine particles in reverse micelles by hydrolysis of titanium tetrabutoxide. Ind. Eng. Chem. Res. **32:** 3014–3019.
25. LAMER, V. & R.J. DINEGAR. 1950. Theory, production and mechanism of formation of monodispersed hydrosols. J. Am. Chem. Soc. **72:** 4847–4854.
26. TERANISHI, T., *et al.* 1997. ESR study on palladium nanoparticles. Phys. Chem. B **101:** 5774–4776.
27. MOORE, J.A., *et al.* 1999. Accuracy of computational hemodynamics in complex arterial geometries reconstructed from magnetic resonance imaging. Ann. Biomed. Eng. **27:** 627–640.
28. LIEPSCH, D.A., *et al.* 1989. Flow visualization studies in a mold of the normal human aorta and renal arteries. J. Biomed. Eng. **227:** 115–222.
29. ETHIER, C.R., *et al.* 1999. Comparisons between computational hemodynamics, photochromic dye flow visualization and magnetic resonance velocimetry. *In* The Hæmodynamics of Arterial Organs—Comparisons of Computational Predictions with *In Vivo and In Vitro* Data. X.Y. Xu & M.W. Collins, Eds. Pineridge Press.
30. MOORE, J.A., *et al.* 1999. Accuracy of computational hemodynamics in complex arterial geometries reconstructed from magnetic resonance imaging. Ann. Biomed. Eng. **27:** 32–41.
31. SMITH, A. & N. RAMACHANDRAN. 1998. Flow field measurements in cast acrylic models using the matched index of refraction technique. NASA MSFC Center Director Discretionary Fund Project Report.

The Measurement of Solute Diffusion Coefficients in Dilute Liquid Alloys

The Influence of Unit Gravity and *G*-Jitter on Buoyancy Convection

R.W. SMITH, B.J. YANG, AND W.D. HUANG

Materials Science and Microgravity Applications Group,
Department of Mechanical and Materials Engineering,
Queen's University, Kingston, Ontario, Canada

ABSTRACT: Liquid diffusion experiments conducted on the MIR space station using the Canadian Space Agency QUELD II processing facility and the microgravity isolation mount (MIM) showed that *g*-jitter significantly increased the measured solute diffusion coefficients. In some experiments, milli-*g* forced vibration was superimposed on the sample when isolated from the ambient *g*-jitter; this resulted in markedly increased solute transport. To further explore the effects arising in these long capillary diffusion couples from the absence of unit-gravity and the presence of the forced *g*-jitter, the effects of a 1 milli-*g* forcing vibration on the mass transport in a 1.5 mm diameter long capillary diffusion couple have been simulated. In addition, to increase understanding of the role of unit gravity in determining the extent to which gravity can influence measured diffusion coefficient values, comparative experiments involving gold, silver, and antimony diffusing in liquid lead have been carried out using a similar QUELD II facility to that employed in the QUELD II/MIM/MIR campaign but under terrestrial conditions. It was found that buoyancy-driven convection may still persist in the liquid even when conditions are arranged for a continuously decreasing density gradient up the axis of a vertical long capillary diffusion couple due to the presence of small radial temperature gradients.

KEYWORDS: liquid diffusion; gravity; G-jitter

INTRODUCTION

A knowledge of diffusion processes in liquids is of fundamental importance in studying and modeling the mass transport arising when multicomponent liquids are processed; for example, the solidification of an alloy from its melt and the redistribution of alloying elements.[1] The development of a numerical model to achieve optimal control of a commercial casting process, such as that involved in continuous casting, requires accurate diffusion data.[2] Despite such importance, currently available

Address for correspondence: R.W. Smith, Materials Science and Microgravity Applications Group, Department of Mechanical and Materials Engineering, Queen's University, Kingston, Ontario K7L 3N6, Canada. Voice: 613-533-2753; fax: 613-533-6610.
 smithrw@post.queensu.ca

Ann. N.Y. Acad. Sci. 1027: 110–128 (2004). ©2004 New York Academy of Sciences.
doi: 10.1196/annals.1324.011

diffusion data, where they exist at all, may not be accurate, the data obtained in two studies often differing by a factor of two! This is primarily due to the influence of buoyancy-driven convection in the experimental systems used to obtain such diffusion coefficients in terrestrial laboratories.[3] Most of these experiments have used the long capillary diffusion couple and, in order to reduce convective transport, the diameters of the capillaries have been progressively decreased. However, it has been reported that if the diffusion capillary is less than about 1 mm in diameter, then the rate of diffusion is reduced, the so-called "wall effect".[4] Thus, it is conventional to write[5]

$$D_{effective} = D_{intrinsic} + D_{buoyancy} + D_{wall\ effect} + D_{thermal\ (Sorét\ effect)}. \quad (1)$$

It has been supposed that if a diffusion couple is isothermally processed in microgravity, then $D_{buoyancy}$ and $D_{thermal}$ will be absent, and only $D_{intrinsic}$ and $D_{wall\ effect}$ should be present. Hence, by choosing capillaries of different diameters, an estimate of $D_{wall\ effect}$ might be made and an accurate value of $D_{intrinsic}$ obtained. Furthermore, if the diffusion couples are processed at a number of temperatures, then the temperature dependence of the transport process taking place could be determined. This can provide clues to the mechanism(s) by which diffusion occurs since most theories of diffusion mechanisms involve temperature considerations (see APPENDIX).

Fortunately, it has been shown that an orbiting platform can offer an environment with much reduced buoyancy-driven transport and so permit the capture of more precise diffusion data. Frohberg *et al.*[6] demonstrated the usefulness of microgravity for liquid diffusion experiments by measuring the unique diffusion coefficients of the isotopes of tin. However, since experimental opportunities under such reduced-gravity conditions are very limited, it is important to understand the effects of gravity on the measurement of solute diffusion coefficients in liquids in order to obtain solute diffusion coefficients with acceptable accuracy, even in a terrestrial laboratory.

The most important conclusion obtained by Frohberg from his systematic study of self- and inter-diffusion in liquid Sn in the 1980s was that, compared to the diffusion data obtained from ground experiments, the diffusion coefficients measured under microgravity were far smaller and the width of the scattering band of the data was extremely narrow.[6] However, not all liquid diffusion experiments done on the ground demonstrated such data scatter. For example, the experimental results in the terrestrial study of self-diffusion in liquid Sn by Bruson and Gerl are seen to be close to Frohberg's microgravity results.[6] However, there are some differences in the two experimental designs. The semi-infinite couple and long capillary technique were used by Frohberg *et al.*, whereas the double-infinite couple and shear cell technique were selected by Bruson and Gerl. This suggests that convection was better controlled in the terrestrial experiments by Bruson and Gerl, as compared with the equivalent by Frohberg *et al.*, because of the physical design of the former experiments. A lesson all experimentalists might learn from this example is that comparative experiments should be done under closely similar conditions if significant differences in the product are to be detected.

In order to explore critically the influence of gravity on various physical processes, the Canadian Space Agency (CSA) developed the microgravity isolation mount (MIM) as a serviced platform capable of isolating the experimental facility from the disturbing effects of *g*-jitter; that is, the multidirectional, multifrequency, and amplitude component of the reduced gravity environment of an orbiting laboratory. The

Canadian microgravity community was then invited to propose experiments to be done on the Russian space station MIR, using the MIM as the serviced platform. As a result, mounted on MIM, Queen's University experiments in liquid diffusion (QUELD II) operated on MIR for more than two years and processed a series of long capillary diffusion-couple samples. These diffusion samples have been analyzed and solute diffusion coefficients for the various alloy systems at selected temperatures have been obtained.[7-9] The data from these reduced-gravity diffusion experiments, some where even g-jitter was suppressed, provide datum points from which it is possible to examine the effects of the gravity on the diffusion of solute in a number low-melting-point liquids by conducting comparative terrestrial experiments. This paper reports studies of the effects of gravity in two magnitude ranges: (1) that of the g-jitter experienced by a diffusion couple being processed in a low-Earth-orbit laboratory with the MIM operating in three modes, isolating (g-jitter suppressed), locked (ambient steady-g plus g-jitter), and forcing (isolating condition with superimposed forcing acceleration); and (2) the ambient (unit) gravity of a terrestrial laboratory.

INFLUENCE OF FORCED AND AMBIENT *G*-JITTER ON SOLUTE DIFFUSION IN LOW-EARTH ORBIT

The advantages offered by orbiting laboratories are reduced by the presence of g-disturbances invariably present on these space platforms because convective effects associated with relatively high g-levels and/or g-jitter may induce concentration distortions with respect to the ideal zero-g field. An "inactive" low-Earth orbiting laboratory will normally experience quasisteady acceleration due to gravity gradient and drag effect that ranges from 1 to $10\mu g$. *G*-jitter is a spectrum of larger amplitude periodic accelerations, with frequencies ranging from subHerz to 100 Hz, which are excited by other time-dependent forces on an active platform, such as crew activities, on-board machinery, and impulsive transients due to thrusters firings and mass dumping.

During the earlier QUELD II/MIM/MIR campaign, a series of diffusion couples were processed with the MIM operating in its three modes; namely, isolating, locked, and forcing. It was found that in all the alloy systems examined, a much reduced diffusion coefficient value was obtained if the MIM was used in the isolating mode.[5] In the case where lead was the solvent and gold the solute, the results shown in FIGURE 1 were obtained. It can be seen that at all the temperatures examined, the suppression of g-jitter brings about a reduction in the measured D. Conversely, when a small periodic forcing vibration was applied, a large increase in D resulted. The following is an attempt to simulate how g-jitter and the small forcing vibration might achieve the observed effects.

The Simulation

As seen in FIGURE 1, the MIM was used to provide the three gravity conditions, residual g + g-jitter, residual g only, and residual g + forced oscillation in Pb–Au long-capillary diffusion couples of dimensions 1.5 mm diameter and 40 mm length.

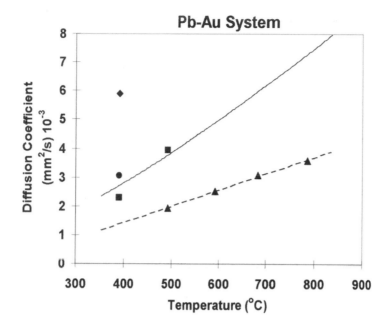

FIGURE 1. Diffusion coefficients of gold in lead (three MIM modes):[5] ▲, QUELD-II, isolated, 1.5 mm; ■, QUELD-II, latched, 1.5 mm; ●, QUELD-II, latched, 3.0 mm; ◆, QUELD-II, forced, 0.1 Hz, 3.0 mm; —, QUELD-I and QUESTS.

In the simulation, the g-jitter was applied in the x-direction, with the MIM in forcing mode; residual gravity was assumed to be $10^{-5} g_0$, as shown in FIGURE 2.

The g-jitter regime of MIM forcing mode is shown in FIGURE 3. It can be seen that 1 mg ($10^{-3} g_0$) is applied parallel to solvent–alloy interface for 0.1 scc, followed by 0.1 mg($10^{-4} g_0$) in opposite direction for 9.9 sec to return QUELD II to its initial rest position.

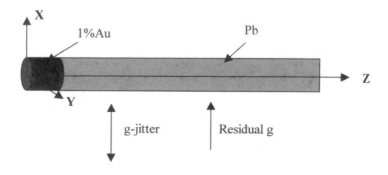

FIGURE 2. Three-dimensional model and gravity condition.

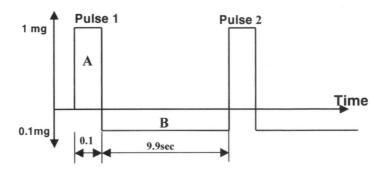

FIGURE 3. The g-jitter regime in MIM forcing mode.

Computed Results

The mathematical and physical models have been applied to two kinds of gravity conditions (forcing mode and isolating mode) to investigate the effect of g-disturbance on the solute distribution in the long capillary diffusion-couple cell samples that were used on the QUELD II/MIM/MIR mission. To excite the distortion of the starting solute profile to investigate the effect of g-jitter, a disturbance was needed; this was obtained by applying a single forcing pulse of g-jitter nonparallel to the solvent/alloy interface plane. FIGURE 4 shows the solute distribution and velocity field at 120 seconds after the condition of the g-jitter regime illustrated in FIGURE 3 has been applied; that is, following the first exciting pulse at 45 degrees to the sample axis, the remaining pulses are applied perpendicular to the axis. FIGURE 5 shows the solute distribution and velocity field at 120 seconds under the condition that a restoring force of 9.8×10^{-5} is applied.

The solute distributions between the two modes are very different, and so is the velocity field for the two modes. The maximum velocity in the forcing mode is two orders of magnitude bigger than that in the isolating mode at the same annealing time. The initial velocity in the forcing mode is of the order of 10^{-4} m/sec, but decreases with increasing annealing time because the density gradient along the axial direction is reduced gradually with the time. We find that a small whirlpool exists around the interface of the alloy/solvent interface region in the longitudinal section. The convection in the isolating mode seems to have little influence on the solute profile, thus a uniform solute concentration in the radial direction is demonstrated.

The averaged solute at each sample cross-section was obtained at 120 sec, and the distributions under the two modes are plotted in FIGURE 6. It is seen that the forcing mode accelerates the solute transport significantly.

Another simulation was done for the case with the g-jitter as shown in FIGURE 3 but when a single pulse of $10^{-1} g_0$ was first applied parallel to solvent/alloy interface for one second. This pulse disturbed the uniform solute profile, so the following forced g-jitter can play a role in expanding the disturbed region of solute. FIGURE 7 shows distributions of velocity and solute at one second during the period of time when the $0.1 g_0$ pulse was applied, and FIGURE 8 shows that the convection continues making the solute distribution more distorted, although only the earlier g-jitter is applied after the first second. However, this solute disturbance decreases with increasing anneal time.

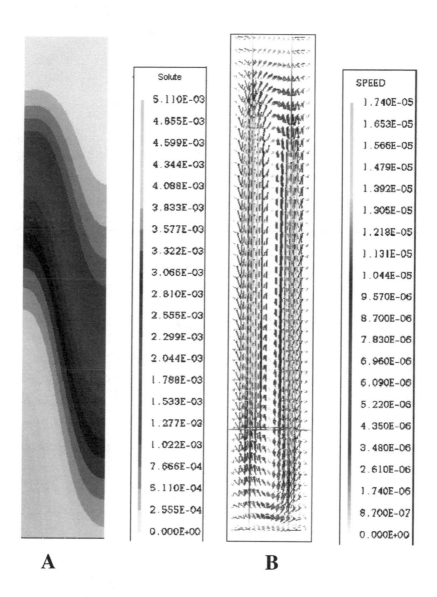

FIGURE 4. Calculated results under the forcing mode at 120 seconds: **(A)** solute distribution and **(B)** velocity field.

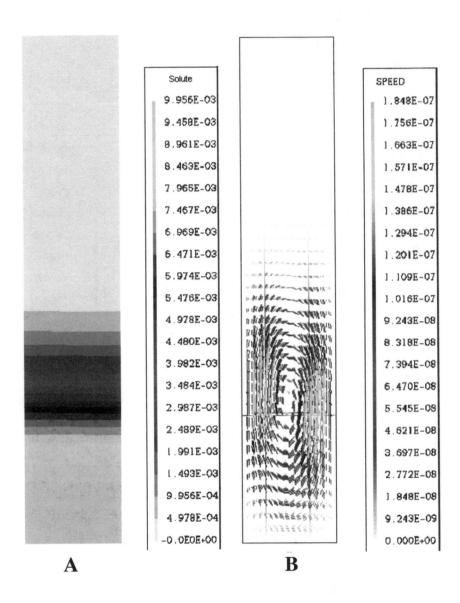

FIGURE 5. Calculated results under the isolating mode at 120 seconds: **(A)** solute distribution and **(B)** velocity field.

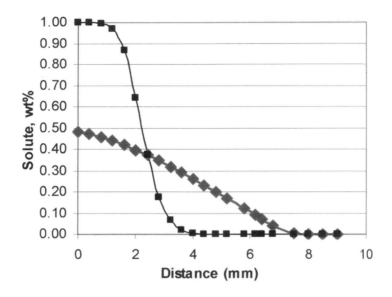

FIGURE 6. Comparison of solute profiles at 120 seconds for forcing and isolating modes: ◆, forcing mode; ■, isolating mode.

It was concluded that, for any MIM condition, convection makes little contribution to the solute distribution when the maximum velocity is reduced to be less than 0.1 mm/sec for a cell diameter of 1.5 mm. This work is now being extended to other alloys and sample geometries.

TERRESTRIAL WORK

To compare the effect of unit gravity with the results from previous liquid diffusion experiments in low Earth orbit, some of which are shown in FIGURE 1, comparative experiments involving gold, silver, and antimony diffusing in liquid lead were conducted using the QUELD II ground base unit.

Experimental Results

Three alloy systems were selected: Pb–1 wt% Au, Pb–1 wt% Ag, and Pb–1 wt% Sb; a 2 mm (diameter), 30 mm (length) specimen was used and the experimental procedures given in the APPENDIX were applied.

The measured diffusion coefficients are shown in TABLES 1, 2, and 3, together with the experimental results obtained previously in the QUELD II/MIM/MIR program under reduced gravity conditions. It can be seen that the diffusion coefficients measured terrestrially are larger than those obtained in the reduced gravity of low Earth orbit. In fact all the terrestrially measured diffusion coefficients shown here are the "effective" or "apparent" diffusion coefficients, since they may contain not only the contribution of the atomic diffusion caused by solute concentration gradients, but also other solute transport contributions; for example, from buoyancy-driven convection, Marangoni free surface convection, and thermotransport. However, since there were no free surfaces in the diffusion specimens, there would be no Marangoni convection present and the near isothermal conditions would render any thermotransport insignificant. As a result, it is believed that in the experimental setup used to obtain the terrestrial data, except for the atomic diffusion, the only extra contribution to the mass transport was buoyancy-driven convection and it is this that makes the measured diffusion coefficients obtained on the ground larger than those obtained in space and this is probably present in all the diffusion experiments conducted on the ground in the absence of strong externally applied damping forces; for

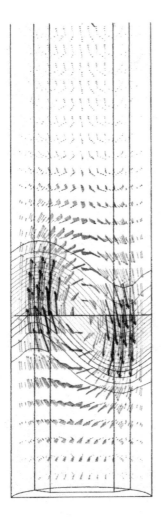 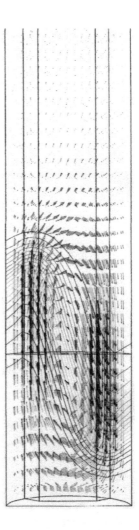

FIGURE 7. Velocity vector and solute distribution at one second during application of forcing pulse of $0.1\,g_0$.

FIGURE 8. Velocity vector and solute distribution at 58 seconds after applying $0.1\,g_0$ pulse.

TABLE 1. Impurity–diffusion coefficients of gold (1 wt%) in lead

T (°C)	$D \times 10^{-3}$ (mm^2/sec) microgravity	$D \times 10^{-3}$ (mm^2/sec) terrestrial
390	—	4.26
492	1.93	5.69
592	2.51	4.45
684	3.07	6.36
785	3.58	9.21

TABLE 2. Impurity–diffusion coefficients of silver (1 wt%) in lead

T (°C)	$D \times 10^{-3}$ (mm^2/sec) microgravity	$D \times 10^{-3}$ (mm^2/sec) terrestrial
390	3.80	7.49
492	4.97	7.80
592	5.88	9.13
684	7.04	9.85
785	8.10	11.20

TABLE 3. Impurity–diffusion coefficients of antimony (1 wt%) in lead

T (°C)	$D \times 10^{-3}$ (mm^2/sec) microgravity	$D \times 10^{-3}$ (mm^2/sec) terrestrial
390	2.74	3.90
492	—	4.06
542	4.21	—
592	—	4.72
684	5.91	6.19
785	6.75	7.36

example, by using magnetic fields. However, one might still wonder why it should exist in these ground experiments with conditions supposed to provide a stabilizing density gradient.

In these comparative diffusion experiments on the ground, the diffusion samples were set with their long axis vertical; that is, the capillary was parallel to the gravity direction. In order to realize this condition, the QUELD II ground based unit was set up with the furnace core axis vertical and with the samples loaded from below. In a temperature verification of the performance QUELD II for the lead solvent sample,[7] it was found that in the horizontal position at $1\,g$ and without any sample in place there was a small temperature gradient along the sample axis, the temperature being lowest at the open sample insertion end. This temperature gradient was found to increase as the setting experimental temperature increased; about 2°C/cm at 400°C. Because of the convection when the empty furnace is set vertically, it is believed that the temperature gradient would be larger. FIGURE 9 A shows the schematic temperature profile along the axis of the furnace; the temperature profile along the axis of an actual sample was found to be much less due to the thermal-smoothing effected by the conductive specimen crucible and metallic sample sheaths and the blocking of the open end of the furnace with an insulating sample support rod.

It is noted that, with a positive vertical temperature gradient, convection does not start in a fluid until the temperature gradient exceeds some critical value. However, according to Gershuni and Zhukhovitskii,[9] a Rayleigh instability should not occur in these diffusion experiments because of the small capillary diameter used.

Thus, even though massive convective cells may not be generated by the vertical temperature gradient, extra mass transport may arise from any horizontal temperature gradients because this convection mode can start instantly without needing to exceed a critical value. FIGURE 9 B shows the heat flow directions around the diffusion sample. The heat flows radially into the sample from the outside to the center and from the upper end, but out from the lower end. Hence, there are almost always horizontal temperature gradients in diffusion samples. Since the temperature at the top of the sample is higher than that at the bottom, the radial heat flow into the sample is less, so that any horizontal temperature gradient near the bottom of the sample is larger than that near the top.

For simplicity in the following analysis, instead of a horizontal temperature gradient, we define the temperature difference between the sample axis and surface, ΔT_R. FIGURE 9 C is a schematic of the change of ΔT_R from the bottom of the sample to the top.

Because of this radial temperature difference, there is always a certain residual convection, for which the velocity V_c is proportional to the radial Grashof number, defined by[10]

$$(G_r)_T = \frac{g \alpha_T R_s^3}{\nu} \Delta T_R,\tag{2}$$

where g is the gravity acceleration, α_T is the thermal expansion coefficient of the diffusion liquid at the experimental temperature, R_s is the sample radius, and ν the viscosity of the experimental liquid. Camel and Favier[11,12] proposed the following expression for a cylindrical configuration:

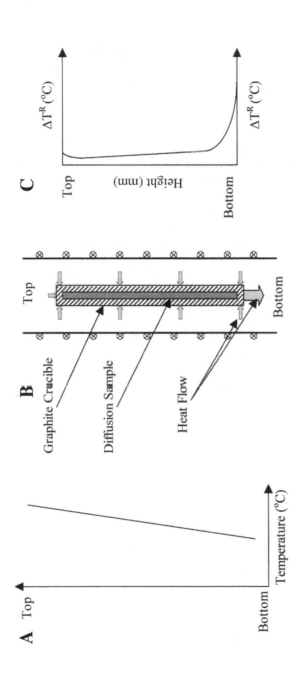

FIGURE 9. Schematic representations of the temperatures and heat flow in a vertical diffusion couple sample: (**A**) temperature profile along the axis of the furnace; (**B**) directions of the heat flow around the diffusion sample; and (**C**) change in temperature difference between the sample axis and the sample surface as a function of position.

$$V_c = 0.024 \frac{g \alpha_T R_s^2}{\nu} \Delta T_R. \tag{3}$$

However, Bejan and Tien[13] developed a different expression for the convection velocity. For the case of steady state and constant temperature gradient, the velocity field can be calculated analytically for a chemically homogeneous melt in an infinite cylinder. Here, the velocity component perpendicular to the capillary axis is shown to be zero whereas the velocity parallel to the capillary axis is given by[13]

$$V(x, y) = \frac{g \alpha_T R_s^2}{8 \nu} \left(1 - \frac{x^2 + y^2}{R_s^2} \right) \Delta T_R. \tag{4}$$

From Equations (2), (3), and (4), it can be seen that when the capillary axis is parallel to the gravity vector the radial temperature gradient is the major reason why the convection occurs. It also can been seen that for a chemically homogeneous melt, the residual convection velocity is proportional to the gravity field and the radial temperature difference. The space diffusion experiments carried out in QUELD II, which was mounted on MIM and operated on MIR, only experienced a $5 \mu g$ ($5 \times 10^{-6} g_0$) steady state acceleration.[10] Thus, compared with the residual convection in the terrestrial diffusion experiments, the residual convection in the space diffusion experiments was negligible. This means that, although the experimental procedures under microgravity conditions may not reveal the truly intrinsic values of any particular diffusion coefficient, that is, only that due to the atomic diffusion for particular alloy systems, they give values that must be much closer to the ideal value than those from terrestrial measurements.

TABLE 4 gives the differences between the diffusion coefficient measured terrestrially and in space at different temperatures for the different alloy systems. It is seen that the average diffusion coefficient difference for gold diffusion in lead is larger than that for silver diffusion in lead and antimony diffusion in lead. This can be explained as follows.

Let D_{exp} represent an experimentally measured diffusion coefficient that contains a contribution due to convection. D_{exp} can be calculated from the mean residual convection velocity V_c and the convection-free diffusion coefficient D using the Taylor relationship,[14]

TABLE 4. Differences between the diffusion coefficients measured under terrestrial conditions and those measured under microgravity conditions ($\times 10^{-3}$ mm^2/sec)

Temperature (°C)	Gold in Lead	Silver in Lead	Antimony in Lead
390	—	3.69	1.16
492	3.76	2.83	—
592	1.94	3.25	—
684	3.29	2.81	0.28
785	5.69	3.10	0.61
Average	3.66	3.14	0.68

$$D_{exp} = D\left(1 + A\left(\frac{R_s V_c}{D}\right)^2\right), \qquad (5)$$

where $A = 0.30$ is a value especially calculated for this velocity field.[15] Because the residual convection in the space diffusion experiments is negligible, it is reasonable to use the diffusion coefficient measured under microgravity conditions D_{mic} as the convection-free diffusion coefficient in Equation (5). From the preceding equations, one can deduce the proportionality:

$$D_{exp} - D_{mic} \propto \Delta T_R^2. \qquad (6)$$

As already noted, the comparative experiments described here were performed with the denser part of the diffusion couple in the lower half of the samples. For gold diffusion in lead, the alloy part was located at the bottom, but for silver diffusion in lead and antimony diffusion in lead, the alloy parts were located at the top of the capillary. From FIGURE 8C, it can be surmised that ΔT_R at the bottom of the sample is much bigger than that at the top. Then, from Equation (6), it can be concluded that the difference between the diffusion coefficient measured under terrestrial conditions and that measured under microgravity conditions for gold diffusion in lead would be larger than that for silver diffusion in lead or antimony in lead. Thus, the experimental results do fit with theoretical predictions.

However, the preceding theoretical analyses cannot explain why the average diffusion coefficient difference for antimony diffusion in lead is much smaller than that for silver diffusion in lead. This is because they are based on the presumption that the liquids are chemically homogeneous. Of course, in diffusion experiments the liquids are not homogeneous, since each is intended to have a concentration gradient along the sample axis.

Hart has dealt with this problem for a configuration confined by two parallel vertical planes separated by a distance H (see FIGURE 10).[16] The fluid is initially linearly stratified and the superimposed velocity–temperature–solute fields are set up by imposing a quasistatically increasing temperature difference across the plate at $x = \pm H/2$. This configuration could be extrapolated to the cylindrical configuration by making $H = R_s$, the characteristic distance.

For the experiments performed here, the solute Rayleigh number can be written as follows:

$$Ra_s = -\frac{g\beta R_s^4}{D\nu}\frac{\partial C}{\partial z}, \qquad (7)$$

where β is the solutal coefficient, $b = |(\rho - \rho_{Pb})/\rho_{Pb}|$, D is the diffusion coefficient, and C is the concentration. The velocity V_N shown in FIGURE 11 is standardized by $g\alpha_T \Delta T_R R_s^2/\nu$. This means that

$$V_C = \frac{g\alpha_T \Delta T_R R_s^2}{\nu}V_N. \qquad (8)$$

The viscosity of lead at 684°C is $1.14\times10^{-3}\,cm^2/sec$.[17] In initial stage, it is reasonable to assume that $\partial C/\partial z$ is approximately 10^{-2}/cm around the interface of the diffusion couple. Thus, using the data in TABLES 1, 2, 3, and 4, we calculated the following values:

$$Ra_s(Ag) = -9.8\times10^2$$

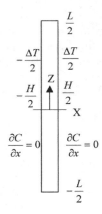

FIGURE 10. Hart's geometry for the problem.[16]

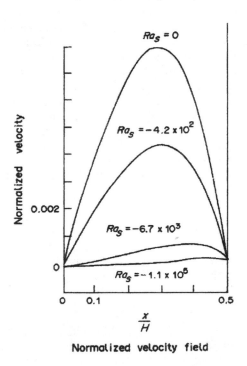

FIGURE 11. Hart's normalized velocity field.[16]

$$Ra_s(\text{Sn}) = -5.9 \times 10^3.$$

From FIGURE 11, it can be shown that the maximum normalized velocity V_N in the (Pb–Ag)–Pb diffusion couple near the interface between the alloy slug and the pure metal is more than three times larger than that in the (Pb–Sb)–Pb diffusion couple. Because the radial temperature difference ΔT_R is same for these two systems, from Equation (**8**), it is obvious that the maximum convection velocity V_C in the (Pb–Ag)–Pb diffusion couple near the interface is more than three times larger than that in the (Pb–Sb)–Pb diffusion couple. This explains why the average diffusion coefficient difference for antimony diffusion in lead was found to be much smaller than that for silver diffusion in lead.

According to Hart,[16] the diffusion specimen may be stabilized by layering. Because the isoconcentration lines are only slightly tilted, in the diffusion zone the lines of equal density can be horizontal, and therefore, convection flow can be suppressed. Since the axial concentration gradient decreases as the diffusion process continues, the ability to stabiles decreases and the convective flow increases. However, when the axial concentration gradient is smaller, the same convective flow causes a smaller effect on the measured value of the diffusion coefficient. Here, the very small differences between the diffusion coefficients measured under terrestrial conditions and those measured under microgravity conditions for antimony diffusion in lead suggests the correctness of the Hart theory prediction.

CONCLUSIONS

To examine the effect of the gravity in the liquid diffusion experiments, performed in the QUELD II/MIM/MIR campaign, two sets of experiments were carried out; the results obtained permit the following conclusions:

1. A computational fluid dynamics simulation confirmed that the microgravity isolation mount was particularly effective in suppressing the mass flow induced by ambient *g*-jitter on the MIR space station when used in *isolating mode* and of producing significant mass transport when used in *forcing mode*.

2. Comparative terrestrial experiments using the QUELD II ground base unit showed. that the diffusion coefficients of gold or silver in liquid lead measured on the ground are much larger than those obtained under microgravity conditions. This means that buoyancy-driven convection exists in the liquid during ground-based diffusion experiments. In fact, convective transport appears to be the dominant transport mode in all $1g$ experiments in Pb–Au, less so in Pb–Ag, but only offers a small proportion of the total transport in Pb–Sb. It is suggested that radial temperature gradients in the diffusion samples cause the buoyancy-driven convection and so enhance solute transport thereby increasing the measured (apparent) diffusion coefficient values. Furthermore, since the difference between the microgravity and $1g$ values for D in the Pb–Sb diffusion couples was found to be small, it is suggested that the convective transport component arising from radial temperature gradients falls as the density of the solute becomes increasingly smaller than that of the solvent.

ACKNOWLEDGMENTS

This work is sponsored by the Canadian Space Agency (CSA) through Contract 9F007-006041/001/SR.

REFERENCES

1. CHALMERS, B. 1967. Principles of Solidification. John Wiley & Sons, Inc.
2. CAMPBELL, J. 1991. Castings. Butterworth-Heinemann Ltd.
3. FROHBERG, G. & K.H. KRAATZ. 1986. Microgravity experiments on liquid self and inter-diffusion. Symposium, Norderney, 27.
4. MALMEJAC, Y. & G. FROHBERG. 1987. Chapters V. *In* Fluids Sciences and Materials Sciences in Space. H.U. Walter, Ed. Springer-Verlag.
5. SMITH, R.W., X. ZHU, M.C. TUNNICLIFFE, *et al.* 2002. The influence of gravity on the precise measurement of solute diffusion coefficients in dilute liquid metals and met-alloids. Ann. N.Y. Acad. Sci. **974:** 57–67.
6. BRUSON, A. & M. GERL. 1980. Phys. Rev. B **21**(12): 5447.
7. SMITH, R.W. 2000. Result of microgravity experiments—final report. PW & GS File No. 9F007-4-6028/01-ST "Diffusion in Liquid"-QUELD Project 4-0026.
8. SMITH, R.W. 1998. Micrograv. Sci. Technol. **XI**(2): 78.
9. GERSHUNI, G.Z. & E.M. ZHUKHOVITSKII. 1976. Convective Stability of Incompressible Fluid. Keter, Jerusalem.
10. PRAIZEY, J. P. 1998. Int. J. Heat Mass Transfer **32**(12): 2385.
11. CAMEL, D. & J.J. FAVIER. 1984. J. Cryst. Growth **67:** 57.
12. CAMEL, D. & J.J. FAVIER. 1986. J. Physique **47:** 1001.
13. BEJAN, A. & C.L. TIEN. 1978. Int. J. Heat Mass Transfer **21:** 701.
14. TAYLOR, G. 1953. Proc. Roy. Soc. A **219:** 186.
15. MATHIAK, G. & G. FROHBERG. 1999. Cryst. Res. Technol. **34:** 181.
16. HART, J.E. 1971. J. Fluid Mech. **49:** 279.
17. GRIGORIEV, I.S. & E.Z. MEILIKHOV. 1997. Handbook of Physical Quantities. CRC Press Inc.

APPENDIX

The Materials Processing Facility

QUELD II is a semi-automated version of Queen's University experiments in liquid diffusion (QUELD I), which flew in 1992 as a NASA STS middeck experiment, and which was used to manually process 12 long-capillary diffusion couples. An operator was still required to manually load the furnaces. For its two processing channels, QUELD II uses a variant of the three-zone furnace and quench system designed by the Queen's University Group and space-proven via two STS flights. The astronaut/cosmonaut time involved in the loading and reloading of samples is minimal (2–3 minutes) since the processing variables are introduced via a multisample ROM data card inserted into QUELD II when starting a new batch of samples. Upon completion of the diffusion anneal, split spring-loaded aluminum quench blocks are brought into contact with the sample causing the liquid specimen to freeze rapidly in a radial manner.

QUELD II is designed to operate with the Canadian Space Agency microgravity isolation mount (MIM) as its physical support and services platform. When on an orbiting platform, the MIM is designed to isolate the materials processing facility from much of the *g*-jitter that intrudes into the steady reduced gravity environment of approximately 5×10^{-6} unit gravity. This permits experiments to be pursued in low earth orbit without the complications brought about by the significant g-jitter usually present. In addition, the MIM can be used in *forcing mode* to introduce small accelerations into the *isolating* condition. The performance specifications of the MIM can be found elsewhere.[5]

Experimental Procedures

The long-capillary method[3,4] was selected for use with QUELD II samples. Typically, a 1.5–3.0 mm diameter × 40 mm length specimen was maintained at a fixed temperature for a given time to permit the solute in the 2 mm of 1–2 wt%-alloy slug attached to one end of the solvent rod to diffuse into the body of the specimen. The diffusion anneal period was selected to ensure that the semi-infinite diffusion condition was not violated; that is, no solute had moved to the far-end of solvent charge by the end of the anneal period.

If a cold sample is inserted into a furnace held at the test temperature, the temperature of the furnace is quenched and gradually climbs back to its design temperature. This results in the slow heating of the sample. It was found in the earlier QUELD I experiments that this situation could be largely overcome by having the furnace at a predetermined "superheat" temperature so that, when inserted, the sample would quench the furnace to the design temperature and so would assume the test temperature in a few seconds. Thus, immediately upon sample insertion, the furnace temperature was reduced to the design temperature and maintained at that value for the diffusion anneal period.

The specimen was then quenched and chemically analyzed using regular atomic absorption spectrophotometry techniques (AAS) in order to determine the solute distribution along the specimen. This solute distribution was then used to calculate the *D* value for the particular alloy, anneal temperature, and MIM condition used.

Diffusion Couple Preparation

In the past, it has been presumed that the solid–solid interface between the alloy slug and the solvent capillary should be planar, perpendicular to the major axis of the sample, and complete before the start of the diffusion period (i.e., no cavities or inclusions present). Recent numerical modeling suggests that some limited departure from this extreme requirement may be tolerated without substantial change in the measured diffusion coefficient. The Queen's Group had worked with many methods before selecting that used in these experiments, namely, the solvent rods are first produced by casting followed by swaging to the desired diameter, cut to length, and then placed in a chill mold such that their ends project into a horizontal gallery. Then the dilute alloy is poured into the mold. Following casting, the finished specimen is cut from the alloy gallery by electro-die-machining (EDM). Careful selection of liquid alloy superheat and mold temperature permits a clean planar interface to be obtained between the solvent rods and the solidified alloy.

Theories/Models of Diffusion in Liquids

The following table of theories and models is reproduced from Reference 5 with permission from the *Annals of the New York Academy of Sciences*.

Diffusion Coefficient–Temperature Relationship	Author(s)
$D = D_0 \exp(-Q/RT)$, where Q is the activation energy, R is the gas constant, and D_0 is a system constant	conventional description of data
$D = kT/6\pi R\eta$, where R is the particle radius and η is the viscosity of the medium	Stokes–Einstein Theory
$D = B\sqrt{T}\exp(-bV^*/V_f)$, where V^* is the critical volume associated with diffusing atom, V_f is the free volume of the liquid, and B and b are constants	model of critical volume
$D = AT^2$, where A is a system constant	fluctuation theory—Swalin 1959
$D = -a + bT$, where a and b are system constants	fluctuation theory—Reynik 1969
$D = \alpha\sqrt{T}(9.385T_m/T)^{-1}$, where T_m is the melting point and α is a system constant	hardsphere models
$D = A^1 T^m$, where $m = 1.7$–2.3 and A^1 is a system constant	molecular dynamics

Effects of Gravity on ZBLAN Glass Crystallization

DENNIS S. TUCKER,[a] EDWIN C. ETHRIDGE,[b]
GUY A. SMITH,[c] AND GARY WORKMAN[c]

[a]SD71, NASA/NSSTC, Huntsville, Alabama, USA

[b]SD46, Marshall Space Flight Center, Huntsville, Alabama, USA

[c]University of Alabama in Huntsville, Huntsville, Alabama, USA

ABSTRACT: The effects of gravity on the crystallization of ZrF_4–BaF_2–LaF_3–AlF_3–NaF glasses have been studied using the NASA KC-135 and a sounding rocket. Fibers and cylinders of ZBLAN glass were heated to the crystallization temperature in unit and reduced gravity. When processed in unit gravity the glass crystallized, but when processed in reduced gravity, crystallization was suppressed. A possible explanation involving shear thinning is presented to explain these results.

KEYWORDS: non-oxide glass; microgravity; ZBLAN; crystallization

INTRODUCTION

Heavy-metal fluoride glasses have been of interest to glass researchers for more than 25 years. One class that has consistently shown the most promise as an optical fiber is that of ZrF_4–BaF_2–LaF_3–AlF_3–NaF, normally referred to as ZBLAN glass.[1] ZBLAN fiber optics are used in a number of applications, such as fiber amplifiers, lasers for cutting, drilling, and surgery and they show promise in applications, such as nuclear radiation, resistant lengths, and non-linear applications.[2] Intrinsic and extrinsic processes limit light propagation at low powers in ZBLAN.[3] Intrinsic processes include band-gap absorption, Rayleigh scatter, and multiphonon absorption. Extrinsic processes include impurities, such as rare-earth and metal ions, and crystallites formed during preform processing and fiber drawing. The theoretical loss coefficient for ZBLAN is 0.001 dB/km at two micrometers. The achievement of this lower limit is hindered by both intrinsic and extrinsic processes. Varma et al.[4] stated that they felt that the devitrification is due to a narrow working range and low viscosity at the drawing temperature. Varma et al.[4,5] found that crystallization of certain tailored ZBLAN glasses is inhibited in a reduced gravity environment. Anselm and Frischat[6] doped ZBLAN glass with Ag and remelted this material in reduced gravity. They found that this glass was very homogeneous away from the crucible walls, a fact that they attributed to the suppression of convective transport processes under weightlessness.

Address for correspondence: Dennis S. Tucker, SD71, NASA/NSSTC, 320 Sparkman Drive, Huntsville, Alabama 35805-1912, USA. Voice: 256-961-7588; fax: 256-961-7148.
  dr.dennis.tucker@nasa.gov

Ann. N.Y. Acad. Sci. 1027: 129–137 (2004). ©2004 New York Academy of Sciences.
doi: 10.1196/annals.1324.012

EXPERIMENTAL

In the first series of experiments, ZBLAN was obtained from commercial vendors (Infrared Fiber Systems Inc., Silver Springs, MD; Galileo Electro-Optical Corp., Sturbridge, MA) and a research laboratory (Lucent Technologies, Murray Hill, NJ) in fiber form. Fiber was prepared and flown on board the KC-135 aircraft and on board a suborbital rocket. Fibers were heated to the crystallization temperature in unit gravity and in reduced gravity. The temperature of the experiments was chosen because it corresponds to nose of the TTT curve (370°C)[7] and the crystallization peak from differential thermal analysis (375°C).[8] For the KC-135 flights, 20 samples of fiber from each vendor were heated in the furnace at 400°C for 20 sec during the low-gravity portion of the parabola. This was repeated on the ground for comparison. The fibers were heated to 400°C for 6.5 min during the low-gravity portion of the suborbital rocket flight. Again, this process was repeated on the ground for comparison. Details of the fiber preparation, furnace construction, and furnace operation are reported elsewhere.[9,10] Optical and electron microscopy and X-ray diffraction was performed on the processed fibers. Experiments using two different ampoule orientations with respect to the horizontal were performed to determine if fiber sticking to the ampoule wall could induce crystallization. This was not observed. An experiment was also performed to determine if the reduced gravity samples reached the same temperature as the ground tested samples. ZBLAN with type-S thermocouples attached was reflown on the KC-135 and compared with ground samples. Both sets of samples reached the crystallization temperature in the same time. During these experiments it was observed that the sample temperature was actually 365°C to 370°C, this being very close to the temperature of maximum crystallization.

In the second set of experiments, samples were prepared at Lucent Technologies. Platinum coils were dipped in molten ZBLAN glass and allowed to cool. The glass resulted in a cylinder of ZBLAN glass within the coil. This coil was then used as the heating element. A Pt/Rh thermocouple was attached in the center of the glass cylinder.

FIGURE 1 shows the ZBLAN heater assembly in the sample mounting fixture. At all times the glass was protected from water vapor. The heater assembly was attached to the sample mounting fixture inside a glove bag that had repeatedly been evacuated and back filled with dry nitrogen until the R.H. in the bag was 0%. The assembly was then mounted in a chamber, also in the glove bag, that allowed continuous viewing with a video camera. Video was recorded during the flights and during ground runs. Rather than look at only one parabola, one sample per day was flown. The optical plus digital magnification of the samples were approximately 90×. The samples were held at 210°C, which is well below the glass transition temperature, T_g (265°C), and heated rapidly to 340°C during the low gravity portion of the parabola. The average time spent at 340°C during this interval was 11 sec. The sample was then cooled to 210°C before entering the high gravity portion of the parabola. This was repeated until crystallization was complete. On the ground, identical intervals were followed for comparison. Samples were examined with optical and electron microscopy and X-ray diffraction.

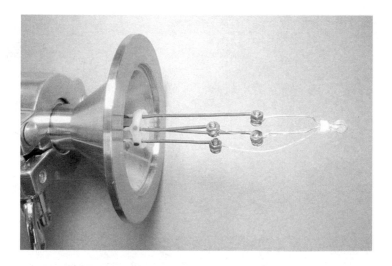

FIGURE 1. ZBLAN furnace fixture.

RESULTS

A scanning electron micrograph of a typical unit gravity processed fiber is shown in FIGURE 2. A close-up of a cross section of a fracture surface of a similar fiber is shown in FIGURE 3. Note the crystal formation. This was observed in all samples processed in unit gravity. Evidence of crystallization was not present for any samples processed in reduced gravity. A typical fiber is shown in FIGURE 4. FIGURE 5 shows a fracture surface of a flight sample that did not indicate evidence of crystallization. The particles on the surface are debris from the fracture surface. X-ray diffraction analysis of the ground processed samples showed the crystal structure to be β-BaZrF$_6$. This is a metastable crystal with an orthorhombic structure. X-ray diffraction analysis of the flight samples showed an amorphous structure showing no crystallinity.

The second series of experiments using the ZBLAN glass cylinders yielded interesting results. FIGURE 6 shows frames from video recordings of the crystallization of two flight samples and one ground sample. These frames were grabbed from the seventh thermal cycle for each sample. In the top frame, is flight sample F4, which does not show any evidence of crystallization at this point. What appear to be crystals are actually pores. The second frame shows ground sample G1. Lamellar crystals are evident in this frame. There is also quite a bit of crystallization occurring around the sample perimeter. The third frame is flight sample F3, which shows a large amount of surface crystallization. This was due to the presence of humidity when the samples were prepared. Upon reviewing the video for each sample, F4 did not completely crystallize until the completion of 29 cycles (parabolas). However, both G1 and F3 crystallized completely by the ninth cycle. It was also observed that once the crystals were observable, growth proceeded at approximately the same rate for each sample.

FIGURE 2. Scanning electron micrograph of ZBLAN fiber processed in unit gravity.

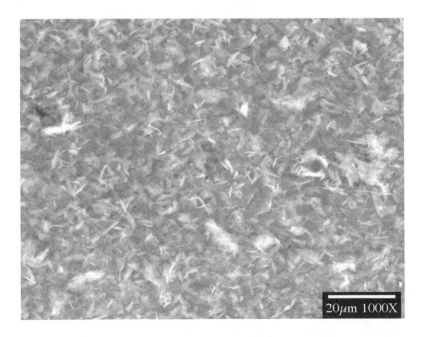

FIGURE 3. Scanning electron micrograph of fracture surface processed in unit gravity.

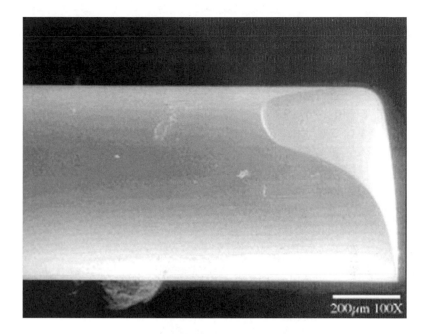

FIGURE 4. ZBLAN fiber processed in microgravity.

FIGURE 5. Scanning electron micrograph of fracture surface of ZBLAN fiber processed in microgravity.

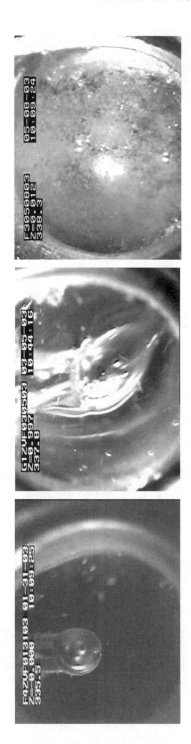

FIGURE 6. Flight sample F4, ground sample G1, and flight sample F3 after seven thermal cycles.

DISCUSSION

The consistency of the observed results in the parabolic flights and sounding rocket flight indicates that gravity does play a role in the crystallization of ZBLAN glass. A possible explanation for enhanced glass formation and reduced crystallization in reduced gravity has been advance by one of the authors (E.C.E.), and a study to investigate this mechanism is under way.[11] It is known that flow in undercooled polymer melts initiates crystallization.[12] It is also known that extrusion processing of glass-ceramics, glass-forming melts catalyzes the nucleation and growth of crystals.[13] A number of glass-forming liquids have been shown to exhibit shear thinning (pseudoplastic) behavior with one order of magnitude lowering of the viscosity. This is attributed to structural rearrangements in the liquid and, in particular, to the orientation of anisometric, chain like, flow units.[14] For example, in phosphate melts, evidence of anisotropic behavior in sheared glass melts is indicated at viscosities less than 10^6 dP·sec, attributed to the "orientation of the phosphate tetrahedral chains".[15] Lithium disilicate glass in the viscosity range 10^9 to 10^{12} dP·sec has as much as one order of magnitude variation in viscosity attributed to shear thinning. Shear thinning has also been observed in ZBLAN class glasses.[16] In the viscosity range of 10^5 to 10^7 dP·sec, a glass with lithium substituted for sodium showed a one order of magnitude shear thinning effect.[17] Another glass with 5% lead exhibited a 50% increase in viscosity of the liquid above the melting point (less than 1 dP·sec) at very low shear rates.[18] To explain the induced crystallization in high viscosity lithium disilicate glass melts simulation extrusion experiments, Gutzow[13] stated "A possible cause for induced crystallization is the reduction in viscosity (and thus kinetic factors) governing both nucleation and growth in melt crystallization."

If one assumes that shear thinning does occur in a glass melt, one can predict the effect on crystallization. Viscosity is the only directly measurable kinetic parameter used in nucleation and growth equations. In the classical treatment of crystallization by Turnbull[19] the nucleation rate, I, and crystal growth rate, U, are both inversely proportional to viscosity, h, with the viscosity term appearing in the preexponential factor.

$$I = \frac{k_n}{\eta} \exp\left[-\frac{b\alpha^3\beta}{T_r(\Delta T_r)^2} \right],$$

$$U = \frac{k_n'}{\eta}[1 - \exp(-\beta\Delta T_r)],$$

where T_m is the melting temperature, T is the absolute temperature, and ΔT_r is the reduced undercooling $(T_m - T)/T_m]$. The kinetic constants k_n and k_n', nucleus shape factor, b, and dimensionless parameters related to the liquid–crystal interface tension, α, and entropy of fusion, β, are described by Turnbull.[19]

The fraction of glass crystallized, X, with time at a given temperature was described by Uhlman[20] to be a function of the rate of nucleation, the third power of the growth rate, and the fourth power of time.

$$X = \frac{\pi\iota}{3}IU^3t^4.$$

Under conditions of shear thinning, the effective viscosity decreases with increasing shear rate so that the viscosity can be expressed as a function of shear rate,

$$\eta = \eta(\varepsilon).$$

The crystallization parameters, such as the nucleation rate, are also functions of shear rate

$$I(\varepsilon) = \frac{k_n}{\eta(\varepsilon)} \exp\left[-\frac{b\alpha^3\beta}{T_r(\Delta T_r)^2}\right].$$

Low-g processing is known to greatly reduce convection, which reduces shear in the liquid. Thus, the viscosity would be higher in these circumstances when compared to samples processed in unit gravity, reducing the nucleation and growth rates. For an increase in viscosity by a factor of two, the nucleation and growth rates are halved, but the fraction crystallized is reduced by a factor of 16. Since shear in liquid occurs as a result of fluid flow and fluid flow is greatly reduced in low-gravity, we have crystallization equations that are affected by gravitational effects. If the viscosity in low-g (low shear) conditions is effectively higher, then crystallization can be suppressed in liquids that exhibit shear thinning.

ACKNOWLEDGMENTS

The authors acknowledge partial financial support by NASA Headquarters for the project entitled "Mechanism for the Crystallization of ZBLAN" awarded under NRA-98-HEDS-05 Microgravity Materials Science, NASA. The authors also thank Dr. Refik Kortan of OFS Fibers for his help in sample preparation and X-ray diffraction

REFERENCES

1. BOEHM, L., K.H. CHUNG, S.N. CRICHTON & C.T. MOYNIHAN. 1987. Crystallization and phase separation in fluoride glasses. Infrared Optical Materials and Fibers V, SPIE 843, Bellingham, Washington, 10–13.
2. TRAN, D.C., G.H. SIEGEL & B. BENDOW. 1984. Heavy metal fluoride glasses and fibers: a review. J. Lightwave Tech. **LT-2**(5): 121–138.
3. BANSAL, N.P., A.J. BRUCE, R.J. DOREMUS & C.T. MOYNIHAN. 1984 Crystallization of fluorozirconate glasses. Mater. Res. Bull. **19**: 522–590.
4. VARMA, S., S.E. PRASAD, I. MURLEY, *et al.* 1991. The role of statistical design in microgravity materials research. Proc. Spacebound **91**: 248–249.
5. VARMA, S., S.E. PRASAD, I. MURLEY, *et al.* 1992. Use of microgravity for investigation phase separation and crystallization in heavy metal fluoride glass. Proc. Spacebound **92**: 109–114.
6. ANSELM, L. & G.H. FRISCHAT. 2000. Early crystallization stages in a heavy metal fluoride glass prepared under normal and weightless conditions. Phys. Chem. Glasses **41**: 32–37.
7. BUSSE, L.E., G. LU, D.C. TRAN & G.H. SIEGEL, JR. 1985. A combined DSC/optical microscopy study of crystallization in fluorozircanate glasses upon cooling from the melt. Mat. Sci. Forum **5**: 219–228.
8. NAKAO, Y. & C.T. MOYNIHAN. 1991. DSC study of crystallization and melting of ZrF_4–BaF_2–LaF_3–AlF_3–NaF glasses. Mat. Sci. Forum **67/68**: 187–195.

9. TUCKER, D.S., G.L. WORKMAN & G.A. SMITH. 1995. Microgravity processing of ZBLAN glass. AIAA Proc. **95:** 3784–3790.
10. TUCKER, D.S., G.L. WORKMAN, G.A. SMITH & S. O'BRIEN. 1996. Effects of microgravity on ZBLAN optical fibers utilizing a sounding rocket. SPIE Proc. 2809, 23–32.
11. ETHRIDGE, E.C. & D.S. TUCKER. 1999. Mechanisms for the crystallization of ZBLAN. Proposal funded by NASA in response to NRA-98-HEDS-05.
12. PENNINGS, A.J., J.M.A. VANDERMARK & H.C. BOOJI. 1970. Hydrodynamically induced crystallization of polymers from solution: II. The effect of secondary flow. Kolloid. Z.u.Z.f. Polym. **236:** 99–111.
13. GUTZOW, I., B. DURSCHANG & C. RUSSEL. 1997. Crystallization of glass forming melts under hydrostatic pressure and shear stress: Part II Flow induced melt crystallization: a new method of nucleation catalysis. J. Mater. Sci. **32:** 5405–5411.
14. DEUBENER, J. & R. BRUCKNER. 1997. Influence of nucleation and crystallization on the rheological properties of lithium disilicate melts. J. Non-Cryst. Solids **209:** 96–111.
15. HABEK, A. & R. BRUCKNER. 1993. Direct connection between anisotropic optical properties, polarizability and rheological behaviour of single-phase glass melts. J. Non-Cryst. Solids **209:** 225–236.
16. WASHE, R. & R. BRUCKNER. 1986. The structure of mixed alkali phosphate melts as indicated by their non-newtonian flow behavior and optical birefringence. Phys. Chem. Glass **27:** 87–89.
17. WILSON, S.J. & D. POOLE. 1985. Viscosity measurements on fluorozirconate glasses. Mat. Sci. Forum **6:** 665.
18. HASZ, W.C., S.N. CRICHTON & C.T. MOYNIHAN. 1988. Viscosity temperature dependence of ZrF_4-based melts. Mat. Sci. Forum **32/33:** 589–594.
19. TURNBULL, D. 1969. Under what conditions can glass be formed? Contemp. Phys. **10:** 473–488.
20. UHLMANN, D.R. 1972. A kinetic treatment of glass formation. J. Non-Cryst. Solids **7:** 337–348.

Space Radiation Transport Properties of Polyethylene-Based Composites

R.K. KAUL,[a] A.F. BARGHOUTY,[a] AND H.M. DAHCHE[b]

[a]NASA-Marshall Space Flight Center, Huntsville, Alabama, USA

[b]Department of Chemistry, Roanoke College, Salem, Virginia, USA

ABSTRACT: Composite materials that can serve as both effective shielding materials against cosmic-ray and energetic solar particles in deep space, as well as structural materials for habitat and spacecraft, remain a critical and mission enabling component in mission planning and exploration. Polyethylene is known to have excellent shielding properties due to its low density, coupled with high hydrogen content. Polyethylene-fiber reinforced composites promise to combine this shielding effectiveness with the required mechanical properties of structural materials. Samples of polyethylene-fiber reinforced epoxy matrix composite 1–5 cm thick were prepared at the NASA Marshall Space Flight Center and tested against a 500 MeV/nucleon Fe beam at the HIMAC facility of NIRS in Chiba, Japan. This paper presents measured and calculated results for the radiation transport properties of these samples.

KEYWORDS: ionizing radiation; space radiation; shielding; shielding effectiveness; HZE; transport; simulation; transmission; polyethylene; hybrid materials; polyethylene-based composites; composites; galactic cosmic rays; solar energetic particles; deep-space missions; micrometeorite shields; Mars missions; multi-functional materials; fluences; dose; dose-equivalent; linear energy transfer

INTRODUCTION

One of the most significant technical challenges to long-duration deep-space missions is that of protecting the crew from harmful and potentially lethal exposure to ionizing radiation. Energetic, high charge galactic cosmic-ray ions (GCR) and solar energetic particles (SEP) constitute the two main sources of this intense radiation environment. Protection against GCR and SEP radiation fields on a manned Mars mission, for example, is vital both during transit and while on the surface of the planet because of the duration of the mission and the lack of sufficient protection from the thin Martian atmosphere. The development of multifunctional materials that can serve as integral structural members of the space vehicle and at the same time provide the necessary radiation shielding for the crew is both mission enabling as well as cost effective. Additionally, by combining shielding and structure the total vehicle mass can be reduced for propulsion system designs.

Address for correspondence: A.F. Barghouty, NASA-Marshall Space Flight Center, SD-46, MSFC, AL 35812, USA. Voice: 256-544-0238; fax: 256–544-7754.
nasser.barghouty@msfc.nasa.gov

Ann. N.Y. Acad. Sci. 1027: 138–149 (2004). ©2004 New York Academy of Sciences.
doi: 10.1196/annals.1324.013

Hybrid laminated composite materials having both ultrahigh modulus polyethylene (PE) and graphite fibers in epoxy and/or PE matrices could meet the above mission requirements. PE fibers have excellent physical properties, including the highest specific strength of any known fiber. Moreover, the high hydrogen content of polyethylene makes this material an excellent shielding material against GCR/SEP radiation. When such fibers are incorporated in a PE matrix, the shielding effectiveness increases. Boron may be added to the matrix resin or used as a coating to further increase shielding effectiveness because of its ability to slow thermal neutrons produced in the interactions of GCR/SEP ions with the shielding materials. These materials may also serve as micrometeorite shields, since PE has high impact energy absorption properties.

In this paper we report on an ongoing effort at the NASA Marshall Space Flight Center to design, fabricate, and test such multifunctional composites. A brief review of the mechanical properties of the composites is followed by a description of the simulation of their shielding effectiveness against GCR and SEP radiation fields. Exposure data from an HIMAC Fe beam at 500 MeV per nucleon and comparison with calculations are then presented, followed by a discussion of our preliminary results.

POLYETHYLENE-BASED COMPOSITES

Reinforced composite materials enjoy significant property advantages that make them excellent candidates for use in aircraft and spacecraft structural applications. Properties such as specific tensile strength, specific tensile modulus, fatigue resistance, damage tolerance, and design flexibility, all make these materials very attractive for aerospace applications. Additionally, a wide range of fiber reinforcement types and matrix resin systems are available to the engineer and designer for application specific use.

The fibers under consideration in this study are PE and graphite; both are high-strength and high-modulus fibers. FIGURE 1 shows a comparison of the specific tensile strength and specific tensile modulus of several common reinforcing fibers. Note that the PE fiber has the highest specific tensile strength, or strength per unit weight, of any reinforcing fiber and a specific modulus approximately equivalent to that of graphite and boron fibers.

PE fibers have the additional advantage in their ability to shield against cosmic radiation; that is, high energy and high charge, highly penetrating, ionizing radiation. Although the high hydrogen content in the PE fiber is responsible for its shielding effectiveness, another added benefit in using PE fibers for radiation shielding is that hydrogen also acts to slow, or thermalize, fast neutrons because of their large collision cross section. Such neutrons are produced in GCR/SEP ions collisions with shielding materials. Thus, added radiation protection can be realized when using PE as a matrix with PE fibers. TABLE 1 shows the hydrogen content for select materials.

Boron may be added to the matrix resin to further improve the shielding effectiveness of these materials. Boron attenuates thermal neutrons that have been slowed down by hydrogen and this, in turn, can reduce the level of captured gamma rays. Values of the hydrogen content of borated polyethylene are listed in TABLE 1 for

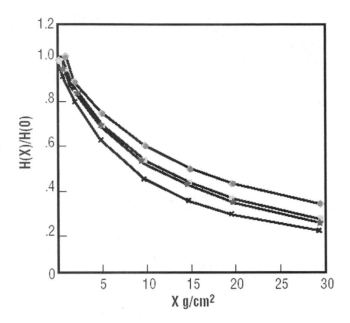

FIGURE 1. Specific tensile strengths of select reinforcing fibers: ➡➊➡, polytetraluo-roethylene; ➡■➡, polyimide; ➡▲➡, polysutine; ✱, polyethermide; ◆, polyethylene. Note the high values of PE fibers for both modulus and tensile strengths.

comparison. Relative to the materials listed in TABLE 1, all of which are considered effective shielding materials, PE enjoys significantly higher hydrogen content.

Epoxy resin systems are widely used as matrix resins in advanced composite systems. These resins also have substantial hydrogen content, making them suitable candidate materials for radiation shielding. A large database exists on the mechanical properties of composites made with these resins. The data shown in FIGURE 2 are taken from studies on the use of both lunar and Martian regolith and regolith/epoxy combinations for radiation shielding. The data show the superior performance of

TABLE 1. Hydrogen content of select materials

Material	Number of H atoms per $cm^3 \times 10^{22}$
Solid hydrogen (4.2 K)	5.7
Water	6.7
Lithium hydride (LiH)	5.9
Pure polyethylene (PE)[a]	8.89
5% Borated PE	6.6
Lithium polyethylene	5.44

[a]Hydrogen content of PE fiber is the same as that of pure PE.

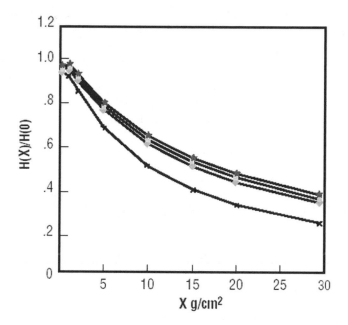

FIGURE 2. Attenuation of dose $H(X)$ as function of depth X in the material for polymeric shielding materials: ◆, lunar regolith; ■, regolith 10% epoxy; ★, regolith 20% epoxy; ✳, epoxy. Note the superior performance of PE. (Reproduced from Ref. 1 with permission.)

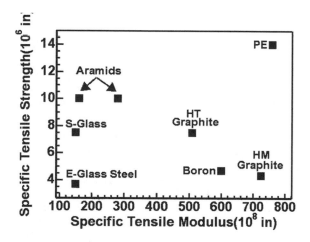

FIGURE 3. Attenuation of dose $H(X)$ as function of depth X in lunar regolith, regolith/epoxy combinations, and epoxy alone, showing the added shielding benefit of incorporating epoxy as matrix resins.

epoxy in the attenuation of radiation. Similar comparisons of various polymeric materials are presented in FIGURE 3 and show PE to be more effective than other polymers in dose attenuation.

A recent accelerator-based study[3] suggests that the addition of modest amounts of PE (or similar materials) to the shielding material of the interior of the International Space Station (ISS), which in its low Earth orbit (LEO) enjoys some magnetospheric and atmospheric shielding, can be effective in reducing the radiation dose from GCR ions to crew members and systems. This study uses the same detection system as the ISS study, as well as similar beams at similar energies, but applies the HIMAC results to the deep-space radiation environment. Simulations in the two studies, however, differ in some details. In this study nuclear fragmentation cross sections, for example, are estimated semi-empirically but are assumed to be energy dependent. Full GCR and SEP spectra are used in this study but no trapped particle component. Calculation of dose and dose-equivalents in both studies are similar.

TRANSPORT SIMULATIONS

The calculations presented here are meant only to benchmark the radiation shielding effectiveness of the proposed PE-based composites relative to pure PE. In these simulations, we assume the radiation field is made up of GCR ions from protons ($Z = 1$) through nickel ($Z = 28$), ranging in kinetic energy from 1 MeV to 1 TeV per nucleon. Solar minimum conditions, where GCR flux is at maximum, are assumed. GCR flux as a function of energy and relative abundance is taken to be that of nominal[4] GCR spectra at 1 AU, appropriate for radiation-shielding simulations. To simulate a worst-case scenario, an unusually intense SEP field (like that associated with the October 1989 solar particle event) is added to the GCR field even though this addition amounts to less than 7% of the total fluence[5] for the simulated flux, energies, and material depths considered here. GCR ions are fully stripped, whereas heavy SEP ions are only partially stripped.

For simulation purposes, nominal parameters; for example, charge state and relative abundance, are used here for the SEP field ions, consistent with recent detailed simulations[6] as well as observations by the solar anomalous and magnetospheric particle explorer (SAMPEX)[7,8] and advanced composition explorer (ACE).[9] The anomalous component of GCR (or ACR) is ignored since its integrated contribution is quite small[10,11] at energies above a few hundreds MeV per nucleon.

The simulations are done in one dimension, but a two-dimensional version of the simulation code is being developed and will be used in the course of the investigation. Although the GCR field is, for the most part, isotropic, two and three-dimensional calculations are more critical for the whole spacecraft or surface habitats. Another addition that will be incorporated, but is not included here in the preliminary simulations, is the neutron field. Contribution of this field to the total dose is expected to become significant[12] (i.e., greater than 5–10%) only for PE-based composites at depths exceeding 25–30 g/cm². Hence, we present simulations up to depths of 25 g/cm².

Given the current uncertainties in a number of the transport calculation parameters; for example, nuclear fragmentation cross sections,[13] contributions of pions and other hadrons are ignored since their yields are quite small relative to the total GCR yield over the energy range of interest; they are certainly smaller than those uncertainties.[14,15] Contributions from target fragments and nuclear energy losses[16] related to charge-changing nuclear interactions are included, even though both contributions are also relatively small. (Energy or momentum losses become more important for two- and three-dimensional simulations, whereas target fragment contribution is relevant for radiation-fatigue analysis of the shielding materials.)

The assumed radiation environment in the simulations consists of solar-minimum GCR plus the October 1989 SEP event at the Earth orbit. No magnetospheric or atmospheric attenuation of the *primary* GCR/SEP flux is assumed. For each ion species at all energy points we apply the continuity equation[17] taking into account species transformations and continuous energy losses due to both atomic (ionization), as well as nuclear collisional losses. In its simplest form, the equation reads

$$\frac{\partial J_i}{\partial \lambda} = -\frac{J_i}{\lambda_i} + \sum_j \frac{J_i}{\lambda_{ij}} + \frac{\partial}{\partial E}[W_i J_i], \tag{1}$$

where J_i is the flux (in units of particles/m²·sec·sr·MeV/nucleon) of species (i), λ is the depth or areal density in g/cm² of the material, λ_i is the mean-free path of species (i) in the material. The latter is related to the mass density of the material and the total charge-changing cross sections (σ_r) of species (i) in that material as follows:

$$\lambda^{-1} = \frac{N_A \sigma_r}{A}, \tag{2}$$

where A is the atomic number of the material and N_A is Avogadro's constant. Similarly, $(\lambda_{ij})^{-1}$ is related to the charge-changing cross section σ_{ij} for species (j) to produce (e.g., by fragmentation) species (i). Finally, E is the kinetic energy of the particle and W_i is the energy-dependent stopping power (in MeV/nucleon per g/cm²) for species (i) in the shielding material.

Energy-dependent nuclear reaction cross sections are taken from Reference 18 and those for nuclear fragmentation from References 19 and 20. Light-ion ($Z \leq 4$) production cross sections are taken from Reference 21. Procedures to calculate stopping powers are taken from various sources cited collectively in Reference 22. The calculations proceed by numerically integrating Equation (1) from a depth of $\lambda = 0$ g/cm² (no shielding) to the required depth, making sure the differential depth (or slab) is small enough so that multiple nuclear interactions in the slab are unlikely (typically of the order of 10^{-2} g/cm²). After each differential slab, the fluxes of all the primary, secondary, tertiary, and so on, nuclei species are updated and stored.

The energy of each species is also adjusted due to ionization and nuclear collisional losses over the slab. Conservation laws for mass, charge, energy, and momentum are checked at the beginning and end of each differential slab. Once the required depth is achieved, the marching algorithm is stopped and the fluxes at that depth and for all species recorded and updated.

The integral linear-energy transfer (LET) spectrum $S(L)$ as a function of LET L in the absorbing material (assumed here to be water) is calculated from the transported flux through the shielding material from a lower limit 0 in stopping power to an infinite upper limit as follows:

$$S(L) = \int_0^\infty J(W')dW'. \qquad (3)$$

For our purposes the lower limit of W is taken to be $1\,\text{MeV/nucleon}$ per g/cm^2 and the upper limit $10^5\,\text{MeV/nucleon}$ per g/cm^2. The dose, in units of energy per mass, J/kg or Gray ($1\,\text{Gy} = 100\,\text{rad}$) in the absorbing material is then calculated from the LET spectrum $S(L)$ using

$$H = \int S(L)LdL. \qquad (4)$$

The dose equivalent, H_q, in Sievert (Sv) is then

$$H_q = \int S(L)Q(L)LdL, \qquad (5)$$

where $Q(L)$ is the LET (or stopping-power) dependent quality factor;[23] which helps to separate the radiation types according to both energy imparted onto the material as well as energy absorbed by the material (hence, different radiation types of the same energy deposition may produce different ionizing effects due to different energy absorption in the material). We use the dose-equivalent measure to benchmark the shielding effectiveness of our samples against pure PE.

FIGURE 4 shows calculated dose equivalents, relative to pure PE, at 0-cm depth in the absorbing material (water) for Composite #1 (70%vol PE fibers + 30% epoxy), Composite #2 (40% PE fibers + 30%C fibers + 30% epoxy), Composite #3 (70% PE fibers + 20% epoxy + 10% B powder), and aluminum as functions of depth in the shielding material. The figure suggests that Composites #1 and #3 have very similar shielding capabilities but come closer to the radiation-shielding performance of pure PE than Composite #2. The dose-depth curve for Composite #4 (not shown), which is made up of Composite #1 interlaced with a thin layer of 70% B + 30% epoxy, is very similar to that for Composite #1.

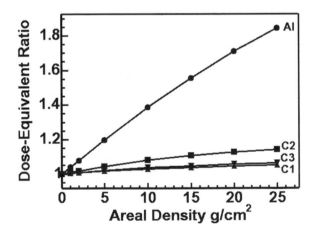

FIGURE 4. Dose-equivalent ratios normalized to pure PE for the three composites C1–C3, as well as Al.

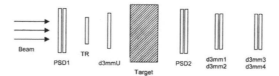

FIGURE 5. Configuration of the detection system. Respective dimensions × areas are: TR, 300 mm × 300 mm²; PSD1 and PSD2, 1 mm × 1,500 mm²; d3mmU and d3mm1-d3mm4, 3 mm × 450 mm². (Reproduced from Ref. 3 with permission.)

Based on FIGURE 4, of the three samples simulated for their radiation shielding effectiveness, the C1 design was chosen to be fabricated and beam-tested at the HIMAC in Chiba, Japan. Beam runs were conducted during January and February of 2003. Currently at the NASA MSFC, other designs are being considered as well and are at various stages of development, in addition to comprehensive mechanical testing of the C1 design.

SAMPLE FABRICATION

The fabrication of test specimens for this study was performed at the NASA-Marshall Space Flight Center NCAM facilities. The thickness range of the test specimens was determined by radiation transport modeling results as outlined above. Three specimens of thickness 1, 3, and 5 cm were selected for radiation exposure at the HIMAC facility.

The PE/epoxy material was cut into 4″ × 10″ size plies. The plies were hand laid to the desired thickness and autoclave cured at 240°F and 100 psi pressure.

BEAM EXPOSURE SETUP

The detection system used in this particular study (see FIGURE 5) is characterized by large acceptance threshold.[24] For the exposure data presented below this has the effect of smearing detection of light fragments (especially protons) into larger fragments, resulting in fluence measurements with low-resolution peaks for fragments with charge up to $Z = 10$. The effect is negligible for fragments with $Z > 10$.

TABLE 2. Sample beam-exposure data for C1

Thickness g/cm²	Measured ratio of Fe flux behind C1 to beam	Calculated ratio of Fe flux behind C1 to beam
1	0.84 ± 0.07	0.79
3	0.61 ± 0.05	0.57
5	0.49 ± 0.04	0.43

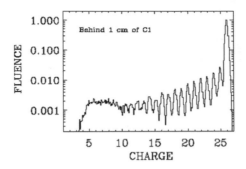

FIGURE 6. Measured fluences for charges 1 through 26 behind 1 cm of C1 from a 500 MeV/nucleon Fe beam. In this figures the high-acceptance threshold of the detection system results in poorly resolved light fragment ($Z \leq 10$) peaks—unlike the higher Z peaks, which are clearly seen.

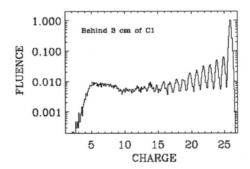

FIGURE 7. As FIGURE 6 but behind 3 cm of C1.

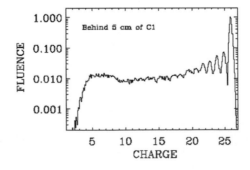

FIGURE 8. As FIGURE 6 but behind 5 cm of C1.

BEAM EXPOSURE DATA

TABLE 2 shows the sample result of exposing C1 to the HIMAC Fe beam of 500 MeV/nucleon. This sample result shows the effectiveness of C1 in attenuating the primary Fe beam by 50% using a thickness of only about 5 cm. Note that, although the data analyses are still preliminary and this attenuation property is only part of the shielding effectiveness of any candidate material, the C1 design, in both simulations and in accelerator-based tests, appears to possess superior shielding properties. FIGURES 6–8 below show fluence measurements for charges $Z = 1$–26 behind 1 cm, 3 cm, and 5 cm of C1 after bombardment with a 500 MeV/nucleon Fe beam.

Comparison with calculations is made complicated by the large-acceptance character of the detection system used. However, a preliminary simulation in which all particles are assumed to be well resolved is presented below. A qualitative comparison between simulation and data is useful and can still be made for large Z fragments, since these fragments contribute the most to the ionizing (proportional to Z^2) effects of high-energy and high-charge particulate radiation (see FIGURE 9).

DISCUSSION

We have presented simulations and accelerator-based test results of the shielding effectiveness of an advanced composite design based on PE fiber in an epoxy matrix resin. This study is part of ongoing efforts at NASA-MSFC to design, fabricate, and test multifunctional materials for use in deep-space mission planning and designs.

The high hydrogen content of this composite gives it superior transport properties against high-energy and high-charge particulate radiation fields that populate the deep-space radiation environment. The PE-based composite was beam-tested against a 500 MeV/nucleon Fe beam at the HIMAC facility in Chiba, Japan.

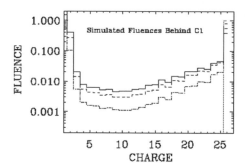

FIGURE 9. Simulated fluences of charges 1 through 26 behind 1 cm (*lower histograms*), 3 cm (*middle*), and 5 cm (*top*) of C1 from the HIMAC Fe beam (vertical) of 500 MeV/nucleon. In contrast to the measurements presented in FIGURES 6–8, calculated fluences for all charges are assumed to be highly resolved, hence the sharp peak for protons not seen (separately) by the detection system.

Accelerator data suggest that this new design is able to attenuate the primary beam by 50% in about 5 cm. In addition, fluence data are able to depict the efficiency and degree to which this composite can reduce the expected radiation dose equivalent by primary-ion fragmentation and spallation. Coupled with high specific strength and modulus, this PE-based composite appears to promise true multifunctionality for deep-space exploration.

Efforts at NASA-MSFC, and elsewhere, appear to point to the potential utility and benefits of PE fiber-based composites as both structural and radiation shielding materials, and, thus, as deep-space mission enabling technology.

ACKNOWLEDGMENTS

This work is supported by NASA internal grant at MSFC (NRA-02-OBRP-02) and by NASA Grant NAG8-1901 (The Transport Consortium). The authors thank J. Miller, L. Heilbronn, and C. Zeitlin (Lawrence Berkeley National Laboratory, University of California, Berkeley) for sample exposure at the HIMAC facility, as well as for supply and analyses of exposure data. The authors extend their special thanks to Y. Iwata and the support staff at the HIMAC facility of the National Institute of Radiological Sciences, Chiba, Japan and NIRS for their gracious hospitality.

REFERENCES

1. WILSON, J.W., *et al.* 1997. Galactic and solar cosmic ray shielding in deep space. NASA Tech. Paper #3682. NASA, Washington, DC.
2. KIM, M.Y., *et al.* 1998. Comparison of Martian meteorites and Martian materials for galactic cosmic rays. NASA/TP-1998-208724. NASA, Washington, DC.
3. MILLER, J., *et al.* 2003. Benchmark studies of the effectiveness of structural and internal materials as radiation shielding for the International Space Station. Rad. Res. **159:** 381.
4. BADHWAR, G.D. & P.M. O'NEIL. 1996. Galactic cosmic radiation model and its application. Adv. Space Res. **17:** 1.
5. REAMES, D.V. 1991. Particle acceleration in solar flares: observations. *In* Particle Acceleration in Cosmic Plasmas. AIP vol. 264. G.P. Zank & T.K. Gaisser, Eds.: 213. AIP, New York.
6. BARGHOUTY, A.F. & R.A. MEWALDT. 2000. Simulation of charge-equilibration and acceleration of solar energetic ions. *In* Acceleration and Transport of Energetic Particles Observed in the Heliosphere. AIP vol. 528. R.A. Mewaldt, *et al.*, Eds.: 73. AIP, New York.
7. OETLIKER, M., *et al.* 1997. The ionic charge of solar energetic particles with energies of 0.3–70 MeV per nucleon. Astrophys. J. **477:** 495.
8. MAZUR, J.E., *et al.* 1999. Charge states of solar energetic particles using the geomagnetic cutoff technique: SAMPEX measurements in the November 6, 1997 solar particle event. Geophys. Res. Lett. **26:** 173.
9. MOBIUS, E.M., *et al.* 1999. Energy dependence of the ionic charge state distribution during the November 1997 solar energetic particle event. Geophys. Res. Lett. **26:** 145.
10. JOKIPII, J.R. 2001. Acceleration and transport of energetic charged particles in space. Astrophys. Space Sci. **277:** 15.
11. CUMMINGS, A.C., E.C. STONE & C.D. STEENBERG. 2002. Composition of anomalous cosmic rays and other heliospheric ions. Astrophys. J. **578:** 194.

12. WILSON, J.W., *et al.* 1991. Transport method and interactions for space radiation. NASA Ref. Pub. No. 1257. NASA, Washington, DC.
13. TSAO, C.H., R. SILBERBERG & A.F. BARGHOUTY. 2001. Cosmic ray sources and source composition. Astrophys. J. **549:** 320.
14. BARGHOUTY, A.F. & G. FAI. 1991. Pion yield and pion spectra from high energy nuclear collisions. Nucl. Phys. A **75:** 726.
15. GLAGOLEV, V.V., *et al.* 2001. Fragmentation of relativistic oxygen nuclei in interactions with a proton. Eur. Phys. J. A. **11:** 285.
16. TSAO, C.H., *et al.* 1995. Energy degradation in cosmic-ray nuclear spallation reactions: relaxing the straight ahead approximation. Astrophys. J. **451:** 275.
17. GAISSER, T.K. 1990. Cosmic Rays and Particle Physics. Cambridge University Press, Cambridge.
18. SIHVER, L., *et al.* 1993. Total reaction and partial cross sections in proton–nucleus and nucleus–nucleus reactions. Phys. Rev. C **47:** 1257.
19. SILBERBERG, R., C.H. TSAO & A.F. BARGHOUTY. 1998. Updated partial cross sections of proton–nucleus reactions. Astrophys. J. **501:** 911.
20. TSAO, C.H., R. SILBERBERG & A.F. BARGHOUTY. 1998. Partial cross sections of nucleus–nucleus reactions. Astrophys. J. **501:** 920.
21. CUCINOTTA, F.A., *et al.* 1996. Light ion components of the galactic cosmic rays: nuclear interactions and transport theory. Adv. Space Res. **17:** 277.
22. ADAMS, J.H., JR. 1992. Cosmic radiation: constraint of space exploration. Rad. Measur. **20:** 397.
23. NATIONAL COUNCIL ON RADIATION PROTECTION AND MEASUREMENTS. 1990. The relative biological effectiveness of radiations of different quality. NCRP Report No. 104. NCRP, Bethesda.
24. MILLER, J. & C. ZEITLIN. Private communications.

The Study of Devitrification Processes
in Heavy-Metal Fluoride Glasses

IAN R. DUNKLEY,[a] REGINALD W. SMITH,[a] AND SUDHANSHU VARMA[b]

[a]Materials Science and Microgravity Applications Group,
Department of Mechanical and Materials Engineering,
Queen's University, Kingston, Ontario, Canada

[b]Supan Technologies In., Orleans, Ontario, Canada

ABSTRACT: Heavy-metal fluoride glasses are very promising optical fiber
materials because of their predicted ultralow loss and long transparency
range. Although conventional silica fibers have attained their theoretical minimum loss of 0.15 dB/km, fluoride glasses have the potential to yield losses of
only 0.001 dB/km. Fluoride glasses also exhibit transparency into mid-IR frequencies, a region inaccessible to silica fibers. However, this group of glasses is
very unstable to devitrification during both bulk glass synthesis and fiber-
drawing. This instability has limited their commercial exploitation to a small
niche market in the laser industry. The ZBLAN glass ($53ZrF_4$–$20BaF_2$–$4LaF_3$–
$3AlF_3$–$20NaF$) is the most promising of these materials since its fiber-drawing
region lies on the edge, or possibly just outside its crystallization region. It is
believed that additional research into understanding the nucleation mechanics
involved in the devitrification of fluoride glasses will lead to the development
of technology to suppress such nucleation, or at least minimize the associated
crystallization temperature region, allowing high optical quality fibers to be
produced. It has recently been demonstrated that a microgravity environment
can suppress devitrification in ZBLAN glass preform preparation, and that
devitrification may be reduced when preparing ZBLAN terrestrially in a containerless facility. It is believed that the role of viscosity is critical in the devitrification mechanism of ZBLAN glass and in determining the optimum fiber-
drawing temperature. Unfortunately, viscosity data for fluoride glasses are
only available above the melting point and around the glass transition. A piezoelectric viscometer has been developed and is being used to determine the missing viscosity data in the fiber-drawing and crystallization temperature regions.
Shear thinning of the glasses and/or the application of hydrostatic pressure on
the glasses have been recently proposed to be responsible for devitrification
during fiber-drawing at $1g$ and in reduced gravity. The study we report here
is to explore the extent to which such a proposal is realistic.

KEYWORDS: fluoride glass; containerless processing; devitrification; ZBLAN

Address of correspondence: Reginald W. Smith, Department of Mechanical and Materials
Engineering, Queen's University, Kingston, ON, Canada K7L 3N6. Voice: 613-533-2753;
fax: 613-533-6610.
smithrw@post.queensu.ca

Ann. N.Y. Acad. Sci. 1027: 150–157 (2004). ©2004 New York Academy of Sciences.
doi: 10.1196/annals.1324.014

INTRODUCTION

Fluoride glasses are attractive materials for infrared optical fibers, but they easily develop microcrystals during bulk glass synthesis and fiber-drawing. Processing these materials in a microgravity environment has recently been shown to suppress this undesirable devitrification, allowing high optical quality glasses to be prepared.[1-4] These highly desirable optical properties initially promised to place fluoride glasses in a prominent role in laser applications, frequency-up converters, non-linear optical communications, transoceanic and transcontinental links, and so forth.[5] However, the instability of fluoride glasses in the vitreous state has resulted in their vast potential remaining largely untapped. The majority of fluoride glasses studied have crystallization temperatures (T_x) below their fiber-drawing region, making it impossible to create fibers of high optical quality, but ZBLAN glass $(53ZrF_4-20BaF_2-4LaF_3-3AlF_3-20NaF)$ may have a narrow region, just below its T_x, in which fibers can be drawn, thus making it the most encouraging of the heavy-metal fluoride glasses.

The use of optical fibers based on silica is well established in many industries, but their optical performance is limited by some of the intrinsic properties associated with that group of glasses. The transparency region of silica-based fibers is short and restricted to the near-IR; furthermore, their high attenuation makes it necessary to amplify the signal frequently in long-haul communication systems. In contrast, ZBLAN fibers are transparent over a much wider spectral region (0.2–6.0 mm wavelength) and their predicted attenuation approaches the theoretical minimum.[6]

The role of viscosity is critical when investigating nucleation mechanisms, but unfortunately, due to the tendency of fluoride glasses to devitrify, the viscosity data for ZBLAN glass is incomplete. Many studies have investigated the viscosity of ZBLAN[7-9] and other fluoride glasses;[10-13] however, the only reliable data produced has been at temperatures below the glass transition and above the melting point.

As in the case of most glasses,[14] the viscosity of ZBLAN glass decreases rapidly with increasing temperature. However, this change is even more pronounced in fluoride glasses and occurs at a far lower temperature. The consequence of this rapid viscosity change with temperature is that the optimum fiber-drawing range is narrow and the risk of entering the crystallization region is heightened.

It has recently been demonstrated that remelting ZBLAN glasses in a microgravity environment can suppress devitrification.[1-4] Several explanations to account for this behavior have yet to receive universal support. It is believed that further research into understanding the nucleation mechanism of fluoride glasses will lead to the development of technology to suppress its nucleation, or at least minimize its crystallization region, allowing optical fibers to be produced.[15]

DEVITRIFICATION MECHANISMS

The devitrification mechanism of ZBLAN and other fluoride glasses at $1g$ is poorly understood. It was initially believed that a gravity-induced density segregation would cause a composition gradient to occur within the glass with the heavier fluorides (i.e., BaF_2 and LaF_2) sinking to the bottom, thus shifting the glass into a composition region that favors crystallization over the vitreous state.[1] However, this explanation appears to be contradicted by studies on rare-earth fluoride-based glasses, which have shown larger glass-forming regions are attained when using mixtures containing heavier rare-earths.[5]

SHEAR THINNING

Tucker et al.[16] offered an explanation based on shear thinning to account for the 1 g crystallization of ZBLAN glass. Heavy metal fluoride glasses are considered non-Newtonian fluids and consequently exhibit pseudoplastic flow behavior; that is, their viscosity decreases with increasing shear rate. It has been proposed that, at 1 g, convective flow within the glass preform is sufficient to cause shearing of the glass, thus lowering its intrinsic viscosity. The lowered viscosity (or increased fluidity) causes an increase in the rate of diffusion within the glass, allowing crystallites to more readily nucleate and grow, which suggests[16] that a decrease in the viscosity by a factor of two would double both nucleation and crystallization rates and, consequently, increase the fraction crystallized by a factor of 16. However, the extent to which convection can produce a substantial shear on the glass in this temperature regime has yet to be established.

HYDROSTATIC PRESSURE

The influence of hydrostatic pressure on the terrestrial devitrification of fluoride glasses is currently being explored. It has been shown that the thermodynamic barrier to nucleation and crystal growth can be lowered by pressurizing a vitreous melt.[17–19] However, exerting this hydrostatic pressure also causes an increase in the viscosity of the melt yielding the opposite effect, thus, counteracting the thermodynamic barrier reduction achieved by pressurization. Gutzow et al.[17] claim that in the Li_2O–SiO_2 system the two effects essentially offset each other and any increase or decrease in the system pressure has a negligible influence on nucleation. In cordierite glass ($2MgO$–Al_2O_3–$5SiO_2$), it has been shown that pressure suppresses nucleation and crystal growth under both low pressure (0–2 kbar) and high pressure (greater than 5 kbar) conditions.[19] However, the growth rate of quartz from vitreous silica is increased with the application of hydrostatic pressure[20] over the range 5–40 kbar. Chason et al. suggest that the application of hydrostatic pressure decreases the atomic transport coefficients (i.e., fluidity, diffusivity, and crystal growth rate) in ionic solids resulting in the suppression of devitrification with pressure in these materials. In contrast, more covalent glasses, such as fused silica, undergo the opposite effect because an increase in hydrostatic pressure enhances the transport properties. Since a fluoride glass network would be more ionic in nature, it would most likely behave more similarly to cordierite under pressure; that is, a suppression of nucleation with the application of hydrostatic pressure. However, the explanation of why pressure suppresses nucleation in more ionic glasses and promotes it in others has been established entirely on observations made on silica-based and/or oxygen containing glasses. Fluoride glasses may behave differently, and the influence of pressure on the devitrification of ZBLAN glasses cannot be ruled out until such studies on halide glasses have been made.

THE PIEZOELECTRIC VISCOMETER

There are many different techniques and even more sophisticated commercially available machines that can be used to measure the viscosity of a glass. It is of great importance when selecting the method to take into account the nature of the material being studied (i.e., its predicted viscosity, its elasticity, its reactivity, the temperature dependence of its viscosity, the required accuracy, etc). Many materials, including glasses, exhibit a substantial change in viscosity with temperature. So much so, that it is usually necessary to employ several different techniques and viscometers to measure the full viscosity range of the material.

Although it may be possible to fill-in the missing viscosity data (see FIGURE 1) with reduced gravity experiments using several different techniques, it would be advantageous to use just a single viscometer. Our recently developed piezoelectric viscometer has been designed to complete the viscosity data for ZBLAN glass without need for a second piece of equipment. It also provides for a compact size and low energy measurement system; characteristics that are critical for reduced gravity experiments (see FIGURE 2).

The use of ultrasonic pulses to measure viscosity is by no means novel. The technology has been employed to measure the viscosity in dextran fermentation[21] and as a viscosity control device in an air-conditioning compressor.[22] The temperature and

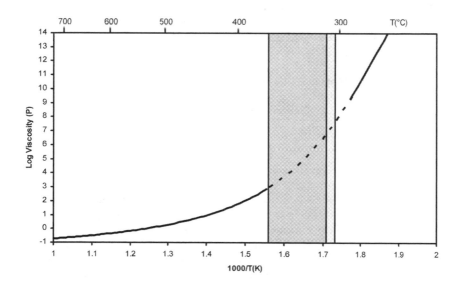

FIGURE 1. The viscosity of ZBLAN glass. Beam bending, isothermal rotational data, and rotational data while cooling at 60°C/min:[8,9] ▨, crystallization region; ☐, fiber drawing region; ——, published viscosity data; ▬ ▬, unknown viscosity data.

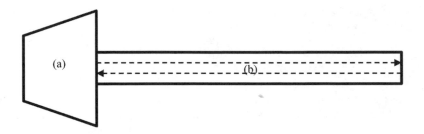

FIGURE 2. Schematic of the piezoelectric viscometer. A pulse is emitted by a transducer (*a*) and travels through the sample glass rod (*b*) until it is reflected back to the transducer by the end glass–air interface.

pressure dependence of the elastic modulus of ZBLAN glass has also been recently determined using a piezoelectric transducer.[23] These examples also show successful measurements on materials with viscosities in the range 10^{-2}–10^{14} P.

The piezoelectric viscometer functions by giving off an ultrasonic pulse that travels through the fluoride glass rod until it is reflected back to the transducer by the glass-rod end/air interface. An ultrasonic detector attached to the piezoelectric transducer then calculates the *time of flight* of the ultrasonic pulse. The time of flight and the viscosity of the glass are both functions of its elastic modulus and density. Therefore, the viscosity of the glass can be determined from the changes in the time of flight of the ultrasonic pulse at various test temperatures.

In our experimental method, a buffer rod was added to the apparatus between the transducer and the ZBLAN glass sample to permit the transducer to operate at room temperature (25–30°C), since even high temperature transducers can only function at the specimen temperatures used in this study for a few seconds at a time. Here, the buffer rod is made of Pyrex glass, with an identical diameter as the ZBLAN rod to optimize ultrasonic transmission between the two materials. A couplant (see below) is applied between the buffer and ZBLAN glass rods and between the transducer and buffer rod to reduce interfacial reflection.

A spring-loaded *jig* with a silver tube is used to ensure that transducer, buffer rod, and ZBLAN rod remain in close contact. The silver tube also assists in obtaining a long isothermal region in the furnace.

A photograph of the piezoelectric viscometer is presented in FIGURE 3. The inset is of a typical signal response showing (from left to right) a large initial peak representing the buffer rod reflection, the middle peak the combined buffer rod, sample rod reflection, and a small double reflection peak at the far right.

The idea of exploiting ultrasound for use in viscosity measurement is based on the observation that a planar ultrasonic pulse passing through a fluid, at normal incidence, will experience a change in propagation velocity as the properties of the fluid are changed.[21] The longitudinal ultrasonic velocity is related to the fluid properties by,

$$V_L = \sqrt{\frac{K}{\rho}}, \tag{1}$$

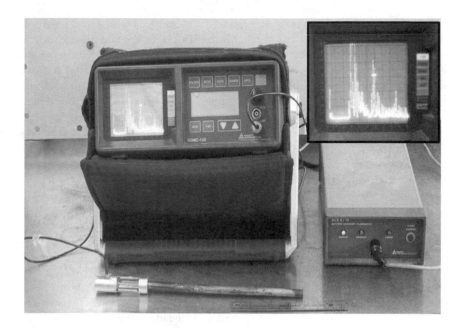

FIGURE 3. Photograph of the piezoelectric viscometer Prototype II.

where K is the bulk modulus and ρ is the density. For highly viscous fluids, however, it is necessary to modify Equation (1) by incorporating the shear modulus, G,[21]

$$V_L = \sqrt{\frac{K + \dfrac{4G}{3}}{\rho}}. \tag{2}$$

The shear modulus is a function of viscosity and allows the correlation of longitudinal speed of sound with viscosity. This complex relationship is developed fully elsewhere.[24]

COUPLANT

The couplant placed between the buffer rod and the sample ZBLAN rod allows transmission of the ultrasonic pulse through the two glasses. Various materials have been employed as couplants during terrestrial experiments; however, a fully satisfactory material has only recently been determined. It is important that the couplant has an acoustic impedance similar to that of the buffer and sample rods and that it adequately wet the two glass surfaces with a thin layer so that its contribution to the pulse time of flight is negligible. The couplant must also maintain these characteristics from room temperature to 400°C. The temperature requirement is the most difficult criterion to meet. The use of diffusion pump oil had yielded the best results of all oils and commercially available couplants examined; it allowed for measurements in excess of 300°C on benign test samples and ZBLAN measurements up to 240°C.

The use of low-melting metals as couplants was examined with mixed results. In three separate experiments, a thin sheet of metal (In, Ga, or Zn) was placed between the buffer and sample rods and the apparatus was heated above the melting point of the particular metal. The apparatus was cooled slowly, allowing the metal couplant to solidify. Ultrasonic transmission was observed when the In and Zn metal couplants were in the solid phase (Ga did not solidify), but failed upon melting. This is due to the inability of the liquid metal couplant to wet the glass surfaces. Consequently, these metals can only be used as couplants up to their melting temperatures. That criterion for Zn does not in itself constitute a problem, but in order to apply the Zn couplant it is necessary to heat the apparatus to temperatures in excess of 420°C. This causes the sample to crystallize and prevents reliable measurements from being made.

The mixed successes of metal couplants provided the inspiration for the use of a plastic couplant. Polyethylene terephthalate (PET) provides the best properties for use as a couplant at a very reasonable price since it can be attained from almost any soft drink or water bottle.

The couplant was applied by heating the system to 280°C, which is slightly above the PET melting point, and allowing the entire apparatus to cool slowly to room temperature. After the couplant has been applied, it is capable of transmitting ultrasonic pulses in both solid and liquid phases from room temperature up to, and likely exceeding, 420°C. This is undoubtedly sufficient for viscosity measurement on fluoride glasses, although other glass systems involving higher temperature ranges would likely require a different polymer couplant. No other polymer has been tested to date for this application.

SUMMARY

The preparation of high optical quality ZBLAN glass cannot begin until the devitrification process is fully described and techniques to suppress its occurrence are developed. The role of viscosity in devitrification is significant, and a reliable model to describe crystallization cannot be advanced until the full viscosity spectrum of ZBLAN glass is determined. A piezoelectric viscometer has been developed to measure the missing viscosity data through terrestrial and reduced gravity experiments.

ACKNOWLEDGMENTS

This research is part of the ongoing microgravity applications studies of Professor Reginald Smith. Funding for this project was provided by the Centre for Research in Earth and Space Technologies (CRESTech) and by our corporate partner Supan Technologies, Inc. (STI), through a Canadian Space Agency contract. Ultrasonic equipment was generously provided by Alcan Inc.

REFERENCES

1. VARMA, S., S.E. PRASAD, A. AHMAD & T.A. WHEAT. 2001. Effect of gravity on crystallization in heavy metal fluoride glasses processed on the T-33 parabolic flight aircraft. J. Mat. Sci. **36:** 4551–4559.

2. VARMA, S., S.E. PRASAD, A. AHMAD & T.A. WHEAT. 2001. Effect of microgravity on crystallization in heavy metal fluoride glasses processed on the CSAR-I sounding rocket. J. Mat. Sci. **36:** 2027–2035.

3. VARMA, S., S.E. PRASAD, A. AHMAD & T.A. WHEAT. 2002. Effect of microgravity on optical degradation in heavy metal fluoride glasses processed on the CSAR-II sounding rocket. J. Mat. Sci. **37:** 2591–2596.

4. TUCKER, D.S. 1997. Effects of gravity on processing heavy metal fluoride fibers. J. Mat. Res. **12:** 2223–2225.

5. DREXAGE, M.G. 1985. Heavy-metal fluoride glasses. Treatise Mater. Sci. Techn. **26:** 151–243.

6. TUCKER, D.S. 1998. ZBLAN continues to show promise. <http://science.nasa.gov/newhome/headlines/msad05feb98_1.htm>.

7. CRICHTON, S.N., R. MOSSADEGH & C.T. MOYNIHAN. Viscosity and crystallization of ZrF_4-based glasses.

8. HASZ, W.C., S.N. CRICHTON & C.T. MOYNIHAN. 1988. Viscosity temperature dependence of ZrF_4-based melts. Mat. Sci. Forum **32-33:** 589–594.

9. KOIDE, M., K. MATUSITA & T. KOMATSU. 1990. Viscosity of fluoride glasses at glass transition temperature. J. Non-Cryst. Solids **125:** 93–97.

10. ZHANG, G., J. JIANG, M. POULAIN, *et al.* 1999. Viscosity of fluoride glasses near the fiber drawing temperature region. J. Non-Cryst. Solids **256/257:** 135–142.

11. NAFTALY, M., A. JHA, E.R. TAYLOR & K.C. MILLS. 1997. Viscosity measurement in halide melts. J. Non-Crystal. Solids **213/214:** 106–112.

12. MATUSITA, K., K. MANABU & T. KOMATSU. 1992. Viscosity of fluoride glasses at glass transition temperature. J. Non-Crystal. Solids **140:** 119–122.

13. WILSON, S.J. & D. POOLE. 1985. Viscosity measurements on fluorozirconate glasses. Mat. Sci. Forum **6:** 665–672.

14. DOREMUS, R.H. 1973. Glass Science. John Wiley & Sons, New York.

15. DUNKLEY, I.R., R.W. SMITH, S. VARMA & M. SAVAS. 2002. Devitrification of fluoride glasses. Exp. Meth. Micrograv. Mat. Sci. **14:** 45–50.

16. TUCKER, D.S. & E.C. ETHRIDGE. 2001. Explanation of the effects of gravity on crystallization of ZrF_2–BaF_2–LaF_3–AlF_3–NaF. J. Mat. Res. **16:** 3027–3029.

17. GUTZOW, I., B. DURSCHANG & C. RUSSEL. 1997. Crystallization of glass forming melts on hydrostatic pressure and shear stress Part I: crystallization catalysis under hydrostatic pressure: possibilities and limitations. J. Mat. Res. **32:** 5389–5403.

18. GUTZOW, I., C. RUSSEL & B. DURSCHANG. 1997. Crystallization of glass forming melts on hydrostatic pressure and shear stress Part II: flow induced melt crystallization: a new method of nucleation catalysis. J. Mat. Res. **32:** 5405–5411.

19. CHASON, E., & M.J. AZIZ. 1991. Effect of pressure on crystallization kinetics of cordierite glass. J. Non-Cryst. Solids **130:** 204–210.

20. FRATELLO, V.J., J.F. HAYS & D. TURNBULL. 1980. Dependence of growth rate of quartz in fused silica on pressure and impurity content. J. Appl. Phys. **51**(9): 4718–4728.

21. ENDO, H., K. SODE & I. KARUBE. 1990. On-line monitoring of the viscosity in dextran fermentation using piezoelectric quartz crystal. Biotech. Bioeng. **36:** 636–641.

22. OYAMADA, T., Y. INOUE & M. MIZUMOTO. 2001. *In situ* ultrasonic viscosity measurement inside of an air-conditioning compressor. Lubricat. Eng. **57**(10): 28–32.

23. HWA, L.G., Y.J. WU, W.C. CHAO & C.H. CHEN. 2002. Pressure and temperature dependence of the elastic properties of a ZBLAN glass. Mat. Chem. Phys. **74:** 160–166.

24. VAN WAZER, J., J. LYONS, K. KIM & R. COLWELL. 1963. Viscosity and Flow Measurement: A Laboratory Handbook of Rheology. Interscience Publishers, New York.

Effect of Magnetic Field on the Crystalline Structure of Magnetostrictive TbFe$_2$ Alloy Solidified Unidirectionally in Microgravity

TAKESHI OKUTANI, YOSHINORI NAKATA, AND HIDEAKI NAGAI

National Institute of Advanced Industrial Science and Technology,
AIST Tsukuba Central 5, Tsukuba, Japan

ABSTRACT: We performed unidirectional solidification experiments on TbFe$_2$ alloy in a static magnetic field in microgravity of $10^{-4}g$ for 10 sec obtained by a 490 m free fall of the Japan microgravity center (JAMIC). When the magnetic field strength was increased from zero to 4.5×10^{-2} T during unidirectional solidification in microgravity, a [1 1 1] crystallographic alignment dominated, and the maximum magnetostriction constant increased from 1,000 ppm to 4,000 ppm. For unidirectional solidification in normal gravity, the maximum magnetostriction constant remained at 2,000 ppm with increasing magnetic field. The columnar structure grows and orients along the magnetic field. TbFe$_2$ crystals grow in microgravity predominantly in the same direction as the magnetic field.

KEYWORDS: TbFe$_2$; magnetostrictive material; unidirectional solidification; magnetic field; microgravity

INTRODUCTION

The compound TbFe$_2$ has drawn much attention due to its excellent magnetostrictive properties[1,2] and offers a promising material for linear actuators and magnetic sensors. Excellent magnetostrictive materials must consist of large and oriented crystals, preferably a single crystal. However, single TbFe$_2$ crystals are very difficult to grow due to double peritectic structure formation during solidification. It has been reported that homogeneous melt remained and nucleation was suppressed during solidification in short-term microgravity.[3] This means there is some possibility of synthesizing large crystals in microgravity. A magnetic field applied to a melt is known to suppress convection during the solidification, thereby reducing segregation. Much research on the effect of a magnetic field during Czochralski and Bridgman crystal growth has been reported.[4–9] Kim[6] showed that the convective mass transfer was inhibited, but that it was very difficult to totally suppress it in a weak magnetic field. However, this research was done in normal gravity. In microgravity, small forces, such as weak magnetic fields, may affect the metallurgical structure and/or crystalline growth direction of magnetic materials, even above the Curie temperature,

Address for correspondence; Takeshi Okutani, National Institute of Advanced Industrial Science and Technology, Tsukuba Central 5, Tsukuba, Ibaraki 305-8565, Japan. Voice: 81 29-861-4737; fax: 81 29-861-6289.
  okutani-takeshi@aist.go.jp

Ann. N.Y. Acad. Sci. 1027: 158–168 (2004). ©2004 New York Academy of Sciences.
doi: 10.1196/annals.1324.015

at which magnetic materials change to paramagnetic from ferromagnetic. In this paper, we report the synthesis of large and oriented TbFe$_2$ crystals in microgravity and/or a static magnetic field.

EXPERIMENTAL

Microgravity experiments were performed in microgravity circumstances of $10^{-4}g$ for 10sec obtained by the free-fall facility of the Japan microgravity center (JAMIC).[10] A unidirectional solidification system was adopted for the drop system in the JAMIC. FIGURE 1 shows a schematic diagram of the unidirectional solidification apparatus. The sample could be heated by infrared heaters from room temperature to

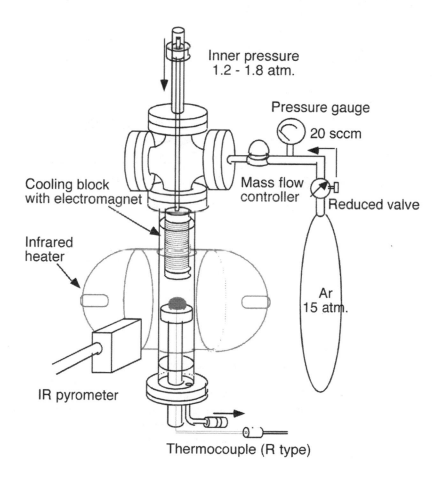

FIGURE 1. Schematic diagram of the unidirectional solidification apparatus for a microgravity experiment.

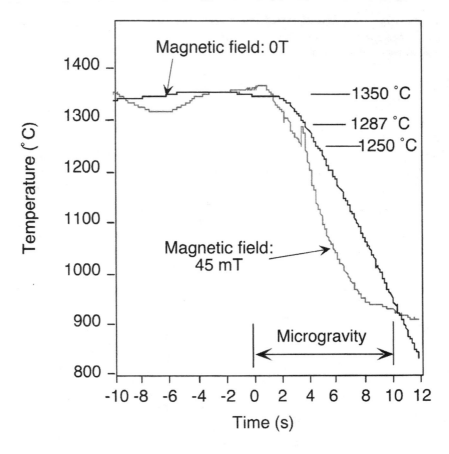

FIGURE 2. Experimental cooling curve for solidification in static magnetic field in microgravity and normal gravity. (Reprinted from Ref. 11, with permission from Elsevier.)

1,550°C within 60 seconds. Unidirectional solidification was performed by cooling with an iron block, that was excited by an electromagnetic coil for the static magnetic field experiment. The maximum magnetic field was about 4.5×10^{-2} T. Santoku Metal Co. provided $TbFe_2$ matrix that had been prepared by melting in an induction furnace at 1,400°C in a 0.1 MPa argon gas atmosphere. For solidification, the sample was cut into a 10 mm diameter × 3 mm thick disk. FIGURE 2[11] shows the cooling curve for solidification in microgravity. Infrared heaters first heated the sample from room temperature to 1,350°C within 40 seconds. On reaching the temperature of 1,350°C (at which sample had melted), the system was dropped. After 2.0 sec dropping, heating was stopped and at the same time sample was cooled by an iron block. In microgravity, the sample temperature decreased from 1,350°C to 1,000°C within 10.0 sec. About 40°C supercooling and recalescence was subsequently observed during solidification in the magnetic field.

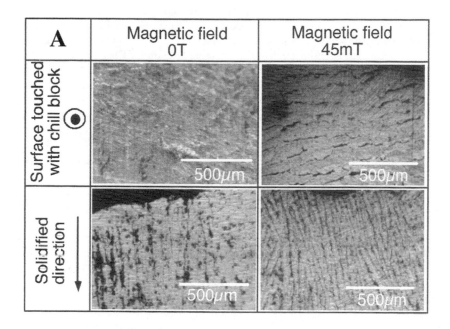

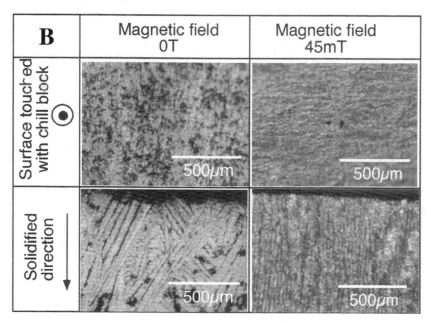

FIGURE 3. Macrostructure of unidirectionally solidified TbFe₂ in **(A)** microgravity and **(B)** normal gravity. (Reprinted from Ref. 11, with permission from Elsevier.)

To investigate which crystallographic orientation of the $TbFe_2$ columnar crystal, solidified unidirectionally in magnetic field in microgravity, is affected by the direction of cooling or magnetic field, the $TbFe_2$ melt was solidified unidirectionally in a magnetic field inclined 30 degrees to the direction of cooling. The microstructures of the solidified products were examined by scanning electron microscopy (SEM) and optical microscopy (OM), and the effect of Fe_{3p} and Tb_{5f} orbitals for various inclinations of the solidified sample was examined by photoelectron spectroscopy.

Solidified samples were evaluated by the following methods: SEM and OM for structure, electron probe micro analysis (EPMA) for composition, and X-ray diffraction (Cu–Kα radiation, 45 keV 30 mA) for crystal structure. The magnetostriction measurement system (Tamakawa Co., Ltd) by a three-terminal capacitance method was used to measure the magnetostriction constant.

RESULTS AND DISCUSSION

FIGURE 3 A and B[11] depict macrostructures of the $TbFe_2$ sample solidified in 0 T and 4.5×10^{-2} T magnetic fields in microgravity and normal gravity. For solidification in microgravity, the samples solidified in 0 T and 4.5×10^{-2} T were columnar in structure. The sample in 4.5×10^{-2} T was slightly larger than that in 0 T. An aligned columnar structure was observed for the sample in 4.5×10^{-2} T, whereas a slightly disordered columnar structure with defects of several microns was observed for the sample in 0 T. For solidification in normal gravity, a randomly aligned columnar structure was observed for the sample in 0 T. The sample solidified in 4.5×10^{-2} T had the same columnar structure alignment as that solidified in microgravity, but the columnar size was only a few microns—smaller than that solidified in microgravity.

FIGURE 4 A and B[11] illustrate the X-ray diffraction patterns of the sample produced in microgravity and normal gravity. XRD analysis was performed for the face that contacted the magnetic cooling chill-block. The $(0\,2\,2)$ and $(1\,1\,3)$ peaks were observed in the sample solidified in normal gravity. The intensities of each peak were on the same order, between 0 T and 4.5×10^{-2} T. For solidification in 0 T and microgravity, the $(0\,2\,2)$ and $(1\,1\,3)$ peaks were also observed, but the $(2\,2\,2)$ peak increased considerably for the sample solidified in a 4.5×10^{-2} T magnetic field. These results indicate that a crystallographic alignment along the [1 1 1] direction was obtained by unidirectional solidification in microgravity and a static magnetic field. Shi et al.[12] and others have reported that a crystallographic alignment along [1 1 0] direction[12] and [1 1 2] texture[13–15] has been obtained in unidirectionally solidified (Tb–Dy)Fe_2 alloys.

FIGURE 5 shows the relationship between the intensities of peaks $(0\,2\,2)$ of $2\theta(Cu–K\alpha) = 34°$, $(1\,1\,3)$ of $2\theta(Cu–K\alpha) = 40.5°$, and $(2\,2\,2)$ of $2\theta(Cu–K\alpha) = 42.5°$ of the $TbFe_2$ and magnetic field applied during solidification in microgravity and normal gravity. The XRD analysis for all samples was conducted for the 10 mm diameter face that contacted the magnetic cooling chill-block. Therefore, the change in intensity semiquantitatively represents the crystal growth rate. For solidification in normal gravity, $(1\,1\,3)$ and $(2\,2\,2)$ peaks have almost the same intensity, 310 and 85 arbitrary-units, respectively, for all magnetic fields. The $(0\,2\,2)$ peak decreases slightly for all magnetic fields. However, the intensities of all peaks are less than a three-

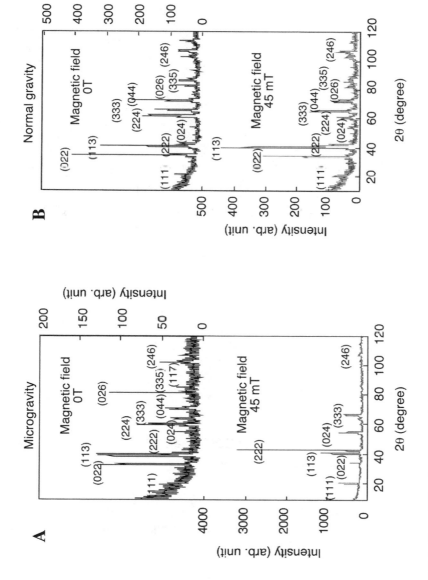

FIGURE 4. X-ray diffraction pattern of TbFe₂ samples produced by unidirectional solidification in static magnetic field in (**A**) microgravity and (**B**) normal gravity. (Reprinted from Ref. 11, with permission from Elsevier.)

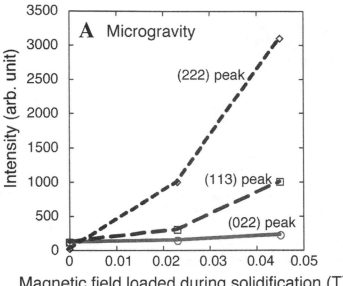

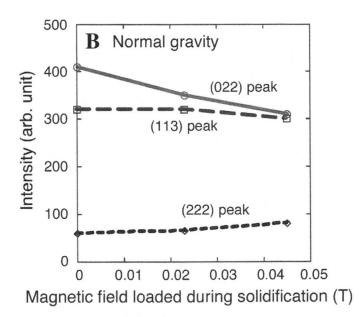

FIGURE 5. Relationship between intensity of TbFe$_2$ and magnetic field loaded during solidification in **(A)** microgravity and **(B)** normal gravity: 2θ(Cu–Kα) of (2 2 2) peak, 42.5°; 2θ(Cu–Kα) of (1 1 3) peak, 40.5°; and 2θ(Cu–Kα) of (0 2 2) peak, 34°.

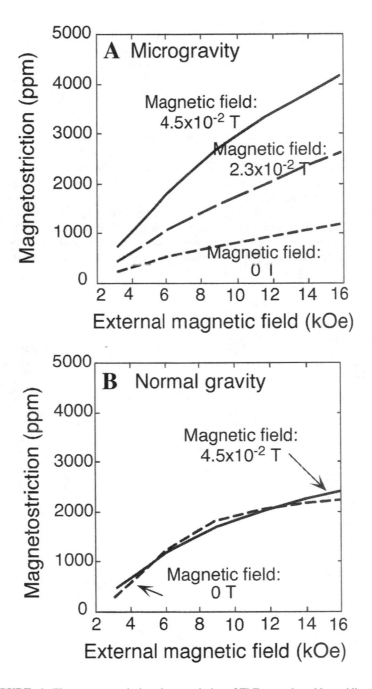

FIGURE 6. The magnetostrictive characteristics of TbFe₂ produced by unidirectional solidification in static magnetic field under (**A**) microgravity and (**B**) normal gravity.

digit number of arbitrary-units, and the three peaks are thought not to change in the range of magnetic fields from 0 mT to 45 mT. On the other hand, with solidification in microgravity, the (2 2 2) peak increases dramatically with increasing magnetic field. The (1 1 3) peak also increases moderately with increasing magnetic field. The (0 2 2) peak does not change with the magnetic field. These results show that $TbFe_2$ crystals grow in the [1 1 1] direction in microgravity and a magnetic field. The small crystal growth in the [1 1 0] direction is observed in the same unidirectional solidification as the growth in the [1 1 1] direction. We believe that this result indicates some part of the magnetic field was not perpendicular to the face that contacted the magnet, because a 10-mm diameter bar magnet with heterogeneous direction and magnetic field strength was used in these experiments. We can, therefore, conclude that the [1 1 1] direction of a $TbFe_2$ crystal is grown by unidirectional solidification in microgravity and a magnetic field.

FIGURE 6 illustrates typical magnetostriction curves of $TbFe_2$ produced in microgravity and normal gravity. Magnetostriction of the sample solidified in microgravity has strong magnetic field dependence; however, magnetostriction of the sample produced in normal gravity does not have the same dependence. Furthermore, for

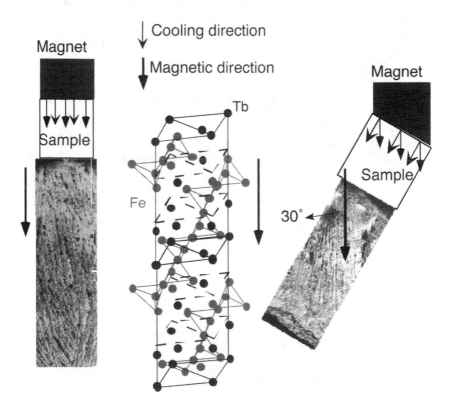

FIGURE 7. Dependence of cooling and magnetic field direction on dominant direction of crystal growth.

solidification in microgravity, the maximum magnetostriction constant increased from 1,000 ppm to 4,000 ppm, when the strength of the magnetic field increased during solidification. For unidirectional solidification in normal gravity, however, the maximum magnetostriction remained at 2,000 ppm despite any increase in magnetic field. These results demonstrate that unidirectional solidification by using a static magnetic field in microgravity yields high-performance TbFe$_2$ magnetostrictive materials.

To investigate which crystallographic orientation of TbFe$_2$ columnar crystals, solidified unidirectionally in a magnetic field in microgravity, is affected by the direction of cooling or magnetic field, the TbFe$_2$ melt was solidified unidirectionally in a magnetic field inclined 30 degrees to the direction of cooling. FIGURE 7 shows that the columnar structure grows and orients along with the magnetic field. TbFe$_2$ crystals grow predominantly in the same direction as the magnetic field in microgravity.

CONCLUSIONS

Columnar crystal-aligned TbFe$_2$ with size of 10 mm diameter $\times$ 3 mm thick has been produced by unidirectional solidification in static magnetic field and microgravity. The TbFe$_2$ alloy unidirectionally solidified in microgravity has a [0 1 1] and [1 1 3] crystallographic alignment and the crystallographic alignment changes from [0 1 1] and [1 1 3] to [1 1 1] as the magnetic field is increased. This increase in magnetic field during solidification also markedly increases the magnetostrictive properties of the materials.

ACKNOWLEDGMENTS

This study was carried out as a part of Ground-based Research Announcement for Space Utilization promoted by Japan Space Forum.

REFERENCES

1. CLARK, A.E. & H. BELSON. 1972. Giant room-temperature magnetostrictions in TbFe$_2$ and DyFe$_2$. Phy. Rev. B **5**: 3642–3644.
2. CLARK, A.E. 1980. Magnetostrictive rare earth–Fe$_2$ compounds. *In* Ferromagnetic Materials, Vol. 1. E.P. Wohlfarth, Ed.: 531–581. North-Holland Publishing Company.
3. OKUTANI, T., H. MINAGAWA, H. NAGAI, *et al.* 1999. Synthesis of crystalline materials with high quality under short-time microgravity. Ceramic Eng. Sci. Proc. **20**(4): 215–226.
4. MOTAKEF, S. 1990. Magnetic field elimination of convective interference with segregation during vertical-bridgman growth of doped semiconductors. J. Cryst. Growth **104**: 833–850.
5. ELURLE, D.T.J. & R.W. SELIES. 1994. Handbook of Crystal Growth, Vol. 2. 261. Elsevier Science, Amsterdam.
6. KIM, K.M. 1982. Supression of thermal convection by transverse magnetic field. J. Electrochem. Soc. **129**: 427–429.
7. MATTHESEN, D.H., M.J. WARGO, S. MOTAKEF, *et al.* 1987. Dopant segregation during vertical bridgman-atockbarger growth with melt stabilization by strong axial magnetic fields. J. Cryst. Growth **85**: 557–560.

8. BECLA, P., J.-C. HAN & S. MOTAKEF. 1992. Application of strong vertical magnetic fields to growth of II–VI pseudo-binary alloys: HgMnTe. J. Cryst. Growth **121:** 394–398.
9. KANG, J., Y. OKANO, K. HOSHIKAWA & T. FUKUDA. 1994. Influence of a high vertical magnetic field on Te dopant segregation in Sb grown by the vertical gradient freeze method. J. Cryst. Growth **140:** 435–438.
10. NAGAI, H., F. ROSSIGNOL, Y. NAKATA, *et al.* 1998. Wetting of molten indium under short-time microgravity. Mater. Sci. Eng. A **248:** 206–211.
11. MINAGAWA, H., K. KAMADA, T. KONISHI, *et al.* 2001. Unidirectional solidification of TbFe$_2$ alloy using magnetic field in microgravity. J. Magn. Magn. Mater. **234:** 437–442.
12. SHI, Z., Z. CHEN, X. WANG, *et al.* 1997. Directionally solidified Tb$_x$Dy$_{1-x}$(M$_y$Fe$_{1-y}$)$_{1.9}$ giant magnetostrictive materials and their applications in transducers. J. Alloy Comp. **258:** 30–33.
13. VERHOVEN, J.D., E.D. GIBSON, O.D. MCMASTERS & H.H. BARKER. 1987. The growth of single crystal terfenol-D crystals. Metal. Trans. **18A:** 223–231.
14. VERHOVEN, J.D., E.D. GIBSON, O.D. MCMASTERS & G.E. OSTENSON. 1990. Directional solidification and heat treatment of terfenol-D magnetostrictive materials. Metal. Trans. **21A:** 2249–2255.
15. MEI, W., T. UMEDA, S.Z. ZHOU & R. WANG. 1995. Magnetostriction of grain-aligned Tb$_{0.3}$Dy$_{0.7}$Fe$_{1.95}$. J. Alloy Comp. **224:** 76–80.

Ground-Based Diffusion Experiments on Liquid Sn–In Systems Using the Shear Cell Technique of the Satellite Mission Foton-M1

SHINSUKE SUZUKI, KURT-HELMUT KRAATZ, AND GÜNTER FROHBERG

Institute for Materials Science and Technology,
Technical University of Berlin, Berlin, Germany

ABSTRACT: This study reported in this paper was aimed at testing the shear cell that was developed for the satellite mission Foton-M1 to measure diffusion coefficients in liquid metals under microgravity (μg)-conditions. Thick Layer diffusion experiments were performed in the system Sn90In10 versus Sn under $1g$-conditions. For this system several μg-diffusion results are available as reference data. This combination provides a low, but sufficiently stable, density layering throughout the entire experiment, which is important to avoid buoyancy-driven convection. The experimental results were corrected for the influences of the shear-induced convection and mixing after the final shearing, both of which are typical for the shear cell technique. As the result, the reproducibility and the reliability of the diffusion coefficients in the ground-based experiments were within the limits of error of μg-data. Based on our results we discuss the necessary conditions to avoid buoyancy-driven convection.

KEYWORDS: diffusion; liquid metal; shear cell; microgravity; ground experiment; Sn–In; Foton

NOMENCLATURE:

c	concentration of In (at%)
c_0	initial concentration of In in the thick layer (at%)
D	diffusion coefficient
ΔD	statistic error in D including temperature error
$\Delta D'$	standard deviation in D values obtained from the four capillaries
g	gravitational acceleration
h	initial thickness of the thick layer
H	height of a cell (3 mm)
P_L	pressure of liquid metal in the lower reservoir
P_U	pressure of liquid metal in the upper reservoir
t_{diff}	diffusion time
T_{diff}	diffusion temperature (°C)
v_0	shear velocity
x	distance from edge of diffusion sample in the gravity vector direction
$\overline{x^2}$	mean square diffusion depth
$\overline{x^2}_{\text{meas}}$	measured $\overline{x^2}$
$\overline{x^2}_{\text{average}}$	additional mean square diffusion depth from concentration averaging
$\overline{x^2}_{\text{shear}}$	additional mean square diffusion depth from shear-induced convection
ρ	density

Address for correspondence: Shinsuke Suzuki, Institute for Materials Science and Technology, Technical University of Berlin, Sekr.PN2-3, Hardenbergstr.36, D-10623 Berlin, Germany. Voice: 49-(0)30-314-23038; fax: 49-(0)30-314-23035.
shinsuke@physik.tu-berlin.de

Ann. N.Y. Acad. Sci. 1027: 169–181 (2004). ©2004 New York Academy of Sciences.
doi: 10.1196/annals.1324.016

INTRODUCTION

Exact diffusion data of liquid states are very much in demand from both physical and metallurgical points of view. Since the Spacelab missions, microgravity conditions have been considered an effective environment in which to measure diffusion in liquid states because this avoids buoyancy-driven convection. Various measurement methods are applied to diffusion experiments. The long capillary method delivered good results for single-component systems in the FSLP,[1] the D1,[2] and the D2[3] missions. On the other hand diffusion measurements on multicomponent systems are often not satisfactory because of the disturbance by segregation during solidification. Better results for multicomponent systems are expected from experiments that use the shear cell technique. Such experiments have been performed under microgravity conditions.[4,5]

FIGURE 1 shows a schematic illustration of a diffusion experiment with the shear cell technique. Before the start of the diffusion phase the two different sample materials are kept separate (1)–(2). By an initial shearing (3) the diffusion pair is formed and the diffusion phase (4) starts. At the end of the diffusion phase, the diffusion pair is separated into many small cells (5)–(6) before the cooling phase starts. Because of this function, the shear cell technique has several advantages compared with the long capillary technique:

1. no time–temperature correction for diffusion is necessary during the heating and the cooling phase;
2. there is a possibility of homogenization before the diffusion phase starts (e.g., in case of polyphasic alloys);
3. no correction of diffusion coefficients for the effect of volume expansion and shrinkage is necessary; and
4. there is no influence of segregation during solidification.

On the other hand, the shear cell technique has also disadvantages:

1. it has a complicated mechanical design and operation;
2. it involves shear-driven convection; and
3. the concentration inside a cell is averaged.

The solutions to these problems are described below.

In our working group, several μg-shear cell experiments were performed in an AGAT-furnace (in-house construction) during the Russian Foton-12 mission.[5,6] For the subsequent mission, Foton-M1, in October 2002, 24 μg-shear cell diffusion experiments were prepared with an improved AGAT-furnace. However, during the launch the rocket failed. In 2005, a reflight of Foton-M1 is planned with nearly the same experimental program. Meanwhile, we concentrate on ground (1g) experiments and testing.

Out of the program, thick layer diffusion experiments in the system Sn90In10–Sn were chosen. For this kind of system several μg-diffusion results are available as reference data. Furthermore, this combination provides a stable density layering that can suppress buoyancy-driven convection according to the results of numerical simulations[7] (and Equation (5), below). It is essential for density layering stability that the density is a monotonic function of the concentration in the composition

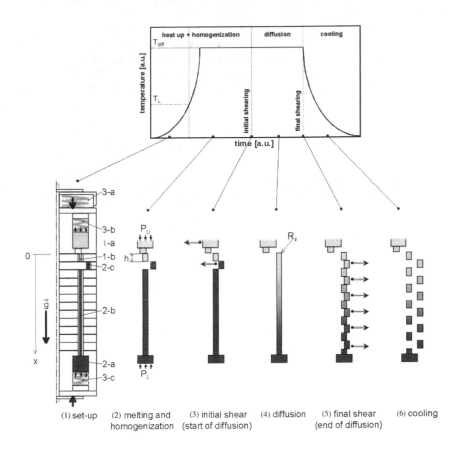

FIGURE 1. Operational diagram of the path of a diffusion experiment illustrated with the FOTON shear cell. The figure shows the setup in this study for a ground-based experiment involving diffusion from a Sn90In10 thick layer to a Sn semi-infinite space. One of four capillaries is visible. (1) Setup: solid Sn90In10 samples for the thick layer (1-b) and for the upper reservoir (1-a), solid Sn-samples for the capillary (2-b), for the intermediate cell (2-c), and for the lower reservoir (2-a). Graphite felts are used in the reservoirs (3-b,c) to provide additional liquid pressure and for compressing the disks to tighten the capillary (3-a). All of them are set inside the shear cell unit. The intermediate cell is set apart from the diffusion axis. (2) Melting and homogenization phase: filling the capillaries, except for the intermediate cell, is completed by the pressure P_U and P_L generated by the graphite felts in the reservoirs. The initial thickness of the thick layer is h. (3) Initial shear process: the upper reservoir (1-a) is disconnected. (4) Diffusion phase: all cells are aligned on the diffusion axis. Hence, filling the intermediate cell is completed by the pressure from the lower reservoir. The curvature radius R_C of the liquid sample at the corners is kept small by the pressure P_L. (5) Final shear process: the capillaries are separated into 20 cells by rotation of every second disk. (6) Cooling phase: the cells are completely separated.

range of the experiment during the entire diffusion time, provided that the density gradient is in the direction of the g-vector.

The experimental results were corrected because of shear-induced convection and the mixing inside the cells after the final shearing, both of which are typical for the shear cell technique. The reproducibility and the reliability of the ground-based experiments were investigated and compared with reference μg-data. Based on our results we discuss the efficiency of a stable density layering to suppress buoyancy-driven convection.

FOTON SHEAR CELL

The shear cell used for the ground experiments was basically the same as that used for the mission Foton-12.[5] Slight improvements were made for the mission Foton-M1 relating to the filling technique, the mechanics of the reservoirs, and the DC-motor of the rotating unit. FIGURE 2 shows a photograph of the Foton shear cell unit. Representative shear cell data are listed in TABLE 1. For the shear cell design the following were taken into account.

1. Physical and chemical points of view:
   - one dimensional setup (simple treatment of thick layer diffusion, see FIG. 1),
   - reduction of shear-induced convection,
   - reduction of Marangoni convection (sufficient pressure on the liquid metal in the capillaries $P_L \approx 20 \text{kPa}$),
   - no chemical reaction of the capillary material with the sample.
2. Technical points of view:
   - sufficient stability of temperature up to 1,000°C,
   - minimization of the weight and the size (for microgravity applications),
   - guaranty of complete filling by using reservoirs,

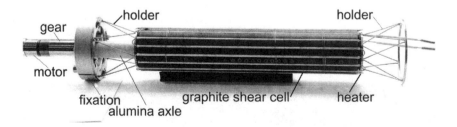

FIGURE 2. Foton shear cell unit. The graphite shear cell consists of an outer tube, an inner tube, disks, reservoirs, and an axle. The shear cell is heated by the wound Mo wire isolated by ceramic tubes. Ceramic fiber sheets (not visible) are wound around the shear cell for thermal isolation and to damp mechanical shocks at launch. The holders on both ends are for fixing the shear cell inside the vacuum chamber. The motor rotates the inner tube, and hence, every second disk.

TABLE 1. Representative data of the Foton shear cell

Shear cell unit	$350 \times \phi 60$ mm, graphite
Process atmosphere	vacuum ($p < 1 \times 10^{-2}$ Pa)
Weight	600 g with a motor
Motor	DC motor-tachometer combination
	(1727 U 012 C 001 G, Faulhaber GmbH and Co. KG)
Temperature range	up to 1,000°C
Shearing unit	20 graphite cells with 4 capillaries $\phi 1.5 \times 3$ mm
Pressure on the liquid metal in the capillaries (under 1g operation)	P_U (upper side), 5 kPa (for filling)
	P_L (lower side) 20 kPa (for filling and diffusion phase)
Thermal isolation	ceramic fiber

- highly precise machining with a tolerance of about 10μm to avoid leakage and misalignment of the capillaries,
- smooth rotation of the initial and final shearing by a computer controlled DC motor with a tachometer output,
- low friction force between disks (compression force, 2N).

EXPERIMENTAL

The experimental procedure was the same as that for the experiments in the improved AGAT-furnace facility (Foton-M1 mission). The 1g diffusion experiments were performed by using the Foton shear cell installed in a vacuum chamber that has the same dimensions of a cartridge chamber in the AGAT-furnace.

The experiment type was diffusion from a thick layer of Sn90In10 into pure Sn (FIG. 1). The diffusion axis was arranged vertically. According to the density diagram of Sn–In alloy (see FIG. 5, below), the higher density sample Sn was set in the lower part (FIG. 1[2-a, b, and c]) of the capillary and the lower density sample Sn90In10 was set in the upper part (FIG. 1[1-a and b]). The volume of samples at setup was 20% smaller than the capillary capacity to avoid leakage due to thermal expansion. Each sample part was weighed before insertion in order to check, after the experiment, the initial thickness h of the thick layer sample. All four capillaries were filled with samples with the same composition from the same charge and the experiments were performed simultaneously so that reproducibility could be discussed on that basis.

After evacuating the chamber, the furnace was heated to the diffusion temperature. After a one-hour homogenization phase, the initial shear process was performed at a shear speed $v_0 = 0.5$ mm/sec to start the diffusion process. At the end of the diffusion time, the final shear was started at 0.5 mm/sec again. The vacuum pressure was about 4×10^{-3} Pa during the experiment. The diffusion temperatures (diffusion times) were 275°C (28,800 sec), 400°C (21,600 sec), 600°C (21,600 sec), and 800°C (14,400 sec).

After the diffusion experiments each sample was pressed out of the cell mechanically, weighed, and dissolved in aqua regia for analyses by atomic absorption spectroscopy (AAS, Varian SpectrAA300).

EXPERIMENTAL RESULTS AND DISCUSSION

Diffusion Profile and Measured Mean Square Diffusion Depth

We noticed that the rotation of the cell was not hindered during the shear processes in any of our experiments. When we opened the shear cell after the diffusion experiments, no leakage of the samples from the cells was detected. All of the individual samples had almost the same weight as the completely filled mass, with a standard deviation of 0.6%. Some of the AAS concentration profiles are shown in FIGURE 3.

The concentration curves $c(x,t)$ for the diffusion experiments were fitted to the thick layer solution,

$$c(x, t) = \frac{c_0}{2}\left[\text{erf}\left(\frac{h+x}{\sqrt{4Dt}}\right) + \text{erf}\left(\frac{h-x}{\sqrt{4Dt}}\right)\right]. \tag{1}$$

In case of additive quasidiffusive convective contributions, the equation is also a solution, but for the effective diffusion coefficient D_{eff} instead of D (the atomic diffusion coefficient). In these calculations, the initial concentration c_0 of In in the thick layer was 10 at% for all experiments. The fitting parameters are h and the product Dt. Dt can be interpreted as half of the measured mean square diffusion depth $\overline{x^2}_{meas}$,

$$\overline{x^2}_{meas} = 2Dt. \tag{2}$$

The values of $\overline{x^2}_{meas}$ obtained are shown in TABLE 2.

The values of capillary C at 400°C are somewhat different to those of the others. This is a result of a smaller value of h than normal, caused by insufficient liquid pressure in the upper reservoir. This means that h in the thick layer (FIG. 1[1-b]) was smaller than 3 mm. However, with the correct h this has no influence on D. The fitted value for h is identical with that determined by weight control. In the other cases this deviation was avoided by correctly setting the pressure on the liquid.

Each profile obtained was smooth and agreed well with the fitting function (coefficient of determination $r^2 > 0.995$). This means that there was probably no indication of buoyancy-driven convection.

The profile can be approximately described by a thin layer profile using

$$c(x, t) = \frac{c_0 h}{\sqrt{\pi D(t + t_0)}} \exp\left(\frac{-x^2}{4D(t + t_0)}\right), \tag{3}$$

where $t_0 = h^2/6D$ is the time offset at $t = 0$ that gives the same square penetration depth as the real start profile. In the case of diffusion-like convective contributions, it is the solution for the effective diffusion coefficient D_{eff} instead of D. Because of the linearity of the relation between x^2 and $\ln c$, Equation (3) can be used to show the absence of large non-diffusion-like convection (see FIGURE 4).

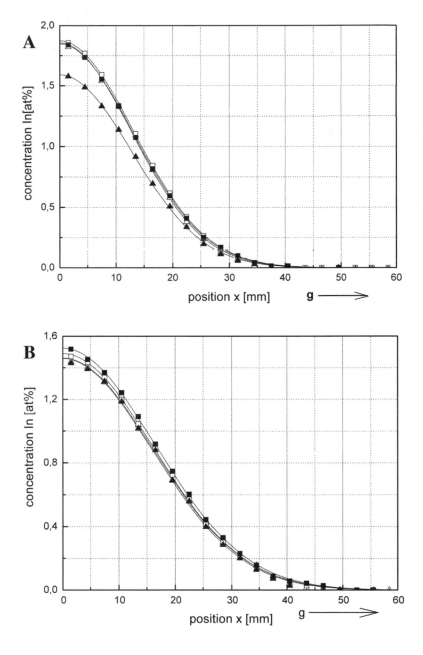

FIGURE 3. Concentration profiles obtained in the capillaries at **(A)** 400°C and **(B)** 600°C: △, capillary A; □, capillary B; ▲, capillary C; ■, capillary D. In the profiles at 400°C the initial thickness of the thick layer was lower than 3 mm.

TABLE 2. Measured mean square diffusion depth and diffusion coefficient obtained

T_{diff} [°C]	t_{diff} [sec]	Capillary	$\overline{x^2_{meas}}{}^a$ [10^{-4}m^2]	D [10^{-9}m^2/sec]
275	28,800	A	1.41	2.44
		B	1.44	2.48
		C	1.43	2.45
		D	1.38	2.37
400	21,600	A	1.57	3.64
		B	1.65	3.82
		C	1.59	3.67
		D	1.62	3.77
600	21,600	A	2.59	5.90
		B	2.54	5.87
		C	2.49	5.75
		D	2.49	5.75
800	14,400	A	2.31	8.04
		B	2.44	8.48
		C	2.45	8.49
		D	2.72	9.46

[a]Uncorrected measured mean square diffusion depth.

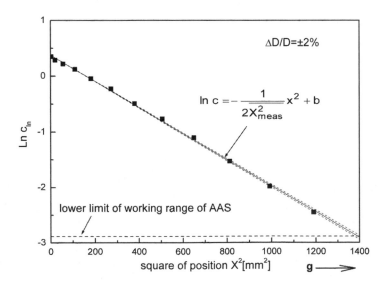

FIGURE 4. Typical linear plot of a concentration profile (capillary A at 800°C).

Determination of the Diffusion Coefficient D by Corrections

The measured mean square diffusion depth $\overline{x^2}_{meas}$ is influenced by two systematic error sources both of which are typical for the shear cell technique. As a result, $\overline{x^2}_{meas}$, and hence D, become larger than the pure diffusional $\overline{x^2}$ or D, respectively. One of the reasons is that the shear-induced convection causes volume exchanges between neighboring cells. This problem has been already discussed elsewhere.[8,9] The increment $\overline{x^2}_{shear}$ caused by shear-induced convection was measured by short-time diffusion experiments on SnIn–Sn at 250°C under low gravity conditions on a parabolic flight with the Foton shear cell (Suzuki *et al.*, to be published). The measurements showed that $\overline{x^2}_{shear}$ is about $2 \times 10^{-7} m^2$ at 250°C. As a result of a more detailed analysis, we assume that the shear-induced convection depends mostly on the cell geometry and the shear speed, rather than on the physical characteristics of the sample materials. Thus, $\overline{x^2}_{shear}$ can be taken as $2 \times 10^{-7} m^2$ for all diffusion temperatures.

Another reason is the mixing of the concentration profiles inside the cells. The mixing is induced by the final shear process, by segregation effects during solidification, or by integrating analyses as AAS or ICP. If the averaged values are plotted at the position of the cell centers, this causes an enlargement of $\overline{x^2}$. If the diffusion time is long enough ($Dt \gg h^2$, applicable also to this study), the increment $\overline{x^2}_{average}$ by this averaging effect can be simply described as $H^2/12$,[10] which depends only on the cell geometry. For the Foton shear cell, $H^2/12$ is $7.5 \times 10^{-7} m^2$, since H is 3 mm.

Diffusion coefficients can be corrected by using the following formula:

$$D = \frac{\overline{x^2}_{meas} - \overline{x^2}_{shear} - \overline{x^2}_{average}}{2t_{diff}} = \frac{\overline{x^2}_{meas} - \overline{x^2}_{shear} - H^2/12}{2t_{diff}}. \tag{4}$$

The corrected diffusion coefficients are shown in TABLE 2. The reference μg-data of the D1-mission and the statistical analyses of the diffusion coefficients in this study are shown in TABLE 3.

Garandet *et al.*[11] analyzed all expected error sources in their diffusion measurements in the Foton-12 flight. For the Sn–SnIn1 interdiffusion experiment at 300°C

TABLE 3. Diffusion coefficients of reference μg-data compared to that for 1g data obtained in this study

T [°C]	D1 Mission $D_{\mu g}$ [$10^{-9} m^2$/sec]	This Study $D^a \pm \Delta D^b$ [$10^{-9} m^2$/sec]	This Study $\Delta D/D$ [%]	Deviation from μg Value $(D_{1g} - D_{\mu g})/D_{\mu g}$ [%]
275	2.27	2.44±0.07	2.8	+6.9
400	3.43	3.73±0.11	3.0	+8.5
600	5.78	5.82±0.14	2.4	+0.7
800	8.73	8.62±0.62	7.2	−1.3
average	—	—	3.9	+3.7

[a] Average values of D obtained from the four capillaries A–D.
[b] Full statistical errors from four experiments, including temperature errors.

they assumed an absolute maximum error of 9% for $\Delta D/D$ (2% from temperature, 6% concentration measurement, and 1% from g-jitter induced convection). No standard deviation was given for that case, but from four other identical experiments in the Al–AlNi1 system the mean scattering of the D values can be derived as $\Delta D/D = 4.6\%$,[6] that is, about half the maximum value. Since the Foton shear cell has been improved, we think that the errors in our present 1γ-experiments are lower, also for the concentration measurements; but there may have been contributions from buoyancy convection.

In TABLE 3 the average D values and the standard deviations $\Delta D'$ are derived for each set of four values of this study for each temperature. Since the $\Delta D'$ values only represent the reproducibility under nearly equal external conditions, all statistical errors should be included, except those from the temperature measurement. Hence, for the total statistical error ΔD in TABLE 3 we have additionally included a standard temperature deviation, estimated at 1% on the basis of the relation $D = AT^2$. Thus, we arrive at relative total statistical errors of $\Delta D/D = 2.4$–7.2% (average 3.9%), which is indeed lower than in the Foton-12 mission. To obtain information about the possible systematic errors, for example, due to buoyancy convection, we compare our values with those from μg-experiments (TABLE 3). Two of the $1g$ values are still systematically larger than the μg values, but very near to them (within 9%). The same holds for the deviations. This may indicate, that there was still some contribution from buoyancy convection, but possibly also from other (unknown) sources. Nevertheless, the D values at 600°C and 800°C coincide with the μg values within the limits of error, which are of course larger than for the μg experiments.

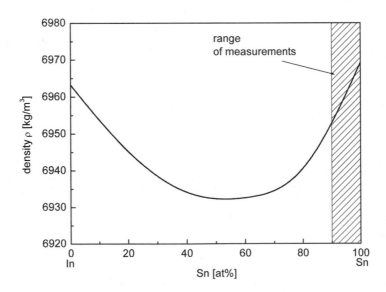

FIGURE 5. Density diagram of Sn–In alloy at 250°C.[12]

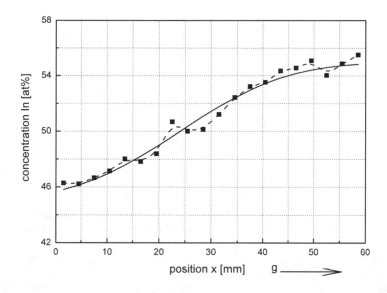

FIGURE 6. Example of an interdiffusion profile (between In55Sn45–In46Sn54 at 275°C) that is disturbed by convection.

Effect of Density Layering

The results of the present study indicate that buoyancy-driven convection was largely suppressed in these $1g$ experiments by using the condition $d\rho/dx > 0$. Strong buoyancy-driven convection is normally caused by a density gradient antiparallel to the g-vector; that is, if $d\rho/dx < 0$ (x-axis parallel to the g-vector). Since

$$\frac{d\rho}{dx} = \frac{\partial\rho}{\partial c}\frac{\partial c}{\partial x} + \frac{\partial\rho}{\partial T}\frac{\partial T}{\partial x} \qquad (5)$$

and $\partial\rho/\partial T$ is normally negative, the condition $d\rho/dx$ can be met by a negative $\partial T/\partial x$ (temperature gradient upward) and/or a positive contribution from the solutal part in Equation (**5**). Liquid diffusion experiments are normally designed to meet both conditions.

To demonstrate the significance of the relation (**5**) if $d\rho/dx < 0$, an interdiffusion experiment in the system InSn45 versus InSn54 was performed at 275°C under unfavorable conditions. The result is shown in FIGURE 6. Since according to the density diagram for Sn–In alloy[12] (see FIGURE 5), the variation in density is small in the diffusion zone and there is always a density minimum at its center, there must be always an unstable region in the diffusion zone. As a consequence, the concentration profile shows typical wave-like deviations from the error function, resulting from convection rolls, and the analysis gives an effective diffusion coefficient $(8\times10^{-9}\,\mathrm{m^2/sec})$ that is larger than the μg data that have been measured to date in that system (less than $2.5\times10^{-9}\,\mathrm{m^2/sec}$).[2]

Since it is difficult to arrange the temperature gradient exactly antiparallel to the g-vector, we have to consider horizontal temperature gradients, say $\partial T/\partial y$. Such gradients produce forces that also try to initiate convection rolls. It can be shown that there are always convection rolls, but a sufficiently positive solutal density gradient can strongly reduce the velocity of the rolls, so that their influence on the determination of the real atomic diffusion coefficient remains below a certain desired level. If the condition can be fulfilled, at least in the main part of the diffusion zone, depends on the material parameters in (**5**). We are presently working on this problem. It can normally not be met in the capillary parts away from the diffusion zone, since there c is constant and dT/dx should be small.

From these results, a stable density layering can be considered an effective method to suppress the buoyancy-driven convection. For stable density layering, it is important to avoid the region with negative values of $(\partial\rho/\partial c)\cdot(\partial c/\partial x)$ during the entire diffusion time. A more exact quantitative discussion concerning the necessary density layering arrangement to avoid the buoyancy-driven convection is a future subject. More exact thermodynamical data of the samples are required for this discussion.

If there are free surfaces on the sample liquid, the difference of the surface tension due to the concentration gradient can cause Marangoni convection. Müller[13] mentioned that Marangoni convection makes the reproducibility of the diffusion measurements worse. However, it is obvious that the density layering can create counter forces (also in the case of free surfaces) that can significantly reduce Marangoni forces. This is not possible under μg conditions, when Marangoni convection is active in the cases of free surfaces, but cannot be canceled because density layering cannot be activated as it can under $1g$ conditions. Hence, in the case of Marangoni convection, a stable density layering on the ground can be more advantageous than under μg conditions. This is the goal of experiments that are currently being investigated by Müller-Vogt *et al.* at the University of Karlsruhe.

CONCLUSIONS

In this study we tested the shear cell, designed for the μg experiments onboard the Foton-M1/2 mission, for its applicability in ground experiments. For the diffusion pair SnIn10–Sn we measured diffusion coefficients close to the μg values with a small data scatter, indicating that sufficient density layering can largely reduce buoyancy-driven convection. This should also be possible for other systems with stable material parameters (see (**5**)). We are presently working on the details of the necessary conditions, since the method of a sufficient stable density layering can avoid expensive μg experiments, if no high-accuracy data are wanted (say up to 10% error). Furthermore, we are working together with Müller-Vogt *et al.* to check if Marangoni convection can also be suppressed by density layering under $1g$.

ACKNOWLEDGMENTS

We thank the German Ministry of Education and Research, BMBF/ German Aerospace Center (DLR), especially Ms. M. Roth, for the financial support under national registration number 50WM0048. Meaningful discussions with Dr. G. Müller-Vogt, Dipl.-Phys. R. Rosu at the University of Karlsruhe and Dr. A. Griesche in HMI Berlin are gratefully acknowledged.

REFERENCES

1. FROHBERG, G. 1986. Thermophysical properties. *In* Fluid Sciences and Materials Sciences in Space. H.U. Walter, Eds.: 425. Springer-Verlag, Berlin.
2. FROHBERG, G., K.-H. KRAATZ, H. WEVER, *et al.* 1989. Diffusion on liquid alloy under microgravity. Defect Diffusion Forum **66**: 295–300.
3. FROHBERG, G. 1995. Diffusion in liquids: scientific results of the German Spacelab Mission D-2. WPF/DLR, 275–287.
4. YODA, S., H. ODA, T. OIDA, *et al.* 1999. Measurement of high accurate diffusion coefficient in melt of semiconductor and metal by using shear cell method. J. Jpn. Soc. Micrograv. Appl. **16**(2): 111–118.
5. GRIESCHE, A., K.-H. KRAATZ, G. FROHBERG, *et al.* 2001. The shear cells used in the AGAT facility onboard FOTON-12. Proc. First International Symposium on Microgravity Research and Applications in Physical Sciences and Biotechnology. ESA SP-454, January 2001. Sorrento, Italy, September 10–15, 2000. 985–992.
6. GARANDET, J.P., P. DUSSERRE, J.P. PRAIZEY, *et al.* 2000. Measurement of solute diffusivities in liquid metal alloys within the AGAT facility during the FOTON 12 mission. Micrograv. Space Station Utilization **1**: 29–34.
7. MATHIAK, G., G. FROHBERG & W.A. ARNOLD. 1996. Numerical simulations of convective flows concerning liquid diffusion experiments. Proc. Second European Symposium on Fluids in Space, Naples, Italy, April 22–26, 369–400.
8. ARNOLD, W. & D. MATTHIESEN. 1995. Numerical simulation of the effect of shearing on the concentration profile in a shear cell. J. Electrochem. Soc. **142**(2): 433–438.
9. GRIESCHE, A., K.-H. KRAATZ, G. FROHBERG, *et al.* 2001. Liquid diffusion measurements with shear cell technique study of shear convection. Proc. First International Symposium on Microgravity Research and Applications in Physical Sciences and Biotechnology. ESA SP-454, January 2001. Sorrento, Italy, September 10–15, 2000. 497–503.
10. SUZUKI, S., K.-H. KRAATZ & G. FROHBERG, *et al.* 2004. Investigation of the convection in a shear cell diffusion experiment (parabolic flight PFC31 pre-experiment for the FOTON-M1 mission). To be published.
11. GARANDET, J.P., G. MATHIAK, V. BOTTON, *et al.* 2004. Reference microgravity measurements of liquid phase solute diffusivities in tin and aluminum based alloys. Int. J. Thermophys. **25**: 249.
12. BERTHOU, P.-E. & R. TOUGAS. 1970. The densities of liquid In–Bi, Sn–In, Bi–Sb, and Bi–Cd–Ti alloys. Metal. Trans. **1**: 2978–2979.
13. MÜLLER, H. 2001. Untersuchung von konvektiven Zusatztransporten bei der Messung von Diffusionskoeffizienten mittels der Scherzellenmethode. Ph.D. Thesis. University of Karlsruhe, Karlsruhe. [German].

Microgravity Experiments on Boiling and Applications

Research Activity of Advanced High Heat Flux Cooling Technology for Electronic Devices in Japan

KOICHI SUZUKI AND HIROSHI KAWAMURA

Department of Mechanical Engineering,
Tokyo University of Science, Noda, Chiba, Japan

ABSTRACT: Research and development on advanced high heat flux cooling technology for electronic devices has been carried out as the Project of Fundamental Technology Development for Energy Conservation, promoted by the New Energy and Industrial Technology Development Organization of Japan (NEDO). Based on the microgravity experiments on boiling heat transfer, the following useful results have obtained for the cooling of electronic devices. In subcooled flow boiling in a small channel, heat flux increases considerably more than the ordinary critical heat flux with microbubble emission in transition boiling, and dry out of the heating surface is disturbed. Successful enhancement of heat transfer is achieved by a capillary effect from grooved surface dual subchannels on the liquid supply. The critical heat flux increases 30–40 percent more than for ordinary subchannels. A self-wetting mechanism has been proposed, following investigation of bubble behavior in pool boiling of binary mixtures under microgravity. Ideas and a new concept have been proposed for the design of future cooling system in power electronics.

KEYWORDS: microgravity boiling; MEB; narrow channel; self-wetting fluid; heat transfer enhancement; cooling technology

INTRODUCTION

Recently, electronic devices, such as IC chips and CPU, have been getting smaller and the processing speed and memory capacity have been increasing. Consequently, power and the thermal emission from the electronic devices have increased considerably. According to an IC manufacturer, the operating power of a personal computer (PC) will be more than 500 W ten years from now. In the near future, conventional cooling methods using air or single-phase fluid will not be useful in practice, an advanced new cooling system is urgently required for practical use in high power electronic devices.

Heat transfer with phase change removes a larger quantity of heat from a hot body than a single-phase fluid. In particular, boiling is the superior heat transfer

Address for correspondence: Koichi Suzuki and Hiroshi Kawamura, Department of Mechanical Engineering, Tokyo University of Science, Noda, Chiba, 278-8510 Japan. Voice: 81-4-7124-1501; fax: 81-4-7123-9814.
suzuki@rs.noda.tus.ac.jp   kawa@rs.noda.tus.ac.jp

Ann. N.Y. Acad. Sci. 1027: 182–195 (2004). ©2004 New York Academy of Sciences.
doi: 10.1196/annals.1324.017

phenomenon over any other thermal transport methods. For this reason, boiling heat transfer is expected to be used as a cooling method for future electronic devices. This technique has been effective for microgravity use and has been studied for long time for the purpose of thermal management and fluid control in space.

The absence of gravity presents severe limitations on heat and mass transfer driven by a gravity effect. However, it is very effective for understanding the mechanism of heat and mass transfer phenomena. For an example, boiling bubbles are difficult to detach from a hot surface and the heat transfer coefficient decreases to very low values; however, Marangoni convection appears in the liquid–vapor interface and is a very important factor for bubble behavior on a heating surface. Therefore, the results of boiling experiments in microgravity are considered to yield important information for industrial technology about heat and mass transfer fields.

This paper reviews recent boiling experiments in microgravity, including experimental data on critical heat flux (CHF), two phase flow of liquid and vapor in a narrow channel, and boiling of binary mixtures. Then, research activity is presented that relates to applications of microgravity experiments on boiling heat transfer to cooling technology of future electronic devices in Japan.

CRITICAL HEAT FLUX AND BUBBLE BEHAVIOR IN POOL BOILING UNDER MICROGRAVITY

In his book,[1] J. Straub expressed the view: "A study of critical heat flux (CHF) is a difficult task, because of destruction of heater surface, thus failure of whole systems." However, it could be easy to observe CHF visually by burnout, using a thin plate heater placed in the pool, if sufficient electric power for burnout is supplied during a microgravity period. Of course, the heater is destroyed.

Boiling bubbles are more difficult to detach from a heating surface in microgravity than on Earth, and part of the surface is easily dried out so that the surface burns out rapidly. Because of the difficulty of bubble detachment, heat transfer coefficient and CHF decrease. In a previous prediction of CHF, proposed by Kutateladze[2] and Zuber,[3] critical heat flux decreases proportionally to one-quarter power of gravity. Their predictions have been widely accepted in pool boiling heat transfer. In the early stage of microgravity studies, pool boiling experiments were conducted by Seigel,[4] and by Merte and Clark.[5] Their results showed that critical heat fluxes roughly agreed with the existing theories.[2,3] Di.Marco and Grassi[6] summarized the main features of pool boiling experiments in microgravity early on in microgravity experiments. However, in recent microgravity experiments on pool boiling, Straub indicated that the critical heat fluxes are 100–300 percent higher than existing theories suggest.[7]

The heat transfer performance of boiling strongly depends on bubble behavior on the heating surface, where bubbles detach or not. Furthermore, bubble behavior depends on several factors, such as wetting of the solid surface by liquid, size of heater surface and the configuration, thermocapillary flow activated on the bubble interface, inertia of bubbles, and g-jitter. These are very important factors for boiling heat transfer research. In particular, it is well known that the wetting of a solid surface by a liquid generates a strong effect on heat transfer performance and boiling

FIGURE 1. Bubble behavior and burn out in subcooled pool boiling of water at 10 K of liquid subcooling under microgravity (0.01 g–0.04 g): (**A**) just before burnout and (**B**) at burnout. (Reproduced from Ref. 9, with permission.)

characteristics. Wetting has generally been evaluated in terms of only the contact angle of a liquid droplet with the solid surface and no inclusion of a wetting term has been suggested in the heat transfer equations. Wetting might be an eternal problem in boiling heat transfer.

Oka and Abe[8] showed that the bubbles detached in pool boiling under high quality microgravity of $10^{-5}g$, using a long period drop facility, JAMIC. This may be a reason why the critical heat flux is higher in microgravity than widely accepted predictions.[2,3]

In our microgravity experiment on subcooled pool boiling of water, performed during a parabolic flight of aircraft,[9] stainless steel plates with 20 mm length, 5 mm width, and 0.1 mm thickness were physically burned out by AC electric power, and the bubble behavior visually observed. No detachment of bubbles from the heating surface was observed until the heating plate burned out, despite the low quality of microgravity with g jitter, 0.01–0.04 g, as shown in FIGURES 1 and 2. The critical heat fluxes obtained were 200–400 percent higher than suggested by existing theories,[2,3] the same as the results of many experiments.[7] In the experiment, the heating power was added at about constant heating rate in microgravity until the heater was burned

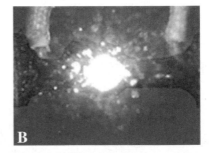

FIGURE 2. Bubble behaviors and burn out in subcooled pool boiling of water at 40 K of liquid subcooling under microgravity (0.01 g–0.04 g): (**A**) just before burnout and (**B**) at burnout. (Reproduced from Ref. 9, with permission.)

out. The burnout heat fluxes (CHF) were obtained from the heating power at burnout. It is considered that, in the microgravity experiment, the boiling bubbles expand rapidly by quick heating and some liquid layers remain between the expanding bubbles and the heating surface. Ohta indicated the existence of microlayers in his microgravity experiment on pool boiling.[10,11] Rapid evaporation of the liquid layer and the microlayer may be closely related to the higher critical heat flux in microgravity.

BOILING IN A NARROW GAP UNDER MICROGRAVITY

According to the bubble behavior obtained in our experiment,[9] subcooled pool boiling of water was performed on the ground under the condition that the bubbles were attached to the heating surface with a transparent glass plate placed over the heating surface. the same critical heat fluxes as found in the microgravity were obtained on the ground by adjusting the clearance between the plate and the heating surface.[12] The clearances were 4 mm for 10 K and 1 mm for 40 K of liquid subcooling.

The critical heat flux was measured with respect to the heating time until burnout by increasing the heating power. The critical heat flux decreased gradually for about 100 seconds and then became constant, as shown in FIGURE 3.[13] In this figure, the heat flux zone modified by the Lienhard and Dhir study on hydrodynamic prediction of finite heating surface is indicated for reference.[14] The constant critical heat flux agreed well with those calculated from existing theories[2,3] and the Lienhard and Dhir heat flux ranges. The liquid layer under the boiling bubbles is considered to decrease for the long heating time because of slower expansion of bubbles.

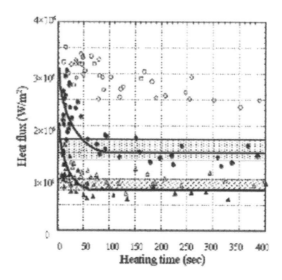

FIGURE 3. Time dependence of burnout heat flux in pool boiling of water with bubble holder under ground conditions: subcooling at 40 K without (○) and with (●) cover; subcooling at 10 K without (△) and with (▲) cover; Dhir–Zuber theory subcooling at 40 K ([ ]) and 10 K ([ ]), 0.01 g–0.04 g. (Reproduced from Ref. 13, with permission.)

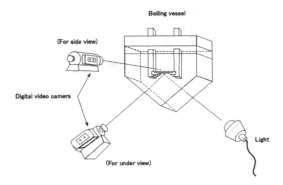

FIGURE 4. Experimental setup for optical observation of contact area of boiling bubbles with heating surface in pool boiling. (Reproduced from Ref. 13, with permission.)

In additional experiments we tried to observe optically the contact area of boiling bubbles with heating surface using a transparent heating surface, as shown in FIGURE 4.[13] The heating surface is an ITO coated soda-lime glass of 2 mm thickness and the bubble holder is placed over the heater surface. DC heating power was applied to the ITO heater. Dry areas are indicated as bright parts by complete reflection, and the liquid contact area is dark because of penetration by the beam. The effect of heating time on contact area of bubbles is shown in FIGURE 5 as an example. The contact area of boiling bubbles in quick heating is smaller than that for long time heating at same heat flux. The experimental result suggests that the higher critical

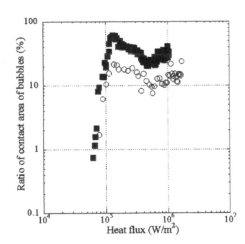

FIGURE 5. Comparison of contact area of bubbles with heating surface for heating time in subcooled pool boiling of water equipped bubble holder at 10 K of liquid subcooling: subcooling at 10 K; ○, quick heating; and ■, long time heating; with cover. (Reproduced from Ref. 13, with permission.)

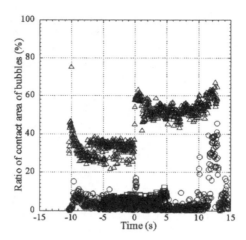

FIGURE 6. Progressive change of contact area of bubbles with heating surface in subcooled pool boiling of water under microgravity: ○, 72 V = 2.7×10^5 W/m²; △, 108 V = 5.2×10^5 W/m²; □, 144 V = 9.5×10^5 W/m². (Reproduced from Ref. 13, with permission.)

heat flux is caused by rapid evaporation, the liquid layer remaining under the expanded bubbles in short period microgravity.

In drop-shaft experiments with the high quality microgravity provided by JAM-IC, constant heating power was applied to the ITO heating surface from solid batteries. The heating began at 10 seconds before the drop and the liquid subcooling was about 40 K. As shown in FIGURE 6, the contact area increases dramatically at the start of microgravity, then it becomes constant.[13] At high subcooling of liquid in microgravity, large coalesced bubbles remain just over the heating surface and absorb the boiling bubbles lifting from the heating surface. It was suggested by Merte that generated bubbles coalesced and hovered just over the heating surface and absorbed bubbles detached from heating surface in microgravity.[15]

For flow boiling of water in a straight channel with narrow gap, Ohta investigated bubble behavior and local heat flux using a precisely processed transparent heating surface in microgravity resulting from parabolic flight of an aircraft.[16] The gap size was 2 mm, the subcooling of liquid was 0–14 K, and the mean flow velocity was 0.06 m/sec. The surface temperature rose at the start of microgravity, 0.03 g, and a large dry patch appeared in the middle and downstream of the channel. The local heat flux decreased in microgravity; however, it occasionally increased because of quenching effects by the subcooled liquid.

WETTING OF LIQUID WITH HEATING SURFACE UNDER MICROGRAVITY CONDITIONS

In a binary mixture, such as volatile alcohol and water, the contact angle decreases and wetting improves with increase of alcohol concentration.[17] In our previous experiments on pool boiling of binary mixtures using a platinum wire, the maximum critical heat flux was observed at each concentration.[18]

Abe, Oka, and Mori demonstrated enhancement of heat transfer in pool boiling of ethanol and water binary mixture under a high quality microgravity condition of $10^{-4}g$ and 10-second drop shaft facility of JAMIC.[19]

Recently, many studies and analyses have concerned liquid supply, bubble growth, bubble behavior, and bubble dynamics in microgravity. For example, the research papers in previous conferences by Abe,[20] Straub,[21] Dhir,[22] and Fujita.[23] Maruyama showed a microscopic treatment of bubble motion and evaporation, including wetting of droplet on the heating surface.[24]

RESEARCH AND DEVELOPMENT OF ADVANCED HIGH HEAT FLUX COOLING TECHNOLOGY FOR ELECTRONIC DEVICES BY BOILING PHENOMENA

An application of microgravity boiling was investigated for high heat flux cooling of high power electronic devices as an international research project promoted by the Japan Space Utilization Promotion Center (JSUP) and New Energy and Industrial Technology Development Organization (NEDO) in 2000. The investigation indicated that boiling in microgravity was strongly effective in the development of the high heat flux cooling technology for high power electronic devices.

In 2002, NEDO started new research project: "Project of Fundamental Technology Development for Energy Conservation". In this project, the research and development on advanced high heat flux cooling technology was carried out for the next generation of power electronics. Professor Ohta, H., Kyusyu University, is a project leader and four institutes are involved. The main goal of this project is to establish a useful model cooling system for high power electronic packages with $300\,W/cm^2$ of maximum cooling heat flux under steady operation.

High Heat Flux Cooling by Microbubble Emission Boiling

In highly subcooled flow boiling, heat flux increases beyond the critical heat flux, emitting many microbubbles from coalesced bubbles on the heating surface in transition boiling.[25,26] This boiling is called microbubble emission boiling (MEB). In experiments, subcooled flow boiling of water was performed in horizontal rectangular channels with 1 mm, 3 mm, and 5 mm height, 14 mm width, and 150 mm length, shown in FIGURE 7.[27] The heating surface, size 10 mm × 10 mm, is placed on the bottom surface of the channel. The heating block (copper) is composed of upper straight and lower conical parts, and the top of the heating block is considered a cooling surface for electronic elements. Heat is applied to the bottom surface of the block by electric heaters. The pressure in the boiling section is held at atmospheric conditions.

Microbubble emission boiling occurred at higher liquid subcooling, 40 K, for the channels with narrow gap and heat flux increased considerably beyond that for ordinary critical heat fluxes, as shown in FIGURE 8.[27] Very high heat flux, $5\,MW/m^2$ ($500\,W/cm^2$) was obtained, even in the narrow channel with 1 mm height. This is five times the present cooling limit for a power electronic package. However, once flow

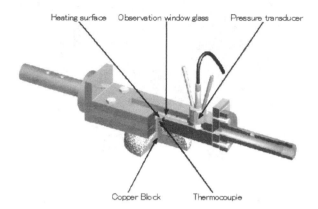

FIGURE 7. Cut view of test section, rectangular channel, and heating block. (Reproduced from Ref. 27, with permission.)

of cooling liquid was stopped, large bubbles emerged, as shown in the microgravity experiments on boiling in the narrow gap[16] and burnout may occur, as indicated by subcooled pool boiling with bubble holder in FIGURES 1 and 2. The experimental results indicate that microbubble emission boiling is expected for the super high heat flux cooling of high power electronic elements, but a self-automatic wetting aid should be employed to avoid emergence dry out.

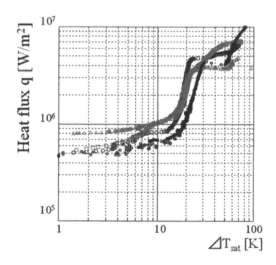

FIGURE 8. Boiling characteristics of subcooled flow boiling of water with microbubble emission in rectangular channels with small height: liquid subcooling, 40 K; liquid velocity, 0.5 m/sec; △, 5×14 mm; ○, 3×14 mm: □, 1×14 mm; and ○, 17×12 mm rectangular channels. (Reproduced from Ref. 13, with permission.)

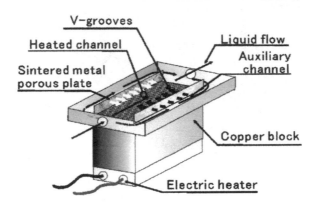

FIGURE 9. Heating surface with microgrooves and subliquid supply channels. (Reproduced from Ref. 28, with permission.)

Enhancement of Heat Transfer by a Heating Surface with Microstructure

Ohta presented a new heating surface concept for an electronics package.[28] The heat transfer section is composed of a fine grooved surface in the main channel with 2 mm height and auxiliary sub-channels for liquid supply placed on both sides, as shown in FIGURE 9. The heat transfer section and the subchannels are separated by porous metal layers. With increasing heat flux, the mode of heat transfer changes to nucleate boiling from forced convection of liquid flow and generated vapor bubbles grow large. A microclearance is formed between the bubble and the triangular groove. Liquid is introduced into the microgaps by capillary effect from the sub-channels through the porous sections.

In the experiment, stable evaporation was observed and dry out did not occur. CHF increased 30–40 percent higher than normal values for R113, as shown in FIGURE 10.[28] This is one of the successful wetting technologies applying the capillary effect that has focused on microgravity conditions.

Influence of Marangoni Convection on Bubble Behavior

In microgravity, Marangoni convection is one of the strong factors that controls bubble motion and boiling performance. Abe investigated bubble behavior generated from a spot heater and Mrangoni convection using a particle tracking method for a saturated 1-butanol–water mixture of 1.5 wt% under high quality microgravity at JAMIC.[29]

In the pool boiling of 1-butanol–water mixture, two generated bubbles detached immediately from the heater without coalescing in the strong Marangoni convection into the contact spot of bubble with heater. In another experiment, using a transparent heating surface, it was observed that liquid flowed into the bottom of the bubble by Marangoni convection. Marangoni convection is probably driven by the concentration gradient resulting from the temperature difference.

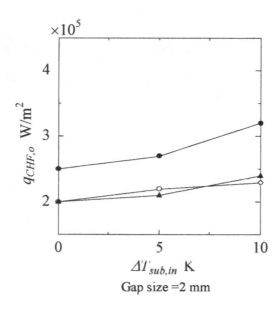

FIGURE 10. Comparison of critical heat flux by heat input between flat and grooved surface R-113: $P = 0.1$ MPa, $u_m = 0.032$ m/sec, $s = 2$ mm; ○, flat surface with plane separating plates; ●, grooved surface with porous separating plates; ▲, grooved surface with plane separating plates. (Reproduced from Ref. 28, with permission.)

Surface tension, generally, decreases with rising temperature. Marangoni flow, activated upward along the liquid–vapor interface, as illustrated in FIGURE 11 A. The dry surface then spreads and is burned out. This is a popular dryout mechanism. However, polyhydric alcohols, such as 1-butanol, exhibit an opposite surface tension property; that is, the surface tension increases with rising temperature.[30] This is very important for boiling phenomena. The temperature of liquid near the liquid–bubble interface of the heating surface is higher than that at the top of the bubble. For 1-butanol–water mixtures, the surface tension is higher at the bottom of bubble than at the top, and the Marangoni flow activates toward the heating surface, then the liquid flows into the boundary of liquid, vapor, and solid, as illustrated in FIGURE 11 B. The dry area would not extend or bubble would be pushed up by the liquid flow. As a result, dryout would be automatically avoided.

According to experimental results for binary alcohol–water mixtures, Abe and Kawamura named the wetting characteristic *self-wetting* and the liquid *self-wetting fluid*.[29] This is one of the new concepts for the development of advanced high heat flux cooling of electronic devices.

The critical heat flux with respect to the concentration of volatile liquid in binary mixtures was measured in subcooled pool boiling at 40 K of subcooling, using a thin stainless steel plate for a heating surface on the ground. The critical heat flux was obtained by applying electric power to the heating surface at the burnout. The test was conducted for several binary mixtures. A specific concentration exists at maximum CHF. For a 1-butanol–water mixture of 3 wt%, the maximum CHF is 160–200 percent higher than water (1.6–2 times as much as that of water). We consider that the data show a strong effect of self-wetting on CHF. In subcooled flow, boiling of a 1-butanol–water mixture of 3 wt% and an ethanol–water mixture of 10 wt% in a rectangular channel, microbubble emission boiling (MEB) occurred, and the maximum heat flux was about 10 percent higher than that of water, as shown in FIGURE 12. In this case, MEB was very silent compared to the case of water, although the heat flux increased considerably. Thus, Kawamura and Suzuki named the MEB, *mild MEB*.

Control of Bubble Behavior by the Application of an Electric Field

Di Marco and Walter Grassi investigated effects of an electric field on a single bubble on the surface under microgravity performed at the drop shaft facility, JAM-IC.[31] With increasing electric field intensity, the detaching diameter of bubble reduced and the bubble moved with higher velocity in the liquid pool, even in absence of gravity. The experimental results are very interesting for the enhancement

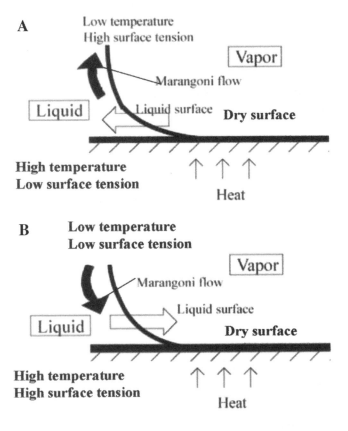

FIGURE 11. Concept of wetting in boiling heat transfer induced by Marangoni convection: **(A)** poor wetting and **(B)** self-wetting. (Reproduced from Ref. 29, with permission.)

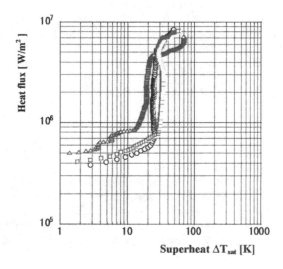

FIGURE 12. Comparison of boiling of binary mixtures and water in a horizontal rectangular channel at 40 K of liquid subcooling and 0.5 m/sec of liquid velocity: channel, 5 mm H × 14 mm W; ○, water–ethanol (10 wt%); △, water; □, water–1-butanol (3 wt%).

of boiling heat transfer in microgravity and high heat flux cooling technology for electronic devices.

CONCLUSIONS

It is very high expensive to perform microgravity experiments. However, it is very important for science and technology, and the results obtained could contribute greatly to the development of industrial technology. For example, the absence of gravity has provided valuable information about Marangoni convection to fluid science and boiling heat transfer. The research and development introduced in this paper are useful for future technology suggested by the application of microgravity science.

For the cooling of small electronic chips and elements, heat and mass transfer are performed in a small- or a microchannel. It has been understood that Marangoni convection and capillary flow are very strong factors for the boiling heat transfer in a narrow or a small channel, enhancing their heat transfer performance.

The following experimental results have been obtained in the NEDO project since 2002. A microgrooved surface with subliquid channels automatically supplies liquid to the heating surface in boiling condition by a capillary effect of boiling bubbles and microgrooves, and CHF increases 30–40 percent higher than usual for R-113. The surface tension of binary mixtures of polyhydric alcohol and water increases with rising temperature, and a self-wetting property was discovered. In subcooled flow boiling of a self-wetting fluid, "mild MEB" occurs in transition boiling and the maximum heat flux reaches $8 \times 10^6 \text{W/m}^2$, 10 percent higher than in water. These experimental results give a new concept for advanced high heat flux cooling technology of future electronic devices.

ACKNOWLEDGMENTS

Since 2002 the new NEDO project introduced in this paper has been carried out by Professor Ohta, project leader, and four institutes. The institutes are Kyushu University, Tokyo University of Science, National Institute of Advanced Industrial Science and Technology, and Toshiba Corporation.

We, all the members of this project, express our deep appreciation for the promotion and support of NEDO. Also, we greatly appreciate useful advice and suggestions from Professor M. Ishizuka (Toyama Pref. University), Dr. M. Mochizuki (Fujikura Corporation), Professor J. Straub (Munich Technical University), Professor Di Marco and Professor W. Grassi (Pisa University) for the project.

REFERENCES

1. STRAUB, J. 2001. Boiling heat transfer and bubble dynamics in microgravity. Adv. Heat Trans. **35:** 119–120.
2. KUTATELADZE, S.S. 1952. Heat transfer in condensation and boiling. USAEC. Rep. AEC-tr-3770.
3. ZUBER, N. 1958. On stability of boiling heat transfer. Trans. ASME. J. Heat Trans. **80:** 711–720.
4. SIEGEL, R. 1967. Effect of reduced gravity on heat transfer. In Advances in Heat Transfer, Vol. 4. T.F. Irvine, Ed.: 144–228. Academic Press.
5. MARTE, H. & J.A. CLARK. 1964. Boiling heat transfer to cryogenic fluid at standard, fractional, and near-zero gravity. Int. J. Heat Mass Trans. **86:** 351–359.
6. DI MARCO, P. & W. GRASSI. 2000. Pool boiling in microgravity: accessed results and open issues. Proc. Third European Thermal Science Conference, Vol. 1. 81–90.
7. STRAUB, J., M. ZELL & B. VOGEL. 1991. Boiling under microgravity conditions. Proc. First European Symposium on Fluid in Space. Ajaco, France. ESA SP-353. 269–297.
8. OKA, T., Y. ABE, Y. MORI & A. NAGASHIMA. 1994. Pool boiling heat transfer in microgravity-experiments with CFC-113 and water using drop-shaft facility. Trans. JSME 60-557: 3093–3100.
9. SUZUKI, K., H. KAWAMURA, et al. 1997. Burnout in subcooled pool boiling of water under microgravity. Proc. Xth European and VIth Russian Symposium on Physical Science in Microgravity. St. Petersburg, Russia. 366–369.
10. OHTA, H., K. KAWASAKI, et al. 1996. Nucleate pool boiling heat transfer under reduced gravity condition. Micrograv. Q. **6**(2–3): 114–120.
11. OHTA,H., M. KAWAJI, et al. 1998. Heat transfer in nucleate pool boiling under microgravity condition. Proc. 11th Intl. Heat Transfer Conference, Vol. 2: 401–406.
12. SUZUKI, K., H. KAWAMURA, et al. 1999. Experiments on subcooled pool boiling of water in microgravity (observation of bubble behavior and burnout). Proc. 5th ASME-JSME Joint Thermal Engineering Conference. No.6420. CD-ROM.
13. SUZUKI, K., M. SUZUKI, S. TAKAHASHI, et al. 2003. Bubble behavior in subcooled pool boiling of water under reduced gravity. Proc. Space Technology and Applications International Forum 2003 (STAIF2003). 142–148.
14. LIENHARD, J.H. & V.K. DHIR. 1973. Hydrodynamics prediction of peak pool boiling heat fluxes from finite bodies. J. Heat Trans. **95:** 152–158.
15. LEE, H.S. & H. MERTE. 1999. Pool boiling mechanisms in microgravity. Proc. Intl. Conference on Microgravity Fluid Physics and Heat Transfer. 126–135.

16. OHTA, H., H. WATANABE, et al. 1999. Gravity effect on flow boiling heat transfer in narrow gaps. Proc. 5th ASME-JSME Thermal Engineering Joint Conference. No. 6421. CD-ROM.

17. SEFIAN, K. & L. TADRIST. 2003. Experimental investigation on wetting of binary volatile sessile drop. Proc. Space Technology and Applications International Forum 2003 (STAIF2003). 247–254.

18. HOVESTREUDT. J. 1963. The influence of the surface tension difference on the boiling of mixtures. J. Chem. Eng. Sci. **18:** 631–639.

19. ABE, Y., T. OKA, Y.H. MORI & A. NAGASHIMA. 1994. Pool boiling of a non-azerotropic binary mixture under microgravity. Int. J. Heat Mass Trans. **37:** 2405–2413.

20. ABE, Y. & A. IWASAKI. 1999. Single and dual vapor bubble experiments in microgravity. Proc. Int. Conference on Microgravity Fluid Physics and Heat Transfer. 55–61.

21. STRAUB, J. 2002. Origin and effect of thermocapillary convection in subcooled boiling: observations and conclusions from experiments performed at microgravity. Ann. N.Y. Acad. Sci. **974:** 348–363.

22. SINGH, S. & V.K. DHIR. 1999. Effect of gravity, wall superheat and liquid subcooling on bubble dynamics during nucleate boiling. Proc. Intl. Conference on Microgravity Fluid Physics and Heat Transfer. 103–113.

23. BAI, Q. & Y. FUJITA. 1999. Numerical simulation of bubble growth in reduced-gravity. Proc. Intl. Conference on Microgravity Fluid Physics and Heat Transfer. 136–143.

24. MARUYAMA, S., T. KIMURA & M.C. LU. 2002. Molecular scale aspects of liquid contact on solid surface. J. Therm. Sci. Eng. **10**(6): 23–29. The Heat Transfer Society of Japan.

25. SUZUKI, K., K. SAITO, T. SAWADA & K. TORIKAI. 2000. An experimental study on microbubble emission boiling. Proc. 3rd European Thermal Science Conference, Vol. 1. 81–90.

26. SUZUKI, K., H. SAITOH & K. MATSUMOTO. 2002. High heat flux cooling by microbubble emission boiling. Ann. N.Y. Acad. Sci. **974:** 364–377.

27. SUZUKI, K., H. KAWAMURA, et al. 2003. High heat flux cooling for electronic devices by subcooled flow boiling with microbubble emission. Proc. 6th ASME-JSME Joint Thermal Engineering Conference. TED-AJ03-106. CD-ROM.

28. OHTA, H., Y. SHINMOTO, T. OHNO & A. OKAMOTO. 2002. Development of high-performance space cold plate by improved liquid supply for flow boiling in narrow gaps. J. Therm. Sci. Eng. **10**(6): 39–44. The Heat Transfer Society of Japan.

29. ABE, Y. & A. IWASAKI. 2002. Study on the boiling bubble of self-wetting fluid in microgravity. Proc. Thermal Engineering Conference. Thermal Engineering Division. JSME. 291–292.

30. VOCHTEN, R. & P. PETRE. 1973. Study of the heat of reversible adsorption at the air-solution interface II, experimental determination of the heat of adsorption of some alcohols. J. Coll. Interf. Sci. **42**(2): 320–327.

31. DI MARCO, P. & W. GRASSI. 2000. Development of high heat flux cooling technology by micro-scale boiling. NEDO Report: International Research Project on Energy Saving #3.

Momentum Effects in Steady Nucleate Pool Boiling During Microgravity

HERMAN MERTE, JR.

Mechanical Engineering Department, University of Michigan, Ann Arbor, Michigan, USA

ABSTRACT: Pool boiling experiments were conducted in microgravity on five space shuttle flights, using a flat plate heater consisting of a semitransparent thin gold film deposited on a quartz substrate that also acted as a resistance thermometer. The test fluid was R-113, and the vapor bubble behavior at the heater surface was photographed from beneath as well as from the side. Each flight consisted of a matrix of three levels of heat flux and three levels of subcooling. In 26 of the total of 45 experiments conditions of steady-state pool boiling were achieved under certain combinations of heat flux and liquid subcooling. In many of the 26 cases, it was observed from the 16-mm movie films that a large vapor bubble formed, remaining slightly removed from the heater surface, and that subsequent vapor bubbles nucleate and grow on the heater surface. Coalescence occurs upon making contact with the large bubble, which thus acts as a vapor reservoir. Recently, measurements of the frequencies and sizes of the small vapor bubbles as they coalesced with the large bubble permitted computation of the associated momentum transfer. The transient forces obtained are presented here. Where these arise from the conversion of the surface energy in the small vapor bubble to kinetic energy acting away from the solid heater surface, they counter the Marangoni convection due to the temperature gradients normal to the heater surface. This Marangoni convection would otherwise impel the large vapor bubble toward the heater surface and result in dryout and unsteady heat transfer.

KEYWORDS: pool boiling; microgravity; momentum effects

NOMENCLATURE:

D	bubble diameter
E	surface energy of bubble
f	frequency
F	force
m	mass of bubble
n_t	total number of bubble sizes considered
t	time
V	velocity

Greek Symbols

ρ	density
σ	surface tension

Subscripts

b	bubble
i	index for size
v	vapor

Address for correspondence: Herman Merte, Jr., Mechanical Engineering Dept., 2026 G.G. Brown Building, University of Michigan, Ann Arbor, MI 48109-2125, USA. Voice: 734-764-5240; fax: 734-647-3170.
  merte@umich.edu

Ann. N.Y. Acad. Sci. 1027: 196–216 (2004). ©2004 New York Academy of Sciences.
doi: 10.1196/annals.1324.018

INTRODUCTION

The relatively recent availability of long-term, high quality microgravity associated with space flight provides opportunities for the study of pool boiling under this condition, resulting in insights into its behavior not encountered in Earth gravity. As has been extensively and well demonstrated over the years, the phenomena of nucleate or bubble boiling, including both pool and flowing, are highly complex, resisting attempts to describe or predict their behavior under the variety of parameters to which they may be subjected in applications. Considerable progress has indeed been made in this regard and is expected to continue in light of the continuing advances in measurement and computational capabilities. However, it should be recognized that the lack of adequate understanding of the mechanisms involved severely inhibits the capacity for modeling, and therefore, for computation. Hence, the necessity, at times, for experiments that are accompanied and followed by analytical activities for purposes of confirmation.

Pool boiling experiments were conducted on five space shuttle flights as part of the NASA Get Away Special (GAS) program during the period 1992–1996. Detailed descriptions of the hardware, procedures and general results are given elsewhere.[1,2] A brief description of certain unexpected behaviors encountered is repeated below for convenience, as well as a description of the hardware in the following section. The objective of this paper is to present some initial quantitative results for one of these unexpected behaviors obtained from measurements to date. Needless to say, were some of these unexpected behaviors "expected", the design of the experiments and the corresponding instrumentation might have been quite different.

Two identical facilities resulted from the development of the experiment, a prototype and a flight version. Each flight experiment consisted of three levels of input heat flux and three levels of initial liquid subcooling, for a total of nine different test runs per flight. The nominal variable parameters associated with each flight are given in TABLE 1. These were identical in the first three flights except for minor changes in timing of the camera speeds. These three flights are of special significance in that any questions as to the reproducibility of the unexpected behaviors are answered. The first and third are identical experiments with identical hardware; the first and second (or second and third) are identical experiments with different hardware of identical construction. For the fourth and fifth flights, an opportunity was taken to increase the levels of subcooling and to decrease the imposed input heat flux, respectively. Viewing the first three flights as one, one experiment in each flight employed the same

TABLE 1. Nominal variable parameters used for pool boiling experiments

Shuttle Flight	Initial Bulk Liquid Subcooling (°C)	Imposed Heat Flux to Film Heater (W/cm^2)	Hardware
STS-47	11.1, 2.8, 0.3	8, 4, 2	prototype
STS-57	11.1, 2.8, 0.3	8, 4, 2	flight
STS-60	11.1, 2.8, 0.3	8, 4, 2	prototype
STS-72	22.2, 16.7, 11.1	8, 4, 2	flight
STS-77	11.1, 2.8, 0.3	2, 1, 0.5	flight

parameters as used in the third, fourth, and fifth flights in order to confirm reproducibility within the last two flights.

In 26 of the total of 45 experiments (9 × 5), conditions of what can be termed steady-state pool boiling were surprisingly achieved during the two minutes generally allocated for each experiment. In all of these boiling was associated with the formation, growth, and motion of relatively small vapor bubbles, giving rise to average heat transfer coefficients often greater than values obtained in Earth gravity, as presented elsewhere.[3,4] In the remaining 19 of the 45 experiments, a continuous temperature rise of the heater surface occurred resulting from dryout. Dryout of the heater surface took place in all cases with the highest input heat flux of $q'' = 8\,\text{W/cm}^2$, except when the largest bulk liquid subcooling, 22.2°C, was applied during the STS-72 flight. As described by Lee et al.,[4] highly effective steady-state boiling takes place when a large vapor bubble slightly removed from the heater surface is present to act as a reservoir for the small bubbles growing beneath, thereby inhibiting the onset of dryout. This behavior took place with a combination of moderate heat flux levels, 4 and $2\,\text{W/cm}^2$, and subcoolings below 16.7°C (TABLE 1). Increasing the subcooling to 22.2°C eliminated the formation of a large vapor bubble, but still produced highly effective steady boiling because of thermocapillary effects,[5] with a technical basis presented most recently by Betz and Straub[6] and by Sides.[7]

According to the theories of thermocapillarity, the large vapor bubble described above, which acts as a reservoir to produce steady-state boiling, should be impelled toward the heated surface, resulting in dryout and unsteady behavior. The mechanistic elements that are believed to inhibit this are presented in the following sections, together with order-of-magnitude estimates of the forces involved, as determined from the space experiments.

A description of two additional interesting and unexpected behaviors arising with pool boiling in microgravity are briefly repeated here. The first is what is termed *quasihomogeneous* nucleation, in which slow heating produces nucleation that appears not to take place on the heating surface itself. For all experiments with input heat flux levels of $q'' = 4\,\text{W/cm}^2$ and below shown in TABLE 1 (a total of 33 of the 45 experiments), initial nucleation occurred at various and apparently random locations over the heating surface, and thus could not be identified with a geometric characteristic of the surface itself. The theoretical basis for the influence of system pressure, used to produce the bulk liquid subcooling, and these low levels of heat flux, were accounted for by a modification of classical homogenous nucleation theory, with the details presented elsewhere.[8] What remains to be done to remove the *quasi* term from the title above is a definitive observation of nucleation indeed taking place away from the heater surface. For the remaining 12 (of 45) experiments in TABLE 1 conducted with an input heat flux of $q'' = 8\,\text{W/cm}^2$, nucleation invariably took place at the same physical location on the heater surface, differing only between the two sets of hardware used, and thus was characterized as heterogeneous nucleation. A physical (and theoretical) explanation for the role of the input heat flux in producing these two different (apparently) types of nucleation in microgravity remains to be determined.

The second unexpected behavior here is the extremely dynamic or "explosive" type of vapor bubble growth at low levels of heat flux, generally following the quasi-homogeneous nucleation described above. The standard macroscopic conservation and transport equations successfully describe the vapor bubble growth rates associated with heterogeneous nucleation.[9,10] Dynamic vapor bubble growth, on the other hand, is characterized by liquid–vapor interfaces that are wrinkled and corrugated, leading to the conclusion that some type of instability mechanism is acting. A theoretical basis for this is given elsewhere,[11] although the dynamics of vapor bubble growth under these conditions remain to be modeled.

DESCRIPTION OF EXPERIMENTAL HARDWARE

As can be seen in FIGURES 1 and 2, two heater surfaces are placed on a single flat substrate, with one acting as a backup, installed so as to form one wall of the test vessel having internal dimensions 15.2 cm diameter by 10.2 cm height.

Each heater consists of a 400 Å thick semitransparent gold film sputtered on a highly polished quartz substrate, and serves simultaneously as a heater, with an uncertainty of ±2% in the heat flux, and a resistance thermometer, with an overall uncertainty of ±1.0°C in the mean surface temperature. The heater is rectangular, 19.05 × 38.1 mm (0.75 × 1.5 inch). System subcooling is obtained by increasing the system pressure above the saturation pressure and is controlled and measured with an uncertainty of ±0.345 kPa. Degassed commercial grade R-113 was used because of its electrical nonconductivity, compatible for direct contact with the thin gold film heater.

In light of the propensity for the R-113 to absorb large amounts of gases, considerable efforts were expended to reduce the dissolved gases to the lowest practical level, thereby minimizing the possibility for thermocapillary convection around bubbles

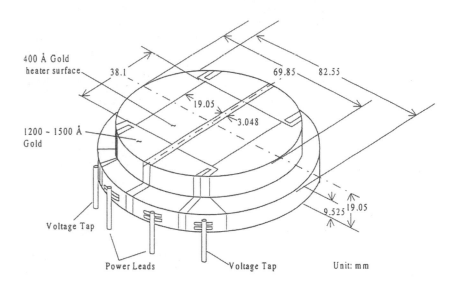

FIGURE 1. Transparent gold film heater/resistance thermometer on quartz substrate.

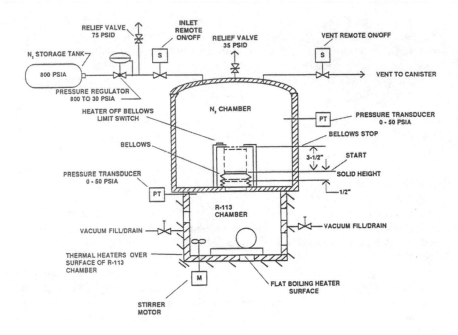

FIGURE 2. Schematic of test vessel.

containing mixtures of gas and vapor.[6] The R-113 was purified and degassed by a double distillation process, a molecular sieve and freezing on a highly convoluted surface at liquid nitrogen temperature levels, while continuously pumping out at a low pressure. The degassing was deemed adequate when the measured vapor pressure conformed to the equation of Mastroianni *et al.,*[12] to within the accuracy of the standard laboratory instruments ($\pm0.06°C$ and $\pm170\,Pa$) over the temperature range used.

Photographs of the boiling process were obtained simultaneously from the side and from beneath the heater surface at framing rates of 10 and 100 fps, with a 16-mm cine camera. An example is given in FIGURE 3.

ANALYSIS

FIGURE 4 is a schematic of the representative photograph in FIGURE 3 in which a relatively large vapor bubble is maintained off of the heater surface, in opposition to the thermocapillary forces that otherwise would move this bubble toward the heated surface, producing dryout as described in the previous section. This opposing force is ascribed to the momentum of small bubbles nucleating and growing beneath large bubbles, arising from the conversion of surface energy to kinetic energy as the small bubbles combine with the large one. Holding the large bubbles away from the heater surface permits liquid inflow beneath, producing the resulting steady nucleate boiling observed.

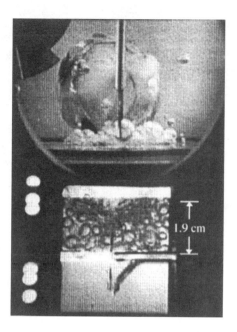

FIGURE 3. Representative photograph of steady nucleate pool boiling in microgravity: STS-60 Run #2; $q'' = 4\,\text{W/cm}^2$, $\Delta T_{\text{sub}} = 11.1°\text{C}$; frame #1329; time, 51.23 sec.

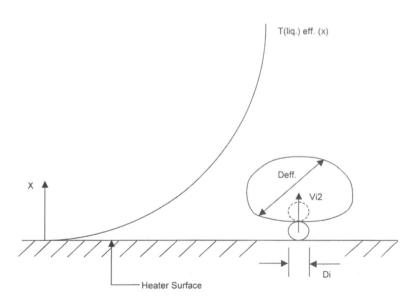

FIGURE 4. Model for sustaining a large vapor bubble away from heater surface in microgravity to counter thermocapillarity.

From measurements of the size of the small bubbles as they combine with the larger ones, and the frequency of this process, it is possible to estimate the order of magnitude of the forces involved, which at some future time might be compared with estimates of the associated thermocapillary forces.

For a vapor bubble of diameter D_i at the moment of combination with a large bubble, at which point it "disappears" from the field of view, the surface energy per bubble is given by

$$E_i = \sigma_i \pi D_i^2. \tag{1}$$

Assuming that this energy is converted completely (ideally) to kinetic energy when the two bubbles combine, then

$$KE_i = \frac{1}{2} m_{b_i} V_i^2 = \frac{1}{2}\left(\rho_{vi} \times \frac{1}{6}\pi D_i^3\right) V_i^2 = \sigma_i \times \pi D_i^2 \tag{2}$$

or

$$V_i^2 = \frac{12\sigma_i}{\rho_{v_i} D_i} \equiv \overline{V}_i^2. \tag{3}$$

The force contribution from bubbles of size i, neglecting any initial velocity, is

$$F_i = \frac{d}{dt}(m_i V_i) = \dot{m}_i \overline{V}_i. \tag{4}$$

The total force for all bubble sizes is

$$F_i = \sum_{i=1}^{n_t} F_i = \sum_{i=1}^{n_t} \dot{m}_i \overline{V}_i, \tag{5}$$

where n_i is the total number of bubble sizes considered. The mass flow rate of bubbles of size D_i is

$$\dot{m}_i = \overline{f}_i \times (\text{mass/bubble}) = \overline{f}_i \times \left(\rho_{v_i} \times \frac{\pi D_i^3}{6}\right), \tag{6}$$

where $\overline{f}_i$ is the mean bubble frequency of size i. Substituting Equations (3) and (6) into (5),

$$F_i = \pi \sum_{i=1}^{n_t} \overline{f}_i \times \left(\frac{\sigma_i \times \rho_{v_i}}{3}\right)^{1/2} \times D_i^{5/2}. \tag{7}$$

It was noted that the vapor bubble generation rate at any particular physical location on the heater surface was never faster than the film framing rate used here, 10pps, which means that all "disappearing" bubbles of size D_i could be observed and counted with reasonable certainty. That the bubble generation rate is this low in microgravity appears reasonable, since these bubbles are between the large *sink* bubble and the heating surface, where the liquid is unlikely to be significantly subcooled. This was confirmed by liquid thermistor temperature measurements 1mm above the center of the heater.

It was found to be adequate to count the disappearing bubbles within only four size ranges i in each frame, with diameters between 2–4mm, 4–6mm, 6–8mm, and 8–10mm and then use the mean diameter within each of these ranges in Equation (7). The mean frequencies of each of these four size ranges were determined by averaging each over ±1 second (±10 frames) for each time of interest, and then advancing one frame at a time over the entire period of interest. Computation of the time-varying forces induced by the disappearing bubbles on the large hovering bubbles were then carried out using (7).

RESULTS

The variation of the mean heater surface temperature and the derived heat transfer coefficient are shown in FIGURE 5 for the entire experiment period of two minutes for STS-47 Run # 9. This plot is taken directly from Merte *et al.*[1] Following the steady nucleate boiling up to 80 seconds the large vapor bubble resulting from the virtually saturated liquid condition contacted the opposite side of the test vessel, producing the subsequent dryout and associated heater surface temperature rise.

The lift-off force induced by the disappearing or engulfed bubbles, computed from (7), is shown in FIGURE 6 for the steady 70–80 second time interval. Note that this varies over approximately 3.8–5.6 N, which can be considered essentially constant if some smoothing of the small-scale oscillations takes place.

The frequencies of the engulfed bubbles corresponding to the four size ranges measured are plotted in FIGURE 7. On comparing the force contributions of each of these size ranges to the total shown in FIGURE 6, it becomes obvious that the major contributions come from bubbles in the two smaller size ranges, 2–4 mm and 4–6 mm. This was also true for the majority of the additional experiments whose bubble measurements have been completed, and for which only the results corresponding to FIGURES 5 and 6 are given below.

The four pairs of FIGURES 5 and 6, 8 and 9, 10 and 11, and 12 and 13, all have the same input level of heat flux, with the latter three differing from that of FIGURES 5 and 6 in having a higher subcooling level of 3°C (vice 0.3°C). We restate here that in comparing the resulting behaviors and their differences, FIGURES 5, 8, and 12 and the corresponding FIGURES 6, 9, and 13 were obtained with the *prototype* experimental hardware, whereas FIGURES 10 and 11 were obtained with the identically

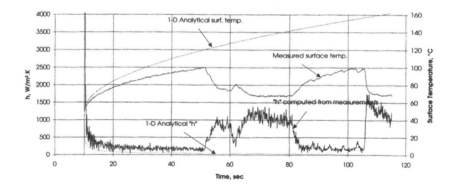

FIGURE 5. Mean heater surface temperature and heat transfer coefficient: STS-47 Run #9, $q'' = 2\,\mathrm{W/cm^2}$, $\Delta T_{sub} = 0.3°C$.

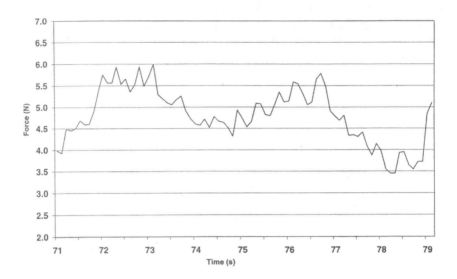

FIGURE 6. Lift-off force induced by engulfed bubbles: STS-47 Run #9, $q'' = 2\,\mathrm{W/cm^2}$, $\Delta T_{sub} = 0.3°C$.

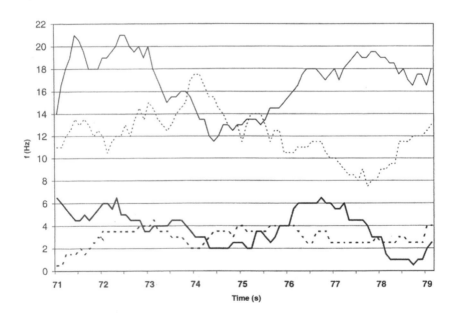

FIGURE 7. Frequency of the engulfed bubbles in FIGURE 6 for the four size ranges measured: ——, 2–4 mm; ······, 4–6 mm; ——, 6–8 mm; · - · , 8–10 mm; STS-47 Run #9.

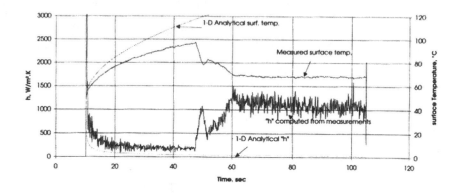

FIGURE 8. Mean heater surface temperature and heat transfer coefficient: STS-47 Run #6, $q'' = 2\,\mathrm{W/cm}^2$, $\Delta T_{\mathrm{sub}} = 3°\mathrm{C}$.

designed *flight* experimental hardware. Thus, on comparing FIGURES 8 and 9 with the corresponding FIGURES 12 and 13, the reproducibility is excellent for the initial nucleation heater surface temperature (and the related time to the onset of nucleation), the subsequent steady heater surface temperature, and the derived mean heat transfer coefficient. Comparisons between the liftoff forces, varying over 4.0–1.8 N in FIGURE 9 and 2.5–0.8 N in FIGURE 13 suggest that they are reasonably reproducible, when viewing the dynamic and somewhat chaotic motions of the larger *hovering* bubbles.

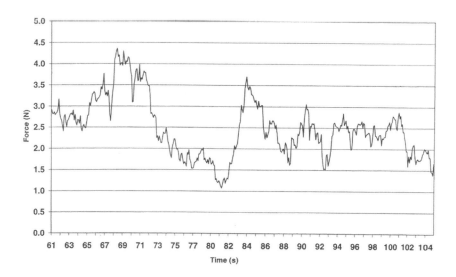

FIGURE 9. Liftoff force induced by engulfed bubbles: STS-47 Run #6, $q'' = 2\,\mathrm{W/cm}^2$, $\Delta T_{\mathrm{sub}} = 3°\mathrm{C}$.

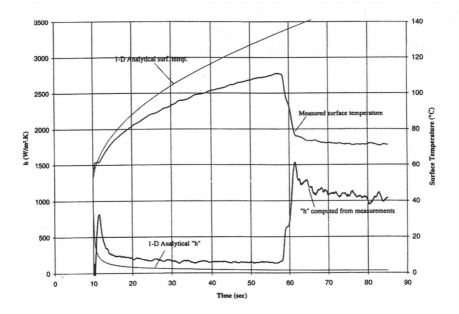

FIGURE 10. Mean heater surface temperature and heat transfer coefficient: STS-57 Run #6, $q'' = 2\,\text{W/cm}^2$, $\Delta T_{\text{sub}} = 3°\text{C}$.

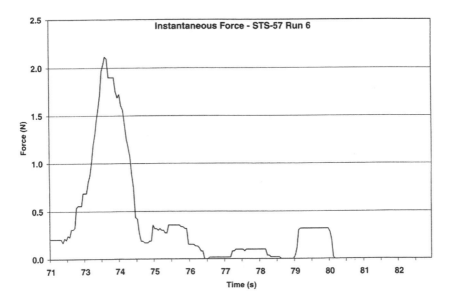

FIGURE 11. Lift-off force induced by engulfed bubbles: STS-57 Run #6, $q'' = 2\,\text{W/cm}^2$, $\Delta T_{\text{sub}} = 3°\text{C}$.

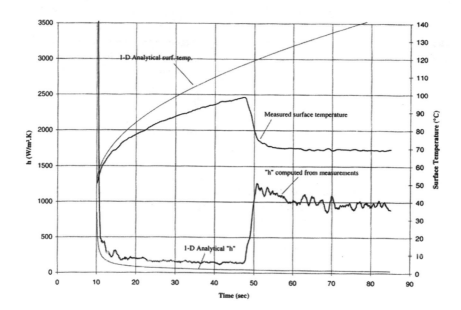

FIGURE 12. Mean heater surface temperature and heat transfer coefficient: STS-60 Run #6, $q'' = 2\,\mathrm{W/cm}^2$, $\Delta T_{sub} = 3°C$.

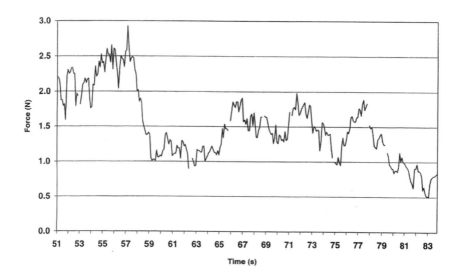

FIGURE 13. Lift-off force induced by engulfed bubbles: STS-60 Run #6, $q'' = 2\,\mathrm{W/cm}^2$, $\Delta T_{sub} = 3°C$.

This is to be contrasted with the larger variability in FIGURE 11, with the *flight* hardware. This difference is associated with the longer time to nucleation in FIGURE 10 and the corresponding higher heater surface superheat at nucleation. This leads to a much more dynamic initial vapor bubble growth which, when combined with the resulting shorter experiment time remaining, is felt to require a longer time period for the violent interfacial instabilities to be dissipated.

As stated already, the variable parameters in the five space shuttle flights consisted of the imposed heater heat flux and the initial bulk liquid subcooling. It is instructive to examine the influence of these on the lift-off forces induced by the engulfed bubbles.

FIGURE 14 presents the mean heater surface temperature and heat transfer coefficient and FIGURE 15 shows the lift-off forces induced by the engulfed bubbles for STS-72 Run #1, with an imposed heat flux $q'' = 8\,W/cm^2$ and initial subcooling $\Delta T_{sub} = 22.2°C$. This is the only experiment among the 15 conducted at this highest heat flux level that produced a large hovering vapor bubble during the allotted experiment time. Because of the high initial liquid subcooling here, it took a considerable period of time for a large vapor bubble to become stable, until the subcooling around this bubble in the vicinity of the heater surface had decreased sufficiently because of condensation. Prior to this, highly subcooled nucleate boiling with small bubbles took place in the immediate vicinity of the heater surface. Of course, with a sufficiently long experimental time available it can be expected that condensation will reduce the bulk liquid subcooling, and the behavior would revert to that observed for lower subcooling levels, to be shown in the following figures. This would occur provided that dryout did not take place, which requires that the vessel be large enough

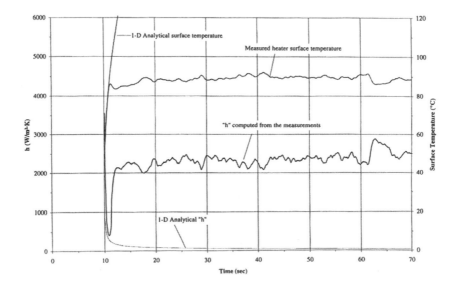

FIGURE 14. Mean heater surface temperature and heat transfer coefficient: STS-72 Run #1, $q'' = 8\,W/cm^2$, $\Delta T_{sub} = 22.2°C$.

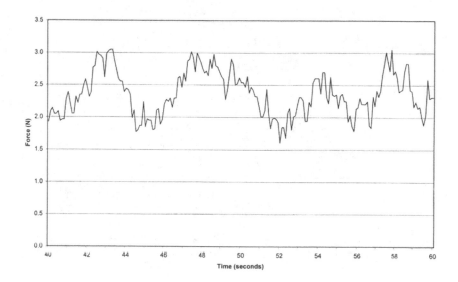

FIGURE 15. Liftoff force induced by engulfed bubbles: STS-72 Run #1, $q'' = 8\,\text{W/cm}^2$, $\Delta T_{\text{sub}} = 22.2\,°\text{C}$.

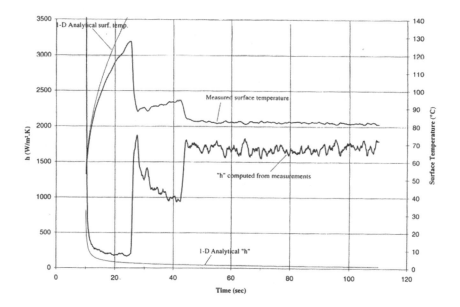

FIGURE 16. Mean heater surface temperature and heat transfer coefficient: STS-57 Run #2, $q'' = 4\,\text{W/cm}^2$, $\Delta T_{\text{sub}} = 11.1\,°\text{C}$.

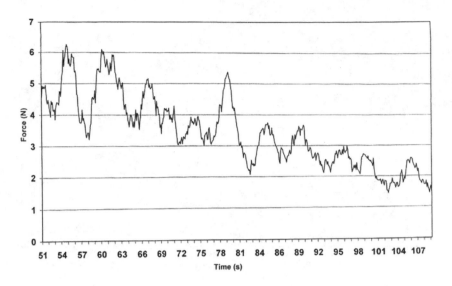

FIGURE 17. Liftoff force induced by engulfed bubbles: STS-57 Run #2, $q'' = 4\,\mathrm{W/cm}^2$, $\Delta T_{\mathrm{sub}} = 11.1^\circ\mathrm{C}$.

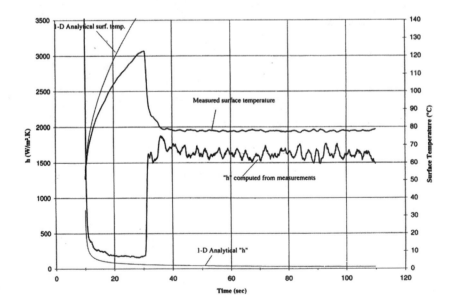

FIGURE 18. Mean heater surface temperature and heat transfer coefficient: STS-60 Run #2, $q'' = 2\,\mathrm{W/cm}^2$, $\Delta T_{\mathrm{sub}} = 11.1^\circ\mathrm{C}$.

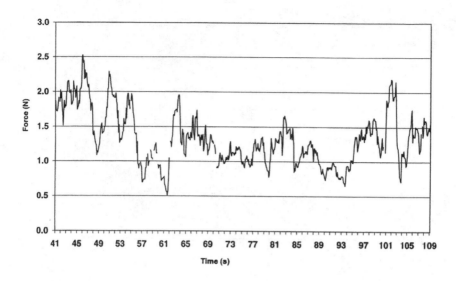

FIGURE 19. Liftoff force induced by engulfed bubbles: STS-60 Run #2, $q'' = 4\,\text{W/cm}^2$, $\Delta T_{\text{sub}} = 11.1°\text{C}$.

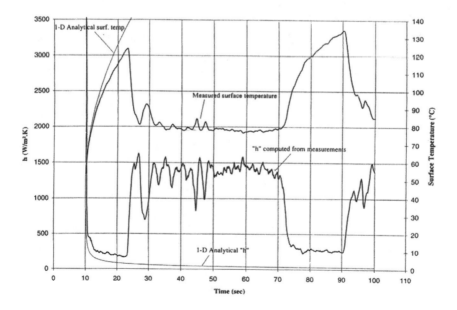

FIGURE 20. Mean heater surface temperature and heat transfer coefficient: STS-57 Run #5, $q'' = 4\,\text{W/cm}^2$, $\Delta T_{\text{sub}} = 3°\text{C}$.

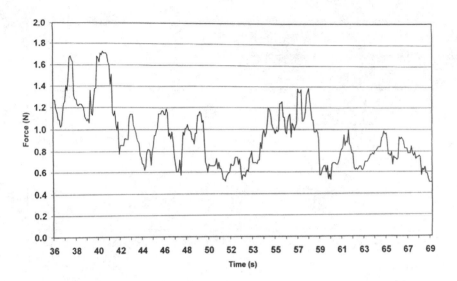

FIGURE 21. Liftoff force induced by engulfed bubbles: STS-57 Run #5, $q'' = 4\,W/cm^2$, $\Delta T_{sub} = 3°C$.

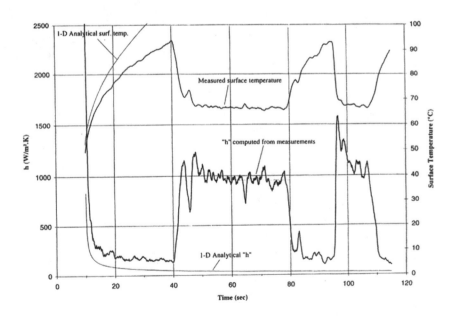

FIGURE 22. Mean heater surface temperature and heat transfer coefficient: STS-60 Run #9, $q'' = 2\,W/cm^2$, $\Delta T_{sub} = 0.3°C$.

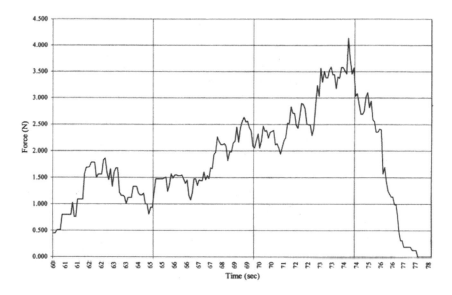

FIGURE 23. Liftoff force induced by engulfed bubbles: STS-60 Run #9, $q'' = 2\,W/cm^2$, $\Delta T_{sub} = 0.3°C$.

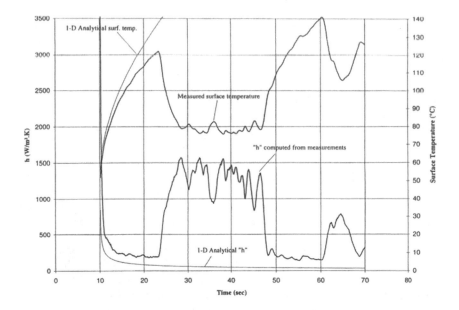

FIGURE 24. Mean heater surface temperature and heat transfer coefficient: STS-57 Run #8, $q'' = 4\,W/cm^2$, $\Delta T_{sub} = 0.3°C$.

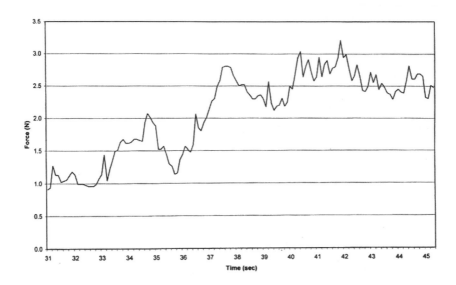

FIGURE 25. Liftoff force induced by engulfed bubbles: STS-57 Run #8, $q'' = 4\,\mathrm{W/cm}^2$, $\Delta T_{sub} = 0.3°\mathrm{C}$.

to prevent bubble contact with the opposing side. It is noted in FIGURE 15 that the liftoff force is virtually constant up to the maximum time shown, when the bulk liquid stirrer was turned on, producing the subsequent behavior seen in FIGURE 14.

The behavior of the experiments at the next lower heat flux level, $q'' = 4\,\mathrm{W/cm}^2$, are shown in FIGURES 16–19 for a subcooling $\Delta T_{sub} = 11.1°\mathrm{C}$ and in FIGURES 20 and 21 for a subcooling $\Delta T_{sub} = 3°\mathrm{C}$. It is noted that, in FIGURES 17, 19, and 21, for all three cases, the induced liftoff forces distinctly decrease with time, as the subcooling level decreases. This was also the behavior at the lower heat flux level, $q'' = 2\,\mathrm{W/cm}^2$ and subcooling $\Delta T_{sub} = 3°\mathrm{C}$ in FIGURES 9 and 13. The exceptional behavior in FIGURE 11 for these parameters is discussed above.

Where the bulk liquid is initially almost at saturation, a quite different behavior takes place. Here the induced lift-off force either remains almost constant, as in FIGURE 6 ($q'' = 2\,\mathrm{W/cm}^2$, $\Delta T_{sub} = 0.3°\mathrm{C}$), or increases with time, as in FIGURE 23 ($q'' = 2\,\mathrm{W/cm}^2$, $\Delta T_{sub} = 0.3°\mathrm{C}$) and in FIGURE 25 ($q'' = 4\,\mathrm{W/cm}^2$, $\Delta T_{sub} = 0.3°\mathrm{C}$).

In interpreting the significance of the levels of the liftoff forces presented above, it is important to keep in mind the assumption made in the development of the analysis by which the forces were computed from the measurements: below Equation **(1)** it was assumed that the surface energy of each *engulfed* or combined small vapor bubble is completely (ideally) converted to kinetic energy. General experience indicates that some efficiency factor should be applied, which at this time is highly subjective, perhaps ranging from 1–50%. It might be anticipated that a considerable expenditure of appropriate experiment effort (accompanied by analysis) will be needed to reduce this uncertainty. On physical and geometric grounds this *efficiency* could be expected to depend on the size of the engulfed bubble, a larger value for smaller bubbles.

CONCLUSION

Among the total of 45 experiments conducted of pool boiling on a flat heater surface in the long-term microgravity of space, 26 somewhat unexpectedly resulted in steady-state pool boiling. Examination of 16-mm motion films showed that 9 of the 26 experiments were associated with the relatively high subcooling levels in which the resulting small vapor bubbles remained in the vicinity of the heater surface, with motion parallel to the surface caused by thermocapillary effects.

In the remaining 17 of 26 experiments, highly effective steady-state boiling took place when a large vapor bubble, slightly removed from the heater surface, was present to act as a reservoir for the small bubbles growing beneath. The momentum transfer of these small bubbles to the large bubble as they combine is believed to provide the force necessary to counter the thermocapillary force that otherwise would move the large bubble toward the heater surface, resulting in dryout. Measurements of the size and frequency of the small bubbles as they are *engulfed* by the large one provide estimates of the so called liftoff force that counters that of thermocapillarity. The maximum possible levels of this force have been determined to be on the order of 0.8–6.0 N for the experimental parameters used.

It was observed that where the initial bulk liquid was near saturation, the total liftoff force tended to increase with time, whereas for the initially subcooled bulk liquid case, the force tended to decrease with time.

ACKNOWLEDGMENTS

The results presented here were obtained using films produced under NASA Contract NAS 3-25812. The assistance of NASA and Dr. Francis P. Chiaramonte, the Project Scientist at the time, are gratefully acknowledged. The author thanks the students, Mr. Eric Preiditsch, Ms. Valerie Toth, and Mr. Shawn Burgdorf for the, at times, tedious task of measurements from the films.

REFERENCES

1. MERTE, H., JR., H.S. LEE & R.B. KELLER. 1995. Report on pool boiling experiment flown on STS-47 (PBE IA), STS-57 (PBE-IB), STS-60 (PBE-IC). NASA Contract NAS 3-25812, Report No. UM-MEAM-95-01, Department of Mechanical Engineering and Applied Mechanics, University of Michigan, Ann Arbor, Michigan.
2. MERTE, H., JR., H.S. LEE & R.B. KELLER. 1998. Dryout and rewetting in the pool boiling experiment flown on STS-72 (PBE-IIB), STS-77 (PBE-IIA). Final Report, NASA Grant NAG-1684, Report No. UM-MEAM-98-01, Department of Mechanical Engineering and Applied Mechanics, University of Michigan, Ann Arbor, Michigan.
3. MERTE, H., JR., H.S. LEE & J.S. ERVIN. 1994. Transient nucleate pool boiling in microgravity—some initial results. Micrograv. Sci. Techn. **VII/2:** 173–179.
4. LEE, H.S., H. MERTE, JR. & F. CHIARAMONTE. 1997. Pool boiling curve in microgravity. J. Thermophys. Heat Transf. **11:** 216–222.
5. LEE, H.S., H. MERTE, JR. & F. CHIARAMONTE. 1998. Pool boiling phenomena in microgravity. Heat Transfer 1998. Proceedings of 11th IHTC, Vol. 2, August 23–28, 1998, Kyongju, Korea. 395–399.
6. BETZ, J. & J. STRAUB. 2002. Thermocapillary convection around gas bubbles. Ann. N.Y. Acad. Sci. **974:** 220–245.

7. SIDES, P.J. 2002. A thermocapillary mechanism for lateral motion of bubbles on a heated surface during subcooled nucleate boiling. J. Heat Transf. **124:** 1203–1206.
8. MERTE, H., JR. & H.S. LEE. 1997. Quasi-homogenous nucleation in microgravity at low heat flux: experiments and theory. J. Heat Transf. **119:** 305–312.
9. LEE, H.S. & H. MERTE. 1996. Spherical bubble growth in uniformly superheated liquids. Int. J. Heat Mass Transf. **39:** 2427–2447.
10. LEE, H.S. & H. MERTE. 1996. Hemispherical vapor bubble growth in microgravity: experiments and model. Int. J. Heat Mass Transf. **39:** 2449–2461.
11. LEE, H.S. & H. MERTE, JR. 1998. The origin of the dynamic growth of vapor bubbles related to vapor explosions. J. Heat Transf. **120:** 174–182.
12. MASTROIANNI, M.J., R.F. STAHL & P.N. SHELDON. 1978. Physical and thermodynamic properties of 1,1,2-trifluorotrichloroethane (R-113). J. Chem. Eng. Data **23:** 113–118.

Development of a High-Performance Boiling Heat Exchanger by Improved Liquid Supply to Narrow Channels

HARUHIKO OHTA, TOSHIYUKI OHNO,
FUMIAKI HIOKI, AND YASUHISA SHINMOTO

Department of Aeronautics and Astronautics, Kyushu University,
Hakozaki, Higashi-ku, Fukuoka, Japan

ABSTRACT: A two-phase flow loop is a promising method for application to thermal management systems for large-scale space platforms handling large amounts of energy. Boiling heat transfer reduces the size and weight of cold plates. The transportation of latent heat reduces the mass flow rate of working fluid and pump power. To develop compact heat exchangers for the removal of waste heat from electronic devices with high heat generation density, experiments on a method to increase the critical heat flux for a narrow heated channel between parallel heated and unheated plates were conducted. Fine grooves are machined on the heating surface in a transverse direction to the flow and liquid is supplied underneath flattened bubbles by the capillary pressure difference from auxiliary liquid channels separated by porous metal plates from the main heated channel. The critical heat flux values for the present heated channel structure are more than twice those for a flat surface at gap sizes 2 mm and 0.7 mm. The validity of the present structure with auxiliary liquid channels is confirmed by experiments in which the liquid supply to the grooves is interrupted. The increment in the critical heat flux compared to those for a flat surface takes a maximum value at a certain flow rate of liquid supply to the heated channel. The increment is expected to become larger when the length of the heated channel is increased and/or the gravity level is reduced.

KEYWORDS: microgravity; flow boiling; cold plate; narrow channel; critical heat flux

NOMENCLATURE:

g/g_e	ratio of gravity to the normal gravity
P	pressure, Pa
q_{CHF}	critical heat flux at local surface position, W/m^2
$q_{CHF,o}$	critical heat flux evaluated from power input, W/m^2
q	heat flux at local surface position, W/m^2
q_0	heat flux evaluated from power input, W/m^2
s	gap size, mm
u_{in}	inlet velocity, m/sec
x	distance from center of heating surface, mm
x_1	concentration of more volatile component
y	distance from bottom of heating block, mm
T_i	interfacial temperature, K

Address for correspondence: Haruhiko Ohta, Dept. Aeronautics and Astronautics, Kyushu University, 6-10-1 Hakozaki, Higashi-ku, Fukuoka 812-8581, Japan. Voice: 81 92 642 3489; fax: 81 92 642 3752.

ohta@aero.kyushu-u.ac.jp

Ann. N.Y. Acad. Sci. **1027**: 217–234 (2004). ©2004 New York Academy of Sciences.
doi: 10.1196/annals.1324.019

T_{bulk} saturation temperature at bulk concentration, K
Greek Symbols
$\alpha_{local,M}$ local value heat transfer coefficient in center of heating surface, W/m^2K
$\Delta T_{sub,in}$ inlet liquid subcooling, K
σ surface tension, N/m
τ time, sec

INTRODUCTION

In recent years, heat generation density of small semiconductor tips has markedly increased, and air cooling is usually applied with the aid of heat spreaders. It is estimated that during the next decade the power density will increase tenfold or hundredfold beyond that required by existing devices. Hence, a delay in the development of cooling technologies could prevent further advance of electronics. In most existing research, however, cooling of a small surface area is the goal, and only the value of heat flux attainable seems to be emphasized. On the other hand, no innovative technology has been developed for the cooling of large-scale semiconductors, referred to as power electronics, despite their many practical applications replacing conventional devices. Incidentally, to promote energy saving and prevent environmental destruction due to CO_2, serious efforts are being made to develop new energy resources. The application of power electronics, for example, to electric conversion systems in power stations would improve their efficiency, assisted by the development of SiC semiconductors, and thus contribute significantly to the resolution of these problems. In the cooling of large-scale semiconductors, problems of heat generation from large areas (large total power) and of thermal energy transportation over a long distance to the cooling section need to be investigated in addition to situations involving high heat flux conditions.

This situation is also true for space applications. The enlarged size of new space platforms requires an increase in both the thermal power (beyond 100 kW) and in the distance of thermal energy transport (to more than 100 m). These requirements are beyond the range for application of existing passive methods, including those of heat pipes, consequently, the application of a pumped loop forming the core of a thermal management system is being extensively investigated. There are two types of pumped loop for transporting thermal energy from cold plates to radiators; namely, single-phase fluid loops, based on forced convection of liquid, and two-phase fluid loops, that handle boiling and two-phase flow. The latter type is subdivided into mechanical pumped loops, capillary pumped loops that use the interfacial pressure difference in, for example, the porous material, and hybrid loops in which these driving forces can be switched depending on thermal load.

High-performance heat exchange in both boiling on a cold plate and condensation in a radiator offers an advantage to two-phase flow loops. Furthermore, because of latent heat transport, lower liquid flow rate reduces the mechanical pump power and also the liquid inventory, which is directly related to the launch weight, as long as the same amount of thermal energy is concerned. In addition, the rate of thermal energy transportation using the two-phase fluid loop can easily be adjusted by simply changing the system pressure via accumulators in response to thermal load varying with power generation from solar cells in orbit, consumption by instrument operation, and battery charging. In the International Space Station (ISS) to be constructed soon, however, only Russian module seems to adopt a two-phase fluid loop as a thermal management system. It was abandoned and single-phase fluid loops are to be introduced in the modules by other agencies, including the Japanese one (JEM). This is because of insufficient knowledge about boiling and the two-phase flow under microgravity conditions that are needed for the design of two-phase fluid loops. In particular, there are almost no data for CHF conditions for the fundamental systems of flow boiling in microgravity, despite its significance to safe operation in the case of unexpected thermal power excursion. An outline of experimental results on boiling in microgravity and additional subjects to be investigated are summarized in reviews elsewhere.[1–4]

In our research, flow boiling heat transfer in narrow gaps is investigated so as to develop high-performance compact cold plates as a component of a two-phase fluid loop. This paper describes the results of experiments in attempting to increase CHF for flow boiling in narrow gaps by the supply of liquid from a direction transverse to the flow by means of capillary force. The scope for the methods to additionally increase CHF and for application to space technology recently became a topic of interest to us.

FLOW BOILING IN NARROW CHANNELS
UNDER MICROGRAVITY CONDITIONS

A narrow heated channel is one of the ideal configurations for heat exchangers because it has larger heat transfer area for a given duct cross sectional area. Research on the boiling heat transfer in narrow gaps has clarified the heat transfer characteristics required for various experimental systems.[5–7] An increase in critical heat flux (CHF) to avoid burnout phenomena is one of goals for this system.

For a system immersed in a liquid pool, heat transfer coefficients increase beyond those for pool boiling with a decrease in gap size under constant heat flux conditions; however, further decrease of gap size results in heat transfer deterioration. The critical heat flux decreases monotonically with decreasing in gap size.[7] The mechanism of a system immersed in the pool is quite ambiguous because the inlet liquid velocity of induced flow in a pool is a function of heat flux supplied to the heating surface. The system of flow boiling is required for a strict definition of inlet liquid conditions.

Under microgravity conditions, there are a few investigations that focus on boiling in narrow channels,[8,9] including those with a variation in the direction of gravity. Three typical bubble behaviors with various gravity effects were confirmed for narrow channels that are vertically oriented,[9] and in which glass heating surfaces are employed to relate the heat transfer data directly to the observed liquid–vapor behavior.[10] On the heat transfer surface, temperature sensors are directly coated to evaluate the distributions of surface temperature and heat flux. Transparent ITO films are attached to the back of the heat transfer side, and uniform heat flux is supplied by Joule heating. The experimental conditions are; test liquid, distilled water; system pressure, $P = 0.093\,\mathrm{MPa}$; inlet velocity, $u_{\mathrm{in}} = 0.06\,\mathrm{m/sec}$; inlet liquid subcooling, $DT_{\mathrm{sub,in}} = 0$, $10\,\mathrm{K}$; gap size, $s = 0.7$–$10\,\mathrm{mm}$; average heat flux, $q_0 = 4\times10^4$–3×10^5.

The heat transfer area is 40mm wide and 70mm long. The opposite surface in the duct is kept unheated. Gravity levels are varied in parabolic flight experiments.

In FIGURE 1, pictures of bubble behaviors at normal (1 g) and microgravity (μ g) and the transition of heat transfer coefficients in the center of heating surface along a parabolic trajectory are shown for three cases: a small gap size and inlet liquid at near saturation state, and its variation with an extremely small gap size or higher inlet liquid subcooling. A few interesting results were obtained: (1) In narrow gaps, heat transfer is, in general, deteriorated by the reduction of buoyancy. The extension of a flattened bubble on the heating surface and then of dry patches underneath it results in unstable heat transfer behavior at high heat flux because of dryout underneath the bubbles. (2) When the gap size is decreased to an extremely small value, heat transfer deterioration is observed regardless of gravity levels, and CHF values

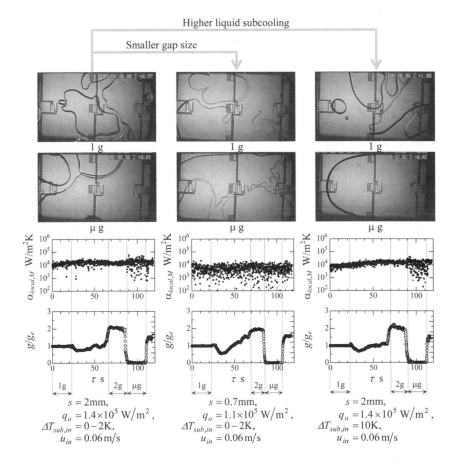

FIGURE 1. Variation of liquid–vapor behavior and heat transfer coefficient by the reduction in gravity gap size and liquid subcooling.

are smaller. The surface temperatures oscillate at high frequency within limited ranges and the averaged values are considered to be quite stable. This is because the enhanced growth rate (i.e., interfacial velocity) of flattened bubbles promotes the quick exchange of locations covered by flattened bubbles and those wetted by bulk liquid penetrating them. Dry patches that extend underneath the flattened bubbles are quenched at high frequency. (3) In microgravity, a small increase of liquid subcooling at the inlet prevents further growth of a large flattened bubble toward the upstream direction and fixed it on the heating surface for a long time, balancing the rate of evaporation with that of condensation in the upstream region. Larger extension of dry patches underneath the large flattened bubble emphasizes the deterioration of heat transfer at higher liquid subcooling.

The reduction in critical heat flux in microgravity is obvious from these results, and the development of methods to increase CHF is desired to be able to use the advantage of a narrow channel heat exchanger under safe operating conditions.

NEW STRUCTURE FOR A HEATED CHANNEL

An outline of the structure and the principle for the proposed narrow channel are shown in FIGURE 2. The channel consists of a main heated channel and auxiliary unheated liquid channels located on both sides. On the heated surface, arrays of V-shaped grooves are machined in a direction perpendicular to the flow. When heat flux is high enough, the main heated channel is filled with flattened bubbles. Dry patches in the microlayers underneath these bubbles are extended by the growth of flattened bubbles. At heat flux near CHF, the central part of the heated channel is dried by the lateral coalescence of the bubbles. The liquid meniscus in a V-shaped groove has a smaller radius near the center location than that at the side, if the same contact angle is assumed under the static liquid–vapor equilibrium. The difference in capillary pressures between the center and the side results in a pressure gradient in the liquid filling the groove, which induces the liquid flow underneath the flattened bubble from the auxiliary unheated channel to the center of the heated channel. The liquid supply in the present structure is enhanced when the heated channel is completely filled by coalesced flattened bubbles that squeeze the liquid meniscus in the grooves.

The heat transfer surface is vertically oriented and is 30 mm wide and 50 mm long in the flow direction. It occupies one end of a copper block 35 mm thick. On the opposite side of the block, electric cartridge heaters are inserted and heat fluxes up to $3.0 \times 10^5 \, \text{W/m}^2$ are applied.

Each groove has an apex angle of 90 deg and is 1 mm deep with 2 mm pitch. To compare the performance of the grooved surface, a flat copper surface without additional channels was prepared. Six thermocouples were inserted in the copper block at depths of 3, 10, and 17 mm from the heat transfer surface, at the center and side locations shown in the figure. An unheated glass plate was attached in parallel with the heating surface, remaining a constant distance of 2 mm or 0.7 mm between the surfaces to realize a narrow channel structure with a rectangular cross section. For the grooved surface, two unheated auxiliary channels were located as liquid feeders,

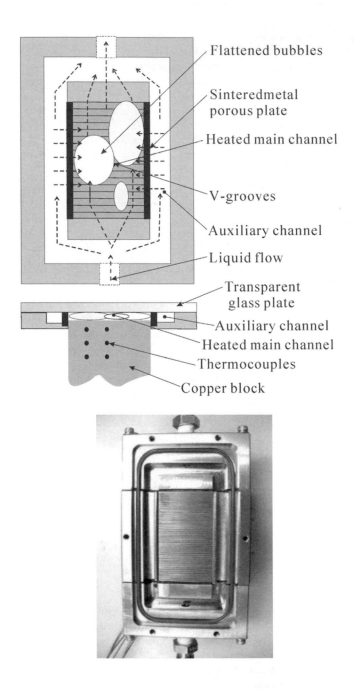

Flattened bubbles

Sinteredmetal porous plate

Heated main channel

V-grooves

Auxiliary channel

Liquid flow

Transparent glass plate

Auxiliary channel

Heated main channel

Thermocouples

Copper block

FIGURE 2. Structure of grooved surface in a narrow gap with auxiliary liquid feeders.

separated by sintered metal plates of porous structure so that bubbles do not penetrate the liquid flow in the auxiliary channels.

Test liquids R113 and FC72 with saturation temperature 47.0°C and 55.7°C, respectively, at a system pressure $P = 0.1$ MPa were employed. The experiments were conducted under the following conditions: heat flux from the electric heater, $q_0 = 8.0 \times 10^4 - 3.3 \times 10^5$ W/m^2; velocity of inlet liquid, $u_{in} = 0.032 - 0.14$ m/sec; and inlet liquid subcooling, $\Delta T_{sub,in} = 0 - 10$ K. Vertical upward flow was realized in experiments on ground.

The test liquid was heated to obtain prescribed inlet liquid subcooling by means of a preheater installed in the test loop and the temperature of inlet liquid was maintained constant until the end of data recording. Heat flux was applied in steps. After the equilibrium state was established in the temperature distribution of the heating block, the e.m.f. of the thermocouples was recoded. Under CHF conditions, transition of e.m.f. due to temperature excursion was also recorded. The heat flux distribution in the copper block was analyzed by two-dimensional transient heat conduction using the measured temperatures and appropriate boundary conditions on the surfaces of the block.

Burnout was detected by the temperature excursion resulting from the increase of heat flux in steps. Once the occurrence of burnout was confirmed, the heating was stopped to cool the heating block. The heating was started again at a heat flux one-step smaller than that resulting in burnout. The increment of heat flux was reduced during this time to determine the CHF values with more accuracy. The critical heat flux $q_{CHF,o}$ is defined as the heat input, corresponding to the heat flux just before that resulting in burnout divided by the nominal heat transfer area ignoring the increase in area due to the existence of grooves. Near CHF conditions, the final increment of heat flux input by the electric heaters was 1.0×10^4 W/m^2 for the early experiments (see FIGURES 3 and 4), whereas it is decreased later to 2.5×10^3 W/m^2 (see later, FIG. 7), which implies the resolution of measured CHF values.

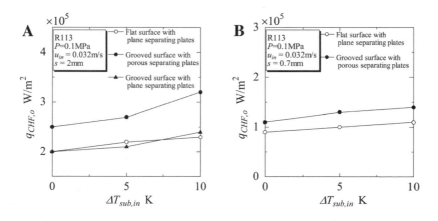

FIGURE 3. Comparison of critical heat flux values evaluated by heat input between flat and grooved surfaces (R113): (**A**) $s = 2$ mm, (**B**) $s = 0.7$ mm.

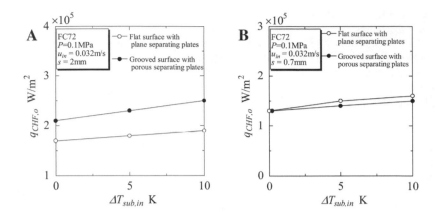

FIGURE 4. Comparison of critical heat flux evaluated by heat input between flat and grooved surfaces (FC72): **(A)** $s = 2\,mm$, **(B)** $s = 0.7\,mm$.

EXPERIMENTAL RESULTS AND DISCUSSION

FIGURE 3 shows the relation between the critical heat flux $q_{CHF,o}$ and inlet liquid subcooling $\Delta T_{sub,in}$ for R113, $s = 2$ and $0.7\,mm$, where $q_{CHF,o}$ is determined from the power input required to reduce the values by $1.0 \times 10^4\,W/m^2$ from the values resulting in temperature excursion (i.e., the heat flux increment). The following were clarified. (1) At $s = 2\,mm$, $\Delta T_{sub,in} = 0\,K$, CHF for the flat surface is $q_{CHF,o} = 2.0 \times 10^5\,W/m^2$, the same value as that calculated from the Zuber correlation for pool boiling[11] because the conflicting effects of the channel confinement and the superimposed liquid flow cancel. At $s = 0.7\,mm$, $\Delta T_{sub,in} = 0\,K$, CHF is decreased to $q_{CHF,o} = 9 \times 10^4\,W/m^2$. (2) For both cases ($s = 2$ and $0.7\,mm$), the effect of liquid subcooling is much smaller than the evaluation by Ivey and Moris.[12] From the heat balance, a saturation state is established at the locations $4.1\,mm$ and $3.0\,mm$, respectively, from the inlet, even for $\Delta T_{sub,in} = 10\,K$ under the heat flux values shown in the figure. (3) For $s = 2\,mm$, CHF values for the grooved surface are 1.3–1.4 times larger than those for the flat surfaces as a result of expected liquid supply in the proposed structure. (4) To confirm the validity of the structure with auxiliary unheated channels as liquid feeders, the porous plates on the grooved surface were replaced by plane metal plates with no holes and the supply of liquid from the auxiliary channels was prevented. The results are shown by black triangle symbols in FIGURE 3 for $s = 2\,mm$. No increase of CHF values from those for the flat surface was observed, and it is concluded that the difference in the surface structure itself cannot contribute to the increase in CHF. (5) At $s = 0.7\,mm$, CHF values for the grooved surface are 1.2–1.3 times larger than those for the flat surface. The increase in CHF for $s = 0.7\,mm$, smaller than that for $s = 2\,mm$, seems to be contradictory to the substantial increase of gap size for $s = 0.7\,mm$ due to the existence of grooves. The gap size is defined as a distance between the unheated plate and the top of the grooved surface.

FIGURE 4 shows the results for FC72. (1) At $\Delta T_{\text{sub,in}} = 0\,\text{K}$, $q_{\text{CHF,o}} = 1.7 \times 10^5\,\text{W/m}^2$ and $1.3 \times 10^5\,\text{W/m}^2$ are obtained for $s = 2\,\text{mm}$ and $0.7\,\text{mm}$, respectively, whereas the predicted value from the Zuber correlation is $1.51 \times 10^5\,\text{W/m}^2$. (2) For gap size $s = 2\,\text{mm}$, CHF on the grooved surface is increased to 1.2–1.3 times that on the flat surface, whereas the trend is reversed for $s = 0.7\,\text{mm}$ under subcooled inlet conditions, and the CHF values for the grooved surface are slightly smaller than those for the flat surface.

To clarify the reason for this contradictory trend, heat flux distribution in the copper block was calculated on the basis of the measured temperature transition. In FIGURE 5, heat flux distributions in the copper block just before CHF conditions, for $s = 0.7\,\text{mm}$, $\Delta T_{\text{sub,in}} = 0\,\text{K}$, are compared for both surfaces. Symmetric heat flux distributions are shown for 18 mm of the total 30 mm width based on a calculated half-zone of distance 9 mm between the thermocouple arrays. (1) For the flat surface, since the dryout already occurred on the center of the heating surface, the heat supplied from the bottom of the block flows toward the sides and a part of it diffuses from the block walls as heat loss before it reaches to the surface. (2) The large difference in the surface heat flux values between the center and side for the flat surface reflects directly the difference in the heat transfer under dried and wetted surface conditions, respectively. (3) For the grooved surface, almost uniform transverse distribution of surface heat flux is obtained, reflecting substantial liquid supply to the center.

In FIGURE 6, the surface heat flux q and the heat flux evaluated from power input q_0 are compared between the center ($x = 0$) and side ($x = 9\,\text{mm}$) for both surfaces for FC72, $s = 0.7\,\text{mm}$, and $DT_{\text{sub,in}} = 0\,\text{K}$. The following were clarified: (1) For the flat surface, the surface heat flux q starts to decrease when dry patches arise at the center and this trend is extended toward the side with an increase in heat flux q_0. (2) For the grooved surface, on the other hand, the values of q for both surface locations start to decrease at the same input heat flux q_0.

With reference to FIGURE 6, the critical heat flux is considered to be the maximum value of the surface heat flux q_{CHF} at the center when q_0 is varied. The values of q_{CHF}

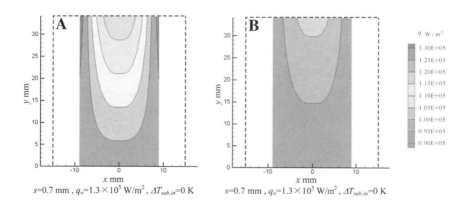

FIGURE 5. Distribution of heat flux in a heating copper block just before burnout (FC72, $s = 0.7\,\text{mm}$): **(A)** flat surface, **(B)** grooved surface.

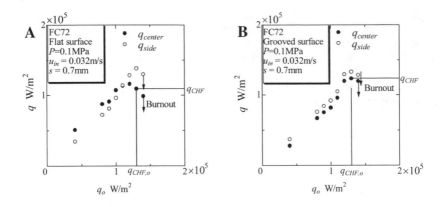

FIGURE 6. Comparison of surface heat flux values with nominal heat flux based on heat input (FC72, $s = 0.7\,mm$): **(A)** flat surface, **(B)** grooved surface.

evaluated are shown in FIGURE 7 for FC72 in the same manner as in FIGURE 3. Even at $s = 0.7\,mm$, a marked increase in q_{CHF} on the grooved surface is observed under the subcooled inlet conditions. (1) For $s = 2\,mm$, q_{CHF} for the grooved surface is larger by a factor 1.7–2.0 than that for the flat surface, regardless of $\Delta T_{sub,in}$. (2) For $s = 0.7\,mm$, it is larger by a factor 1.8–1.9 at $\Delta T_{sub,in} = 5$ and $10\,K$, whereas almost no increase is observed at $\Delta T_{sub,in} = 0\,K$. (3) For $s = 0.7\,mm$, $\Delta T_{sub,in} = 0\,K$, the growth rate of flattened bubble is very large, resulting in their vigorous motion and, thus, the complicated shape of bubble peripheries by the liquid penetration under the imposed liquid flow. The extension of dry patches and their rewetting by the penetrating liquid are repeated at very high frequency under the saturated inlet conditions. Hence, the liquid supply via grooves can no longer promote the quench

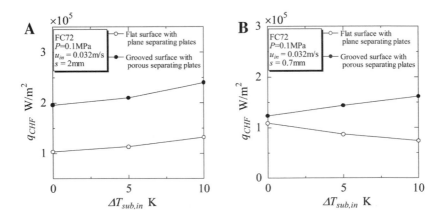

FIGURE 7. Marked increase in CHF on a grooved surface in a narrow gap with auxiliary liquid feeders (FC72): **(A)** $s = 2\,mm$, **(B)** $s = 0.7\,mm$.

process. (4) For $s = 0.7$mm, values of q_{CHF} for the flat surface decreases with increase in $\Delta T_{sub,in}$, which is interpreted as follows. The liquid subcooling prevents the growth and detachment of flattened bubbles by the balance of evaporation rate with that of condensation at the tip located upstream. As a result, large dry patches extend underneath the bubbles.[9] In this case, the liquid supply by the groove structure exerts significant effects on the increase in CHF.

FIGURE 8 shows the transitions of local heat flux at the center ($x = 0$mm) and the side ($x = 9$mm) for both surfaces at $\Delta T_{sub,in} = 0$K. Since experiments for the two surfaces are performed separately, their time scales are adjusted so that the heat flux values at the center are zero at the same instant for both surfaces, as shown by a small circle in FIGURE 8. (1) For the flat surface, the surface heat flux decreases rapidly, whereas it decreases gradually for the grooved surface. The existence of grooves as liquid feeders prevents the dried areas from extending rapidly and from occupying the heat transfer surface. (2) Heat flux tends to decrease with time, and finally becomes negative, except for the case of side location on the grooved surface. Negative values of the y-component of heat flux are possible because heat flows from the bottom to the side of the block just after the dried areas extend to the center of heat transfer surface. For the grooved surface, heat flux at the side never becomes negative because sufficient liquid is supplied from the auxiliary channels, despite local dryout at the center. The transition in the local value and the direction of heat flux is shown in FIGURE 9 at various instants A–D corresponding to those in FIGURE 8. In FIGURE 9, small arrows indicate the direction of heat flux. Once a dry area is extended on the surface, heat flow is directed to the auxiliary liquid channels operating as heat sinks.

In FIGURE 10, the relation between CHF values and inlet liquid velocity is examined. The data at $u_{in} = 0.032$m/sec is almost consistent with those in FIGURE 7, some discrepancy remains due to the scattering. (1) CHF values for the grooved surface are higher than those for the flat surface except the highest inlet liquid velocity tested. (2) With the increase of inlet liquid velocity, critical heat flux for the flat surface

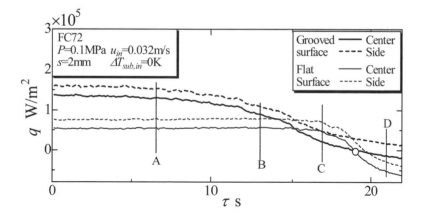

FIGURE 8. Transition of local surface heat flux under CHF conditions.

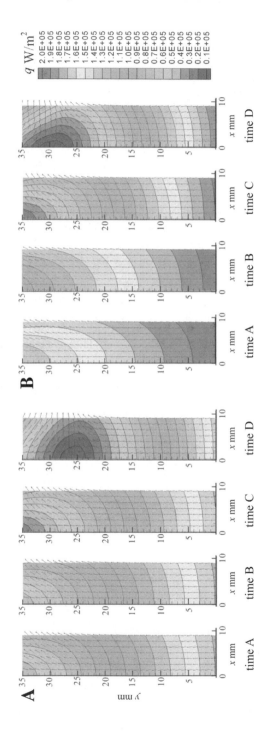

FIGURE 9. Transition of heat flux values and directions in the copper block under CHF conditions (FC72, $P = 0.1\,\text{MPa}$, $s = 2\,\text{mm}$, $\Delta T_{\text{sub,in}} = 0\,\text{K}$): **(A)** flat surface ($q_0 = 1.8 \times 10^5\,\text{W/m}^2$), **(B)** grooved surface ($q_0 = 2.1 \times 10^5\,\text{W/m}^2$).

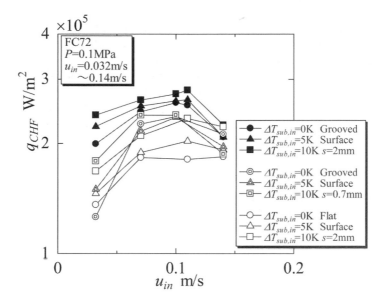

FIGURE 10. Relationship between critical heat flux q_{CHF} and inlet liquid velocity.

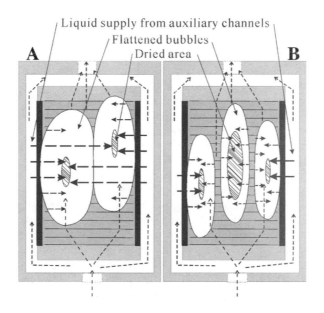

FIGURE 11. Explanation for the negative effect of inlet liquid velocity on CHF: (**A**) low liquid velocity, (**B**) high liquid velocity.

increases as expected, but the trend is saturated at higher velocity. For the grooved surface, however, maximum values of CHF are clearly recognized, and additional increase in inlet liquid velocity reduces the CHF values. (3) At $s = 2$ mm, the increment of CHF values by the grooved surface due to the present structure disappears at about a liquid velocity of 0.14 m/sec or at higher values. This result indicates that the groove structure has its optimum flow rate. The situation is similar for $s = 0.7$ mm.

The reason for the negative effect of inlet velocity on CHF is thought to be as follows. At high flow rate, flattened bubbles become elongated in the flow direction, as depicted in FIGURE 11. Liquid rivulets exist between the elongated bubbles that reduce the liquid supply directly from the auxiliary channels via V-shaped grooves. This implies that the role of liquid supply from the structure with grooves and auxiliary channels is weakened, and the conditions of CHF are determined by the balance between the rates of liquid evaporated and that supplied by the bulk flow. The situation is similar to that for the flat surface in the channel without additional structure. This explains why the increment in CHF decreases with increasing inlet liquid velocity. From observing these phenomena, the existence of grooves prevents the bubble detaching from the surface by the reduction of bubble surface area exposed to bulk flow. The situation becomes serious in the case of high velocity of bulk flow passing over the bubbles partly hidden in the grooves. At higher liquid inlet velocity, the advantage of grooved structure for the increase of CHF might disappear, but CHF values are expected to increase again with further increase in the liquid velocity.

TOPICS CONCERNING FURTHER INCREASE IN CHF AND APPLICATIONS

Increase in CHF by Marangoni Effect

To increase CHF values there are three different approaches: changes in the structure of the heating surface or of the heated duct as discussed above; changes in the conditions of coolant; and changes in the kind of coolant used. The most powerful method for the second approach is the application of microbubble emission boiling (MEB) under highly subcooled liquid conditions. CHF values can be increased by as much as ten times those for saturated nucleate boiling, for example.[13,14] The last approach is very simple and effective because it can be achieved by the addition of various liquids to change the effect due to surface tension. The validity of this approach is confirmed for alcohol–aqueous solutions.[15] At very low concentrations of alcohol, there is a temperature range over which surface tension increases with temperature. The Marangoni force, acting along the surface of micolayer underneath attached bubbles, induces flow from the bulk liquid to the receding front of the microlayer, resulting in an increase in CHF values. The liquid is referred to as a *self wetting* liquid. This feature could be possible as a result of the surface tension variations with both temperature and concentration. FIGURE 12 shows four possible cases, classified by the sign of their surface tension derivatives with respect to temperature and concentration. Interfacial temperature T_i on the surface of microlayer becomes larger than the saturation temperature at a bulk concentration T_{bulk} due to the preferential evaporation of more volatile component.[16] The present authors

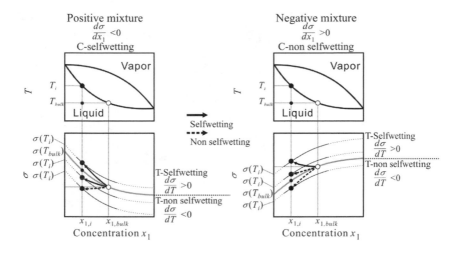

FIGURE 12. Explanation of self-wetting ability of mixtures with surface tension gradient along the liquid–vapor interface of a microlayer.

tried to explain the effects of concentration and temperature gradients on the variation of surface tension, from the periphery to the center along the surface of the microlayer. In the case $d\sigma/dx_1 < 0$ (C self wetting) for a positive mixture (i.e., lower surface tension with higher concentration of more volatile component, x_1) the self wetting nature is observed for $d\sigma/dT > 0$ (T self wetting) and also for $d\sigma/dT < 0$ (T non self wetting) provided that the negative gradient is not too large. In the case of $d\sigma/dx_1 > 0$ (C non self wetting) for negative mixture, the self-wetting nature is observed only for large positive gradient $d\sigma/dT > 0$, and the mixture becomes non-self wetting for $d\sigma/dT < 0$.

Design of the Thin Cold Plate Using the Same Principle as for Increasing CHF

The unheated liquid channels arranged on both sides of a heated channel are operated as temporary heat sinks once dryout occurs and they prevent the temperature of heat sinks from increasing excessively for a certain time before safety measures are devised. In contrast with this advantage, the structure has an inherent difficulty in extending the width of heating surface area as is required for practical applications. One solution is to locate the auxiliary liquid channels behind the main heated channel and supply the liquid through multiple sintered metal plates connected between both channels at an optimum distance. To extend this idea, keeping the thickness of the channel small, a corrugated sintered plate was introduced between the channels to supply liquids at a fixed distance, as shown in FIGURE 13. To prevent an increase in the temperature of the subcooled liquid in the auxiliary channel, an insulating material, such as a teflon sheet, is attached to the single side of the porous metal sheet facing the heated channel. The sheet has holes for outlet of cold liquid and for the connecting pins of both channels. The ratio of cross section areas for both channels can easily be changed by means of the shape of the corrugated sheet.

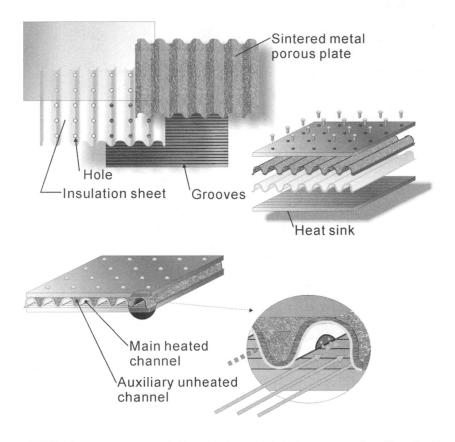

FIGURE 13. An example of thin cold plate with built in structure of auxiliary liquid channels.

Requirements From the Design of a Space Solar Power System

Recent requirements for the development of new energy resources created the concept of a space solar power system (SSPS) where the electric power generated from solar cells is transmitted via microwave to the surface of the Earth. An idealized unit of the system generates electric power at 1 GW, which is the same as that from a unit of a nuclear power plant. To reduce the weight and size of SSPS satellites, and also to simplify the system, a generator/transmitter combined module with light condensing mirrors was proposed. The structure has an inherent difficulty because it cannot directly radiate most of the incident solar energy; the rest of the energy converted to electric power, directly from areas on the modules, and the use of powerful heat transportation system is required for their radiation at the periphery of the satellite. From the analysis of thermal management for an assumed 10 GW model,[17] it was deduced that the transportation of waste heat is quite difficult for the 1 GW model, because huge amounts of heat generated in the central part of the satellites cannot be transported along a radius of 1 km by two-phase loops, even though water is used as working fluid. There are two possible solutions to this problem. One solution is to separate the 1 GW power satellite into several segments and to construct an assembly of ten 100 MW class satellites. On the other hand, if sunlight is condensed at high magnification, the satellite diameter reduction is inversely proportional to the square root of the magnification. The method is accompanied by a marked increase in heat flux from the generators and requires the development of technologies to achieve heat transfer at high heat flux from a large extension of cold plate area. There are many investigations into the development of cooling systems for small semiconductors of high heat generation density, but almost none into technologies that handle both high heat flux and a large heat transfer area. The structure mentioned in the preceding sections has the ability to supply cold liquid uniformly to a large area.

CONCLUSIONS

An increase in critical heat flux for flow boiling in a narrow channel was attempted by the development of a new structure for a heating surface assembly. In the channel between parallel plates, a heating surface with V-shaped grooves in the transverse direction to the flow was installed, and the test liquid was supplied at the bottom of flattened bubbles via grooves from unheated auxiliary liquid channels located in parallel with the main channel. The following were clarified. (1) CHF values increase to about twice those for a flat surface without additional liquid supply. (2) The CHF value for the present structure takes a maximum at an optimum inlet liquid velocity. (3) The auxiliary channels operate as heat sinks to prevent rapid temperature excursion under CHF conditions.

The proposed structure of narrow channels has the following additional advantages: (1) An increase in CHF is also expected for longer heated channels without the resulting serious vapor blanketing downstream. (2) There are optimum designs for the shape and dimensions of grooves and optimum operating conditions to achieve the maximum increment in CHF. (3) The structure is valid for an increase in CHF under microgravity conditions, where the decrease in departure frequency and the extension of flattened bubbles on the heating surface are emphasized.

An increase in CHF values for a large heat transfer area is desired for waste heat management on large space platforms. There are three different approaches to this subject: a new cold plate structure to promote liquid supply or to prevent the emergence of dry patches in nucleate boiling, as discussed here; the application of high subcooled liquid to suppress bubble generation or realize microbubble emission boiling; and the use of mixtures with self-wetting ability. The challenge to establishing high-performance cooling systems stems not only from space applications, but also from applications on the ground.

ACKNOWLEDGMENT

The present investigation was conducted under the New Energy and Industrial Technology Development Organization (NEDO) program, "Project of Fundamental Technology Development for Energy Conservation", 2002–2004. The authors appreciate support from Mr. Katsuharu Ohtsubo, Dr. Hiroya Shida, and Osamu Ogawa of the Energy Conservation Technology Department in NEDO, and collaboration on the present project with Professors Hiroshi Kawamura and Koichi Suzuki of Tokyo University of Science, Dr. Yoshiyuki Abe of the National Institute of Advanced Industrial Science and Technology, and Dr. Hideo Iwasaki of Toshiba Corporation.

REFERENCES

1. STRAUB, J., M. ZELL & B. VOGEL. 1990. Pool boiling in a reduced gravity field. Proc. 9th Intl. Heat Transfer Conf. **1:** 91–112.
2. MERTE, H., JR. 1990. Nucleate pool boiling in variable gravity. Low-gravity fluid dynamics and transport phenomena. Prog. Astronaut. Aeronaut. **130:** 15–69.
3. DI MARCO, P. 2000. Pool boiling in microgravity: assessed results and open issues. Proc. 3rd European Thermal Sciences Conference. **1:** 81–90.
4. OHTA, H. 2003. Microgravity heat transfer in flow boiling. *In* Advances in Heat Transfer, Vol. 37. 1–76. Academic Press.
5. YAO, S.C. & Y. CHANG. 1983. Pool boiling heat transfer in a confined space. Intl. J. Heat Mass Transf. **26:** 841–848.
6. HUNG, Y.H. & S.C. YAO. 1985. Pool boiling heat transfer in narrow horizontal annular crevices. J. Heat Transf. **10:** 656–662.
7. FUJITA, Y., H. OHTA. & S. UCHIDA. 1987. Nucleate boiling heat transfer in vertical narrow space. Proc. 2nd ASME- JSME Thermal Engineering Joint Conference. **5:** 469–476.
8. BRUSSTAR, M.J., H. MERTE, JR. & R.B. KELLER. 1995. Relative effect of flow and orientation on the critical heat flux in subcooled forced convection boiling. NASA Rep. UM-MEAM-95-15.
9. OHTA, H., Y. SHINMOTO & K. MATSUNAGA. 2002. Effect of gravity on flow boiling in narrow ducts and enhancement of CHF values. Proc. 12th Intl. Heat Transfer Conf. **3:** 725–730.
10. OHTA, H., H. WATANABE, T. SABATO, *et al.* 1999. Gravity effect on flow boiling heat transfer in narrow gaps. Proc. 5th ASME-JSME Thermal Engineering Joint Conference. CD-ROM. AJTE 99-6421.
11. ZUBER, N., M. TRIBUS & J.W. WESTWATER. 1961. The hydrodynamic crisis in pool boiling of saturated and subcooled liquids. Int. Devel. Heat Transf. **II:** 27.
12. IVEY, H.J. & D.J. MORIS. 1962. On the relevance of the vapor liquid exchange mechanism for subcooled boiling heat transfer at higher pressure. UK Atomic Energy Authority. AEEW-R. 137.
13. SUZUKI, K., H. SAITO & K. MATSUMOTO. 2002. Heat flux cooling by microbubble emission boiling, microgravity transport process in fluid, thermal, biological and materials sciences. Ann. N.Y. Acad. Sci. **974:** 364–377.
14. SUZUKI, K., S. TAKAHASHI & H. OHTA. 2003. A study on subcooled pool boiling of water: contact area of bubbles heating surface in heating process. Proc. Microgravity Transport Process in Fluid, Thermal, Biological and Materials Sciences III. ECI-MTP-03-55. CD-ROM.
15. ABE, Y. & A. IWASAKI. 2003. Microgravity experiments on phase change of self-wetting fluids. Proc. Microgravity Transport Process in Fluid, Thermal, Biological and Materials Sciences III. ECI-MTP-03-18. CD-ROM.
16. OHTA, H & Y. FUJITA. 1994. Nucleate pool boiling of binary mixtures. Heat Transfer 1994. Proc. 10th Intl. Heat Transfer Conf. **5:** 129–134.
17. OHTA, H., S. TOYAMA, H. KAWASAKI, *et al.* 2003. On the feasibility of heat removal from generator/transmitter units for assumed 10 MW space solar power system by using two-phase flow loop with latent heat transportation. Proc. 54th Intl. Astronautical Congress. IAC-03-R.2.02. CD-ROM.

Numerical Simulation and Experimental Validation of the Dynamics of Multiple Bubble Merger During Pool Boiling Under Microgravity Conditions

H.S. ABARAJITH,[a] V.K. DHIR,[a] G. WARRIER,[b] AND G. SON[c]

[a]University of California, Los Angeles, California, USA

[b]Mechanical and Aerospace Engineering Department,
University of California, Los Angeles, California, USA

[c]Sogang University, South Korea

ABSTRACT: Numerical simulation and experimental validation of the growth and departure of multiple merging bubbles and associated heat transfer on a horizontal heated surface during pool boiling under variable gravity conditions have been performed. A finite difference scheme is used to solve the equations governing mass, momentum, and energy in the vapor liquid phases. The vapor–liquid interface is captured by a level set method that is modified to include the influence of phase change at the liquid–vapor interface. Water is used as test liquid. The effects of reduced gravity condition and orientation of the bubbles on the bubble diameter, interfacial structure, bubble merger time, and departure time, as well as local heat fluxes, are studied. In the experiments, multiple vapor bubbles are produced on artificial cavities in the 2–10 micrometer diameter range, microfabricated on the polished silicon wafer with given spacing. The wafer was heated electrically from the back with miniature strain gage type heating elements in order to control the nucleation superheat. The experiments conducted in normal Earth gravity and in the low gravity environment of KC-135 aircraft are used to validate the numerical simulations.

KEYWORDS: pool boiling; bubble merger; microgravity; multiple bubbles; force on bubble; levelset method

NOMENCLATURE:

A_0	dispersion constant, J
c_p	specific heat at constant pressure, kJ/(kg K)
D	liftoff diameter of the bubble, m
g_e	gravitational acceleration at Earth level, m/s^2
g	gravitational acceleration at any level, m/s^2
H	step function
h	grid spacing for the macro region
h_{ev}	evaporative heat transfer coefficient, W/(m^2 K)
h_{fg}	latent heat of evaporation, J/Kg
K	interfacial curvature, m^{-1}
l_0	characteristic length, m
M	molecular weight

Address for correspondence: Vijay K Dhir, 48-121 EnGR IV, 420 Westwood Plaza, University of California, Los Angeles, CA 90095, USA. Voice: 310-825-8507; fax: 310-206-2302.
vdhir@ucla.edu

Ann. N.Y. Acad. Sci. 1027: 235–258 (2004). ©2004 New York Academy of Sciences.
doi: 10.1196/annals.1324.020

$\vec{m}$	evaporative mass rate vector at interface, $kg/(m^2 \, sec)$
p	pressure, Pa
q	heat flux, W/m^2
R	radius of computational domain, m
$\bar{R}$	universal gas constant
R_0	radius of dry region beneath a bubble, m
R_i	radial location of the interface at $y = h/2$, m
r	radial coordinate, m
T	temperature, K
t	time, sec
t_0	characteristic time, l_0/u_0, sec
u	velocity in r direction, m/sec
$\vec{u}_{int}$	interfacial velocity vector, m/sec
u_0	characteristic velocity,
$\dot{m}_{micro}$	evaporative mass rate from microlayer, kg/sec
V_c	volume of a control volume in the microregion, m^3
v	velocity in y direction, m/sec
Z	height of computational domain, m
z	vertical coordinate normal to the heating wall, m
α	thermal diffusivity, m^2/sec
β_t	coefficient of thermal expansion, 1/K
δ	liquid thin film thickness, m
δ_T	thermal layer thickness, m
$\delta_\varepsilon(\phi)$	smoothed delta function
φ	apparent contact angle, deg
ϕ	level set function
θ	dimensionless temperature, $(T - T_s)/(T_{wall} - T_s)$
κ	thermal conductivity, W/mK
μ	viscosity, Pa sec
ν	kinematic viscosity, m^2/sec
ρ	density, kg/m^3
σ	surface tension, N/m
Γ	mass flow rate in the micro layer, kg/sec
ΔT_s	heating wall superheat, K

Subscripts

l, v	liquid and vapor phases
r, z, t	$\partial/\partial r, \partial/\partial y, \partial/\partial t$
s, wall	saturation, wall
int	interface

INTRODUCTION

Boiling, the most efficient mode of heat transfer, is employed in various energy conversion systems and component cooling devices. Boiling has been studied extensively. The process allows accommodation of high heat fluxes at low wall superheats. The process is very complex and understanding it imposes severe challenges. Prior to incipient boiling, at low heat flux, heat transfer is controlled by natural convection. At higher heat flux, heat transfer is controlled by bubble dynamics. Initially during partial nucleate boiling, discrete bubbles form on the heater surface. At moderately high heat flux, bubbles start merging laterally as well as vertically. The transition from partial nucleate boiling to fully developed nucleate boiling is indicated by lateral and vertical merger of bubbles.

Because of space constraints, in space applications, boiling heat transfer is the preferred choice for power production. Applications of boiling heat transfer in space applications include thermal management, fluid handling, and control and power systems. The key factors that need to be addressed for space systems based on the Rankine cycle are the boiling heat transfer coefficients and the critical heat flux under reduced gravity conditions.

Keshock and Siegal[1] observed that the bubbles grow larger and show longer growth periods before becoming detached from the heater surface under reduced gravity conditions because of the reduced buoyancy force acting on the bubbles. Merte[2] and Lee and Merte[3] reported the results of pool boiling experiments conducted in the space shuttle for a surface similar to that used in drop tower tests. Subcooled boiling was found to be unstable during long periods of microgravity conditions. It was concluded that subcooling has negligible influence on the steady state heat transfer coefficient.

Straub, Zell, and Vogel[4,5] conducted a series of nucleate boiling experiments using thin platinum wires and a gold-coated flat plate as heaters under low gravity conditions in the flights of ballistic rockets and in KC 135 aircraft. For a flat plate heater with R12 as the test liquid, boiling curves similar to those at normal gravity cases were obtained. Using R113 as the test liquid, rapid bubble growth and large bubbles were observed. However, neither the bubble growth rate nor the bubble diameter on departure were given. For subcooled boiling under microgravity conditions, they observed a reduction of up to 50% in heat transfer coefficient, in comparison with the normal gravity cases.

Qiu *et al.*[6] conducted experimental studies on the growth and detachment mechanisms of a single bubble and multiple bubble mergers during parabolic flights of the KC-135 aircraft. Experiments were carried out under normal and reduced gravity conditions for various wall superheats and liquid subcoolings at a system pressure varying from 1 atm to 1.13 atm Artificial cavities were made on a polished silicon wafer. The wafer was heated on the backside to control nucleation superheat. Bubbles were produced on artificially etched cavities at the middle of the wafer in degassed and distilled water. Gravity in these experiments was not always constant, but varied with time.

Experimental investigations of boiling heat transfer for space applications impose serious restrictions in terms of the duration of the experiments, maintaining gravity conditions, the number of experiments that can be conducted, the size of the experimental apparatus required, and other features. Thus, numerical simulation can play an important role in predicting the boiling heat flux under microgravity conditions.

Several previous attempts have been made to model bubble growth on a heated wall and the bubble departure processes. Lee and Nydahl[7] calculated the bubble growth rate by solving the flow and temperature fields numerically from the momentum and energy equations. They used the formulation of Cooper and Llyod[8] for microlayer thickness. However, they assumed a hemispherical bubble and wedge shaped microlayer and, thus, they could not account for the shape change of the bubble during growth.

Zeng *et al.*[9] used a force balance approach to predict the bubble diameter at departure. They included the surface tension, inertial force, buoyancy, and the lift force created by the wake of a previously departed bubble. However, there was empiricism involved in computing the inertial and drag forces. The study assumed a power law profile for growth rate and the coefficients were determined from the experiments.

Mei et al.[10] studied the bubble growth and departure time, assuming a wedge shaped microlayer. They also assumed that the heat transfer to the bubble was only through the microlayer, which is not correct for both subcooled and the saturated boiling. The study did not consider the hydrodynamics of the liquid motion induced by the growing bubble and introduced empiricism through the shape of the growing bubble. Welch[11] published a scheme, using a finite volume method and an interface tracking method. The conduction in the solid wall was also taken in to account. However, the microlayer was not modeled explicitly.

Son et al.[12] numerically simulated single bubble growth during nucleate boiling by using the *level set method*. This method has been applied to adiabatic incompressible two-phase flow by Sussman et al.[13] and to film boiling near critical pressures by Son and Dhir.[14] Singh and Dhir[15] obtained numerical results for low gravity

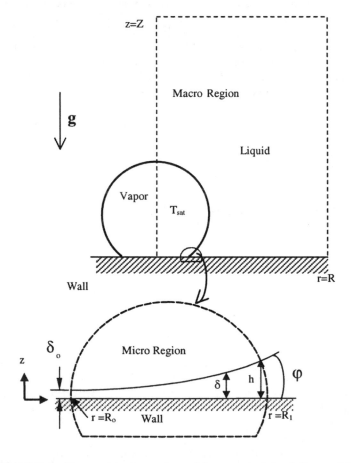

FIGURE 1. Macro- and microregions of the numerical simulation model.

conditions by exercising the numerical simulation model of Son *et al.*[12] when the liquid is subcooled. The computation domain was separated into two regions, micro- and macroregions. The interface shape, position, velocity, and temperature field in the liquid were obtained from the macroregion by solving the conservation equations. For the microregion, lubrication theory was used, which included the disjoining pressure in the thin liquid film. The solutions for the microregion and macroregion were matched at the outer edge of the microlayer (see FIGURE 1).

Abarajith and Dhir[16] studied the effects of contact angle on the growth and departure of a single bubble on a horizontal heated surface during pool boiling under normal gravity conditions. The contact angle was varied by changing the Hamaker constant that defines the long-range forces. They also studied the effect of contact angle on the microlayer and macrolayer heat transfer rates. It was shown that the predicted diameter at departure, normalized to the corresponding values for a contact angle of 90° for water and PF5060, when plotted against contact angle, fell on the same curve.

A complete numerical simulation and experimental validation of the growth, merger, and departure of multiple bubbles on a horizontal heated surface during pool boiling under low gravity conditions is described here. The effects of reduced gravity conditions, spacing and arrangement of the cavities, wall superheat, and liquid subcooling on the bubble diameter, interfacial structure, bubble merger time, and departure time, as well as local heat flux, are to be studied. The main objective is to set a road map to a unified correlation for boiling heat flux that encompasses all these independent and dependent variables.

EXPERIMENTS

The experimental apparatus for pool boiling experiments is shown schematically in FIGURE 2. The system configuration is same as that used elsewhere.[6] It consists of a test chamber, bellows, and a nitrogen (N_2) chamber. Three glass windows are installed on the walls of the test chamber for visual observation. To control the system, pressure transducers are installed in the test chamber and in the N_2 chamber. The test surface for studying nucleate boiling is installed at the bottom of the test chamber. In the vicinity of the heater surface a rake of six thermocouples are placed in the liquid pool to measure the temperature in the thermal boundary layer; another rake is placed in the upper portion of the chamber to measure the bulk liquid temperature. Distilled, filtered, and degassed water is used as the test liquid. Two video cameras operating at 250 frames/second are installed at an angle of 90° to record the boiling processes. The liquid temperature and pressure in the chamber are controlled according to the set-points established by the operator on board. A three-component accelerometer is installed on the frame on which the setup is mounted.

A polished silicon wafer was used as the test surface for the nucleate boiling experiments. From manufacturer specifications, the roughness of the bare polished wafer was less than 5 Å. The contact angle for the liquid–surface combination was measured before the experiment by taking photographs of the preplaced drops of the test liquid on the heater surface.

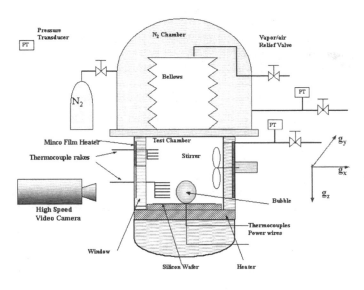

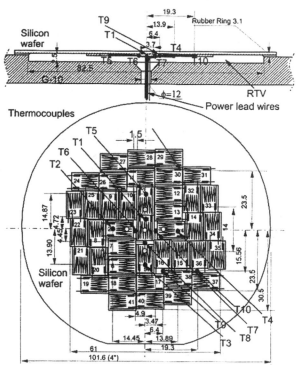

FIGURE 2. Schematic of the experimental setup.

At the back of the silicon wafer, strain gages are bonded as heating elements. In the central area, miniature elements of size $2\times2\,mm^2$ are used and are grouped so as to cover small areas of the test surface. In each group a thermocouple is directly attached to the wafer. The heater surface temperatures in various regions are then separately controlled through a multichannel feedback control system. As such, the wall superheat can be maintained constant during an experimental run and can be changed to the desired setpoint. The local heating rate in individual areas is recorded during the experiments. The power lead wires and the thermocouple wires are led out from the hole in the base made from phenolic garolite grade 10 (G-10). The wafer is cast with RTV on the G-10 base. The base in turn is mounted in the test chamber. Cylindrical cavities of predetermined sizes and at specified locations are etched in the wafer using the deep reactive ion etching (DRIE) technique.

NUMERICAL MODEL

Mathematical Development of the Model

The two-dimensional model of Son *et al.*[12] was extended three dimensions to study bubble merger under low gravity conditions. The computation domain was separated into two regions—microregion and macroregion—as shown in FIGURE 1. The microregion is a thin film that lies underneath the bubble; the macroregion consists of the bubble and its surroundings. Both regions are coupled and solved simultaneously. The calculated shapes of the interface in the microregion and macroregion are matched at the outer edge of the microlayer.

Assumptions

The assumptions made in the model are:

1. the process is three-dimensional;

2. the flows are laminar;

3. the wall temperature remains constant throughout the process;

4. pure water at atmospheric pressure is used as the test fluid;

5. the thermodynamic properties of the individual phases are insensitive to small changes in temperature and pressure (this assumption is reasonable since the computations are performed for low wall superheat range); and

6. the static contact angle is known (variations of contact angle during advancing and receding phases of the interface are not included).

Microregion

A two-dimensional quasistatic model is used for the microregion and no azimuthal variation is considered. As such, the solution for the microlayer thickness is obtained in the radial direction from the center of the bubble base. This solution is assumed to be valid for all azimuthal positions.

The equation of mass conservation in the microregion is written as

$$\frac{q}{h_{fg}} = -\frac{1}{r}\frac{\partial}{\partial r}\int_0^\delta \rho_l \cdot r u dz, \tag{1}$$

where q is the conductive heat flux from the interface, defined by $k_l(T_{wall} - T_{int})/\delta$, where δ is the thickness of the thin film.

Lubrication theory has been assumed in a manner similar to that used in earlier papers.[17–19] According to the lubrication theory, the momentum equation in the microregion can be written as

$$\frac{\partial p_l}{\partial r} = \mu\frac{\partial^2 u}{\partial z^2}u, \tag{2}$$

where p_l is the pressure in the liquid. Heat conducted through the thin film must match that due to evaporation from the vapor–liquid interface. By using modified a Clausius–Clapeyron equation, the energy conservation equation for the microregion yields

$$\frac{k_l(T_{wall} - T_{int})}{\delta} = h_{ev}\left[T_{int} - T_v + \frac{(p_l - p_v)T_v}{\rho_l h_{fg}}\right]. \tag{3}$$

The evaporative heat transfer coefficient is obtained from kinetic theory as

$$h_{ev} = 2\left[\frac{M}{2\pi\bar{R}T_v}\right]^{1/2}\frac{\rho_v h_{fg}^2}{T_v}, \quad T_v = T_s(p_v). \tag{4}$$

The pressure of the vapor and liquid phases at the interface are related by

$$p_l = p_v - \sigma K - \frac{A_0}{\delta^3} + \frac{q^2}{2\rho_v h_{fg}^2}, \tag{5}$$

where A_0 is the dispersion constant. The second term on the right-hand side of Equation (5) accounts for the capillary pressure caused by the curvature of the interface, the third term is for the disjoining pressure, and the last term originates from the recoil pressure. The curvature of the interface is defined by

$$K = \frac{1}{r}\frac{\partial}{\partial r}\left[\frac{r\frac{\partial S}{\partial r}}{\sqrt{1 + \left(\frac{\partial \delta}{\partial r}\right)^2}}\right]. \tag{6}$$

The combination of the mass conservation, Equation (1), momentum conservation, Equation (2), mass balance and energy conservation, Equation (3) and pressure balance, Equation (5), together with Equation (6) for the curvature for the microregion, yields a set of three nonlinear first order ordinary differential equations:

$$\frac{\partial \delta_r}{\partial r} = -\frac{\delta_r(1 + \delta_r^2)}{r} + \frac{(1 + \delta_r^2)^{3/2}}{\sigma}\left[\frac{\rho_l h_{fg}}{T_v}\left(T_{int} - T_v - \frac{q}{h_{ev}}\right) - \frac{A_0}{\delta^3} + \frac{q^2}{\rho_v h_{fg}^2}\right] \tag{7}$$

$$\frac{\partial T_{int}}{\partial r} = -\frac{q\delta_r}{\kappa_l + h_{ev}\delta} + \frac{3T_v h_{ev}\mu\Gamma}{(\kappa_l + h_{ev}\delta)\rho_l^2 h_{fg}r\delta^2} \tag{8}$$

$$\frac{\partial[\Gamma]}{\partial r} = -\frac{rq}{h_{fg}}. \tag{9}$$

The three differential equations, **(7)**–**(9)**, can be simultaneously integrated by using a Runge–Kutta method, when the initial conditions at $r = R_0$ are given. In present case, the radial location R_l, the interface shape obtained from the micro- and macro-solutions are matched. This is the end point for the integration of these equations. The radius of the dry region beneath a bubble, R_0, is related to R_l from the definition of the apparent contact angle, $\tan\varphi = 0.5h/(R_l - R_0)$.

The boundary conditions for film thickness at the end points are

$$\delta = \delta_0, \quad \delta_r = 0, \quad \Gamma = 0 \quad \text{at } r = R_0$$
$$\delta = h/2, \quad \delta_{rr} = 0 \quad \text{at } r = R_1,$$

(10)

where δ_0 is the interline film thickness at the tip of the microlayer, which is calculated by combining **(3)** and **(4)** and requiring that $T_{int} = T_{wall}$ at $r = R_0$, and h is the spacing of the two-dimensional grid for the macroregion. For a given $T_{int,0}$ at $r = R_0$, a unique vapor–liquid interface is obtained. The static contact angle, φ, for a water–silicon system based on measurements was taken to be 54°.

Macroregion

To numerically analyze the macroregion, the level set formulation[12] for nucleate boiling of pure liquid is used. The interface separating the two phases is captured by ϕ, defined as the signed distance from the interface. The negative sign is chosen for the vapor phase and the positive sign for the liquid phase. The discontinuous pressure drop across vapor and liquid caused by surface tension force is smoothed into a numerically continuous function with a δ-function formulation (see Ref. 13 for details). The continuity, momentum, and energy conservation equations for the vapor and liquid in the macroregion are written as follows:

$$\rho_t + \nabla \cdot (\rho \vec{u}) = 0$$

(11)

$$\rho(\vec{u}_t + \vec{u} \cdot \nabla \vec{u}) = -\nabla p + \nabla \cdot \mu \nabla \vec{u} + \nabla \cdot \mu \nabla \vec{u}^T$$
$$+ \rho \vec{g} - \rho \beta_T (T - T_s) - \sigma K \nabla H$$

(12)

$$\rho c_p (T_t + \vec{u} \cdot \nabla T) = \nabla \cdot \kappa \nabla T \quad \text{for } H > 0$$

(13)

$$T = T_s(\rho_v) \quad \text{for } H = 0.$$

(14)

The fluid density, viscosity, and thermal conductivity of water are defined in terms of the step function H by

$$\rho = \rho_v + (\rho_l - \rho_v)H$$

(15)

$$\mu^{-1} = \mu_v^{-1} + (\mu_l^{-1} - \mu_v^{-1})H$$

(16)

$$\kappa^{-1} = \kappa_l^{-1}H,$$

(17)

where H is the Heviside function, smoothed over three grid spaces as described below

$$H = \begin{cases} 1 & \text{if } \phi \geq 1.5h \\ 0 & \text{if } \phi \leq 1.5h \\ 0.5 + \dfrac{\phi}{3h} + \dfrac{1}{2\pi}\sin\left(\dfrac{2\pi\phi}{3h}\right) & \text{if } |\phi| \leq 1.5h. \end{cases} \qquad (18)$$

The mass conservation equation, (11), can be rewritten as

$$\nabla \cdot \vec{u} = -\frac{\rho_l + \vec{u} \cdot \nabla\rho}{\rho}. \qquad (19)$$

The term on right hand side of (19) is the volume expansion due to the liquid–vapor phase change. From the conditions of the mass continuity and energy balance at the vapor–liquid interface, the following equations are obtained:

$$\vec{m} = \rho(\vec{u}_{\text{int}} - \vec{u}) = \rho_l(\vec{u}_{\text{int}} - \vec{u}_l) = \rho_v(\vec{u}_{\text{int}} - \vec{u}_v) \qquad (20)$$

$$\vec{m} = \frac{\kappa \nabla T}{h_{fg}}, \qquad (21)$$

where $\vec{m}$ is the water evaporation rate vector and $\vec{u}_{\text{int}}$ is the interface velocity. If the interface is assumed to advect in the same way as the level set function, the advection equation for density at the interface can be written

$$\rho_l + \vec{u}_{\text{int}} \cdot \nabla\rho = 0. \qquad (22)$$

Using Equations (18), (20), and (21), the continuity equation, (19), for the macro-region is rewritten as

$$\nabla \cdot \vec{u} = \frac{\vec{m}}{\rho^2} \cdot \nabla\rho. \qquad (23)$$

The vapor produced as a result of evaporation from the microregion is added to the vapor space through the cells adjacent to the heated wall, and is expressed as

$$\left(\frac{1}{V_c}\frac{dV}{dt}\right)_{\text{mic}} = \frac{\dot{m}_{\text{mic}}}{V_c \rho_v}\delta_\varepsilon(\phi), \qquad (24)$$

where V_c is the control volume and $\dot{m}_{\text{mic}}$ is the evaporation rate from the microlayer expressed as

$$\dot{m}_{\text{mic}} = \int_{R_0}^{R_1} \frac{\kappa_l(T_w - T_{\text{int}})}{h_{fg}\delta} r\,dr. \qquad (25)$$

The volume expansion contributed by the microlayer is smoothed at the vapor–liquid interface by the smoothed delta function,

$$\delta_\varepsilon(\phi) = \frac{\partial H}{\partial \phi}. \qquad (26)$$

In the level set formulation, the level set function, ϕ, is used to keep track of the vapor–liquid interface location in terms of the set of points for which $\phi = 0$, and it is advanced by the interfacial velocity while solving the following equation:

$$\phi_l = -\vec{u}_{\text{int}} \cdot \nabla\phi. \qquad (27)$$

To keep the value of ϕ close to that of the signed distance function, $|\nabla\phi| = 1$, ϕ is reinitialized after each time step,

$$\frac{\partial\phi}{\partial t} = \frac{\phi_0}{\sqrt{\phi_0^2 + h^2}}(1 - |\nabla\phi|),\qquad(28)$$

where ϕ_0 is a solution of Equation **(27)**.

The boundary conditions for velocity, temperature, concentration, and level set function for the governing equations, **(11)**–**(14)** are:

$$
\begin{aligned}
u = 0,\ \ v = w = 0,\ \ T_x = 0,\ \ \phi_x = 0 \quad &\text{at } x = 0\\
u_x = v_x = w_x = 0,\ \ T_x = 0,\ \ \phi_x = 0 \quad &\text{at } x = X\\
u = v = w = 0,\ \ T = T_{\text{wall}},\ \ \phi_y = -\cos\varphi \quad &\text{at } y = 0\\
u_y = w_y = v = 0,\ \ T_y = 0,\ \ \phi_y = 0 \quad &\text{at } y = Y\\
u_z = v_z = w_z = 0,\ \ T_z = 0,\ \ \phi_z = 0 \quad &\text{at } z = 0\\
u_z - v_z = w_z = 0,\ \ T_z = 0,\ \ \phi_z = 0 \quad &\text{at } z = Z.
\end{aligned}
\qquad(29)
$$

For the numerical calculations, the governing equations for the micro- and macro-regions are non-dimensionalized by defining the characteristic length, l_0, the characteristic velocity, u_0, and the characteristic time, t_0, as follows:

$$l_0 = \sqrt{\frac{\sigma}{g(\rho_l - \rho_v)}},\qquad u_0 = \sqrt{gl_0},\qquad t_0 = \frac{l_0}{u_0}.\qquad(30)$$

Solution

The governing equations are numerically integrated by following the procedure of Son *et al.*[12]

1. A guess is made for the value of A_0, the Hamaker constant at a given contact angle.
2. The macrolayer equations are solved to determine the value of R_1 (radial location of the vapor–liquid interface at $\delta = h/2$.
3. The microlayer equations are solved with the guessed value of A_0 to determine the value of R_0 (radial location of the vapor-liquid interface at $\delta = 0$.
4. The apparent contact angle is calculated using the equation $\tan\phi = 0.5h/(R_1 - R_0)$ and steps 1–4 are repeated for a different value of A_0 if the values of the given and the calculated apparent contact angles are different.

The computation domain is chosen to be $(X/l_0,\ Y/l_0,\ Z/l_0) = (2, 2, 3)$, so that the bubble growth process is not affected by the boundaries of the computation domain. The initial velocity is assumed to be zero everywhere in the domain. The initial fluid temperature profile is taken to be linear in the natural convection thermal boundary layer, and the thermal boundary layer thickness, δ_T, is evaluated using the correlation for turbulent natural convection on a horizontal plate:

$$\delta_T = 7.14\left(\frac{\nu_l\alpha_l}{g\beta_T\Delta T}\right)^{1/3}.$$

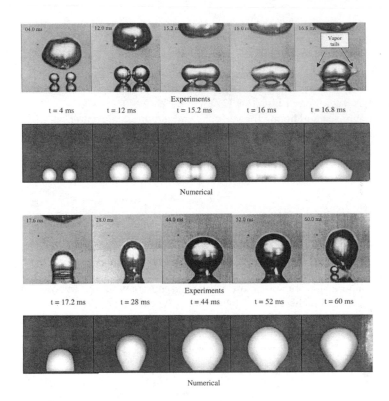

FIGURE 3. Comparison between experimental and numerical bubble shapes in a two-bubble merger for saturated water at $\Delta T_w = 7\,K$, $g = 1\,g_e$.

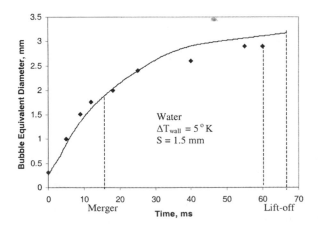

FIGURE 4. Comparison between experimental ($\blacklozenge$) and numerical (—) equivalent bubble diameters in a two-bubble merger for saturated water at $\Delta T_w = 7\,K$, $g = 1\,g_e$.

The mesh size for all calculations is chosen as $98 \times 98 \times 136$. This represents the best tradeoff in calculation accuracy and computing time. The calculations are completed for one quarter of the domain due to the symmetric nature of the problem with respect to the x and z axes.

RESULTS AND DISCUSSION

Numerical simulations were carried out for two, three, five, seven, and nine bubble mergers with water as test fluid at a gravity level of $0.01\,g_e$ and at normal Earth gravity. Bubble dynamics and heat transfer data were analyzed. The effect of bubble merger on heat transfer is discussed.

Two Bubble Merger

FIGURE 3 shows a comparison of numerically calculated and experimentally observed bubble shapes during the merger of two bubbles at normal Earth gravity. The bubbles are placed 1.5 mm apart in saturated water at a wall superheat of 5 K. The shapes obtained from numerical simulation are found to compare well with those observed in the experiments of Mukharjee and Dhir.[20] However, the trapped liquid layer ($15 < t < 17$ msec) calculated from numerical simulations is much smaller than that observed in experiments. A possible reason for the difference could be the existence of some liquid subcooling in the experiments. Numerical simulations were carried out assuming liquid to be saturated. FIGURE 4 shows the variation of equivalent bubble diameter with time. The numerical predictions are again found to match well with the data from experiments. The bubbles merge when they grow to a diameter equal to the spacing between them (i.e., 1.5 mm at 15 msec) and oscillate for a period of time. The merged bubble lifts off at 62 msec.

FIGURE 5 shows comparison of predicted and observed bubble shapes for a two-bubble merger in a low gravity environment. The spacing between cavities was 7 mm. The wall superheat was 5 K, and the liquid subcooling was 3 K. The calculated shapes are in excellent agreement with those observed in experiments. It should also be noted that in this case there was a phase lag between the two bubbles and gravity varied with time. FIGURE 6 A shows the variation of gravity with time during the experiments.[6] FIGURE 6 B shows the variation of equivalent bubble diameter with time. The effect of time-varying gravity is introduced in the model by introducing piecewise linear gravity functions as used in the experiments of Qiu *et al.*[6] The numerical data are found to match well with the experimental data. The bubbles merge when they grow to a diameter equal to the spacing between them; that is, 7 mm at 2.9 sec. Thereafter, the merged bubble lifts off at 3.188 sec. The bubble diameter at departure is much smaller than that would be the case for a single bubble at the prevailing gravity level. The cause for this is the additional lift-off force that develops during the merger. This is elaborated below.

Three Inline Bubble Merger

FIGURE 7 shows the results of numerical simulations for the growth, merger, and departure sequence of three inline bubbles placed 6 mm apart in a domain of

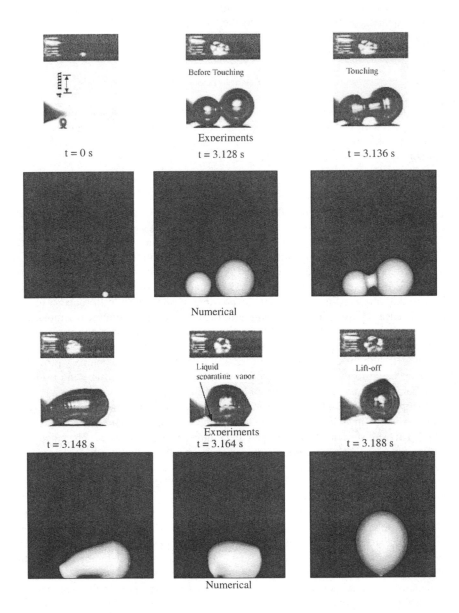

FIGURE 5. Growth, merger, and departure in a two-bubble merger for water at $\Delta T_w = 5\,\mathrm{K}$, $\Delta T_{sub} = 3\,\mathrm{K}$.

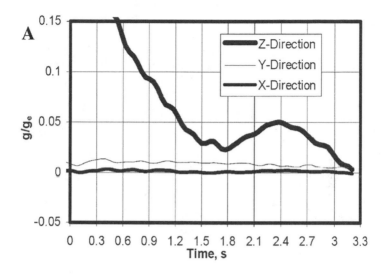

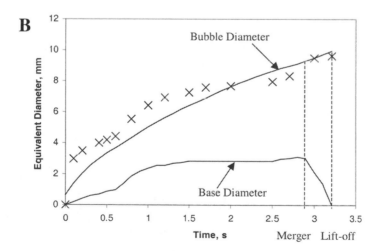

FIGURE 6. The variation of (**A**) gravity level and (**B**) equivalent bubble diameter with time for a two-bubble merger in water.

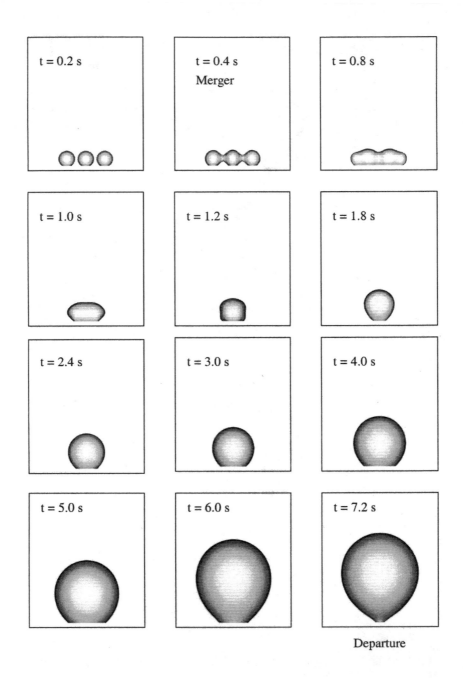

FIGURE 7. Growth, merger, and departure in a three-bubble merger for water at $\Delta T_w = 7\,\mathrm{K}$, $g = 0.01\,g_e$.

$40\,\text{mm} \times 40\,\text{mm} \times 80\,\text{mm}$ at a gravity level of $0.01\,g_e$. The results are for saturated water with a wall superheat of 7 K. FIGURE 8 A shows the corresponding variation of equivalent bubble diameter and bubble base diameter with time. The dotted lines show the data for single bubble under the same conditions. As can be seen in the figure, the bubble merger occurs when the equivalent diameters of the individual bubbles become equal to the bubble spacing. After merger, the bubble base starts to shrink in one direction and at the same time expands in the second direction. This

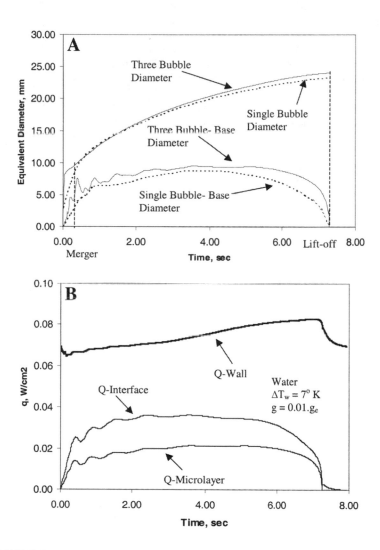

FIGURE 8. Variation of (**A**) equivalent bubble diameter and bubble base diameter and (**B**) heat transfer rate with time for a three-bubble merger placed inline 6 mm apart at a gravity of $g = 0.01\,g_e$.

process of oscillations in bubble shape as surface tension tries to round off the merged vapor mass, causes agitation in the liquid and an increase in the natural convection heat transfer from the wall to liquid. After the merger, expansion, and shrinking process, the merged bubble growth is similar to that of a single bubble. The bubble departs at the same time with same equivalent diameter as a single bubble. FIGURE 8 B shows the heat transfer associated with the three-bubble merger case discussed above. As can be seen in the figure, the wall heat transfer decreases immediately after the merger. This can be attributed to a loss in base area as the merger occurs. Thereafter, the wall heat transfer continuously increases and once again decreases as the bubble base starts to shrink before bubble lift off.

FIGURE 9 shows how the vertical component of normal shear and pressure forces acting on the bubble surface are calculated. FIGURE 10 shows the variation in normal force acting on the vapor mass during merger of three inline bubbles and two bubbles at low gravity. As can be seen in FIGURE 10 A, rapid variations in the normal force acting on the bubbles occur just after the merger. The force is negative, implying that the bubble is being pushed against the wall. The force acting on the bubble becomes positive (i.e., acts upward) before the bubble departs. As a result, the bubble base starts to shrink rapidly. For the two-bubble case, the magnitude of the normal force acting initially on the bubble is much smaller than that of the three-bubble case. The normal force decreases dramatically after merger and reaches zero value within 0.2 sec, causing the merged bubble to lift off. The reduced downward force is reflective of the additional upward acting (lift off) that develops when two bubbles are growing with a phase lag.

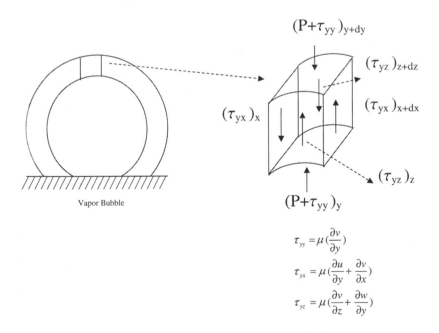

FIGURE 9. Normal force calculations on a bubble surface.

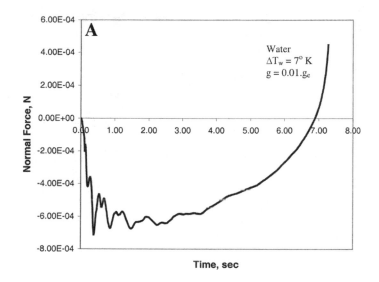

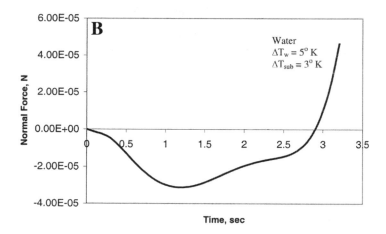

FIGURE 10. The variation of normal force with time: (**A**) for a three-bubble merger for water at $g = 0.01\, g_e$ and (**B**) for a two-bubble merger for water under the same conditions as in FIGURE 6 A.

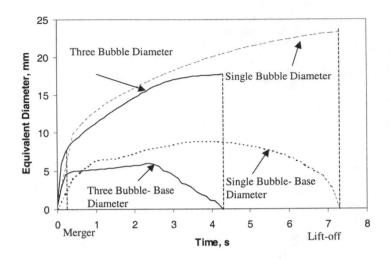

FIGURE 11. Variation of equivalent bubble diameter and bubble base diameter with time for a three-bubble merger placed inline 6 mm apart at a gravity of $g = 0.01\,g_e$. (Water, $\Delta T_w = 7\,K$.)

Bubble Merger in a Plane

FIGURE 11 shows the variation of equivalent bubble diameter and bubble base diameter with time under low gravity conditions for saturated water at a wall superheat of 7 K when three bubbles in a plane merge. The dotted lines show the prediction for single bubble under the same conditions. Unlike the case of inline bubble mergers, the bubble departure diameter and the time period of growth are smaller than that for a single bubble.

FIGURE 12 shows the equivalent bubble diameter and equivalent bubble base diameter for five, seven, and nine bubble mergers in a plane. The graphs show a similar trend during growth, merger, and departure. In all the cases, the individual bubble grows to an equivalent diameter corresponding to the spacing between them. After merger, the shape of the vapor mass continues to oscillate for some time. It should be noted that the bubble diameter at departure for the five- and seven-bubble cases is much smaller than that for a single bubble. However, for the nine-bubble case, it again increases to about the same as that for a single bubble or the three-bubble inline case. FIGURE 13 shows the variation of bubble departure diameter and time period of growth with the number of bubbles. The bubble departure diameter and time period of growth reach a minimum with five bubbles, increases thereafter, and is expected to attain a value equal to that of a single bubble as the number of bubbles increases. FIGURE 14 shows the heat transfer associated with five-, seven-, and nine-bubble mergers. A microlayer contribution of about 30% of the total heat flux is somewhat higher than that for a single bubble. The time-averaged wall heat flux increases as the number of bubbles increases.

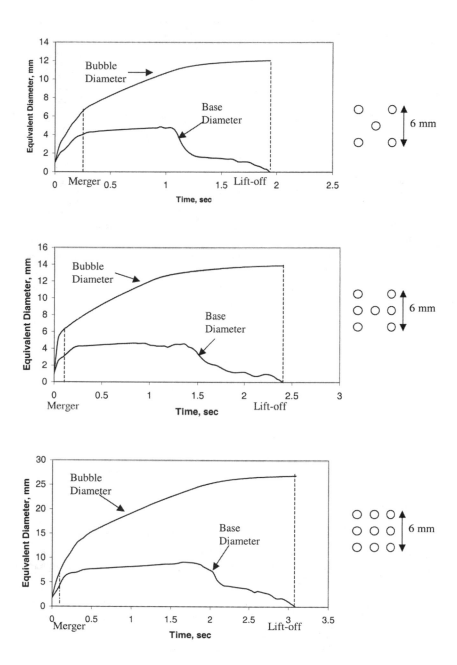

FIGURE 12. Variation of equivalent bubble diameter and bubble base diameter with time for a multiple merger in a plane for water at $\Delta T_w = 7\,\mathrm{K}$, $g = 0.01\,g_e$.

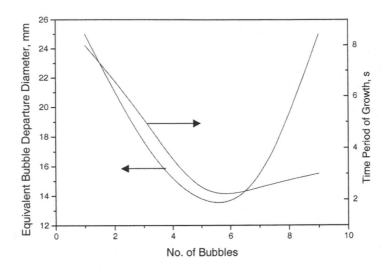

FIGURE 13. Variation of equivalent bubble departure diameter and time period of growth with number of bubbles in plane for water at $\Delta T_w = 7\,\mathrm{K}$, $g = 0.01\,g_e$.

CONCLUSIONS

1. The results of numerical simulations for a two-bubble merger in low gravity are found to be in good agreement with those obtained from experiments.

2. In low gravity, a three-bubble merger in a plane leads to early bubble lift-off and smaller bubble departure diameter, whereas in the case of a three-inline merger, the bubble departure diameter is about the same as for a single bubble.

3. After the force acting on merged vapor mass becomes positive (upward), the bubble base starts to shrink and eventually the bubble lifts off. Prior to the bubble shrinking, the force is negative and pushes the bubble toward the heater.

4. Bubble merger in a plane lead to early lift off of the bubbles under low gravity conditions. Except for nine-bubble merger case, the bubble departure diameter is much smaller than that for a single bubble case. The cause of premature bubble lift off is the additional lift off force that develops due to fluid motion generated during merger process.

5. The time averaged wall heat flux increases as the number of bubbles increases. microlayer contribution to the total heat flux is higher than that for a single bubble.

ACKNOWLEDGMENTS

This work received support from NASA Microgravity Fluid Physics Program.

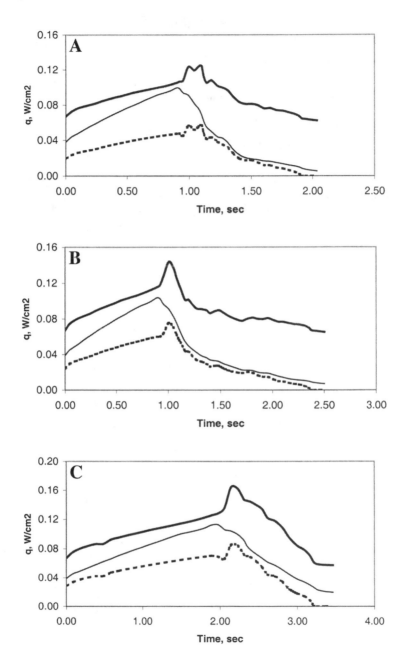

FIGURE 14. Variation of heat transfer rate with time in a multiple merger for water at $\Delta T_w = 7\,\mathrm{K}$, $g = 0.01\,g_e$: (**A**) five-bubble merger; (**B**) seven-bubble merger; (**C**) nine-bubble merger. ——————, q-wall; ‐ ‐ ‐ ‐, q-microlayer; ———, q-interface.

REFERENCES

1. SIEGEL, R. & E.G. KESHOCK. 1964. Effect of reduced gravity on nucleate bubble dynamics in water. J. AIChE **10**(4): 509–516.
2. MERTE, H. 1994. Pool and flow boiling in variable and microgravity. Second Microgravity Fluid Physics Conference, Paper No.33, Cleveland, OH, June 21–23.
3. LEE, H.S. & H. MERTE. 1997. Pool boiling curve in microgravity. J. Thermophys. Heat Transf. **11**(2): 216–222.
4. STRAUB, J., M. ZELL & B. VOGEL. 1992. Boiling under microgravity conditions. Proceedings of 1st European Symposium on Fluids In Space, Ajaccio, France, Nov. 18–22.
5. STRAUB, J. 1994. The role of surface tension for two-phase heat and mass transfer in the absence of gravity. Exp. Therm. Fluid Sci. **9**: 253–273.
6. QIU, D.M., V.K. DHIR, M.M. HASAN & D. CHAO. 2000. Single bubble dynamics during nucleate boiling under low gravity conditions. *In* Microgravity Fluid Physics and Heat Transfer. V.K. Dhir, Ed.: 62–71. Begell House, New York.
7. QIU, D.M., V.K. DHIR, M.M. HASAN & D. CHAO, 2000. Single and multiple bubble dynamics during nucleate boiling under low gravity conditions. Thirty-fourth National Heat Transfer Conference, Pittsburg, Pennsylvania, Aug. 20–22.
8. LEE, R.C. & J.E. NYADHL. 1989. Numerical calculation of bubble growth in nucleate boiling from inception to departure. J. Heat Transf. **111**: 474–479.
9. ZENG, L.Z., J.F. KLAUSNER & R. MEI. 1993. A unified model for the prediction of bubble detachment diameters in boiling systems—1. Pool boiling. Intl. J. Heat Mass Transf. **36**: 2261–2270.
10. MEI, R., W. CHEN & J.F. KLAUSNER. 1995. Vapor bubble growth in heterogeneus boiling—1. Growth rate and thermal fileds. Intl. J. Heat Mass Transf. **38**: 921–934.
11. WELCH, S.W.J. 1998. Direct simulation of vapor bubble growth. Intl. J. Heat Mass Transf. **41**: 1655–1666.
12. SON, G., V.K. DHIR & N. RAMANUJAPU. 1999. Dynamics and heat transfer associated with a single bubble during nucleate boiling on a horizontal surface. J. Heat Transf. **121**: 623–632.
13. SUSSMAN, M., P. SMEREKA & S. OSHER. 1994. A level set approach for computing solutions to incompressible two-phase flow. J. Comput. Phys. **114**: 146–159.
14. SON, G. & V.K. DHIR. 1998. Numerical analysis of film boiling near critical pressure with level set method. J. Heat Transf. **120**: 183–192.
15. SINGH, S. & V.K. DHIR. 2000. Effect of gravity, wall superheat and liquid subcooling on bubble dynamics during nucleate boiling. Microgravity Fluid Physics and Heat Transfer. V.K. Dhir, Ed.: 106–113. Begell House, New York.
16. ABARAJITH, H.S. & V.K. DHIR. 2002. Effect of contact angle on the dynamics of a single bubble during pool boiling using numerical simulations. Proceedings of IMECE2002 ASME International Mechanical Engineering Congress and Exposition, New Orleans.
17. STEPHAN, P. & J. HAMMER. 1994. A new model for nucleate boiling heat transfer. Intl. J. Heat Mass Transf. **30**: 119–125.
18. LAY, J.H. & V.K. DHIR. 1995. Shape of a vapor stem during nucleate boiling of saturated liquids. J. Heat Transf. **117**: 394–401.
19. WAYNER, P.C. 1992. Evaporation and stress in the contact line region. Proceedings of Engineering Federation Conference on Pool and Flow Boiling.
20. MUKHERJEE, A. & V.K. DHIR. 2004. Numerical and experimental study of bubble dynamics associated with lateral merger of vapor bubbles during nucleate pool boiling. J. Heat Transf. In press.

A Study of Subcooled Pool Boiling of Water

Contact Area of Boiling Bubbles with a Heating Surface During a Heating Process

KOICHI SUZUKI,[a] SAIKA TAKAHASHI,[b] AND HARUHIKO OHTA[c]

[a]*Department of Mechanical Engineering,*
Tokyo University of Science, Noda, Chiba, Japan

[b]*Graduate School of Mechanical Engineering,*
Tokyo University of Science, Noda, Chiba, Japan

[c]*Department of Aeronautics and Astronautics,*
Kyushu University, Hakozaki, Higashi-ku, Fukuoka, Japan

ABSTRACT: The contact area of bubbles with a transparent heating surface was optically measured during subcooled pool boiling of water on the ground. In the experiments, boiling bubbles were attached to the heating surface with a bubble holder and nearly reproduced the bubble behavior observed in low gravity. DC power was applied to the ITO heater and increased until the heater surface burned out. In quick heating, that is about 20 second until burnout and equal to the heating time during the low gravity period, the contact area was smaller than that for long time heating at the same heat flux. The experimental results suggest the reason why the critical heat flux in pool boiling is higher than the widely accepted predictions in microgravity. In a drop shaft experiment with constant heating, the contact area increased dramatically at the start of microgravity and became constant. Boiling bubbles coalesced and remained just over the heating surface.

KEYWORDS: subcooled pool boiling; microgravity; contact area; boiling bubbles

INTRODUCTION

Many experimental studies on boiling heat transfer in microgravity have been carried out during the past half-century. As space facilities have developed, the management and control of heat and mass transfer have become one of the most important tasks in support of space systems. On Earth, heat control is also very important as a means of energy saving in industrial fields.

Boiling heat transfer is a superior technology for removing a large quantity of heat from a hot body, but there are some difficulties to its use in electronic cooling devices and in space, for example. Because it is an unstable transition, boiling and dryout can result in considerable damage to the cooling surface of an electronic

Address for correspondence: Koichi Suzuki, Department of Mechanical Engineering, Tokyo University of Science Noda, Chiba, 278-8510, Japan. Voice: 81-4-7124-1501; fax: 81-4-7123-9814.

suzuki@rs.noda.tus-ac.jp

Ann. N.Y. Acad. Sci. **1027**: 259–268 (2004). ©2004 New York Academy of Sciences.

doi: 10.1196/annals.1324.021

device. Furthermore, the heat transfer coefficient is considerably less under microgravity conditions. Nevertheless, it is important to understand the dry out mechanism and bubble behavior in microgravity in order to use boiling phenomena in electronics and in space. In microgravity, there are not only negative factors but also positive aspects to boiling heat transfer development. Thermal convection in a low gravity field is very small. Thus, the lack of gravity is very effective in understanding bubble behavior and dryout of a cooling surface in boiling research.

In many recent microgravity experiments on pool boiling, the critical heat flux (CHF) is 100–300% higher than previous predictions.[1–3] In our low gravity experiments on subcooled pool boiling of water performed during parabolic flight of aircraft, critical heat flux (CHF) values were 100–400% higher than existing predictions and the same as for other experimental data.[3] It has been indicated that, in many cases, in pool boiling, boiling bubbles detach from the heating surface by Marangoni convection in microgravity,[4] and many theoretical treatments of bubble dynamics have reached the same conclusions. However, no bubble detachment was observed in our experiment, although the low gravity was of poor quality, $0.01\,g$–$0.04\,g$ with g-jitter.[5] Bubble detachment shown in many experiments, is an strong reason why CHF is higher than existing predictions, as is also the fact that bubbles remain on the heating surface during low gravity.

In the experiments we describe here, the contact area of bubbles with the heating surface during subcooled pool boiling of water was investigated using a transparent heating surface under the condition that boiling bubbles are attached to the heating surface. The objective was to investigate the reason why the critical heat fluxes are higher than theoretical predictions in microgravity, including drop shaft experiments.

CRITICAL HEAT FLUX IN LOW GRAVITY

FIGURE 1 shows the critical heat flux (burnout heat flux) during subcooled pool boiling of water under low gravity obtained during parabolic flight of aircraft and in ground experiments.[5,6] In low gravity, increasing AC electric power was applied to a thin stainless plate 20 mm long, 5 mm wide, and 0.1 mm thick until burnout occurred. The critical flux values are higher than existing predictions, as shown during recent microgravity experiments.[3]

In the ground experiments, the critical heat flux values were obtained by adjusting of the clearance between bubble holding plate and heating surface. While extending the heating time until burnout occurred, the heat flux gradually decreased until about 80 sec and then remained constant, as shown in FIGURE 2. The constant heat flux values coincide well with values calculated from existing predictions.[7]

According to the time dependence of the critical heat flux, we assume that some liquid layers remain between the rapidly expanding bubbles and the heating surface during the short period heating, 20–30 sec. In low gravity duration and rapid evaporation of the liquid layer the critical heat flux increases, as illustrated in FIGURE 3.

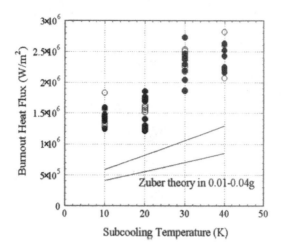

FIGURE 1. Burnout heat flux with liquid subcooling in microgravity (○) and ground (●) experiments. (Reproduced from Refs. 5 and 6, with permission.)

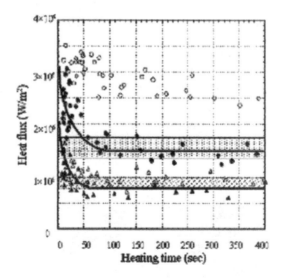

FIGURE 2. Effect of heating time until burnout on critical heat flux in subcooled pool boiling of water: 40 K subcooling without (○) and with (●) cover; 10 K subcooling without (△) and with (▲) cover; Dhir–Zuber theory subcooling at 40 K (▨) and 10 K (▨), 0.01 g–0.04 g. (Reproduced from Ref. 7, with permission.)

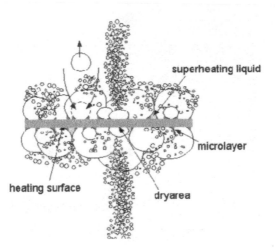

FIGURE 3. Bubble behavior and rapid evaporation of liquid layer during subcooled pool boiling for short period heating in low gravity. (Reproduced from Ref. 7, with permission.)

EXPERIMENTS ON CONTACT AREA OF
BOILING BUBBLES WITH THE HEATING SURFACE

Ground Experiment

To confirm the effect of heating time on critical heat flux, we investigated the contact area of boiling bubbles using a transparent heating surface and the observation system illustrated in FIGURES 4 and 5. The transparent heating surface is composed of soda lime glass and an ITO coating layer, 20 mm long and 5 mm wide. The clearance between holding plate and heating surface is 4 mm for 10 K and 1 mm for 40 K

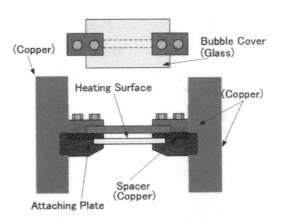

FIGURE 4. Transparent heating surface assembly with ITO coating and bubble holding plate.

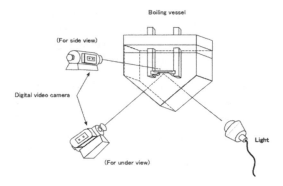

FIGURE 5. System for observing the contact area of boiling bubbles with the transparent heating surface.

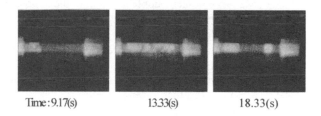

FIGURE 6. High-speed video pictures of the contact area of boiling bubbles with the transparent heating surface during quick heating at 40 K of liquid subcooling in a ground experiment.

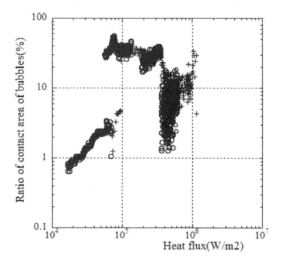

FIGURE 7. Contact area of bubbles for varying heat flux at 10 K of liquid subcooling in a ground experiment: +, quick eating with cover; ○, long time heating with cover.

of liquid subcooling, the same as in the case of the stainless steel heating surface shown in FIGURE 4.

The heating section is placed in the transparent pool vessel shown in FIGURE 5 and DC electric power is applied to the ITO heater. A light beam is applied to the back of the heating surface and a high-speed video camera focused on the light spot. The light beam is reflected at the dry surface of a bubble causing the contact area of the bubble to appear brightest.

An example of the high-speed contact area video pictures of boiling bubbles is shown in FIGURE 6 for quick heating at 40 K of liquid subcooling. The quick heating procedure is same as that used in the low gravity experiment. The long time heating is about 200 sec. The total contact area of boiling bubbles is obtained by summing the areas of the brightest parts and is expressed as a fraction of the heating surface area. Examples of the contact area of boiling bubbles with varying heat flux are shown in FIGURE 7 at 10 K and in FIGURE 8 at 40 K of liquid subcooling. As time passes, the heating power increases and the bubbles spread with larger contact area, shown in FIGURE 6, but a large coalesced bubble slips out of the bubble holder before burnout occurs. For this reason, the contact area of bubbles decreases again, as shown on the right of FIGURES 7 and 8.

No remarkable differences in the heating time are indicated for the data at 10 K of liquid subcooling, but the contact area is likely to be smaller in quick heating than it is in long time heating at same heat flux. At 40 K of liquid subcooling, the contact area in quick heating is smaller than that in long time heating at same heat flux, as shown in FIGURE 8.

The experimental data supports our assumption that the rapid evaporation of the liquid layer remaining between rapid expanded bubbles and the heating surface results in values close to the higher critical heat flux in microgravity. Hence, the heating process is more or less influenced by the bubble behavior and heat flux in microgravity. Of course in many cases, it is easy to accept that liquid is supplied to the heating surface by bubble detachment caused by Marangoni convection in microgravity.

Observation of Contact Area of Boiling Bubbles with the Heating Surface in a Drop Shaft Experiment

The contact area of boiling bubbles with the heating surface was investigated in a microgravity experiment performed by JAMIC. The experimental setup used in the ground experiment, shown in FIGURES 4 and 5, was employed in the drop shaft experiment. Dry batteries were used to heat the ITO surface. The temperature of liquid could not be controlled, but it was about 40 K in liquid subcooling at the start of drop. The heating of the ITO surface was started 10 sec before the drop. Constant heating power was applied to the heating surface. The change in contact area is shown in FIGURE 9. At 5.22×10^5 W/m^2 of applied heat flux, the contact area increased dramatically at the start of microgravity and then became constant.

Thus, the heating process is considered rather similar to long time heating. The heat flux values obtained in the drop shaft experiment are indicated on the boiling curve given by long time heating, as shown in FIGURE 10. Typical data suggest that the bubbles stay on the heating surface in microgravity, but that detachment of

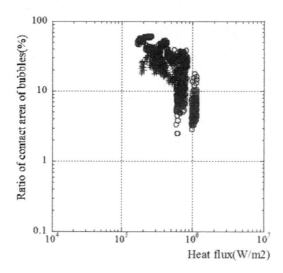

FIGURE 8. Contact area of bubbles for varying heat flux at 40 K of liquid subcooling in a ground experiment: +, quick heating with cover; ○, long time heating with cover.

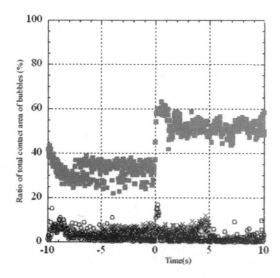

FIGURE 9. Progressive change in contact area of bubbles in microgravity, performed in a drop shaft experiment at JAMIC: ○, $2.82 \times 10^6 \, \text{W/m}^2$; ■, $5.22 \times 10^6 \, \text{W/m}^2$; ×, $8.77 \times 10^6 \, \text{W/m}^2$. (Reproduced from Ref. 6, with permission.)

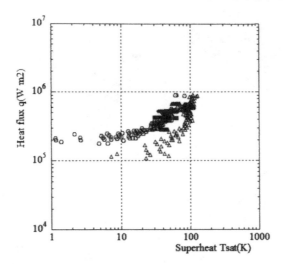

FIGURE 10. Boiling curves for quick heating and long time heating at 40 K of liquid subcooling obtained in a ground experiment under microgravity at 72 V (●), 96 V (■), 108 V (◆), 132 V (▲), and 144 V (○); △, quick heating with cover; ○, long time heating with cover.

bubbles occurs. If the bubbles did not detach from the heating surface, the contact area of bubbles would increase. In this case, generated bubbles would coalesce just above the heating surface as a cloud and absorb small bubbles lifted from the heating surface, as shown in FIGURE 11, an idea proposed by Merte.[8]

To confirm the hovering of coalesced bubbles in microgravity, the contact area of bubbles was investigated using the bubble holder, as shown in FIGURE 11, for various heating power values and clearances between bubble holder and heating surface. The contact area is 50–70% of the heating surface area in many cases, as shown in FIGURE 12. Then the coalesced bubbles at $5.22 \times 10^5 \, \text{W/m}^2$ obtained in the drop shaft experiment are estimated to hover less than about 1.5 mm from the heating surface.

In industrial technology, subcooled flow boiling has been proposed for high heat flux cooling of future electronic devices. The cooling fluid needs to be passed through a narrow channel on the cooling surface of the electronic package. In an emergency, when the fluid flow is stopped, for example, heat is removed from the

Bubble holder

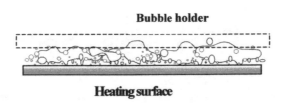

Heating surface

FIGURE 11. Image of bubble behavior in subcooled pool boiling under microgravity. Coalesced bubbles hover over the heating surface. (Reproduced from Ref. 6, with permission.)

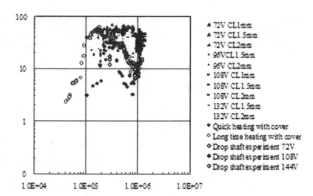

FIGURE 12. Contact area of bubbles in nucleate boiling for various heating powers at 40 K of liquid subcooling in drop shaft and ground experiments.

cooling surface by pool boiling in the narrow space until the electronic system can be shut down. The results of the present experiments are very effective in predicting the cooling limit of heat flux for electronic devices in the case of an urgent shutdown.

CONCLUSIONS

Bubble behavior is very important in boiling heat transfer. Why is the critical heat flux higher than theoretically predicted under short period microgravity in the case that boiling bubbles attach to the heating surface? The contact area of boiling bubbles with heating surface was observed using a transparent heating surface in subcooled pool boiling of water for heating process to understand bubble behavior based on the short period microgravity experiments.

The contact area of bubbles in quick heating is the same as for short period microgravity but smaller than that in long time heating at same heat flux. In short period microgravity, boiling bubbles expand rapidly and some liquid layers remain beneath the bubbles. The rapid evaporation of the liquid layers might be closely related to the higher critical heat flux in microgravity.

In a drop shaft experiment on subcooled pool boiling of water with constant heating power, the contact area of bubbles dramatically increases at the start of microgravity and then becomes constant. Boiling bubbles coalesce and remain just over the heating surface, like a cloud. The distance of the bubble cloud is estimated to be less than about 1.5 mm above the heating surface, according to experimental results from the ground experiment. The experimental results offer effective data for predicting an urgent cooling limit for electronic devices using subcooled flow boiling.

ACKNOWLEDGMENT

Some of the experiments described in this paper were carried under the foundation of Grants-in-Aid for Scientific Research, promoted by Japan Society for Promotion of Science (JSPS) in 2002. The authors greatly appreciate the support of JSPS.

The authors acknowledge the professional assistance of JAMIC for the drop shaft experiment and the work of Mr. Motohiro Suzuki (Toyota Motor Company) and Mr. Junya Kawasoko (Daihatsu Motor Company) for the series of experiments.

REFERENCES

1. KUTATELADZE, S.S. 1952. Heat transfer in condensation and boiling. USAEC. Rep.AEC-tr-3770.
2. ZUBER, N. 1958. On stability of boiling heat transfer. Trans. ASME. J. Heat Transf. **80:** 711–720.
3. STRAUB, J., M. ZELL & B. VOGEL. 1991. Boiling under microgravity conditions. Proc. 1st European Symposium on Fluid in Space. Ajaco, France. ESA SP-353. 269–297.
4. OKA,T., Y. ABE, Y. MORI & A. NAGASHIMA. 1994. Pool boiling heat transfer in microgravity-experiments with CFC-113 and water using drop-shaft facility. Trans. JSME **60-557:** 3093–3100.
5. SUZUKI, K., H. KAWAMURA, et al. 1997. Burnout in subcooled pool boiling of water under microgravity. Proc. Xth European and VIth Russian Symposium on Physical Science in Microgravity. St.Petersburg, Russia. 366–369.
6. SUZUKI, K., H. KAWAMURA, et al. 1999. Experiments on subcooled pool boiling of water in microgravity (observation of bubble behavior and burnout). Proc. 5th ASME-JSME Joint Thermal Engineering Conference. No. 6420. CD-ROM.
7. SUZUKI, K., M. SUZUKI, S. TAKAHASHI, et al. 2003. Bubble behavior in subcooled pool boiling of water under reduced gravity. Proc. Space Technology and Applications International Forum 2003. 142–148.
8. H.S. LEE & H. MERTE. 1999. Pool boiling mechanisms in microgravity. Proc. International Conference on Microgravity Fluid Physics and Heat Transfer. 126–135.

Microgravity Experiments on Phase Change of Self-Rewetting Fluids

YOSHIYUKI ABE,[a] AKIRA IWASAKI,[a] AND KOTARO TANAKA[b]

[a]*National Institute of Advanced Industrial Science and Technology, Tsukuba, Japan*

[b]*Shizuoka Institute of Science and Technology, Fukuroi, Japan*

ABSTRACT: A series of microgravity experiments on self-rewetting fluids has been conducted at the 10-second drop shaft of the Japan Microgravity Center (JAMIC). In all the experiments, 1.5 wt% of 1-butanol aqueous solution were employed as a self-rewetting fluid. The objective of the first experiment was to observe the boiling behavior of two-dimensional adjacent dual vapor bubbles with the aid of a two-wavelength interferometer and tracer particles. A significant difference was observed between a self-rewetting fluid and a normal fluid (CFC-113 in this experiment) in bubble interaction and flow developed along vapor/bubble interface. The second experiment focused on the flow at the bubble/heater contact area and around the three-phase interline, visualized with tracer particles. Differing behavior among three fluids, 1-butanol aqueous solution, CFC-113, and ethanol aqueous solution, was observed. The last microgravity experiment was a demonstration of wickless heat pipes containing three different fluids as a working fluid, 1-butanol aqueous solution, water, and ethanol aqueous solution. The temperature variation of working fluid in the heat pipe was monitored, and the liquid flow returning from the condensation region to the evaporation region was visualized by tracer particles. In addition to microgravity experiments, the performance of conventional heat pipes with 1-butanol aqueous solution was evaluated on the ground, and compared with water heat pipes. Our preliminary results are presented.

KEYWORDS: microgravity; boiling phenomena; heat pipe; Marangoni effect; aqueous solutions; non-azeotropic mixtures

NOMENCLATURE:

C	concentration
U	liquid velocity in the y direction
y	axial direction of heat pipe
z	radial direction of heat pipe
σ	surface tension
η	viscosity

INTRODUCTION

The objectives of nearly a half a century of studying microgravity boiling[1–7] can be classified into two main categories: (1) to obtain heat transfer data for space thermal management applications; and (2) to understand boiling mechanisms, such as

---

Address for correspondence: Yoshiyuki Abe, Energy Technology Research Institute, AIST, Tsukuba, Ibaraki 305-8568 Japan. Voice: 81-29-861-5749; fax: 81-29-861-5709.
 y.abe@aist.go.jp

Ann. N.Y. Acad. Sci. **1027:** 269–285 (2004). ©2004 New York Academy of Sciences.
doi: 10.1196/annals.1324.022

nucleation, bubble growth/coalescence, microlayer/macrolayer, three-phase inter-line, and thermocapillary convective flow around a bubble, by using a microgravity environment.

Since the role of surface tension in phenomena that accompany the liquid–gas interface is more pronounced in microgravity conditions, that environment may pro-vide us a better understanding on the significance of the Marangoni effect in the boil-ing process.[8]

Although most of studies on microgravity boiling have been conducted with sin-gle component fluids, for more than a decade the present authors have focused atten-tion on binary mixtures. The original objective of our interest in binary mixtures was a possibility for passive boiling heat transfer enhancement in microgravity environ-ment. In the case of non-azeotropic compositions, preferential evaporation of more volatile component takes place. Preferential evaporation results in a composition gradient along the liquid/vapor interface, and the surface tension gradient due to the composition gradient induces Marangoni flow. Concerning the so-called positive mixtures, such as the ethanol–water system, in which the more volatile component has a lower surface tension, it is expected that the Marangoni effect induces liquid inflow to the bubble–heater contact area, which may result in an enhancement of boiling heat transfer, especially under microgravity conditions.

A series of microgravity experiments with ethanol–water binary mixtures (11.3 wt% and 27 wt%) were conducted in the drop shaft of the Japan Microgravity Center (JAMIC) in which a high quality microgravity condition of the order of 10^{-4}G was available for 10 sec.[9] The experimental results showed that the nucleate pool boiling heat transfer coefficients in microgravity were considerably enhanced (20% to 60%) compared to those under terrestrial conditions.[10] The appreciable pool boiling heat transfer enhancement for positive binary mixtures in microgravity was also subsequently confirmed by Ahmed et al.[11]

The present authors also pointed out, from observation of a transparent heater surface from the rear side and by interferometry study, that the bubble–heater contact behavior for binary mixtures is totally different to that for single component fluids, such as CFC-113 or water. Single component fluids showed the development and subsequent consumption of a thin microlayer at the bubble–heater contact area in the course of bubble growth. In contrast, the bubble growth of binary mixtures is not accompanied by the development of a microlayer at the bubble–heater contact area but develops a liquid layer of appreciable thickness as a result of vigorous radial liq-uid inflow to the center of the contact area, which then pushes the grown bubble away from the heater surface.[10]

In our next step we tried to visualize both temperature and concentration profiles around a two-dimensional single vapor bubble of binary mixtures with the aid of a two-wavelength interferometer in the microgravity environment available at JAMIC. The development processes of temperature and concentration fields around a single bubble for a CFC-12–CFC-112 mixture (6.2 wt% of CFC-12) were successfully identified by analysis of the interferograms.[12] Although a clear interferogram did not develop around the vapor bubble for a saturated ethanol–water mixture (31 wt% of ethanol), we did note that the bubbles in ethanol–water are always accompanied by a secondary bubble at the bubble base and that the primary bubble is lifted up and pushed away by growth of the secondary bubble, finally leaving the heater surface.[13]

This observation supports the existence of a liquid layer of appreciable thickness, as we noted elsewhere.[10]

The Marangoni effect due to a concentration gradient around the vapor bubble in positive binary mixtures mentioned so far induces liquid inflow to the bubble–heater contact area. Another Marangoni effect due to temperature gradient, in other words a thermocapillary effect, takes place at the liquid–vapor interface. The incipient mechanism of the thermocapillary effect, temperature gradient along the liquid/vapor interface, might be attributable to:

- disjoining pressure at the microregion of the liquid meniscus,[14]

- capillary pressure due to curvature gradient along bubble interface,[15] and

- partial pressure gradient in bubble due to a trace of incondensable gas dissolved in the liquid.[16]

Although discussion on the origin of the thermocapillary effect in boiling process is rather sparse,[17] a number of previous papers point out, mostly from optical procedures, the significance of the thermocapillary effect in the case of subcooled boiling of single component fluids in microgravity or while using a downward facing heater orientation. We also showed the temperature profile developed by the thermocapillary effect in the interferogram for CFC-113 obtained in microgravity.[12] The thermocapillary convective flow developed along the bubble interface directs from the bubble base toward the bubble top, and as a result the bubble is constrained to the heater surface, which is actually the opposite direction to Marangoni flow induced by the concentration gradient for positive mixtures. In the case of highly subcooled conditions or for those mixtures with a very narrow difference between bubble point and dew point, boiling heat transfer enhancement in positive mixtures may not be emphasized because the thermocapillary flow dominates over the Marangoni flow due to concentration gradient. If the surface tension increases with increasing temperature, however, the situation would be dramatically improved.

Although most of fluids show a decrease in the surface tension with increasing temperature, some fluids, such as molten SiO_2, GeO_2, B_2O_3,[18] some binary alloys, and dilute aqueous solutions of higher alcohols, are known to exceptionally show the opposite behavior. The temperature variation of the surface tensions of aqueous solutions of higher alcohols with more than four carbon atoms was found to show minima; the surface tension increases with increasing temperature at higher temperatures, these temperatures giving a minimum decrease with an increase in the number of carbon atoms in the alcohols.[19] In these particular fluids, the thermocapillary effect induces a liquid inflow to the bubble–heater contact area, which is expected to enhance boiling heat transfer. In addition, since the dilute aqueous solutions are non-azeotropic, the Marangoni flow due to concentration gradient becomes appreciable. This function of the Marangoni flow is also expected to induce a liquid inflow to the bubble–heater contact area, since the alcohol rich component preferentially evaporates in dilute composition, although aqueous solutions of higher alcohols are essentially so-called negative mixtures.

The significance, especially in microgravity, of the coupled Marangoni effects caused by this surface tension behavior may be listed as follows:

- spontaneous cooling by liquid supply to hot spot or hot region in liquid free surface,

- emphasized boiling heat transfer enhancement by amplified liquid inflow to the bubble–heater contact area in conjunction with Marangoni flow due to a concentration gradient, and

- spontaneous rewetting by liquid supply in conjunction with Marangoni flow to incidental dry patches appearing on heater surface in the course of nucleate boiling, which prevents the development of dry patches.

The term "self-rewetting" that appears in the title of this paper was coined for these particular characteristics. The present authors have been conducting a series of microgravity experiments at JAMIC focused on self-rewetting fluids. Dual bubble experiments on a self-rewetting fluid are reported elsewhere.[20] TABLE 1 summarizes our classification of fluids from the viewpoint of their surface tension characteristics.

The objective of the dual bubble experiment was to visualize the flow, temperature, and concentration profiles developed around/between two adjacent bubbles with the aid of 200–230μm diameter glass tracer particles and a two-wavelength interferometer. The dual bubbles were nucleated at spot heaters fabricated on a glass substrate, and then grown. The experimental apparatus was basically the same as that described in Reference 12, except for the interferometer. The He–Ne lasers of 544nm and 633nm used previously were replaced with a 437nm solid-state laser and a 780nm laser-diode.

In this experiment, CFC-113 subcooled at 5K and 1.5wt% 1-butanol aqueous solution at the saturation condition were employed. The dual bubbles of CFC-113 immediately coalesced in the growing process, and the coalesced bubble kept attaching to the heater surface. A strong Marangoni flow was then developed along the bubble interface, which yielded a vigorous flow from the bubble base toward the bubble top of up to 50mm/sec. Estimation of the Marangoni number, by using the temperature difference as a measure of subcooling, resulted in 2.6×10^5 in this case. For 1-butanol aqueous solution, on the other hand, the behavior was totally different. A number of small bubbles, typically up to 5mm in diameter, were nucleated and immediately emitted from the heater surface. A moderate flow of 4–10mm/sec toward the bubble base developed and the bubble departure was promoted by a relatively strong flow, up to 20mm/sec, near the three-phase interline toward the bubble–heater contact area. Although the bubble size emitted from the heater surface was considerably smaller than ethanol aqueous solution, we observed the growth of a secondary bubble at the base of primary bubble. We had first observed this in the previous microgravity experiment for ethanol aqueous solution, in 1-butanol aqueous

TABLE 1. Classification of fluids from surface tension behavior

	Single Component	Positive Mixtures	Negative Mixtures	Self-Rewetting
$\dfrac{\partial \sigma}{\partial T}$	< 0	< 0	< 0	> 0
$\dfrac{\partial \sigma}{\partial C}$	—	< 0	> 0	< 0

solution. This observation supports the existence of certain amount of liquid at the bubble–heater contact area.

The role of coupled Marangoni effects in nucleate pool boiling may be appreciable not only in microgravity, but also in the normal gravity condition. It is worthwhile to recall that 2.5–3.5 times higher CHFs (critical heat fluxes) at particular compositions of dilute aqueous solution of 1-butanol and 1-pentanol have been reported.[21] These facts suggest that the self-rewettability caused by the coupled Marangoni effects plays a significant role in the increase in CHF, although Hovestreijdt discussed only the Marangoni effect due to the concentration gradient.[8]

One of the most promising applications of self-rewetting fluids to space thermal management is a wickless heat pipe, by means of coupled Marangoni effects. The idea of wickless heat pipes using ethanol aqueous solutions at non-azeotropic compositions was successfully demonstrated under microgravity conditions, and the Marangoni effect due to concentration gradient of the working fluid was shown to provide a motive force that was strong enough to return the condensate to the evaporation region of heat pipes.[22] Although the attention of the previous study was only focused on the Marangoni effect caused by concentration gradient, the coupled Marangoni effects claimed in the present study are expected to generate a rather strong motive force for returning the working fluid in wickless heat pipes.

This paper describes the fundamental characteristics of boiling behavior of self-rewetting fluids and the experimental results for wickless heat pipes with self-rewetting fluids from viewpoint of possible applications to space thermal management devices. The self-rewetting characteristics may allow an improvement in the boiling limit and the capillary limit in conventional heat pipes, even in the terrestrial conditions. Preliminary results of the performance evaluation of conventional heat pipes with self-rewetting fluids as a working fluid are also introduced. In addition to heat transfer studies, a thorough study of the precise surface tension measurements for self-rewetting fluids with the aid of the differential capillary method is ongoing work by our colleagues. The existence of a minimum in the surface tension variation against temperature is confirmed in 1-butanol aqueous solution.[23] It should be noted that, from exactly the same point of view as that of the present authors, Zhang and Chao proposed models to predict the bubble detachment diameters, the nucleate boiling heat transfer coefficient, and the CHF for these particular aqueous solutions.[24]

EXPERIMENTAL

Two different microgravity experiments were conducted at the drop shaft in JAMIC. Conventional heat pipe performance tests were conducted on the ground.

Single Bubble Experiment

The point of this experiment was to observe the flow developed around the bubble–heater contact area. To this end, a pair of 45° prisms was employed, as shown schematically in FIGURE 1. As illustrated, an indium tin oxide (ITO) transparent spot heater was fabricated on one of the prism surfaces and nickel was plated on the surface of the other prism as a mirror. One can observe the bubble/heater contact from the side and the behavior of the tracer particles around the bubble/heater contact area

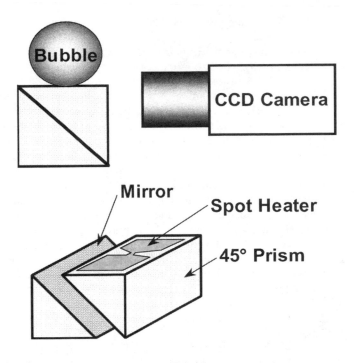

FIGURE 1. Principle of observation in a single bubble experiment.

from the bottom simultaneously by using the prism heater. Four chromel-alumel sheathed thermocouples of 0.25 mm in diameter were immersed in the boiling cell to monitor the temperature profile; one of them was placed as close as possible to the spot heater.

Three different fluids, CFC-113, 20 wt% of ethanol aqueous solution, and 1.5 wt% 1-butanol aqueous solution, were used in this experiment.

Wickless Heat Pipe Experiment

A glass tube 10 mm inner diameter, 1 mm thick, and 165 mm long was employed as a wickless heat pipe. A transparent ITO heater was fabricated on the outer surface of the closed end of the tube, 50 mm long. Four chromel-alumel sheathed thermocouples of 0.25 mm diameter were put in the heat pipe at the inner wall to monitor the temperature distribution of the working fluid, and one thermocouple was attached to the heat pipe outer wall to monitor the wall temperature in the evaporation region. Tracer particles, 100 μm diameter hollow glass spheres, were mixed in 4 cm^3 of working fluid. FIGURE 2 illustrates schematically the experimental setup. The transparent wickless heat pipe was previously heated by the ITO heater with a slight inclination; that is, bottom heating mode. The same experimental conditions continued during microgravity, and temperature variation and the motions of bulk liquid and the tracer particles were monitored.

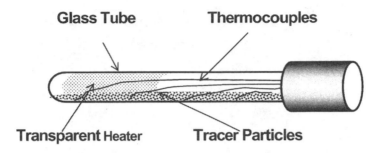

FIGURE 2. Glass tube wickless heat pipe.

The experiments were conducted for three different fluids, water, 20 wt% of ethanol aqueous solution, and 1.5 wt% 1-butanol aqueous solution.

Heat Pipe Performance Test

Conventional copper heat pipes with a composite wick, groove plus fine copper wire, were manufactured with water and 1.5 wt% 1-butanol aqueous solution as a working fluid. The performance tests were conducted on ground for comparison. The same length (250 mm) heat pipes were used, but two different sizes, 4 mm and 8 mm diameter, were prepared for each working fluid. The heat pipes contained 0.41 cm^3 and 1.9 cm^3 of working fluid, respectively.

A copper block, 50 mm long with cartridge heaters, was attached to the evaporation region of the heat pipe, and a water-cooled copper jacket 75 mm long was attached to the condensation region (see FIGURE 3). The temperature distribution on the heat pipe wall, the electric power input, the inlet and outlet temperatures at the copper cooling jacket, and the flow rate of cooling water were monitored in the

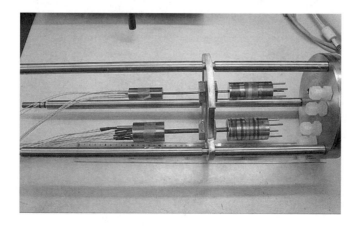

FIGURE 3. Setup of heat pipes.

FIGURE 4. Installation in vacuum chamber.

experiment. The entire apparatus was installed in a vacuum chamber for thermal insulation, as shown in FIGURE 4.

RESULTS AND DISCUSSION

Single Bubble Experiment

FIGURE 5 shows an example of the observation view for CFC-113 from the rear side of the transparent heater. Immediately after nucleation of the vapor bubble, the development of a circular stationary bubble–heater contact area of about 3 mm diameter was observed around the 1 mm × 1 mm spot heater. The tracer particles were first excluded from the spot heater at the inception of boiling, and then accelerated, sliding

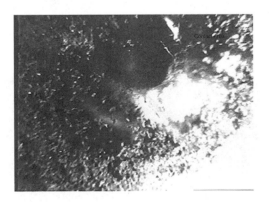

FIGURE 5. Observation from rear side of heater (CFC-113).

from the surrounding area to the contour of the contact area. The tracer particles then suddenly left the contour and moved away along the bubble interface. This inward radial liquid flow, in other words liquid supply, along the heater surface was partly due to counter flow from the Marangoni effect and partly due to consumption of liquid evaporated at the three-phase interline.[25] The maximum velocity was 5 mm/sec near the contact area contour. The temperature of a thermocouple located at the spot heater in the course of microgravity boiling is shown in FIGURE 6. As can be seen in the figure, the temperature was exactly at the saturation temperature without appreciable variation.

In contrast to CFC-113, the experimental results for 1-butanol aqueous solution were entirely different. No stationary bubble–heater contact was maintained (see FIGURE 7) and a vigorous inward radial flow along the heater surface to the spot heater was observed. The fluid velocity along the heater surface reached about 15 mm/sec at the spot heater and 2.5 mm/sec at 3 mm from the spot heater. FIGURE 8 depicts the temperature of a thermocouple at the spot heater. A large amplitude temperature fluctuation implies a vigorous liquid supply to the spot heater caused by the coupled Marangoni effects.

The experimental results for ethanol aqueous solution were similar to those for 1-butanol aqueous solution, but much more moderate. We observed periodic bubble growth and departure with 6–10 Hz frequency, and also periodic radial inward and outward flow synchronized with bubble detachment. The velocity along the heater surface was rather moderate, 0.7 m/sec, compared to prescribed fluids. The temperature of a thermocouple at the spot heater also showed a moderate fluctuation in microgravity boiling (see FIGURE 9), the amplitude and the frequency were both

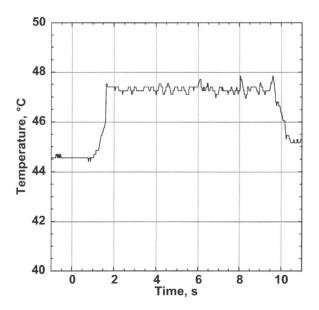

FIGURE 6. Temperature at spot heater (CFC-113).

FIGURE 7. Observation from rear side of heater (1-butanol aqueous solution).

appreciably more moderate compared to 1-butanol aqueous solutions. This different behavior from the coupled Marangoni effect is attributable to the opposite direction of two different Marangoni effects induced by the concentration gradient and the temperature gradient.

Wickless Heat Pipe Experiment

As expected, water did not allow spontaneous liquid return from the condensation region to the evaporation region. Liquid plugs were formed in the glass tube and a

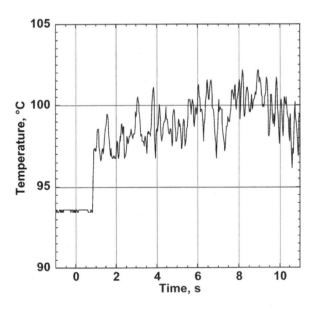

FIGURE 8. Temperature at spot heater (1-butanol aqueous solution).

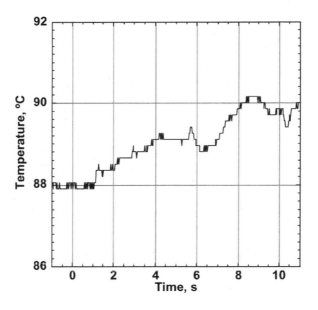

FIGURE 9. Temperature at spot heater (ethanol aqueous solution).

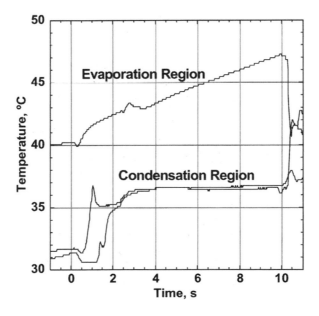

FIGURE 10. Temperature of liquid in a glass tube wickless heat pipe (water).

moderate Marangoni flow was observed at each liquid plug. The plug formation behavior in microgravity depends on the amount of liquid in the tube if the condition is isothermal.[26] FIGURE 10 shows the temperature distribution in the heat pipe in the course of microgravity. As can be seen in the figure, the temperature at the evaporation region monotonically increased. It was also confirmed from the temperature history that the glass tube did not work as a heat pipe.

In the case of ethanol aqueous solution, spontaneous liquid return by the Marangoni effect observed previously[22] was reconfirmed. Immediately after the microgravity condition, the entire inner wall of the glass tube was covered with liquid film, and liquid plugs were formed at both end of the glass tube. The liquid return velocity along the heat pipe wall was 13 mm/sec.

1-Butanol aqueous solution gave rather enhanced liquid return to the evaporation region, and the liquid velocity measured was 25 mm/sec. The temperature distribution in the heat pipe is shown in FIGURE 11, in which the temperatures at the evaporation and the condensation regions are nearly constant during the microgravity condition.

The measured liquid velocity was evaluated by estimating the balance of the shear stress and the surface tension in the estimated liquid film of uniform thickness 0.55 mm at the heat pipe inner wall,

$$\frac{d\sigma}{dy} = -\eta \frac{du}{dz}, \tag{1}$$

where σ is the surface tension, y the axial direction of the heat pipe, η the viscosity, u the velocity, and z the radial direction of heat pipe. Assuming the concentration at the condensation region as the equilibrium vapor composition, and estimating the

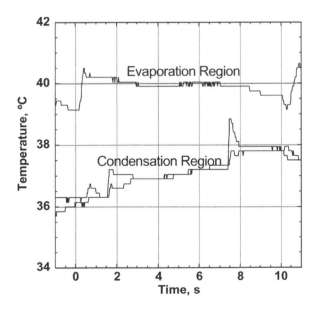

FIGURE 11. Temperature of liquid in a glass tube wickless heat pipe (1-butanol aqueous solution).

thermophysical properties for 1-butanol aqueous solution,[27] the axial mean velocities for ethanol aqueous solution and 1-butanol aqueous solution were estimated to be 13 mm/sec and 21 mm/sec, respectively. Agreement between the measured fluid velocities and estimated values was excellent. Since the estimate of the surface tension for aqueous solutions[27] does not take into account the surface tension anomaly of 1-butanol aqueous solutions, it is likely that the velocity is underestimated. If more precise experimental data for the surface tension of 1-butanol aqueous solution were available, the agreement would be improved.

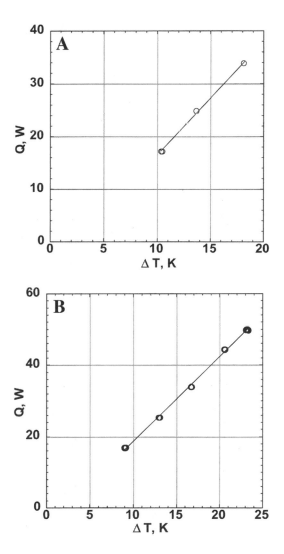

FIGURE 12. Heat transfer characteristics of a conventional heat pipe: (**A**) water, (**B**) 1-butanol aqueous solution.

Heat Pipe Performance Test

Although experimental work on the performance test of conventional heat pipes has just been initiated, we have obtained preliminary results that show an improvement in the performance of 1-butanol aqueous solution heat pipes over water heat pipes. Since the role of surface tension is more pronounced in a smaller scale, the performance improvement was more appreciable for 4mm diameter heat pipes. We have, so far, not reached a definite conclusion for 8mm heat pipes. FIGURE 12 compares the heat transfer capability of 4mm heat pipes as a function of the temperature difference between the evaporation region and the condensation region. The

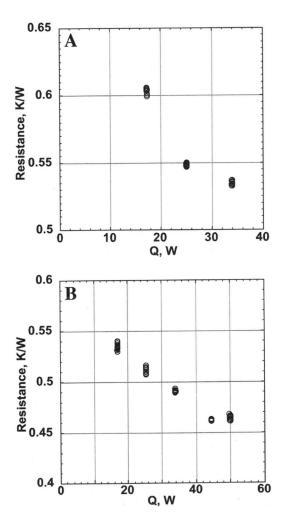

FIGURE 13. Thermal resistance of a conventional heat pipe: **(A)** water, **(B)** 1-butanol aqueous solution.

1-butanol aqueous solution heat pipe did work without dryout up to 50 W, although the water heat pipe dried out around 35 W. A comparison of the thermal resistance is given in FIGURE 13, in which we note a lower thermal resistance for 1-butanol aqueous solution heat pipes.

CONCLUDING REMARKS

The uniqueness of self-rewetting fluids, especially in microgravity applications, was emphasized and a series of microgravity experiments at JAMIC and a preliminary ground-based experiment were described. The single bubble experiment in microgravity qualitatively revealed liquid inflow to the nucleation site caused by coupled Marangoni effects in the self-rewetting fluid. The experiment for single component fluid also gave a new experimental finding, radial liquid inflow along the heater surface to the contour of the stationary bubble–heater contact area; that is, the stationary three-phase interline.

The demonstrative experiment of a glass tube wickless heat pipe with the self-rewetting fluid as a working fluid showed an outstanding practicability in microgravity. Although ordinary positive non-azeotropic binary mixtures, such as ethanol aqueous solution, allow for the operation of wickless heat pipe in microgravity, a considerably faster liquid return was achieved by using the self-rewetting fluid because of coupled Marangoni effects. Such characteristics are believed to improve the dryout limit of wickless heat pipes.

The advantage of self-rewetting fluids is promising, not only under microgravity conditions, but probably also in terrestrial applications. Conventional heat pipes were fabricated with the self-rewetting fluid and their performance was compared experimentally with conventional water heat pipes. Although the experiment has not yet been completed, we have confirmed appreciable improvement in dryout limit and thermal resistance compared to water heat pipes, especially for a smaller size.

The applications of self-rewetting fluids are considered to be extensive, not only in cooling technology, such as spray cooling, liquid film cooling, and quenching, but also in fluidic devices that use the Marangoni effect, such as actuators and optical switches. In addition, the particular behavior characterized by the existence of a surface tension minimum may provide new scientific and technical applications. A calibration in surface tension measurements, which requires temperature independence of surface tension, was successful using 1-butanol aqueous solution.[28]

ACKNOWLEDGMENTS

This work was financially supported by the Japan Space Forum on microgravity work for space applications and also by NEDO on ground-based work for terrestrial applications. The authors received many helpful suggestions for the performance evaluation of heat pipes from Dr. M. Mochizuki of Fujikura Ltd.

REFERENCES

1. SIEGEL, R. 1967. Effect of gravity on heat transfer. *In* Advances in Heat Transfer, Vol. 4. J.P. Hartnett & T.F. Irvine, Jr., Eds.: 143–228. Academic Press, New York.
2. MERTE, JR. H. 1990. Nucleate pool boiling in variable gravity. *In* Low-Gravity Fluid Dynamics and Transport Phenomena. J.N. Koster & R.L. Sani, Eds.: 15–69. AIAA, Washington, DC.
3. STRAUB, J. 2001. Boiling heat transfer and bubble dynamics in microgravity. *In* Advances in Heat Transfer, Vol. 35. J.P. Hartnett & T.F. Irvine, Jr., Eds.: 57–172. Academic Press, San Diego.
4. OHTA, H. 2003. Microgravity heat transfer in flow boiling. *In* Advances in Heat Transfer, Vol. 37. J.P. Hartnett, *et al.*, Eds.: 1–76. Academic Press, San Diego.
5. DI MARCO, P. 2003. Review of reduced gravity boiling heat transfer: European research. J. Jpn. Soc. Micrograv. Appl. **20:** 252–263.
6. KIM, J. 2003. Review of reduced gravity boiling heat transfer: US research. J. Jpn. Soc. Micrograv. Appl. **20:** 264–271.
7. OHTA, H. 2003. Review of reduced gravity boiling heat transfer: Japanese research. J. Jpn. Soc. Micrograv. Appl. **20:** 272–285.
8. HOVESTREIJDT, J. 1963. The influence of the surface tension difference on the boiling of mixtures. Chem. Eng. Sci. **18:** 631–639.
9. MORI, T., *et al.* 1993. Capabilities and recent activities of Japan Microgravity Center (JAMIC). Micrograv. Sci. Tech. **4:** 238–242.
10. ABE, Y., *et al.* 1994. Pool boiling of a non-azeotropic binary mixture under microgravity. Int. J. Heat Mass Transf. **37:** 2405–2413.
11. AHMED, S. & V.P. CAREY. 1998. Effects of gravity on the boiling of binary fluid mixture. Int. J. Heat Mass Transf. **41:** 2469–2483.
12. ABE, Y. & A. IWASAKI. 1999. Observation of vapor bubble of non-azeotropic binary mixture in microgravity with a two-wavelength interferometer. Proc. 5th ASME/JSME Joint Thermal Eng. Conf. AJTE99-6418.
13. ABE, Y. & A. IWASAKI. 2000. Single and dual vapor bubble experiments in microgravity. *In* Microgravity Fluid Physics and Heat Transfer. V. Dhir, Ed.: 55–61. Begell House Inc., New York.
14. WAYNER, JR. P.C., *et al.* 1976. The interline heat-transfer coefficient on an evaporating wetting film. Int. J. Heat Mass Transf. **19:** 487–492.
15. STEPHAN, P. & J. HAMMER. 1994. A new model for nucleate boiling heat transfer. Waerme Stoffuebertragung **30:** 119–125.
16. STRAUB, J. 1994. The role of surface tension for two-phase heat and mass transfer in the absence of gravity. Exp. Therm. Fluid Sci. **9:** 253–276.
17. MAREK, R. & J. STRAUB. The origin of thermocapillary convection in subcooled nucleate pool boiling. Int. J. Heat Mass Transf. **44:** 619–632.
18. JANZ, G.J., *et al.* 1969. Surface tension data. *In* Molten Salts, Vol. 2. NSRDS-NBS **28:** 49–111.
19. VOCHTEN, R. & G. PETRE. 1973. Study of the heat of reversible adsorption at the air-solution interface II. Experimental determination of the heat of reversible adsorption of some alcohols. J. Coll. Interf. Sci. **42:** 320–327.
20. ABE, Y. & A. IWASAKI. 2002. Microgravity experiments on dual vapor bubbles of self-wetting fluid. AIP CP608 (STAIF 2002). 189–196.
21. VAN WIJK, W.R., *et al.* 1956. Heat transfer to boiling binary liquid mixtures. Chem. Eng. Sci. **5:** 68–80.
22. KURAMAE, M. & M. SUZUKI. 1993. Two-component heat pipes utilizing the Marangoni effect. J. Chem. Eng. Japan **26:** 230–231.
23. IWASAKI, E., *et al.* 2003. Measurements of the surface tension and density for 1-butanol. Proc. 24th Jpn. Symp. Thermophys. Properties. 315–317.
24. ZHANG, N. & D. CHAO. 1999. Models for enhanced boiling heat transfer by unusual Marangoni effects under microgravity conditions. Intl. Comm. Heat Mass Transf. **26:** 1081–1090.
25. MITROVIC, J. 1998. The flow and heat transfer in the wedge-shaped liquid film formed during the growth of a vapour bubble. Int. J. Heat Mass Transf. **41:** 1771–1785.

26. TAKAMATSU, H., *et al.* 1999. Stability of annular liquid film in microgravity. Micrograv. Sci. Tech. **12:** 2–8.
27. POLING, B.E., *et al.* 2001. The Properties of Gases and Liquids, 5th edit. McGraw-Hill, New York.
28. OBA, T. & Y. NAGASAKA. 2003. Application of laser induced capillary wave for viscosity measurement. Proc. 24th Jpn. Symp. Thermophys. Properties. 279–281.

Unsteady Near-Critical Flows in Microgravity

V.I. POLEZHAEV, A.A. GORBUNOV, AND E.B. SOBOLEVA

The Institute for Problems in Mechanics,
Russian Academy of Sciences, Moscow, Russia

ABSTRACT: This paper presents analysis of the different time scales associated with unsteady fluid flow phenomena near the thermodynamical critical point and that are typical for experiments carried out in microgravity. A focus of the paper is modeling the initial stage of convection under low and zero gravity on the basis of the two-dimensional Navier–Stokes equations for a compressible gas with the Van der Waals state equation. We also consider a thermoacoustic problem on the basis of three-dimensional linearized equations for an isentropic inviscid gas near the critical point in zero gravity. We compare the heat transfer due to unsteady convection and the piston effect in an enclosure with side heating in zero and low gravity with pure conductivity.

KEYWORDS: unsteady near-critical flow; microgravity

NOMENCLATURE:

V	velocity
T	temperature
p	pressure
ρ	density
η'	viscosity
c'_p	constant-pressure heat capacity
β	compressibility
Ra_r	Rayleigh number
Rv_r	vibrational Rayleigh number
Re	Reynolds number
Pr_r	Prandtl number
M	Mach number
γ	equal to C_p/C_v
p_e	equilibrium pressure
j_l, j_r	local heat flux along the left and right boundaries, respectively
Nu	Nusselt number
ε	reduced temperature, equal to $(T - T_c)/T_c$

INTRODUCTION

Compressible fluid flows in microgravity have features that are absent in the widespread Oberbeck–Boussinesq convection model. Due to the complicated inner structure and a variety of time scales, including differing time scales for convection and thermoacoustic, these problems have not been sufficiently studied, even for a perfect gas. Irregularities of the physical properties and peculiarities of the state

Address for correspondence: V.I. Polezhaev, The Institute for Problems in Mechanics Russian Academy of Sciences, Prospect Vernadslogo 101, B1, Moscow 119526, Russia. Voice: 7-095-434-3656; fax: 7-095-938-2048.

polezh@ipmnet.ru

Ann. N.Y. Acad. Sci. 1027: 286–302 (2004). ©2004 New York Academy of Sciences.

doi: 10.1196/annals.1324.023

equation that are typical for fluid flows near a critical point, add new unsolved problems. During the past 10 years a number of experiments on the orbital complex MIR with Alice-1 and Alice-2 instruments were carried out (see Refs. 1–5 and references cited therein). This research has initiated a number of theoretical works among the international microgravity community, to analyze and explain the observed phenomena, works that are currently in progress. Study of the heat transfer processes in a near-critical fluid in microgravity and on Earth reveals a number of characteristic features of the phenomenon due to the singularity of thermodynamic and transport properties of the fluid near its critical point. In space, the heat exchange process was found to be strongly affected by the external excitation (microgravity environment and spatially excited vibrations of various types). On Earth, due to high compressibility of the fluid, the process develops in a background of strong stratification, which interacts with the developing convection. This paper continues previous work[2–6] on the development of models for near critical fluid flows. A focus of this paper is on two general problems: fast (thermoacoustic) motions for long term duration under zero gravity and unsteady heat transfer under zero and low gravity conditions.

PECULIARITIES OF THE NEAR-CRITICAL FLUID FLOWS

A near-critical fluid is characterized by a temperature, pressure, and density that are close to values at the thermodynamical critical point. Under these conditions the fluid displays specific static and dynamic properties. Static critical properties (asymptotic discrepancy of the constant-pressure heat capacity, coefficients of isothermal compressibility, and heat expansion) relate with the equation of state. The first and second derivatives $\partial p / \partial \rho$ and $\partial^2 p / \partial \rho^2$, at the critical point must be zero (p denotes the pressure and ρ denotes the density). Dynamic critical properties include the anomalistic behavior of transport coefficients; for example, large increased in the thermal conductivity.[7] These peculiar properties lead to certain features in heat and mass transfer that differ from those of a perfect gas. It is known that, in enclosures, the temperature may propagate very fast as a result of the piston effect,[8] obtained numerically in one-dimensional and two-dimensional simulations.[9,10]

Thermal gravity-driven convection is characterized by the Rayleigh number Ra_r and Prandtl number Pr_r. If the external mass force changes quickly or the cavity with the fluid oscillates, vibrational convection is induced; this is described by the vibrational Rayleigh number Rv_r. The nondimensional criteria above include the real parameters of the near-critical fluid and are indicated by the subscript r. For a fluid with density ρ' and viscosity η' in a cavity with size l' and temperature difference Θ' the criteria values are defined by

$$
\begin{aligned}
Ra_r &= \frac{\theta' \beta' g' l'^3 \rho'^2 c_p'}{\lambda' \eta'} \\
Pr_r &= \frac{c_p' \eta'}{\lambda'} \\
Rv_r &= \frac{1}{2} \frac{(A' \omega' \beta' \Theta' l')^2 \rho' c_p'}{\lambda'}.
\end{aligned}
\tag{1}
$$

The fluid is affected by a static mass force g' and oscillates with frequency ω' and amplitude A'—the primes represent dimensioned values. The coefficient of heat expansion η' and constant-pressure heat capacity c_p' increase in the near-critical region and are defined by the equation of state,

$$\beta' = -\frac{1}{\rho'(\partial\rho'/\partial T')_{p'}},$$

$$c_p' = c_v' + \frac{T'}{\rho'^2(\partial p'/\partial T')_{\rho'}^2(\partial\rho'/\partial p')_{T'}}. \tag{2}$$

In this investigation, the equation of state in form of the Van der Waals equation is taken into account, but it may be different. For this form the *real* parameters become as follows (we consider the critical isochor):

$$Ra_r = \frac{2}{3}\varepsilon^{-1}\left(\frac{1}{\gamma} + \frac{\gamma-1}{\gamma}\frac{1+\varepsilon}{\varepsilon}\right)\frac{1}{\lambda}Ra$$

$$Pr_r = \left(\frac{1}{\gamma} + \frac{\gamma-1}{\gamma}\frac{1+\varepsilon}{\varepsilon}\right)\frac{1}{\lambda}Pr \tag{3}$$

$$Rv_r = \frac{4}{9}\varepsilon^{-2}\left(\frac{1}{\gamma} + \frac{\gamma-1}{\gamma}\frac{1+\varepsilon}{\varepsilon}\right)\frac{1}{\lambda}Rv.$$

The last relations include the Rayleigh and Prandtl numbers, Ra and Pr, the vibrational Rayleigh number, Rv, and the ratio of specific heats, γ, that do not depend on the vicinity to the critical point and should be defined in terms of characteristics far from this point (characteristics of the perfect gas),

$$Ra = \theta'g'l'^3\rho'^2\frac{c_{v_0}' + B'}{T_c'\lambda_0'\eta_0'}$$

$$Pr = \frac{(c_{v_0}' + B')\eta_0'}{\lambda_0'} \tag{4}$$

$$Rv = \frac{1}{2}\left(\frac{A'\omega'\Theta'\rho_c'\,l'}{T_c'}\right)^2\frac{c_{v_0}' + B'}{\lambda_0'\eta_0'}$$

$$\gamma = 1 + \frac{B'}{c_{v_0}'}.$$

The subscript 0 denotes values in a perfect gas; $B' = R'/\mu_g'$ is the perfect gas constant R' per unit molecular weight μ_g'. Near-critical features are associated with the parameter $\varepsilon = (T' - T_c')/T_c'$—the temperature difference from the critical point— and with a coefficient of thermal conductivity that depends on ε and is described by $\lambda = 1 + \Lambda\varepsilon^{-\psi}$. Consequently only the temperature parameter ε is responsible for the near-critical properties indicated in *real* nondimensional criteria. In the near-critical region, where $\varepsilon \to \infty$, these criteria strongly increase and asymptotically diverge:

$$Ra_r \sim \varepsilon^{\psi-2} \to \infty, \quad Pr_r \sim \varepsilon^{\psi-1} \to \infty, \quad Rv_r \sim \varepsilon^{\psi-3} \to \infty \ (\psi < 1).$$

The real Rayleigh number Ra_r and vibrational Rayleigh number Rv_r, calculated with the help of relations (3) are compared with the values defined from the experimental temperature variable characteristics of the real fluids (CO_2 and SF_6). The results show good agreement between criteria calculated on the basis of Van der Waals equation of state and their experimental analogies, thus validating the mathematical model applied.[5] One can see that **(3)** and **(4)** demonstrate strong dependence of the near critical processes on gravity, because real criteria may increase by several orders of magnitude and become, under microgravity conditions, the same as on Earth. Most important peculiarities of near critical convection (gravity phenomena in hydrostatics, using state equation and real properties) are taken into account below.

Peculiarities of formulating the problem for computer simulation of fluid flows near a critical point are as follows:

1. the use of thermodynamic and thermophysical property data near the critical point,

2. study of the impact of the state equation and thermophysical properties and requirements for modeling,

3. taking into account and analyzing the impact of non-boussinesq effects,

4. separating slow (convective) and fast (thermoacoustic) fluid flows,

5. taking into account reduced diffusion near the critical point and separating steady state and unsteady processes for data analysis, and

6. developing efficient numerical solution methods for convection of compressible media.

A draft classification of processes that have been studied during recent years[3–6] includes natural (free) convection—ground-based steady state, unsteady, steady state, vibrational convection,[5] and piston phenomena in a microgravity environment. Forced fluid flow under zero gravity may be induced by a temperature drop.[3] Low frequency wave motions during abnormal thermistor heating were also analyzed.[4] Long term high frequency thermoacoustic motion is one of the unsolved problems. We focus on a more detailed analysis of near supercritical unsteady (single-phase) flows as a basis for the more comprehensive area of sub- and supercritical flows near a critical point.

GOVERNING EQUATIONS AND NUMERICAL TECHNIQUE

A hierarchy of models and software that includes the full Navier–Stokes equations for a compressible gas, intermediate (acoustic filtering), as well as the Boussinesq approach has been developed during the past five to seven years. Navier–Stokes equations and the equation of energy for a non-perfect gas with an equation of state having two arbitrary parameters, and the so-called acoustic filtering approximation to low-speed flow are used in this paper (see Refs. 2–6 and references cited therein). A low speed intermediate model based on the Navier–Stokes equations for a compressible gas with acoustic filtering is described below and the results are compared with the full model.

The total pressure p is decomposed into two parts: equilibrium pressure p_e and dynamic pressure p_1. Thus, $p = p_e + 1/\gamma M^2 p_1$, where M is the Mach number. To close the set of equations, the integral mass balance is involved

$$\frac{\partial \rho}{\partial t} = \nabla(\rho \overline{U}) = 0$$

$$\rho \frac{\partial \overline{U}}{\partial t} + \rho(\overline{U}\nabla)\overline{U} = -\nabla p_1 + \frac{1}{Re}\left[2\nabla(\mu \dot{D}) - \nabla\left(\left(\frac{2}{3}\mu - \zeta\right)\nabla \overline{U}\right)\right] + \frac{Ra}{Pr\Theta Re^2}(\rho - \rho_e)\overline{g}$$

$$\rho \frac{\partial T}{\partial t} + \rho(\overline{U}\nabla)T = -(\gamma - 1)T\left(\frac{\partial p_e}{\partial T}\right)_\rho \nabla \overline{U} + \frac{\gamma}{RePr}\nabla(\lambda \nabla T) \qquad (4)$$

$$p_e = p_e(\rho, T)$$

$$\int_V \rho \, dv = \text{constant.}$$

Stratification is described by a linear approximation (subscript e means at equilibrium, $*$ means on the boundary):

$$\rho_e = \rho^*\left(1 + \left(\frac{\partial \rho^*}{\partial p^*}\right)_{T^*}\varepsilon_g \overline{g}(\overline{r} - \overline{r}^*)\right) \qquad (5)$$

$$p_e = p^* + \rho^*\varepsilon_g \overline{g}(\overline{r} - \overline{r}^*).$$

The Reynolds number $Re = \rho_c'U'l'/\eta_0'$, temperature difference $\Theta = \Theta'/T_c'$, and parameter of hydrostatic compressibility $\varepsilon_g = g'l'/B'T_c'$ arise in governing the dimensionless equations (4) and (5), in addition to the parameters mentioned above. To simulate heat and mass transfer in the frame of the model (4) and (5), the equation of state should be defined. We use the Van der Waals equation of state $p_e = \rho T/(1 - b\rho) - a\rho^2$ and (where a and b are constants) to describe a near-critical fluid and the equation of state $p_e = \rho T$ for a perfect gas. This choice allows us to define derivatives $(\partial p_e/\partial T)_\rho$ in the equation for energy (4) and $(\partial \rho^*/\partial p^*)_{T^*}$ in the equation for density stratification (5). The governing equations are solved numerically by means of a complex novel program that uses implicit finite difference methods (SIMPLE-type method and others). The results obtained by using the full model (NS equations with two-scale pressure presentation) and NS equations with *acoustic filtering* during acoustic time τ_c (time for spreading sound over a distance equal to the size of the domain) for VdW gas were compared. The initial state is zero velocity and uniform temperature in the domain ($T - T_c = 1$ K). The density on the upper

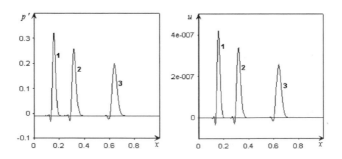

FIGURE 1. Temporal behavior of the dynamic pressure p' and horizontal velocity component u in a central horizontal cross section after time steps $t = 0.2$ (1), 0.4 (2), and 0.8 (3) ($\tau_a = 0.5 \cdot \tau_c$) for a full Navier–Stokes model.

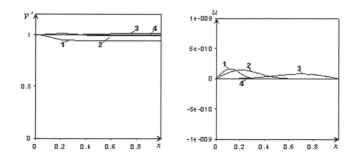

FIGURE 2. Temporal behavior of the dynamic pressure p' and horizontal velocity component u in a central horizontal cross section after 1 (1), 3 (2), 10 (3), and 100 (4) time steps ($\tau_a = 0.5 \cdot \tau_c$) for a model with acoustic filtering.

boundary rises to the critical value ρ_c. Subsequently, small disturbances due to the pressure drop $\Delta p = \rho_c \cdot c^2$, where c is the velocity of sound, spread. The initial conditions are:

$$\overline{U} = 0, \qquad T - T_c = 1\,\text{K}, \qquad \rho = \rho_e.$$

On the left boundary, the pressure drop is $p' = p_e' + \Delta p'$.

FIGURE 1 shows the modeling results on the basis of the full model—motion of the small disturbances. FIGURE 2 shows the analogous results on the basis of the model with acoustic filtering. One can see that this model does not describe sound waves—pressure and velocity distributions do not have local areas of disturbance. Sound oscillation damping during a number of time steps, and averaging the results shows only the long term behavior of fluid flow. However, comparison of the results from models with and without acoustic filtering, for slow (convective) motions, shows that the full model with a two-scale pressure presentation, provides an efficient means to calculate slow motions, with a time step of about 10^3–$10^4\tau_c$, thus validating the model for acoustic filtering—long-term regimes with $t \gg \tau_c$ and low convective motions. The full model with a two-step pressure presentation is universal enough; however, the problem of modeling high frequency disturbances, together with convection, which is of special interest for near critical phenomena, still remains.

ISENTROPIC FLOWS UNDER ZERO GRAVITY
RESULTING FROM MINOR EQUILIBRIUM DISTORTIONS

Near-critical, high-frequency thermoacoustic motions may be of specific interest in zero gravity. We describe these below for the three-dimensional case of with adiabatic boundaries. Isentropic gas movement under zero gravity conditions is determined by the following dimensionless system of equations:

$$\rho M^2 A_1(\rho)\left(Sh\frac{\partial \overline{V}}{\partial t} + (\overline{V}, \nabla)\overline{V} \right) + A_2(\rho)\nabla\rho = 0 \tag{6}$$

$$Sh\frac{\partial \rho}{\partial t} + (\overline{V}, \nabla)\rho + \rho(\nabla, \overline{V}) = 0. \tag{7}$$

The variables t, x, y, and z denote time and Cartesian coordinates. The functions $\bar{V}(t, x, y, z)$ and $\rho(t, x, y, z)$ represent velocity and density in the medium. The coefficients $A_1(\rho)$ and $A_2(\rho)$ depend on the gas law, M is the Mach number, and Sh is the Struchal number. Equation (6) determines the medium movement and results from known equations for an inviscid compressible gas,[12] supposing the spatial and temporal constancy of entropy. In this case, the energy balance equation for the gas is reduced to the equation of continuity (7). For a perfect gas this is

$$A_1 = A_1^{(p)} = 1, \qquad A_2 = A_2^{(p)} = T^{(p)}, \tag{8}$$

and for a Van der Waals (VdW) gas it is

$$A_1 = A_1^{(v)} = 4\gamma_0(3-\rho)^2, \qquad A_2 = A_2^{(v)} = 9(4\gamma_0 T^{(v)} - \rho(3-\rho)^2), \tag{9}$$

where γ_0 is the adiabatic index of a perfect gas and $T^{(p)}$ and $T^{(v)}$ are the corresponding adiabatic temperatures determined by the density value ρ.

We suppose that gas is concentrated in a rectangular parallelepiped (see FIGURE 3),

$$\Pi = \{0 < x < l_x, 0 < y < l_y, 0 < z < 1\}. \tag{10}$$

Isentropic equilibrium inside the parallelepiped Π is determined by

$$\bar{V} = 0, \qquad \rho = \rho_e = \text{constant}, \tag{11}$$

where the adiabatic temperatures for a perfect gas and a VdW gas are, respectively,

$$T_e^{(p)} = (\rho_e)^{\gamma_0 - 1}, \qquad T_e^{(v)} = \left(\frac{2\rho_e}{3 - \rho_e}\right)^{\gamma_0 - 1}, \tag{12}$$

supposing the critical temperature and critical density of a VdW gas to be the scale factors. Following Reference 13, the system of equations (6) and (7) can be linearized in the neighborhood of mechanical equilibrium (11) and (12).

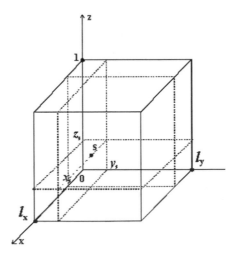

FIGURE 3. The coordinate system and area Π occupied by gas.

Suppose

$$\bar{V}' = \bar{V}'(t, x, y, z) = (V_x', V_y', V_z')$$

$$\rho' = \rho'(t, x, y, z) \tag{13}$$

$$T' = T'(t, x, y, z)$$

are infinitesimal distortions of velocity, density, and temperature, respectively, in **(11)** and **(12)**. By substituting the distorted values of unknown variables into **(6)** and **(7)**, the system of equations for distortions **(13)** can be derived as

$$\rho_e M^2 A_1(\rho_e) Sh \frac{\partial \bar{V}'}{\partial t} + A_2(\rho_e)\nabla\rho' = 0 \tag{14}$$

$$Sh\frac{\partial\rho'}{\partial t} + \rho_e(\nabla, \bar{V}') = 0, \tag{15}$$

for each point $(x, y, z) \in \Pi$. Equation **(14)** corresponds here to the movement equation **(6)**, whereas Equation **(15)** corresponds to the continuity equation **(7)**. Note that, because of **(12)**, the temperature and density distortions are linked by the relations:

$$\text{for a perfect gas} \quad T' = T'_{(p)} = (\gamma_0 - 1)\frac{T_e^{(p)}}{\rho_e}\rho' \tag{16}$$

$$\text{for a VDW gas} \quad T' = T'_{(v)} = \frac{3(\gamma_0 - 1)T_e^{(v)}}{\rho_e(3 - \rho_e)}\rho'. \tag{17}$$

Suppose the harmonic distortions of **(13)**, that is put

$$V_x' = f_x e^{-i\omega t}e_0, \qquad V_y' = f_y e^{-i\omega t}e_0, \qquad V_z' = f_z e^{-i\omega t}e_0$$

$$\rho' = f_0 e^{-i\omega t}e_0, \qquad e_0 = e^{i(xk_x + yk_y + zk_z)}, \tag{18}$$

where ω is the oscillation frequency, and k_x, k_y, k_z are wave numbers, corresponding to coordinate directions. Substituting **(18)** into the system of equations **(14)** and **(15)**, the characteristic equation can be derived:

$$D = \begin{vmatrix} -\rho_e M^2 A_1 Sh\omega & 0 & 0 & k_x A_2 \\ 0 & -\rho_e M^2 A_1 Sh\omega & 0 & k_y A_2 \\ 0 & 0 & -\rho_e M^2 A_1 Sh\omega & k_z A_2 \\ \rho_e k_x & \rho_e k_y & \rho_e k_z & -Sh\omega \end{vmatrix} = 0. \tag{19}$$

Note that, regardless of the number of spatial dimensions, a pair of eigenfrequencies exists,

$$\omega = \pm\omega_t, \qquad \omega_t = \frac{1}{ShM}\sqrt{\frac{A_2}{A_1}}k, \qquad k^2 = k_x^2 + k_y^2 + k_z^2. \tag{20}$$

It is also to be noted that, except in the one-dimensional case, a zero value of eigenfrequency exists; this being single in the two-dimensional case and double in the three-dimensional case. Stationary solutions of the system of equations **(14)** and **(15)** may thus exist in the two- and three-dimensional cases.

It can be shown from **(20)** that, if the wave properties are similar, the ratio of eigenfrequencies for perfect gas and VdW gas has the form

$$\frac{\omega_t^{(v)}}{\omega_t^{(p)}} = \sqrt{\frac{A_2^{(v)} A_1^{(p)}}{A_1^{(v)} A_2^{(p)}}}, \tag{21}$$

and is determined solely by gas law peculiarities.

Using particular solutions of (19) (see, e.g., Ref. 14) that are latent vectors related to the characteristic constants, a complete solution of Equation (19) can be derived

$$V_x' = [-k_x\sqrt{A_2}(C_1 e^{it\omega_t} - C_2 e^{-it\omega_t}) + k_z C_3] e_0$$

$$V_y' = [-k_y\sqrt{A_2}(C_1 e^{it\omega_t} - C_2 e^{-it\omega_t}) + k_z C_4] e_0$$

$$V_z' = [-k_y\sqrt{A_2}(C_1 e^{it\omega_t} - C_2 e^{-it\omega_t}) - k_x C_3 - k_y C_4] e_0 \tag{22}$$

$$\rho' = \rho_e Mk\sqrt{A_1}(C_1 e^{it\omega_t} + C_2 e^{-it\omega_t}) e_0.$$

Here C_1 and C_2 are arbitrary constants that correspond to latent vectors with non-zero characteristic constants, and C_3 and C_4 determine the effect of half-space with zero characteristic constant.

To find the values of the arbitrary constants suppose (see (18)) the entry conditions

$$V_x'|_{t=0} = a_x e_0$$

$$V_y'|_{t=0} = a_y e_0$$

$$V_z'|_{t=0} = a_z e_0 \tag{23}$$

$$\rho'|_{t=0} = a_0 e_0,$$

where the numbers a_x, a_y, a_z, and a_0 are, respectively, the initial amplitudes of velocity distortion and density distortion components. Substituting entry conditions (23) into Equation (22), a system of linear algebraic equations can be derived for the unknown quantities C_1, C_2, C_3, and C_4:

$$-k_x\sqrt{A_2}C_1 + k_x\sqrt{A_2}C_2 + k_z C_3 = a_x$$

$$-k_y\sqrt{A_2}C_1 + k_y\sqrt{A_2}C_2 + k_z C_4 = a_y$$

$$-k_x\sqrt{A_2}C_1 + k_z\sqrt{A_2}C_2 - k_x C_3 - k_y C_4 = a_z \tag{24}$$

$$\rho_e Mk\sqrt{A_1}C_1 + \rho_e Mk\sqrt{A_1}C_2 = a_0.$$

It possible to distinguish the set of real solutions for certain boundary conditions of the system (14) and (15) from a family of complex solutions (22).

Consider the impenetrability (*non-leaking*) boundary conditions

$$V_x'|_{x=0, l_x} = V_y'|_{y=0, l_y} = V_z'|_{z=0, 1} = 0. \tag{25}$$

Two families of real solutions of the system (14) and (15) can be derived from the complete solution (22). The first family of real solutions is determined by

$$V_x' = (k_x A_0 \cos t\omega_t + A_x)\sin xk_x \cos yk_y \cos zk_z$$

$$V_y' = (k_y A_0 \cos t\omega_t + A_y)\cos xk_x \sin yk_y \cos zk_z$$

$$V_z' = (k_z A_0 \cos t\omega_t + A_z)\cos xk_x \cos yk_y \sin zk_z \quad (26)$$

$$\rho' = -\rho_e M k A_0 \sqrt{\frac{A_1}{A_2}}\sin t\omega_t \cos xk_x \cos yk_y \cos zk_z,$$

The designation

$$A_0 = \frac{a_x k_x + a_y k_y + a_z k_z}{k^2} \quad (27)$$

is used here.

The second family is determined by

$$V_x' = \frac{k_x a_0}{\rho_e M k}\sqrt{\frac{A_1}{A_2}}\sin t\omega_t \sin xk_x \cos yk_y \cos zk_z$$

$$V_y' = \frac{k_y a_0}{\rho_e M k}\sqrt{\frac{A_1}{A_2}}\sin t\omega_t \cos xk_x \sin yk_y \cos zk_z \quad (28)$$

$$V_z' = \frac{k_z a_0}{\rho_e M k}\sqrt{\frac{A_1}{A_2}}\sin t\omega_t \cos xk_x \cos yk_y \sin zk_z$$

$$\rho' = a_0 \cos t\omega_t \cos xk_x \cos yk_y \cos zk_z.$$

Both solutions, (26) and (28) are subject to spectral limitations on distortion wave properties; that is, the relations

$$k_x = \frac{\pi n_x}{l_x}, \; k_y = \frac{\pi n_y}{l_y}, \; k_z = \pi n_z, \; k_z = \pi n_z, \; n_x, n_y, n_z = 0, \pm 1, \pm 2, ..., \quad (29)$$

should be true, resulting from the boundary conditions, (25), linking the wave numbers with the corresponding linear dimensions of the parallelepiped Π in (10).

Analyzing the families (26) and (27) it was noted that both are standing waves of limited amplitude, which probably varies in the time domain. It is also obvious that stationary flows, irrotational vibrations, and general flows, composed of stationary flows and irrotational vibrations, may exist.

Stationary flows are determined by the conditions,

$$A_0 = 0, \quad A_x = a_x, \quad A_y = a_y, \quad A_z = -\frac{1}{k_z}(a_x k_x + a_y k_y) \quad (30)$$

and, according to (26), take the form

$$V_x' = V_x'\big|_{t=0}, \quad V_y' = V_y'\big|_{t=0}, \quad V_z' = V_z'\big|_{t=0}, \quad \rho' = 0, \quad (31)$$

where the entry conditions are

$$V_x'\big|_{t=0} = A_x \sin xk_x \cos yk_y \cos zk_z$$

$$V_y'\big|_{t=0} = A_y \cos xk_x \sin yk_y \cos zk_z$$

$$V_z'\big|_{t=0} = A_z \cos xk_x \cos yk_y \sin zk_z \qquad (32)$$

$$\rho'\big|_{t=0} = 0.$$

The conditions (30) show that stationary flows depend on two parameters, making an independent pair of initial amplitudes of velocity component distortions.

Irrotational vibrations are likely to be generated by both (26) and (28) families. In family (26) the irrotational vibrations occur when the following relations hold:

$$A_x = A_y = A_z = 0, \quad A_0 = \frac{a_z}{k_z}, \quad \left(a_x = \frac{k_x}{k_z}a_z, a_y = \frac{k_y}{k_z}a_z\right). \qquad (33)$$

In this case the solution (26) is transformed into

$$V_x' = V_x'\big|_{t=0}\cos t\omega_t$$

$$V_y' = V_y'\big|_{t=0}\cos t\omega_t$$

$$V_z' = V_z'\big|_{t=0}\cos t\omega_t \qquad (34)$$

$$\rho' = -\frac{\rho_e Mka_z}{k_z}\sqrt{\frac{A_1}{A_2}}\sin t\omega_t \cos xk_x \cos yk_y \cos zk_z,$$

with entry conditions

$$V_x'\big|_{t=0} = \frac{k_x a_z}{k_z}\sin xk_x \cos yk_y \cos zk_z$$

$$V_y'\big|_{t=0} = \frac{k_y a_z}{k_z}\cos xk_x \sin yk_y \cos zk_z \qquad (35)$$

$$V_z'\big|_{t=0} = a_z \cos xk_x \cos yk_y \sin zk_z$$

$$\rho'\big|_{t=0} = 0.$$

It follows from conditions (33) that in this case irrotational vibrations depend on a single parameter, the initial amplitude of some velocity component distortion.

Irrotational distortions can, therefore, be generated solely by initial distortions of velocity (35) in one case, or solely by initial distortions of density or temperature (28) in another case.

Finally, for general flows determined by the family (26) the following conditions should hold:

$$A_0 \neq 0, \quad A_x^2 + A_y^2 + A_z^2 \neq 0. \qquad (36)$$

Therefore, the long term behavior of the near critical disturbances in a cell with adiabatic boundaries, for the limiting case of inviscid Van der Waals gas in zero gravity, can be predicted in detail. Note, that a paper elsewhere[15] contains figures that more clearly illustrate these analytic solutions.

Comparison with unsteady convection (see below) shows that sound velocity is still much faster, even near the critical point, however, boundaries of the validation of the acoustic theory based on the mechanics of continuum media in the vicinity of a critical point needs more detailed study.

UNSTEADY FEATURES OF NEAR-SUPERCRITICAL
CONVECTION IN AN ENCLOSURE

Natural gravity-driven convection in a square cavity with differently heated side walls is a classical problem of convection and was studied for the case of a perfect gas on the basis of a Boussinesq approach, but the steady state case for a near-critical fluid was achieved only recently (see Ref. 6 and references cited therein). The unsteady problem in a square with side heating was solved to date for the case of side heating one wall and isolating the other walls.[11] Below we carry out more detailed analysis of unsteady heat transfer for the case of near supercritical fluid for the problem under low and zero gravity for the same configuration as in Reference 6.

A schematic of the problem is provided in FIGURE 4. The initial state is characterized by zero velocity and uniform temperature in the domain $T^* = 1 + e^*$. The density is changed due to hydrostatic stratification, so that the critical value $\rho^* = 1$ is realized on the upper boundary. The temperature at the left side is rising as $T = T^* + k_h t$ during time $t \in [0, \tau_h]$, after that it is fixed, and the temperature difference Θ on the side walls is constant. Therefore, isothermal conditions on the side walls and adiabatic conditions on the top and bottom walls remain in effect. The governing equations of a non-perfect gas with a two arbitrary parameter equation of state in the approximation for low-speed flow, (4) and (5), are used. The near-critical fluid is described with the help of the van der Waals equation of state. The Rayleigh number, Ra, and Prandtl number, Pr, obtained from the dimensionless governing equations, include the physical characteristics of a perfect gas. The real Rayleigh number, Ra_r, and real Prandtl number, Pr_r, based on the near-critical properties were introduced as Equation (3). Convection arises due to heating the left side for a time.

Fluid flow and heat transfer were studied for CO_2 (critical parameters: $T_c' = 304.15\,K$, $\rho_c' = 0.468\,g/cm3$, $p_c' = 7.387\,MPa$) in a cell with dimension $l' = 1\,cm$ and initial reduced temperature $T' - T_c' = 1\,K$. The left boundary was heated for $0.1\,sec$ with a temperature difference $0.1\,K$. The dimensional values correspond to the

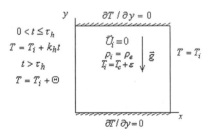

FIGURE 4. Schematic of the problem of near-critical gravity-driven convection in an enclosure with side heating.

non-dimensional parameters $Ra = 10^3$, $\varepsilon^* = 3.3 \times 10^{-3}$, $Re = 3.85 \times 10^4$, $Pr = 1$, $\Theta = 3.3 \times 10^{-4}$, $\gamma_0 = 1.4$, $\varepsilon_g = 2.86 \times 10^{-9}$, $g = (0, -1)$, $\Lambda = 0.028$, $\psi = 0.74$, $\tau_h = 2.85$, (velocity scale, $U' = 28.5$ cm/sec; time scale, $t' = 0.0351$ sec). Therefore, formulation of the problem is similar to that used in Reference 6, but here heating of the left side is 100 times faster and close to the initial temperature drop. Three cases were compared to define the impact of convection and piston effect on heat transfer. The first run, marked by f, corresponds to the full problem with convection due to the gravity,

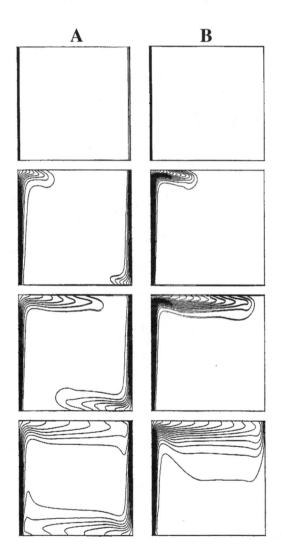

FIGURE 5. Temporal evolution of the temperature field for **(A)** VdW and **(B)** perfect gas in an enclosure with near-critical gravity-driven convection: $t = 2.75$, 34.4, 68.9, and 137 sec.

piston effect, and thermal diffusion. The second run, marked by *pe* is the same but for zero gravity, therefore, only piston effect and thermal diffusion exist. The third run, marked by 0 is characterized by zero gravity and zero velocity. It corresponds to the purely conductivity regime.

The characteristic time for the piston effect may be written as follows:[6,7]

$$\tau_{pe} = \left(\frac{1}{\gamma_0(\gamma_0 - 1)^2} + \frac{1}{\gamma_0(\gamma_0 - 1)} \frac{1 + \varepsilon^*}{\varepsilon^*} \right) \varepsilon^{*2} \frac{1}{1 + \Lambda \varepsilon^{* - \psi}} Re Pr.$$

For this case the dimensional value is $\tau'_{pe} = 2.75 \, \text{sec}$.

FIGURE 5 shows the temporal evolution of a temperature field in the case of gravity-driven convection in VdW and perfect gases after the start of heating. One can see the appearance of a jet-type flow and thermals, upflow on the hot wall and downflow on the cold wall, and formation of convection flow in the enclosure. A question for the analysis below is a comparison of the heat transfer and temporal behavior of convection, piston effect, and thermal diffusion without motion.

FIGURE 6 shows that for initial state $t' = 2.75 \, \text{sec}$, convection impact is negligible: the curves j_l and j_r for the cases with and without gravity are the same and there is uniform heating over the surface. One can see small convection feature only at the end (j_l, curve 1). However, the piston effect enhances heat transfer since the values of j_l and j_r are several times larger than their values for the case without motion. Steady state convection characterizes non-uniform heat distribution, but without gravity heat distribution is uniform in any case—with or without piston effect. Convection and piston effect increase heat transfer in comparison with pure conductivity (see FIGURE 7). The piston effect is significant for zero gravity only during the initial stage, subsequently heat transfer passes to pure conductivity. One can see that the transition period is reduced, not only due to the piston effect, but also due to convection.

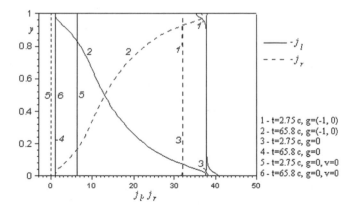

FIGURE 6. Initial and steady state local heat fluxes along the left (j_l) and right (j_r) boundaries under low gravity (1 and 2), zero ground (3 and 4), and in the media without motion (5 and 6).

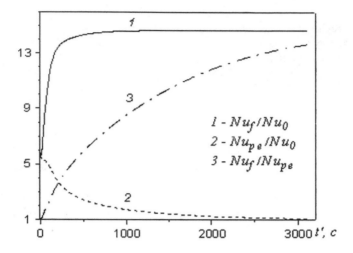

FIGURE 7. Ratio between the average Nusselt numbers on the left boundary for various cases.

Comparison of the heat distribution for the case of a piston effect and for pure conductivity (see FIGURE 8) shows an increase in temperature for the initial state due to piston effect. However, transition to the steady state, when piston effect exists, is

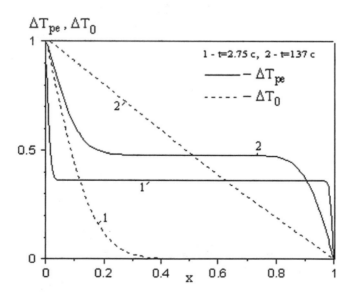

FIGURE 8. Relative addition of the temperature for various times along the central horizontal line for the zero gravity case.

slow in comparison with pure conductivity. Because of this effect, input heat quantity inside the volume is enhanced, but not its distribution inside the volume.

CONCLUSIONS

Unsteady phenomena due to convection and a piston effect and thermoacoustic, with the use of numerical code that was developed for near-critical fluid flows are analyzed. Isentropic thermoacoustic flows over a long time interval in an enclosure, on the basis of three-dimensional linearized equations for the near-critical steady state regimes, are described by analytic solutions. It should be observed in zero gravity regimes to provide a check for model validation.

Comparison of the heat transfer due to unsteady convection, piston effect, and pure conductivity for a VdW gas near its critical point was done for the two-dimensional problem in a square with side heating. Convection and piston effect enhance heat transfer in comparison with pure conductivity; however, the piston effect may be observed only for zero gravity. For long time intervals, heat transfer passes to pure conductivity. One can see that the transition period is reduced due to convection but increased due to the piston effect. The results should help with interpretation of phenomena observed in experiments aboard the MIR station. Study of near-critical convection and heat transfer for microgravity characterization with separation and more detailed registration of the "working processes" aboard the International Space Station are planned.

ACKNOWLEDGMENTS

This work is supported by the Russian Foundation for Fundamental Research (RFBR), Grants 03-01-00682 and 02-02-16995; by the Leading Scientific Schools (Grant 2239.2003.8); and by the Ministry of Science (Grant 0201.06.253).

REFERENCES

1. LAHERRERE, M. & P. KOUTSIKIDES. 1993. Alice, an instrument for the analysis of fluids close to the critical point in microgravity. Acta Astronaut. **29**(10/11): 861.
2. AVDEEV, S.V., A.I. IVANOV, A.V. KALMYKOV, *et al.* 1997. Experiments in the far and near critical fluid aboard MIR station with the use of the "ALICE-1" instrument. Proceedings of the Joint Xth European and VIth Russian Symposium on Physical Sciences in Microgravity. St.-Petersburg, Russia, I. 333–340
3. AVDEEV, S.V., A.I. IVANOV, A.V. KALMYKOV, *et al.* 1997. Experiments for near critical phenomena on Mir station with the use of Alice-1 instrument. Scientific Report. Inst for Problems in Mech.
4. POLEZHAEV, V.I., V.M. EMELIANOV & A.A. GORBUNOV. 1998. Near critical fluids in microgravity: concept of research and new results of convection modeling. J. Jpn. Soc. Micrograv. Appl. 1998: 123–129.
5. ZYUZGIN, A.B., A.I. IVANOV, V.I. POLEZHAEV, *et al.* 2001. Convective motion in near-critical fluids under real zero-gravity conditions. Cosmic Res. **39**(2): 175–186.
6. POLEZHAEV, V.I. & E.B. SOBOLEVA. 2001. Thermal gravity-driven convection near-critical fluid in an enclosure with side heating. Izvestia RAN MZnG 2001, N 3, 143–154. (Russian; transl. as Fluid Dynamics).

7. STANLEY, H.E. 1971. Introduction to Phase Transition and Critical Phenomena. Oxford University Press, London.
8. ONUKI, A., H. HAO & R.A. FERRELL. 1990. Fast adiabatic equilibration in a single-component fluid near the liquid–vapor critical point. Phys. Rev. A **41:** 2256–2259.
9. ZAPPOLI, B. & A. DURAND-DAUBIN. 1994. Heat and mass transport in a near supercritical fluid. Phys. Fluid **6**(5): 1929–1936.
10. MAEKAWA, T., K. ISHII & S. MASUDA. 1998. Temperature propagation and cluster structures in a near-critical fluid. Joint 1st Pan-Pacific Basin Workshop and 4th Japan–China Workshop on Microgravity Sciences Proceedings. J. Jpn. Soc. Microgravity Appl. **15-II:** 130–135.
11. ZAPPOLI, B., S. AMIRODINE, P. CARLES & J. OUAZZANI. 1996. Thermoacoustic and buoyancy-driven transport in a square side-heated cavity filled with a near-critical fluid, J. Fluid Mech. **316:** 53–72.
12. LOITSYANSKY, L.G. 1987. Fluid and Gas Mechanics. Nauka, Moscow.
13. MONIN, A.S. & A.S. YAGLOM. 1965. Statistical Hydromecanics. Part 1. Nauka, Moscow.
14. STEPANOV, V.V. 1953. The Course of Differential Equations. Gostekhteoretizdat, Moscow.
15. POLEZHAEV, V.I., A.A. GORBUNOV & E.B. SOBOLEVA. 2003. Near-critical flows in microgravity. Preliminary Proceedings of the Microgravity Transport Processes in Fluid, Thermal, Biological and Materials Sciences III. Davos, Switzepland. Sept. 14–19, 2003. MTP-03-53.

Thermocapillary Migration of a Drop

An Exact Solution with Newtonian Interfacial Rheology and Stretching/Shrinkage of Interfacial Area Elements for Small Marangoni Numbers

R. BALASUBRAMANIAM[a] AND R. SHANKAR SUBRAMANIAN[b]

[a]National Center for Microgravity Research, NASA Glenn Research Center, Cleveland, Ohio, USA

[b]Department of Chemical Engineering, Clarkson University, Potsdam, New York, USA

ABSTRACT: In this paper we analyze the effects of the following phenomena associated with the thermocapillary migration of a drop. The first is the influence of Newtonian surface rheology of the interface and the second is that of the energy changes associated with stretching and shrinkage of the interfacial area elements, when the drop is in motion. The former occurs because of dissipative processes in the interfacial region, such as when surfactant molecules are adsorbed at the interface in sufficient concentration. The interface is typically modeled in this instance by ascribing to it a surface viscosity. This is a different effect from that of interfacial tension gradients arising from surfactant concentration gradients. The stretching and shrinkage of interfacial area elements leads to changes in the internal energy of these elements that affects the transport of energy in the fluids adjoining the interface. When an element on the interface is stretched, its internal energy increases because of the increase in its area. This energy is supplied by the neighboring fluids that are cooled as a consequence. Conversely, when an element on the interface shrinks, the adjoining fluids are warmed. In the case of a moving drop, elements of interfacial area are stretched in the forward half of the drop, and are shrunk in the rear half. Consequently, the temperature variation on the surface of the drop and its migration speed are modified. The analysis of the motion of a drop including these effects was first performed by LeVan in 1981, in the limit when convective transport of momentum and energy are negligible. We extend the analysis of LeVan to include the convective transport of momentum by demonstrating that an exact solution of the momentum equation is obtained for an arbitrary value of the Reynolds number. This solution is then used to calculate the slightly deformed shape of the drop from a sphere.

KEYWORDS: thermocapillary migration; Newtonian interfacial rheology; small Marangoni numbers; drop; stretching and shrinking of interface

Address for correspondence: R. Balasubramaniam, National Center for Microgravity Research, NASA Glenn Research Center, Mail Stop 110-3, Cleveland, OH 44135, USA. Voice: 216-433-2878; fax: 216-433-3793.
  bala@grc.nasa.gov

Ann. N.Y. Acad. Sci. 1027: 303–310 (2004). ©2004 New York Academy of Sciences.
doi: 10.1196/annals.1324.024

INTRODUCTION

The thermocapillary migration of drops was first investigated theoretically by Young, Goldstein, and Block.[1] When a drop is present in an unbounded liquid that has a uniform temperature gradient, a thermocapillary stress is generated at the interface arising from the variation of interfacial tension with temperature. Typically, the drop is propelled toward the warm portion of the continuous phase. Young *et al.* obtained a result for the steady velocity of a drop, including a contribution from the effect of gravity, in the limit when the Reynolds and Marangoni numbers approach zero, so that the effect of convective transport of momentum and energy can be neglected. To verify their theory, Young *et al.* performed experiments on air bubbles in silicone oils, in which the effects of thermocapillarity and gravity were counterbalanced, so that the bubbles were nearly stationary.

Balasubramaniam and Chai[2] showed that the solution obtained by Young *et al.* is an exact solution of the Navier–Stokes equations for any value of the Reynolds number, when the Marangoni number is equal to zero and the effect of gravity is not present. A discussion of other pertinent literature on the motion of drops due to thermocapillarity can be found in Subramanian and Balasubramaniam.[3]

Our objective in this paper is to consider the effect of Newtonian interfacial rheology and stretching/shrinkage of interfacial area elements on the thermocapillary motion of a drop. In the limit when the Reynolds and Marangoni numbers approach zero, this problem has been addressed by LeVan[4] and discussed by Subramanian and Balasubramaniam.[3] The rheology of the surface needs to be considered in special situations, for example, when surfactant molecules are adsorbed on the interface in sufficient quantity so that an interfacial viscosity can be ascribed to the interface. As a consequence the drop encounters additional resistance to its motion. Scriven[5] has provided the framework for a Newtonian model of the interface, in which the contribution to the surface stress depends linearly on the rate of deformation of area elements on the interface. A detailed discussion of surface rheology models is available in Edwards, Brenner, and Wasan.[6] The second effect that we consider is the consequence of the deformation of interfacial area elements on the energy balance at the interface. Even when a drop moves in a continuous phase fluid that is isothermal, temperature variations can be generated on the interface. This is because elements of surface area grow as they move from the stagnation point to the equatorial region in the forward half of the drop. The increase in area increases the interfacial internal energy, which must be provided by the fluids adjoining the interface. Thus, the neighboring fluids are cooled. In the rear portion of the drop, the situation is opposite, and the adjoining fluids are warmed. Therefore, an interfacial temperature variation can occur, and the consequent thermocapillary stress influences the motion of the drop. This effect has been considered by Harper *et al.*,[7] Kenning,[8] LeVan,[4] and Torres and Herbolzheimer.[9]

One infers from the analysis presented by Subramanian and Balasubramaniam[3] that the thermocapilliary migration velocity of the drop is unaffected by the surface shear viscosity, but is influenced by the surface dilatational viscosity and the effect of stretching and shrinkage of interfacial area elements. The velocity of the drop is reduced as a consequence. We shall revisit the analysis performed by LeVan[4] and Subramanian and Balasubramaniam[3] below and show that their results are valid even when the convective transport of momentum is included in the analysis. The effect of inertia is shown to alter the pressure field, and lead to deformation of the drop. The small change in the shape of the drop from a sphere is obtained using a technique similar to that used by Balasubramaniam and Chai.[2]

FORMULATION

We consider the steady migration of a spherical drop of radius R_0 immersed in a liquid that is of infinite extent. The liquid has a density ρ, viscosity μ, thermal conductivity k, and thermal diffusivity κ. The corresponding properties in the drop are denoted by $\gamma\rho$, $\alpha\mu$, βk, and $\lambda\kappa$. The rate of change of interfacial tension between the drop and the continuous phase liquid is denoted by σ_T, and is assumed to be a negative constant. All the physical properties of the fluids, with the exception of the interfacial tension, are assumed to be constant. A temperature gradient of magnitude G is imposed in the continuous phase liquid. We assume that gravitational effects are not present.

The surface shear and dilatational viscosities in the Newtonian model are denoted by μ_s^* and λ_s^*, respectively. The corresponding dimensionless surface viscosities, scaled by μR_0, are denoted by μ_s and λ_s. The quantity $e_s - \sigma^*$, where e_s is the surface internal energy per unit area and σ^* is the interfacial tension, appears in the energy balance condition at the interface to accommodate the stretching and shrinkage of interfacial area elements. We assume that $e_s - \sigma^*$ is a constant over the drop surface, and define a dimensionless parameter $E_s = -(e_s - \sigma^*)\sigma_T/(\mu k)$.

A reference velocity for motion in the fluids is obtained by balancing the tangential stress at the interface in the continuous phase liquid with the thermocapillary stress. This velocity scale is

$$v_0 = \frac{(-\sigma_T)GR_0}{\mu}. \tag{1}$$

In addition to the property ratios α, β, γ, and λ, the dimensionless interfacial viscosities μ_s and λ_s, and the interfacial internal energy parameter E_s, the dimensionless parameters that are important in determining the motion of the drop are the Reynolds and Marangoni numbers. These parameters are defined using the above velocity scale and the properties of the continuous phase liquid.

$$Re = \frac{\rho v_0 R_0}{\mu} \tag{2}$$

$$Ma = \frac{v_0 R_0}{\kappa}. \tag{3}$$

The problem is analyzed in a reference frame attached to the moving drop. We use a spherical polar coordinate system, with the origin located at the center of the drop. The radial coordinate, scaled by R_0, is r. The polar coordinate measured from the direction of the temperature gradient is θ. The azimuthal coordinate is ϕ; however, the problem posed below for the velocity and temperature fields is independent of ϕ due to axial symmetry about the direction of the imposed temperature gradient. The scaled radial and tangential velocity fields in the continuous phase liquid are denoted by u and v, respectively, and those in the drop by u' and v'. The scaled migration velocity of the drop is denoted by v_∞. These are obtained by dividing the physical velocities by v_0. The scaled pressure fields in the two fluids are p and p' and are obtained by dividing the physical pressure by ρv_0^2. The temperature field in the continuous phase is scaled by subtracting the temperature in the undisturbed continuous phase at the location of the drop and dividing by GR_0.

$$T = \frac{\bar{T} - Gv_0 v_\infty t}{GR_0}, \tag{4}$$

where $\bar{T}$ is the physical temperature and t denotes time. The scaled temperature T' in the drop is defined similarly. The governing equations for the velocity and temperature fields in the continuous phase liquid and the drop are given below:

$$\frac{1}{r^2}\frac{\partial}{\partial r}(r^2 u) + \frac{1}{r\sin\theta}\frac{\partial}{\partial\theta}(v\sin\theta) = 0 \tag{5}$$

$$u\frac{\partial u}{\partial r} + \frac{v}{r}\frac{\partial u}{\partial\theta} - \frac{v^2}{r} = -\frac{\partial p}{\partial r} + \frac{1}{Re}\left[\nabla^2 u - \frac{2u}{r^2} - \frac{2}{r^2}\frac{\partial v}{\partial\theta} - \frac{2v}{r^2}\cot\theta\right] \tag{6}$$

$$u\frac{\partial v}{\partial r} + \frac{v}{r}\frac{\partial v}{\partial\theta} + \frac{uv}{r} = -\frac{1}{r}\frac{\partial p}{\partial\theta} + \frac{1}{Re}\left[\nabla^2 v + \frac{2}{r^2}\frac{\partial u}{\partial\theta} - \frac{v}{r^2\sin^2\theta}\right] \tag{7}$$

$$v_\infty + u\frac{\partial T}{\partial r} + \frac{v}{r}\frac{\partial T}{\partial\theta} = \frac{1}{Ma}\nabla^2 T \tag{8}$$

$$\frac{1}{r^2}\frac{\partial}{\partial r}(r^2 u') + \frac{1}{r\sin\theta}\frac{\partial}{\partial\theta}(v'\sin\theta) = 0 \tag{9}$$

$$u'\frac{\partial u'}{\partial r} + \frac{v'}{r}\frac{\partial u'}{\partial\theta} - \frac{v'^2}{r} = -\frac{1}{\gamma}\frac{\partial p'}{\partial r} + \frac{\alpha}{\gamma}\frac{1}{Re}\left[\nabla^2 u' - \frac{2u'}{r^2} - \frac{2}{r^2}\frac{\partial v'}{\partial\theta} - \frac{2v'}{r^2}\cot\theta\right] \tag{10}$$

$$u'\frac{\partial v'}{\partial r} + \frac{v'}{r}\frac{\partial v'}{\partial\theta} + \frac{u'v'}{r} = -\frac{1}{\gamma r}\frac{\partial p'}{\partial\theta} + \frac{\alpha}{\gamma}\frac{1}{Re}\left[\nabla^2 v' + \frac{2}{r^2}\frac{\partial u'}{\partial\theta} - \frac{v'}{r^2\sin^2\theta}\right] \tag{11}$$

$$v_\infty + u'\frac{\partial T'}{\partial r} + \frac{v'}{r}\frac{\partial T'}{\partial\theta} = \frac{\lambda}{Ma}\nabla^2 T'. \tag{12}$$

The boundary conditions are as follows. Far from the drop, the velocity field is uniform and equal to the migration velocity of the drop (i.e., in the laboratory reference frame, the fluid far away is quiescent). The temperature field far away is the imposed linear field. Therefore, as $r \to \infty$,

$$u \to -v_\infty\cos\theta, \quad v \to v_\infty\sin\theta \tag{13}$$

$$T \to r\cos\theta. \tag{14}$$

From the analysis by LeVan,[4] it can be shown that when the Reynolds and Marangoni numbers are set equal to zero, the normal stress balance is satisfied by a spherical shape of the drop. This is no longer the case when the Reynolds number is nonzero, the case considered here. Assuming the deformation from the spherical shape is small, it can be shown by using a domain perturbation scheme that it is permissible to obtain the leading order velocity and pressure fields by applying the relevant boundary conditions at the fluid–fluid interface at a spherical boundary, designated here by $r = 1$. In the next section, we show how the small perturbation to this shape arising from inertia can be calculated using these leading order fields. At the fluid–

fluid interface, the kinematic boundary condition holds, and the velocity and temperature fields are continuous.

$$u(1, \theta) = u'(1, \theta) = 0 \tag{15}$$

$$v(1, \theta) = v'(1, \theta) \tag{16}$$

$$T(1, \theta) = T'(1, \theta). \tag{17}$$

The tangential stress balance and the heat flux balance at the fluid–fluid interface appear below. It is in these boundary conditions that the effects of surface viscosity and surface internal energy appear.

$$\frac{\partial}{\partial r}\left(\frac{v}{r}\right)(1, \theta) - \alpha\frac{\partial}{\partial r}\left(\frac{v'}{r}\right)(1, \theta)$$

$$= -\frac{\partial T}{\partial \theta}(1, \theta) - 2\mu_s v(1, \theta) - (\lambda_s + \mu_s)\frac{\partial}{\partial \theta}\left[\frac{1}{\sin\theta}\frac{\partial}{\partial \theta}\{v(1, \theta)\sin\theta\}\right] \tag{18}$$

$$\frac{\partial T}{\partial \theta}(1, \theta) - \beta\frac{\partial T'}{\partial r}(1, \theta) = \frac{E_s}{\sin\theta}\frac{\partial}{\partial \theta}[v(1, \theta)\sin\theta]. \tag{19}$$

The migration velocity of the drop is determined from a force balance—at steady state the drop is not accelerating, and hence, the net force acting on it must be zero,

$$\int_0^\pi\left[\frac{\partial}{\partial r}\left(\frac{v}{r}\right)(1, \theta)\sin^2\theta + \left(Re\, p(1, \theta) - 2\frac{\partial u}{\partial r}(1, \theta)\right)\cos\theta\sin\theta\right]d\theta = 0 \tag{20}$$

RESULTS

Exact Solution for Small Marangoni Number

When $Ma \ll 1$, convective transport of energy can be neglected compared with conduction, as a first approximation. In this case, the terms on the left side of Equations (8) and (12) may be ignored. Then, it can be shown that the solution first obtained by LeVan[4] and given in Subramanian and Balasubramaniam[3] for the temperature and velocity fields in the case $Re = 0$ continues to apply for arbitrary values of the Reynolds number and other parameters in the problem. Only the pressure field is modified by the inclusion of inertia. This exact solution is given below:

$$T = \left(r + \frac{B}{r^2}\right)\cos\theta \tag{21}$$

$$T' = (1 + B)r\cos\theta \tag{22}$$

$$u = -v_\infty\left(1 - \frac{1}{r^3}\right)\cos\theta \tag{23}$$

$$v = v_\infty\left(1 + \frac{1}{2r^3}\right)\sin\theta \tag{24}$$

$$u' = \frac{3}{2}v_\infty(1 - r^2)\cos\theta \tag{25}$$

$$v' = 3v_\infty\left(r^2 - \frac{1}{2}\right)\sin\theta \tag{26}$$

$$p = p_\infty + \frac{v_\infty^2}{2}\left[1 - \cos^2\theta\left(1 - \frac{1}{r^3}\right)^2 - \sin^2\theta\left(1 + \frac{1}{2r^3}\right)^2\right] \tag{27}$$

$$p' = A_0 + \frac{9}{8}\gamma v_\infty^2\left[\sin^2\theta(r^4 - r^2) + 2\cos^2\theta\left(r^2 - \frac{r^4}{2}\right)\right] - 15\frac{v_\infty\alpha}{Re}r\cos\theta \tag{28}$$

$$B = \frac{1 - \beta - \Omega}{2 + \beta + \Omega} \tag{29}$$

$$\Omega = \frac{2E_s}{2 + 3\alpha + 2\lambda_s} \tag{30}$$

$$v_\infty = \frac{2}{(2 + 3\alpha + 2\lambda_s)(2 + \beta + \Omega)} = \frac{\Omega}{E_s(2 + \beta + \Omega)}. \tag{31}$$

The constant A_0 in the pressure field inside the drop can be determined only when the interfacial normal stress balance is considered. The velocity field outside the drop is irrotational and the pressure field outside the drop satisfies Bernoulli's equation. The flow field within the drop is the well known Hill's spherical vortex. The temperature and the velocity fields are coupled by the effects of surface viscosity and interfacial internal energy. The surface shear viscosity μ_s^* has no influence on the dynamics. The surface dilatational viscosity λ_s^* and the effect of the interfacial internal energy reduce the migration velocity of the drop. When λ_s and E_s are zero, the results given above reduce to those given by Balasubramaniam and Chai.[2]

Small Deformation of the Drop

We now consider the normal stress balance at the fluid–fluid interface and calculate the slight deformation of the drop from a sphere. Let $R = R_0[1 + f(\theta)]$ denote the shape of the drop. We assume that $f \ll 1$. It will be seen shortly that this requires that the Weber number $We = Re\,Ca \ll 1$. Here Ca is the Capillary number, defined by $Ca = \mu v_0/\sigma_0$, where σ_0 is a reference value for the interfacial tension. For a slightly deformed drop, the normal stress balance at the interface can be written as

$$\tau_{rr} - \tau'_{rr} = \frac{2H\sigma}{Ca} + \frac{2\lambda_s}{\sin\theta}\frac{\partial}{\partial\theta}[v(1, \theta)\sin\theta], \tag{32}$$

where τ_{rr} denotes the dimensionless normal stress at $r = 1$ evaluated from the leading order fields given above, H is the mean curvature of the interface, and σ is the interfacial tension scaled by σ_0. Note that the surface viscosity makes a direct contribution to the normal stress difference (see Scriven[5]). When the deformation from a spherical shape is small,

$$2H \sim 2 - 2f - \frac{1}{\sin\theta}\frac{d}{d\theta}\left(\sin\theta\frac{df}{d\theta}\right). \tag{33}$$

The normal stress balance can then be written as follows:

$$Ca\,Re\left[A_0 - p_\infty + \frac{v_\infty^2}{2}\left(\frac{9}{4}\gamma - 1 + \frac{9}{4}(1-\gamma)\sin^2\theta\right)\right]$$

$$= 2 - [1 - Ca(1+B)\cos\theta]\left[2f + \frac{1}{\sin\theta\,d\theta}\frac{d}{d\theta}\left(\sin\theta\frac{df}{d\theta}\right)\right].$$

(34)

From Equation (34), one can conclude that, for a fixed Reynolds number, $f(\theta) \sim O(Ca)$ as $Ca \to 0$. In this asymptotic limit, the leading order differential equation for $f(\theta)$ is a nonhomogeneous version of the differential equation for the Legendre Polynomial $P_1(\cos\theta)$. Note the absence of any terms proportional to $\cos\theta$ in the inhomogeneity. If such a term is present, it can be shown that it will lead to unboundedness of the solution for f along $\theta = 0$ and π. In fact, if a force balance on the drop had not been used to determine the speed of the drop, the speed can be obtained by requiring the term proportional to $\cos\theta$ in the inhomogeneity to be zero to eliminate such singular behavior. We impose the constraints that the volume of the drop is a constant (equal to $4\pi R_0^3/3$) and that the origin of the coordinates is the center of mass of the drop. Linearization of these constraints for small f leads to the following equations:

$$\int_0^\pi f(\theta)\sin\theta\,d\theta = 0$$

(35)

$$\int_0^\pi f(\theta)\cos\theta\sin\theta\,d\theta = 0.$$

(36)

The solution can be obtained in a straightforward manner, as shown by Brignell.[10] The unknown constant A_0 is determined as part of the solution,

$$A_0 = \frac{2}{Re\,Ca} + p_\infty - \frac{v_\infty^2}{4}\left(1 + \frac{3}{2}\gamma\right)$$

(37)

$$f(\theta) = \frac{3}{32}Re\,Ca\,v_\infty^2(\gamma-1)(3\cos^2\theta - 1).$$

(38)

The dimensionless surface dilatational viscosity λ_s and the surface internal energy parameter E_s affect the deformation of the drop via their influence on v_∞ (given in (31)). Because we have restricted the deformation of the drop to be small, we see that $We = Re\,Ca$ must be a small parameter. Drops less dense than the continuous phase liquid are predicted to contract in the direction of the temperature gradient; that is, they attain an oblate shape. Drops more dense than the continuous phase attain a prolate shape.

CONCLUDING REMARKS

We have shown that a solution in the limit of zero Reynolds and Marangoni numbers for the velocity and temperature fields, obtained elsewhere[4,3] for the thermocapillary migration of a drop accommodating Newtonian surface rheology and the influence of internal energy changes associated with the stretching and shrinkage of surface elements, holds for arbitrary values of the Reynolds number so long as the Marangoni number is set equal to zero. The pressure fields are modified by inertia. The leading order result for the migration speed of the drop is the same as that obtained when the Reynolds number is zero. Furthermore, whereas the drop will be spherical when the Reynolds number is zero, this is no longer the case for non-zero values of the Reynolds number. A perturbation result for the small deformation of the drop from a spherical shape is given in the case where the Capillary number is asymptotically small, for fixed Reynolds number.

REFERENCES

1. YOUNG, N.O., J.S. GOLDSTEIN & M.J. BLOCK. 1959. The motion of bubbles in a vertical temperature gradient. J. Fluid Mech. **6:** 350–356.
2. BALASUBRAMANIAM, R. & A.T. CHAI. 1987. Thermocapillary migration of droplets: an exact solution for small Marangoni numbers. J. Colloid Interface Sci. **119**(2): 531–538.
3. SUBRAMANIAN, R.S. & R. BALASUBRAMANIAM. 2001. The Motion of Bubbles and Drops in Reduced Gravity. Cambridge University Press, London/New York.
4. LEVAN, M.D. 1981. Motion of a droplet with a Newtonian interface. J. Colloid Interface Sci. **83**(1): 11–17.
5. SCRIVEN, L.E. 1960. Dynamics of a fluid interface. Chem. Eng. Sci. **12:** 98–108.
6. EDWARDS, D.A., H. BRENNER & D.T. WASAN. 1991. Interfacial Transport Processes and Rheology. Butterworth-Heinemann, Boston.
7. HARPER, J.F., D.W. MOORE & J.R.A. PEARSON. 1967. The effect of the variation of surface tension with temperature on the motion of bubbles and drops. J. Fluid Mech. **27**(2): 361–366.
8. KENNING, D.B.R. 1969. The effect of surface energy variations on the motion of bubbles and drops. Chem. Eng. Sci. **24:** 1385–1386.
9. TORRES, F.E. & E. HERBOLZHEIMER. 1993. Temperature gradients and drag effects produced by convection of interfacial internal energy around bubbles. Phys. Fluids A **5**(3): 537–549.
10. BRIGNELL, A.S. 1973. The deformation of a liquid drop at small Reynolds number. Quart. J. Mech. and Applied Math. **XXVI**(1): 99–107.

The Formation of Spikes in the Displacement of Miscible Fluids

N. RASHIDNIA, R. BALASUBRAMANIAM, AND R.T. SCHROER

National Center for Microgravity Research,
NASA Glenn Research Center, Cleveland, Ohio, USA

ABSTRACT: We report on experiments in which a more viscous fluid displaces a less viscous one in a vertical cylindrical tube. These experiments were performed using silicone oils in a vertical pipette of small diameter. The more viscous fluid also had a slightly larger density than the less viscous fluid. In the initial configuration, the fluids were at rest, and the interface was nominally flat. A dye was added to the more viscous fluid for ease of observation of the interface between the fluids. The flow was initiated by pumping the more viscous fluid into the less viscous one. The displacement velocity was such that the Reynolds number was smaller than unity and the Peclet number for mass transfer between the fluids was large compared to unity. For upward displacement of the more viscous fluid from an initially stable configuration, an axisymmetric finger was observed under all conditions. However, a needle-shaped spike was seen to propagate from the main finger in many cases, similar to that observed by Petitjeans and Maxworthy for the displacement of a more viscous fluid by a less viscous one.

KEYWORDS: spikes; displacement; miscible fluids

INTRODUCTION

The dynamics of miscible interfaces is an active area of research that has been identified to benefit from experimentation in reduced gravity. The goal is to study the patterns assumed by the interface when one liquid is slowly displaced by another in a cylindrical tube, such that diffusion plays an important role in the dynamics. It has been suggested that nontraditional stresses in the fluids, caused by the steep variation in the concentration of the miscible fluids in the mixing zone, might be important in the dynamics.[1] These effects are overwhelmed by the flow caused by buoyancy under terrestrial conditions.[2] An improved understanding of the dynamics of multiphase porous media flows is deemed essential for progress in the fields of enhanced oil recovery, fixed bed regeneration, hydrology, and filtration.[3–5]

Taylor[6] and Petitjeans and Maxworthy[3] performed fingering experiments for immiscible and miscible fluids, respectively, in capillary tubes. Taylor measured the amount of fluid displaced by injecting air into a horizontal capillary tube, initially filled with a viscous fluid, in order to calculate the amount of displaced fluid that was left behind on the tube wall. However, in this experiment the flow in the interior of

Address for correspondence: Nasser Rashidnia, National Center for Microgravity Research, 21000 Brookpark Road, MS-110-3, NASA Glenn Research Center, Cleveland, OH 44135, USA. Voice: 216-433-3622; fax: 216-433-3793.
  nasser.rashidnia@grc.nasa.gov

Ann. N.Y. Acad. Sci. 1027: 311–316 (2004). ©2004 New York Academy of Sciences.
doi: 10.1196/annals.1324.025

the finger was dynamically unimportant because of the large liquid to gas viscosity ratios. On plotting the amount of fluid left behind at the tube wall, m, versus the capillary number Ca (where $m = 1 - V_m/V_t$ and $Ca = \mu V_t/\sigma$, σ is the surface tension at the interface, V_t is the finger tip velocity, V_m is the displacement velocity, and μ is the viscosity of the displaced fluid), Taylor[6] as well as Cox[7] found a single curve for a number of fluids. This curve increased at very low Ca as $Ca^{1/2}$ and reached an asymptotic value of 0.60 for large Ca. Numerical simulations by Reinelt and Saffman,[8] based on the Stokes equations, agreed very closely with these experiments.

Petitjeans and Maxworthy[3] performed an experiment very much like that of Taylor[6] for the case of miscible fluids. A less viscous fluid (glycerine–water mixture) with viscosity μ_1 and density ρ_1 displaces a more viscous fluid (glycerine) with viscosity μ_2 and density ρ_2 in a vertical capillary tube of diameter d. An axisymmetric finger of the less viscous fluid forms and the speed of the tip of the finger V_t is measured. The mean velocity of displacement V_m was varied between 0.1 and 10 mm/sec and the Atwood number $At = (\mu_2 - \mu_1)/(\mu_2 + \mu_1)$ was varied between 0 and 1. The quantity $m = 1 - V_m/V_t$ was related to the average thickness of the more viscous fluid left behind at the tube wall ($m = t(2 - t)$, where t is the average thickness of liquid 2 scaled by the tube radius) and m was plotted against the Peclet number $Pe = 2V_m d/D$ (where D is the diffusion coefficient between the glycerine–water mixture and pure glycerine), and the gravity parameter $F = g(\rho_2 - \rho_1)d^2/2\mu_2 V_m$. The experiments were conducted in capillary tubes between 1 and 4 mm in diameter. The Peclet number was in the range $400 \leq Pe \leq 120{,}000$. The gravity parameter was in the range $1 \leq |F| \leq 300$. The small values of $|F|$ could be achieved only for large values of Pe. The Reynolds number was $O(1)$ or less. For large values of Pe, m was found to attain an asymptotic value between 0.5 and 0.6 depending on At. Interestingly, for large Pe, large viscosity ratios, and negligible density effects the asymptotic amount of displaced fluid left behind on the tube walls matches the immiscible data. For $m < 0.5$, a thin spike was observed to grow continuously from the tip of the finger. Experiments were performed with positive and negative values for F, where F was positive in the destabilized case, with the heavier fluid on top of the lighter fluid. In all cases, the lighter fluid, which is less viscous, displaced the heavier fluid. The simulations of Chen and Meiburg[9] confirmed much of the findings of Petitjeans and Maxworthy.[3] In the case of miscible flows, a cutoff length is set by diffusive effects rather than surface tension, so that in some sense the Peclet number takes the place of Ca. A fundamental difference between the immiscible and miscible cases is that the miscible flow can never become truly steady. Sooner or later, diffusion will cut off the supply of *fresh* displacing fluid. However, both simulations and experiments show that for large values of Pe, typically exceeding $O(10^3)$, a quasi-steady finger forms that persists for a time of $O(Pe)$ before it starts to decay. Depending on the strength of the gravitational forces, a variety of topologically different streamline patterns can be observed, among them some that leak fluid from the finger tip (a spike) and others with toroidal recirculation regions inside the finger.

The diffusion coefficient of the miscible fluids is a property that is important in the dynamics. Its values are vitally important to the proper design of reduced gravity experiments. In particular, it determines the desirable range for the speeds of the displacing fluids, so that the effect of diffusion is not overshadowed by convective transport of mass. Measurement of the diffusion coefficients for some fluid pairs were reported by Petitjeans and Maxworthy[3] and Rashidnia *et al.*[10]

In this paper, we report the results of experiments that we have performed on the displacement of a less viscous fluid by a more viscous miscible fluid in cylindrical tubes. We found that, depending on the conditions of flow, the advancing finger of the more viscous fluid exhibits spikes, similar to those obtained by Petitjeans and Maxworthy.[3]

EXPERIMENTAL PROCEDURE

Our experiments were performed using silicone oils in a vertical pipette of small diameter. The more viscous fluid also has a slightly larger density than the less viscous fluid. In the initial configuration, the fluids were at rest, and the interface was nominally flat. A dye was added to the more viscous fluid for ease of observation of the interface between the fluids. Flow was initiated by pumping the more viscous fluid into the less viscous one. The displacement velocity was such that the Reynolds number was small compared to unity, and the Peclet number for mass transfer among the fluids was large compared to unity. The gravitational effects are represented by the dimensionless parameter F, defined above.

RESULTS AND DISCUSSION

When the more viscous fluid displaces a less viscous fluid in a vertical tube, we observed the formation of a spike when the fluids are stably stratified. As mentioned, the relative importance of buoyancy to the viscous forces is represented by the dimensionless parameter F. When the value of F is small, gravity is not important in the dynamics, and the shape of the finger and the flow field are expected to be independent of orientation of the tube and comparable to experiments performed in a reduced gravity environment. FIGURE 1 shows the finger shape when a silicone oil of kinematic viscosity 1,000cSt displaces a silicone oil of viscosity 50cSt. The experiment was performed in a tube of 3mm diameter. The displacement speed was 200micron/sec. The value of F was 33.1. The finger achieved an axisymmetric shape that was stable. The tip of the finger was cone-shaped.

FIGURE 2 shows the image captured from an experiment with the same pair of fluids as in FIGURE 1, when the displacement speed was decreased to 5micron/sec. The value of F in this case was 1,323, and the main finger does have an axisymmetric shape. A thin spike can be seen to propagate from the tip of the main finger. The spike was best observed directly while the experiment was being performed; the faint blue color of the leaking high viscosity liquid was clearly visible. We captured the image using S-VHS video. In the video images, the spike can be observed clearly only when the image is processed digitally. FIGURE 3 shows the same image as in FIGURE 2 after digitally processing the image using thresholding. The spike appears to be stable and its diameter is only a small fraction of the diameter of the tube. Petitjeans and Maxworthy also observed spikes that propagate from the main finger in their experiments, where the less viscous fluid displaces a more viscous fluid in a

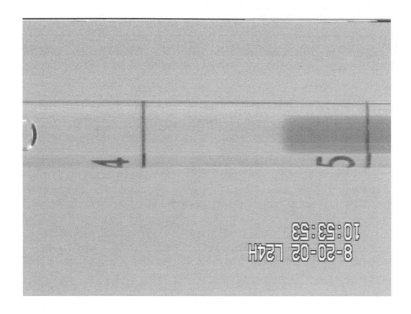

FIGURE 1. Finger shape during the displacement of 50 cSt silicone oil by 1,000 cSt silicone oil. Tube diameter, 3 mm; displacement speed, 200 μm/sec.

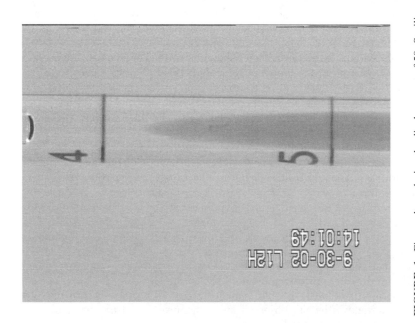

FIGURE 2. Finger shape during the displacement of 50 cSt silicone oil by 1,000 cSt silicone oil. Tube diameter, 3 mm; displacement speed, 5 μm/sec.

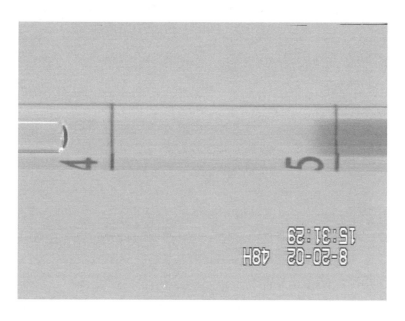

FIGURE 4. Finger shape during the displacement of 50 cSt silicone oil by 1,000 cSt silicone oil. Tube diameter, 3 mm; displacement speed, 1 μm/sec.

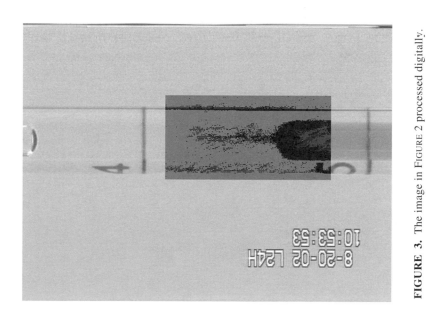

FIGURE 3. The image in FIGURE 2 processed digitally.

small capillary tube. They observed the spikes when the fluids are stably stratified. Our observations are similar to those of Petitjeans and Maxworthy[3] in this respect.

FIGURE 4 shows the results from an experiment when the displacement speed was further decreased to 1 micron/sec. The value of F was 6615. At such a large value of the F parameter, buoyancy dominates the dynamics of the displacement process. The interface between the two fluids was essentially flat. The formation of a spike was suppressed in this case.

Our results, therefore, indicate that spike formation during the displacement of a liquid by a more viscous miscible liquid in a vertical tube occurs when the fluids are stably stratified. When buoyancy is weak compared to the viscous forces, the main finger is axisymmetric and stable, and the finger tip is conical in shape. When the effects of gravity are predominant, the interface is flat. Spikes are formed in an intermediate regime when buoyancy and viscous forces are in optimum balance.

REFERENCES

1. JOSEPH, D.D. & Y. RENARDY. 1993. Fundamentals of Two-Fluid Dynamics. Springer-Verlag.
2. MAXWORTHY, T., E. MEIBURG, R. BALASUBRAMANIAM, *et al.* 2001. The dynamics of miscible fluids: a space flight experiment (MIDAS). AIAA Paper No. 2001-5061, International Space Station Utilization Conference, Kennedy Space Center, Florida, October.
3. PETITJEANS, P. & T. MAXWORTHY. 1996. Miscible displacements in a capillary tube. Part 1 Experiments. J. Fluid Mech. **326:** 37–56.
4. HOMSY, G.M. 1987. Viscous fingering in porous media. Ann. Rev. Fluid Mech. **19:** 271.
5. DAGAN, G. 1987. Theory of solute transport in groundwater. Ann. Rev. Fluid Mech. **19:** 183.
6. TAYLOR, G.I. 1961. Deposition of a viscous fluid on the wall of a tube. J. Fluid Mech. **10:** 161.
7. COX, B.G. 1962. On driving a viscous fluid out of a tube. J. Fluid Mech. **14:** 81.
8. REINELT, D.A. & P.G. SAFFMAN. 1985. The penetration of a finger into a viscous fluid in a channel and tube, SIAM J. Sci. Stat. Comput. **6:** 542.
9. CHEN, C.H. & E. MEIBURG. 1996. Miscible displacements in capillary tubes. Part 2. Numerical simulations. J. Fluid Mech. **326:** 57–90.
10. RASHIDNIA, N., R. BALASUBRAMANIAM, J. KUANG, *et al.* 2001. Measurement of the diffusion coefficient of miscible fluids using both interferometry and Weiners method. Intl. J. Thermophys. **22**(2): 547–555.

Experimental Study of a Constrained Vapor Bubble Fin Heat Exchanger in the Absence of External Natural Convection

SUMITA BASU, JOEL L. PLAWSKY, AND PETER C. WAYNER, JR.

Rensselaer Polytechnic Institute, Troy, New York, USA

ABSTRACT: In preparation for a microgravity flight experiment on the International Space Station, a constrained vapor bubble fin heat exchanger (CVB) was operated both in a vacuum chamber and in air on Earth to evaluate the effect of the absence of external natural convection. The long-term objective is a general study of a high heat flux, low capillary pressure system with small viscous effects due to the relatively large $3 \times 3 \times 40$ mm dimensions. The current CVB can be viewed as a large-scale version of a micro heat pipe with a large Bond number in the Earth environment but a small Bond number in microgravity. The walls of the CVB are quartz, to allow for image analysis of naturally occurring interference fringes that give the pressure field for liquid flow. The research is synergistic in that the study requires a microgravity environment to obtain a low Bond number and the space program needs thermal control systems, like the CVB, with a large characteristic dimension. In the absence of natural convection, operation of the CVB may be dominated by external radiative losses from its quartz surface. Therefore, an understanding of radiation from the quartz cell is required. All radiative exchange with the surroundings occurs from the outer surface of the CVB when the temperature range renders the quartz walls of the CVB optically thick ($\lambda > 4$ microns). However, for electromagnetic radiation where $\lambda < 2$ microns, the walls are transparent. Experimental results obtained for a cell charged with pentane are compared with those obtained for a dry cell. A numerical model was developed that successfully simulated the behavior and performance of the device observed experimentally.

KEYWORDS: vapor bubble; fin heat exchanger

NOMENCLATURE:

A	area of surface element or solid cross-sectional area of quartz, m^2
A_e	total liquid surface area at the cell corners, m^2
A_{quartz}	cross sectional area of solid heat pipe wall
C_l	geometric shape factor, dimensionless
F	view factor, fraction
G	radiative conductance, W/K^4
h	heat transfer coefficient, $W/m^2 \cdot K$, Planck's constant
h_{fg}	latent heat of vaporization/condensation, J/kg
k	thermal conductivity, $W/m \cdot K$
k_{fl}	friction factor, dimensionless
K	liquid film curvature, m^{-1}
K_c	constant film curvature at condenser end, m^{-1}

Address for correspondence: Peter C. Wayner, Jr., Rensselaer Polytechnic Institute, 110 8th Street, Troy, NY, 12180-3590, USA. Voice: 518-276-6199; fax: 518-276-4030.
wayner@rpi.edu

Ann. N.Y. Acad. Sci. 1027: 317–329 (2004). ©2004 New York Academy of Sciences.
doi: 10.1196/annals.1324.027

l	length, m
L	total length of vapor bubble in heat pipe, m
m	parameter related to shape factor, $(12.12/A_{quartz})$
Q	power, W
P	pressure difference, Pa
R_c	radius of curvature, m
T	temperature, °C or K
x	axial distance along the CVB, m

Greek Characters

δ	film thickness of liquid in the CVB, m
ε	total emissivity, dimensionless fraction
ν	kinematic viscosity, m/sec
σ	interfacial tension, N/m, Stefan Boltzmann constant

Subscripts

cell	into the cell
e	evaporator
l	liquid
o	outside
r, rad	radiation
sur	surrounding (air or vacuum chamber wall)
v	vapor

INTRODUCTION

The use of interfacial free energy gradients to control fluid flow naturally leads to simpler and lighter heat transfer systems due to the absence of mechanical pumps. Therefore, passive engineering systems based on this principle are ideal candidates for the space program. One such system is the wickless heat pipe.[1,2] In the Earth environment, these heat pipes usually have very small characteristic dimensions to keep the Bond number small and are known as micro heat pipes. Outside the Earth environment, in microgravity, the characteristic dimension for a small Bond number can be considerably larger. These larger systems should accommodate larger heat fluxes due to smaller viscous losses. However, the basic thermophysical principles controlling these small capillary pressure systems are not well understood. To rectify this deficiency, we initiated a basic research program on a generic system, which we call the constrained vapor bubble (CVB).

The CVB is illustrated in FIGURE 1.[3] The vapor bubble is constrained by fixed walls with liquid separating portions of the bubble from the walls. For a completely wetting system, the liquid will coat all the walls of the chamber. On the other hand, for a finite contact angle system, some of the walls will have only a small amount of adsorbed vapor, which changes the surface properties of the solid–vapor interface. Liquid will fill at least a portion of the corners in both cases. If the temperature $T_2 > T_1$ because of an external heat source and sink, energy flows from end (2) to end (1) by conduction in the walls and by an evaporation, vapor flow, and condensation mechanism. The condensate flows from end (1) to end (2) because of the intermolecular force field, which is a function of the film profile. There is a "pressure jump" at the liquid–vapor interface, $P_l - P_v$, due to the anisotropic stress tensor near interfaces. An intermediate region, that can be approximately isothermal and adiabatic if insulated, connects the regions of evaporation and condensation. On the other hand, as in the current case, if heat losses occur from the complete outer surface, there are only regions of evaporation and condensation controlled by radiation and convection to the surroundings.

For engineering purposes, with regions of both low and high capillary pressures, the CVB discussed herein acts like a fin heat exchanger. In addition, it is also an ideal device to study the fundamentals of transport phenomena controlled by intermolecular forces. Our study is multifaceted: (1) it is a basic scientific study in interfacial phenomena, fluid physics, and thermodynamics; (2) it is a basic study in thermal transport; and (3) it is a study of a passive heat exchanger for use in microgravity. The research is synergistic in that research on a low capillary pressure CVB requires a microgravity environment and the space program needs thermal control systems, like the CVB.

In preparation for a scheduled microgravity flight experiment on the International Space Station during 2005–2006, the CVB was operated in the Earth environment in both a vacuum chamber and in air to evaluate its performance in the absence of natural convection. Within a vacuum chamber to remove external natural convection, the operation of the CVB is significantly influenced by external radiative losses. Since an additional aspect of our experiments is to view the internal liquid film profile using a light microscope, the transmissivity of the walls is important.[3] Although

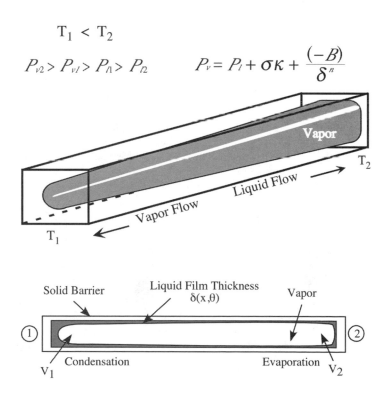

$$T_1 < T_2$$

$$P_{v2} > P_{vl} > P_{l1} > P_{l2} \qquad P_v = P_l + \sigma \kappa + \frac{(-B)}{\delta^n}$$

FIGURE 1. Conceptual diagram of the CVB cell.

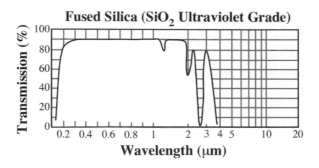

FIGURE 2. Transmissivity of fused silica.

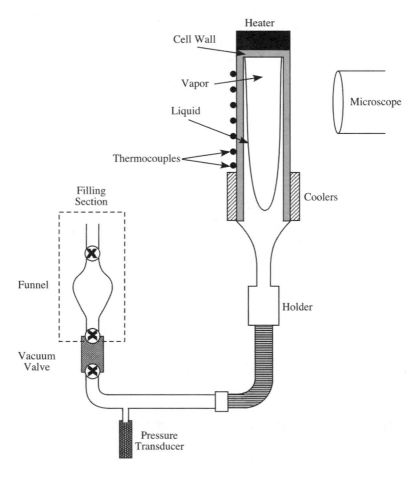

FIGURE 3. Schematic diagram of the experimental vertical CVB system.[7]

quartz has a poor thermal conductivity, it is an ideal substrate for transparent viewing. Its value as an experimental tool stems from the fact that we can see into it and measure the thickness of the fluid layer clinging to the wall surface. The axial variation of the thickness profile gives the pressure gradient for flow of the condensate. That, coupled with the measurement of axial temperature, allows us to completely determine the fluid mechanics and heat transfer characteristics of the device. The effect of wavelength on the transmissivity of UV grade quartz is shown in FIGURE 2. Unfortunately, it is not simple. Ultraviolet grade quartz is transparent in the visual portion of the electromagnetic spectrum, opaque to radiation above a wavelength of four microns and semitransparent between two and four microns. Therefore, the properties of quartz have a dramatic effect on the operation of the CVB. Herein, we report on the experimental measurement of heat transfer to the surroundings from a quartz CVB, with and without a liquid charge, in a vacuum chamber.

A general schematic of the square CVB in the vertical mode for axial symmetry in the Earth environment is presented in FIGURE 3. The device consists of a fused ultraviolet grade quartz cell with inner and outer dimensions measuring 3 mm × 3 mm and 5.5 mm × 5.5 mm × 40 mm. The inside corners of the quartz cell are formed by the fusion of flat plates held at right angles to one another. Therefore, the corners are sharp crevices for liquid flow. The CVB is initially evacuated and then partially filled with the working liquid. A part of the liquid vaporizes to form a vapor bubble constrained within the tube by its own liquid. Operation of the CVB is controlled by temperature and curvature (pressure) gradients created along its length by heating one end and cooling the other. We have been characterizing the CVB system in the Earth gravitational field in preparation for the microgravity flight.[3-8] In these experiments, the system has always been subject to external natural convection. However, in microgravity, natural convection would be absent. Therefore, it is critical to test the CVB in an environment where heat loss from the device may be dominated by radiation, so that we can predict its performance during flight. In this paper we present the results of the first set of experiments on the CVB conducted in the large vacuum chamber at the NASA Glenn Research Center.[7]

BACKGROUND

Both theoretical and experimental work has been done on the operating principles of the non-isothermal CVB heat pipe in a convective environment.[3-8] Bowman *et al.*[9,10] have shown that heat pipe fins, under certain orientations and conditions, can be more efficient than standard solid fins. Peterson[11] and Faghri[2] have written excellent reviews on the operation, performance, and applicability of various forms of miniature and micro heat pipes.

EXPERIMENTAL PROCEDURE

A thoroughly cleaned and dried cuvette assembly was evacuated and partially filled with distilled 99.8%+ GC grade *n*-pentane to give a 22 mm constrained vapor bubble. The system and experiments have been described completely by Basu.[7] The

CVB was operated in a vertical mode, with a resistance heater attached on top using a thermally conductive epoxy to insure axisymmetric operation in the Earth gravitational field. The cooler assembly consisted of a copper block with four thermoelectric coolers, one on each sidewall. Thermocouple beads, embedded 0.5 mm into the quartz, were placed 1.5 mm apart along the axial length of the cell. The T-type thermocouples used in the experiments were calibrated by Omega Engineering Co. to an accuracy of ±0.5°C. To eliminate instrument bias in our measurements we recorded temperature differences between the cell and chamber wall.

A series of experiments were conducted with the assembly inside a vacuum chamber (pressure level of 9.0×10^{-7} Torr) at the Glenn Research Center. The pressure inside the cuvette was monitored using a Druck 900 series pressure transducer. Upon reaching a steady state, the CVB was allowed to operate for one to two hours in each case. The temperature profiles and pressure data were used in conjunction with a three-dimensional SINDA® thermal model[7] to analyze the thermal behavior and performance of the unit. A *dry* evacuated cell was also run in the vacuum chamber as a control and heat inputs were set to achieve the same base temperatures as existed in the CVB with fluid. In addition, interferometry was used to measure the liquid film profile along the axis of the cell.[3–8] However, the interferometry measurements were done outside of the vacuum chamber to show the existence of three operating regions: a dryout zone, an evaporation zone, and a condensation zone. Interferometry could not be used in the vacuum chamber since the video and microscopy equipment could not operate under vacuum conditions.

Since the quartz walls of the CVB are optically clear throughout the operating temperature range (equivalent to $\lambda > 4$ mm), all radiative power exchange with the surroundings occurs from the outer surface of the heat pipe. Simulated (Monte Carlo technique) view factors of the cell to the surrounding space were used to determine the temperature dependent radiative conductance, G, at each axial position of the cell using the equation

$$G = \varepsilon \sigma F_{cell-sur} A. \tag{1}$$

The emissivity was found to be $\varepsilon = 0.85$.

An adsorbed liquid film exists on the flat surfaces of the walls and the thickness of the film is on the order of a few molecular layers. In this region, the resistance to evaporative heat transfer is very high due to the attractive Van der Waals forces and there is little or no evaporation from the liquid. Vapor convection is the main mechanism of heat transfer on the flat surfaces and in the dryout zone. Heat transfer coefficients in the regions dominated by vapor convection were estimated theoretically based on equations for fluid flow in a vertical enclosed tube.[12] However, vapor is generated by the evaporation of liquid in the intrinsic meniscus region near the corners of the cell. An initial value of the liquid heat transfer coefficient at the corners for *each* axial evaporator position was estimated using the liquid film resistance,

$$h_l = \frac{k_l}{\delta_{average}}. \tag{2}$$

The surface area of evaporation, A_e at each 1 mm position of the evaporator region was computed from

$$A_e = 2\pi R_c \Delta x, \tag{3}$$

where $\Delta x = 1\,\text{mm}$. These estimates of the heat transfer coefficient were used in an iterative scheme to best fit the experimental temperature profiles and determine the liquid heat transfer coefficients and the lengths of the dryout, evaporation, and condensation zones within the CVB.

RESULTS AND DISCUSSION

FIGURE 4 shows the measured temperature profiles for a CVB partially filled with pentane that were obtained for five values of the power input. A three-dimensional SINDA® thermal simulation was used to fit the data, predict the temperature and temperature gradient at $x = 0$ (inside of the end wall attached to the heater), and thereby, compute the power input, Q_{cell}, to the CVB. A comparison of the *wet* cell experimental temperature profiles with the corresponding *dry* cell profiles for the same base temperature was made for each power level of the *wet* cell. FIGURE 5 shows the comparison for the lowest power input. In the absence of interferometric film thickness profiles in the vacuum chamber studies, the dryout region for the cell was assumed to be equal to the length of overlap observed, at the heater end, in the temperature profiles of the *dry* and *wet* cells. As the power input to the cell was increased, the overlap of the two profiles also increased. The length of the overlap region increased with increasing power input quickly at first, however it soon reached an asymptote so that beyond a power input of 0.35W, the overlap section length was constant. FIGURE 6 shows the maximum overlap of the two thermal profiles observed. The initial rapid increase in the overlap length followed by the subsequent asymptote, presented in FIGURE 7, demonstrated that heat transfer from the CVB to the external environment and fluid flow associated with evaporation/condensation were controlling the dryout length, or the amount of power that could be transferred via evaporation to the fluid. A similar phenomenon of an asymptotic dryout length within the vertical heat pipe was observed via interferometry for tests con-

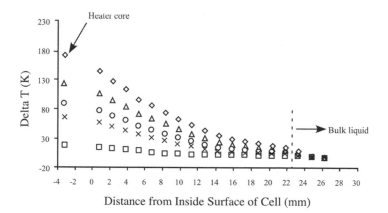

FIGURE 4. Experimental thermal profiles of a 22-mm pentane vapor bubble in the vertical CVB heat pipe, operated inside the vacuum chamber, with the heater on top: $\Diamond$, $Q = 0.50\,\text{W}$; $\triangle$, $Q = 0.35\,\text{W}$; $\bigcirc$, $Q = 0.25\,\text{W}$; $\times$, $Q = 0.19\,\text{W}$; $\square$, $Q = 0.06\,\text{W}$.[7]

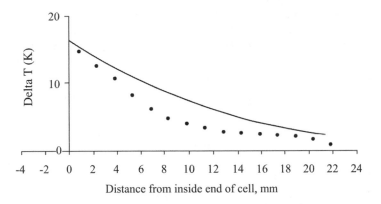

FIGURE 5. Thermal profiles of the *dry* and *wet* CVB heat pipe with a 22-mm bubble in vacuum. $Q_{cell} = 0.06\,W$: ●, experimental ΔT; —, ΔT for dry cell with $T_{heater} = 316.9\,K$.[7]

ducted on a CVB in air.[5,7] A three-dimensional SINDA simulation of each experiment was used to extract a value for the dryout lengths (see Basu[7] for details). Estimation of the dryout length within the *wet* cell was done by assigning vapor convective heat transfer coefficients to nodes in sections suspected to be in the dry region. Evaporative and vapor convective coefficients were attributed to the nodes of those sections exhibiting a sharp deviation in temperature from the dry cell baseline profile (wet regions). The final values of the lengths and heat transfer coefficients for each section were obtained by iterations to best fit the experimental thermal profiles.

Within the confines of the vacuum chamber, the CVB operated in a radiation dominated environment. A theoretical range for the radiative heat transfer coefficients obtained from the operation of the CVB in vacuum is shown in TABLE 1. These

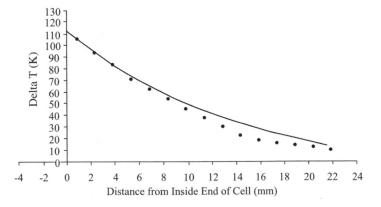

FIGURE 6. Thermal profiles of the *dry* and *wet* CVB heat pipe with a 22 mm bubble in vacuum showing maximum overlap. $Q_{cell} = 0.35\,W$: ●, experimental ΔT; —, ΔT for dry cell with $T_{heater} = 424.8\,K$.[7]

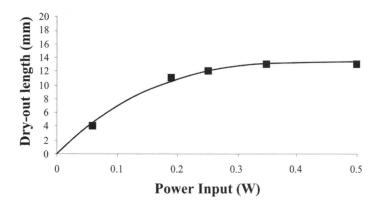

FIGURE 7. Model estimate dryout lengths in the vacuum chamber as a function of the heat input for the vertical CVB heat pipe with a 22-mm pentane vapor bubble: ■, dryout length, mm; —, least squares fit.[7]

coefficients are based on an emissivity of 0.85 for the glass cell. The emissivity was determined by using SINDA to simulate the performance of the dry cell. Previously, we found that the effective heat transfer coefficient for the external surface was significantly higher in an air environment, with $h \approx 20 \, \text{W/m}^2\text{K}$.[8] The dryout lengths for varying power inputs, as determined by the method discussed above, are shown in TABLE 2 and plotted in FIGURE 7. TABLE 2 also contains the lengths for the evaporator and condenser sections of the CVB. FIGURE 8 shows a plot of the estimated liquid heat transfer coefficient, h_l, in the evaporator section of the cell as a function of axial position for the five runs. As the heat input to the cell increased, the liquid receded into the corner and the curvature increased. This provided the additional driving force to supply fluid to the active evaporation zone. An increase in the curvature of the meniscus, translates into a smaller average film thickness in the corner that reduces the resistance to heat transfer. Initially, the dryout length increased rapidly (FIG. 7) with increasing heat load. Beyond $Q_{cell} = 0.35 \, \text{W}$, the dryout length reached a maximum for the given heat pipe configuration. When the power input to the cell was increased further, the overall temperature level along the cell also rose.

TABLE 1. **Range of temperature dependent radiative heat transfer coefficients for the CVB cell in vacuum $\epsilon = 0.85$**

Q_{cell} (W)	Range of h_{or} (W/m²·K)	T Range Experimental (°C)	T_{sur} (°C)
0.04	5.13–5.86	26–53	24.94
0.16	5.21–7.96	29–116	24.97
0.22	5.25–9.06	30–114	25.04
0.32	5.25–11.17	31–191	24.20
0.46	5.35–14.25	34–249	24.88

TABLE 2. Length of the three zones within the vertical CVB heat pipe in vacuum with a 22-mm vapor bubble for varying power input

Heat Pipe Base T (K)	Q_{cell} Experimental (W)	L_{dryout} (mm)	$L_{evaporator}$ (mm)	$L_{condenser}$ (mm)
317.4	0.06	4.0	13.0	5.0
364.5	0.19	11.0	6.0	5.0
387.3	0.25	12.0	5.0	5.0
424.6	0.35	13.0	5.0	4.0
473.7	0.50	13.0	5.0	4.0

However, the coolers kept the temperature of the condenser end relatively constant. Radiation from the surface of the CVB has a significant effect on its performance and a significant portion of the heat input was dissipated before reaching the evaporator section. At the same time, the evaporator region experienced a slight increase in temperature profile with an increase in the power input, leading to higher capillary pumping heads and evaporation rates. Therefore, even though the power input to the CVB was raised, the dryout lengths did not show any further lengthening and the evaporator region ceased to shrink. Undoubtedly, turbulence and oscillations within the CVB also contributed to this phenomenon. The glass walls have a low thermal conductivity and a high emissivity, making conduction of heat down the walls difficult and radiation from the walls easier. The design also demonstrates the need for an adiabatic section to insure the heat pipe dissipates most of its heat via evaporation of the working fluid.

In TABLE 3, a comparison of the radiative heat losses with the evaporative heat transfer capacity of the CVB in a gravitational field is presented. The high radiative

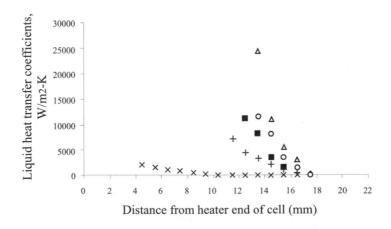

FIGURE 8. Estimated liquid heat transfer coefficients (as a function of axial distance) for a vertical CVB heat pipe with a 22-mm vapor bubble: △, Q_{cell} = 0.50 W; ○, Q_{cell} = 0.35 W; ■, Q_{cell} = 0.25 W; +, Q_{cell} = 0.19 W; ×, Q_{cell} = 0.06 W.[7]

TABLE 3. Evaporative and radiative rate of heat transfer within the vertical CVB heat pipe in vacuum with a 22-mm pentane vapor bubble, gravitational environment

$Q_{cell,x=0}$ Experimental (W)	L_{dryout} (mm)	$Q_{radiative, 0 \leq x \leq 23\,mm}$ (W)	$Q_{boiling+evaporation}$ (W)	$Q_{vapor\,convection}$ (W)
0.06	4.0	0.012	0.030	0.0002
0.19	11.0	0.07	0.088	0.002
0.25	12.0	0.10	0.166	0.008
0.35	13.0	0.16	0.122	0.010
0.50	13.0	0.26	0.142	0.017

power losses (at large dryout lengths) in a convection-free environment are detrimental to the capacity of the device due to evaporation/condensation. Although the larger dimensions of this capillary driven heat pipe (compared to the micro heat pipes) aid our study of the fluid flow properties by reducing viscous effects and enabling interferometric studies, there are some drawbacks. The fluid flow within the vertical CVB heat pipe is subject to instabilities and gravitational pressure heads as a result of the larger dimensions that need to be addressed.

The maximum power input capacity of the 22 mm vapor bubble in the Earth environment, without any dryout region, was estimated to be only 0.01 W based on a numerical extension of the following equation, due to Huang *et al.*,[13,14] for microgravity conditions:

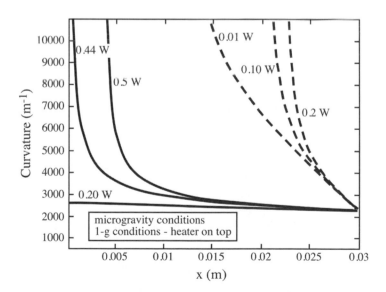

FIGURE 9. Axial curvature gradients as a function of heat input for microgravity and Earth gravity conditions.

$$Q_{max} = \frac{C_l^3 \sigma_l h_{fg} A_{quartz} m^2}{3 v_l k_{fl} K_c^3 \left[12.12 \left(l_e + \frac{1 - \cosh m l_e}{m \sinh m l_e} \right) + A_{quartz} m^2 (L - l_e) \right]}. \tag{4}$$

This is in qualitative agreement with our experiments.

Equation (4) is based on a model in which the major resistance to heat transfer is conduction in the glass and gravitational effects are absent. Therefore, although it is only applicable for an insulated CVB without external losses, it gives a convenient model with which to compare the predicted curvature profiles for microgravity with those in the Earth gravitational field. In FIGURE 9, the curvature profiles at various heat loads for the CVB cell operating in microgravity are compared with those for the Earth environment. FIGURE 9 shows the predicted relative improvement in performance of the CVB in the microgravity environment. The equivalent heat flux for an internal area of $3 \times 3 \times 10^{-6} m^2$ and a power level of 0.5 Watts is $5.5 \times 10^4 W/m^2$. The purpose of the flight experiment is, of course, to compare the performance of the CVB in microgravity with its performance in the Earth environment. Although we used an analytical model above for the purpose of discussion, the data obtained in the flight experiment will be analyzed numerically[7,13] due to the complex transport processes.

CONCLUSIONS

Although the transparency of glass for wavelengths below two microns makes it an ideal material for visual observations of the axial pressure gradient in the liquid, the lack of transparency above four microns makes modeling of the transport processes difficult.

In the absence of natural convection, the operation of CVB is significantly influenced by radiative losses, especially because the capillary limit of the device is approached at relatively low temperature. External radiative losses can dominate internal evaporative heat transfer and become the primary heat dissipation mechanism from the walls of the device.

In the Earth environment, the length of the dryout section initially increases rapidly with an increase in power input and then significantly slows, reaching an asymptote as radiative transfer becomes dominant.

In $1 g$, the 22 mm vapor bubble heat pipe has a small capillary limit (maximum power input with no dryout) since the gravitational head dominates the capillary head, thereby hampering recirculation of the liquid to the heater end. However, in μg, the maximum power input limit for the CVB without reaching dryout is expected to be significantly higher. It is predicted that the device would be able to dissipate at least 60 times more power in a mg environment.

For a practical heat exchanger, these results support the need for an adiabatic region to force the heat to be dissipated via evaporation of the working fluid rather than by radiation or natural convection to the external environment.

ACKNOWLEDGMENT

This material is based on work supported by Northrop Grumman and the National Aeronautics and Space Administration under Grants NAG3-2383 and NAG3-1834. S. Basu acknowledges the contributions of Neil Rowe and Ray Margie. Any opinions, findings, and conclusions mentioned in this paper are those of the authors and do not necessarily reflect the view of Northrop Grumman or NASA.

REFERENCES

1. COTTER, T.P. 1984. Principles and prospects for micro heat pipes. Proc. 5th Intl. Heat Pipe Conf., Tsukuba, Japan. 328–335.
2. FAGHRI, A. 1995. Heat Pipe Science and Technology, 1st edit. Taylor & Francis, Washington DC. 630–631 and 813.
3. DASGUPTA, S., J.L. PLAWSKY & P.C. WAYNER, JR. 1995. Interfacial force field characterization in a constrained vapor bubble thermosyphon. AIChE J. **41:** 2140–2149.
4. KARTHIKEYAN, M., J. HUANG, J. PLAWSKY & P.C. WAYNER, JR. 1998. Experimental study and modeling of the intermediate section of the non-isothermal constrained vapor bubble. ASME J. Heat Transf. **120:** 166.
5. HUANG, J., M. KARTHIKEYAN, P.C. WAYNER, JR. & J.L. PLAWSKY. 2000. Heat transfer and fluid flow in a nonisothermal, constrained vapor bubble. Chem. Eng. Comm. **181:** 203.
6. WANG, Y.-X., J.L. PLAWSKY & P.C. WAYNER, JR. 2001. Optical measurement of microscale transport processes in dropwise condensation. Microscale Thermophys. Eng. **5:** 55.
7. BASU, S. 2002. Experimental Study and Thermal Modeling of the Constrained Vapor Bubble Heat Pipe Operation in a Convection-Free Environment under the Influence of Gravity. Ph.D. Dissertation, Rensselaer Polytechnic Institute, Troy, NY.
8. ZHENG, L., Y.-X. WANG, P.C. WAYNER, JR. & J.L. PLAWSKY. 2003. Microscale transport processes in the evaporator of a constrained vapor bubble. J. Thermophys. Heat Transf. **17:** 166–173.
9. BOWMAN, W.J., J.K. STOREY & K.I. SVENSSON. 1998. Analytical comparison of constant area, adiabatic tip, standard fins and heat pipe fins. J. Thermophys. **13:** 269–272.
10. BOWMAN, W.J., T.W. MOSS, D. MAYNES & K.A. PAULSON. 1999. Efficiency of a constant area, adiabatic tip heat pipe fin. J. Thermophys. **14:** 112–115.
11. PETERSON, G.P. 1994. An Introduction to Heat Pipes. John Wiley & Sons, New York.
12. LIGHTHILL, M.J. 1952. Theoretical considerations on free convection in tubes. Quart. J. Mech. Appl. Math. **VI:** 399–439.
13. HUANG, J., J.L. PLAWSKY, P.C. WAYNER, JR. 2000. Modeling transport processes in a constrained vapor bubble under microgravity conditions. AIP Conference Proceedings 504, Space Technology and Applications International Forum-2000. M.S. El-Genk, Ed.: 261–272.
14. HUANG, J. 1998. Modeling the Transport Processes in a Constrained Vapor Bubble. Ph.D. Dissertation, Rensselaer Polytechnic Institute, Troy, NY.

Surfactant Effect on the Buoyancy-Driven Motion of Bubbles and Drops in a Tube

EISA ALMATROUSHI[a] AND ALI BORHAN[b]

[a]Chemical and Petroleum Engineering Department, U.A.E. University, Al-Ain, U.A.E.

[b]Department of Chemical Engineering, The Pennsylvania State University, University Park, Pennsylvania, USA

ABSTRACT: The effect of surfactants on the buoyancy-driven motion of bubbles and drops in a vertical tube is experimentally examined. The terminal velocities of fluid particles are measured and their steady shapes are quantitatively characterized in systems with various bulk-phase concentrations of surfactant. In the case of air bubbles, the presence of surfactant retards the motion of small bubbles due to the development of adverse Marangoni stresses, whereas it enhances the motion of large bubbles by allowing them to deform away from the tube wall more easily. For viscous drops, the surfactant-enhanced regime of particle motion becomes more pronounced in the sense that the terminal velocity becomes more sensitive to surfactant concentration, whereas the surfactant effect in the surfactant-retarded regime becomes weaker.

KEYWORDS: drops; bubbles; surfactant effect; buoyancy-driven; Marangoni stresses

INTRODUCTION

The motion of drops and bubbles through tubes is of interest in a variety of industrial and technical applications. Much of the earliest interest in this subject was motivated by the suggested analogy between drop motion and the motion of blood cells in the capillaries. These flow systems also arise in many polymer processing operations (e.g., fiber spinning, film blowing, and injection molding) where additives are used to achieve certain required physical/mechanical properties in the final products. Aside from the aforementioned technical applications, the motion of deformable drops through tubes of constant and variable cross section remains of considerable fundamental importance as a pore-scale model for studying the dynamics of two-phase flow through porous media. For example, in the recovery of residual oil from geologic strata at small saturations, the dispersed oil phase resides in the porous matrix in the form of isolated drops that are held fixed in equilibrium configurations by capillary forces. These drops can be mobilized by injecting a second fluid into the pores provided that viscous forces are made sufficiently large compared to interfacial tension forces. The dynamics of droplet motion in the pores, as well as the possible dispersion processes that can occur, strongly influence the overall recovery of the dispersed phase and the pressure drop–volume flux relation, which is the primary

Address for correspondence: Eisa AlMatroushi, Chemical & Petroleum Engineering Department, U.A.E. University, Al-Ain, P.O. Box: 17555, U.A.E.
    almatroushi@uaeu.ac.ae

Ann. N.Y. Acad. Sci. 1027: 330–341 (2004). ©2004 New York Academy of Sciences.
doi: 10.1196/annals.1324.028

function of interest in the macroscopic description of multiphase flow through porous media. Understanding the movement of drops and bubbles in microchannels is also increasingly important in the design and operation of microfluidic devices that involve two-phase flows.

The confined motion of drops and bubbles through Newtonian fluids has been the subject of numerous theoretical and experimental studies, a summary of which is provided by Olbricht.[1] On the experimental side, several investigations have focused on the slow motion of large bubbles in capillary tubes, where the bubble and the tube wall were separated by only a thin layer of suspending fluid.[2-8] The most important result of these experiments is the observation that the film thickness and the drop speed relative to the average bulk fluid velocity become independent of drop size, in good qualitative agreement with the theoretical predictions of Martinez and Udell.[9] However, experimental measurements of the film thickness at small capillary numbers are underpredicted by theoretical analyses, a discrepancy that has been attributed to the presence of surfactant impurities in the experiments.[10-12] The creeping motion of smaller immiscible drops through straight circular tubes has also been the focus of experimental investigations,[13-15] Ho and Leal[13] considered the pressure-driven motion of neutrally buoyant drops through a horizontal circular cylinder whereas Olbricht and Leal[14] examined the effects of a density difference between the drop and the suspending fluid. Ho and Leal[13] measured the variation of drop velocity and extra pressure loss with drop size, bulk velocity, and viscosity ratio. They confirmed the theoretical predictions of a negative extra pressure loss when the viscosity ratio was small. Their experimental data for drop speeds, however, are consistently overpredicted by the analysis of Martinez and Udell.[9] Again, this discrepancy has been attributed to the presence of surfactant impurities in the experiments.[16,17]

Surface-active materials are encountered in many industrial applications involving the flow of droplet dispersions, and are extremely difficult to remove from these systems. Because interfacial tension is often very sensitive to the presence of minute amounts of surface-active impurities, their presence can lead to variations of interfacial tension across free surfaces. It is well established that such interfacial tension gradients play an important role in free-surface flows.[18] Migration of droplets in a temperature gradient, and the larger than expected drag on translating droplets in all but very clean systems are clear examples of this effect. In this paper, we present the results of an experimental study of the steady motion of drops and bubbles through a cylindrical tube in the presence of surfactants. The principal quantities to be examined are the steady shapes and velocities of drops and bubbles.

EXPERIMENTAL PROCEDURE

A schematic illustration of the experimental setup is shown in FIGURE 1. The experimental apparatus was similar to that used by Borhan and Pallinti[19] to study the motion of clean drops in cylindrical capillary tubes. It consisted of a vertical precision-bore glass tube of inside diameter 0.796cm and length 120cm, enclosed by a plexiglas chamber of square cross-section containing an aqueous solution of sodium iodide. The refractive index of the sodium iodide solution was matched with that of the glass tube to minimize optical distortions due to the refraction of light at the outer wall of the tube.

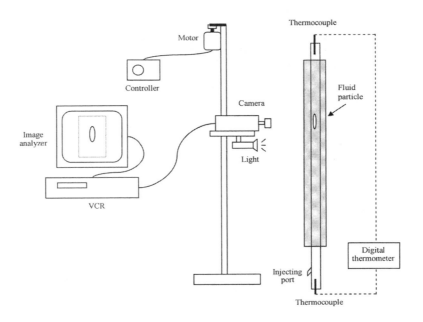

FIGURE 1. Schematic illustration of the experimental setup.

Aqueous glycerol solutions (96.2 wt%), prepared with filtered double-distilled water, were used as the suspending fluid. UCON LB-165 fluid and air were used as drop fluids. The experiments were performed at a constant fluid temperature of 25°C. At this temperature, the suspending fluid was characterized by density of 1,247 kg/m^3 and viscosity of 443 mPa·sec, and the LB-165 fluid had a density of 976 kg/m^3 and viscosity of 100 mPa·sec. Various amounts of 99%+ pure sodium dodecyl sulfate (SDS) supplied by Sigma Chemical Company were dissolved in the suspending fluid to prepare surfactant solutions of various concentrations for the experiments. The interfacial tension of the surfactant-free interface between the two phases was determined to be 8.0 dyn/cm for LB-165 drops, and 65.5 dyn/cm for air bubbles. The equilibrium surface tension of the surfactant solution was determined at the various surfactant concentrations used in the experiments. The critical micelle concentration (CMC) of the surfactant was then estimated from the location of the abrupt change in the slope of the surface tension–surfactant concentration plot, shown in FIGURE 2. Based on these results, the surfactant CMC in the aqueous glycerol solution was estimated to be at (or slightly below) a value of about 8.7 mM corresponding to the concentration of the 0.20 wt% surfactant solution. This value of CMC is in line with previously reported values for aqueous SDS solutions.[20] In all cases, the surfactant solutions were prepared just before the experiments were performed, to avoid possible changes in the properties of the surfactant solution over time. After each set of experiments with the same surfactant solution, the entire system was thoroughly cleaned with filtered double-distilled water, benzene, and acetone, and then dried in air before the start of each new set of experiments.

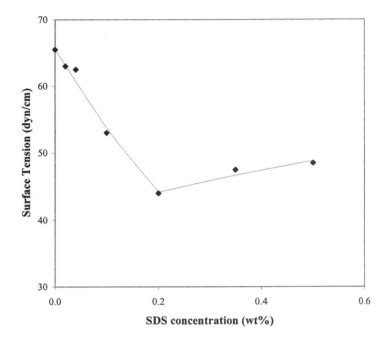

FIGURE 2. Variation of equilibrium surface tension with SDS concentration.

For each experiment, the desired volume of the drop fluid was injected at the symmetry axis of the tube near the inlet region using a micrometer syringe. The density of the drop phase was less than that of the suspending fluid so that the drop moved from the bottom of the tube to the top. A CCD camera, mounted on a moving platform with a variable speed controller, was used to monitor the motion of the drop as it passed through the entire length of the tube, and drop motion was recorded and indexed using a video recorder. To study drop deformations, the recorded images of the drop profile were played back frame by frame and the signal from the video recorder was digitized. A stop-motion filter was applied to the digitized images to remove any jittering caused by the motion of the drop, and Image-Pro image analysis software was used to quantitatively characterize the drop shapes by measuring various geometric features, such as the perimeter of the drop profile in the meridional plane and its maximum axial and radial dimensions. For each experiment, the volume of the axisymmetric drop was also determined more accurately using Image-Pro.

The terminal velocity of the drop was determined by measuring the time required for the drop to travel a specified vertical distance between selected markers on the tube wall. The velocity measurements were made in regions sufficiently far from the inlet of the tube to avoid entrance effects. For each experiment, three independent velocity measurements over different regions of the tube were made. Some experiments were repeated to ensure reproducibility of the results. In all cases, the reported terminal velocity represents the average of at least three velocity measurements, with each measurement having a variation of less than 5% from the reported mean value.

RESULTS AND DISCUSSION

In this section, we present the experimental results in terms of the effect of surfactant concentration on the steady shape and terminal velocity of drops and bubbles. The Reynolds number based on the terminal velocity of the drop or bubble was less than 0.1 in all of the experiments reported here. The Bond number $Bo = \Delta \rho g R^2 / \sigma_{eq}$ varied in the range $3.0 \leq Bo \leq 6.5$, where $\Delta \rho$ and σ_{eq} represent the density difference and equilibrium interfacial tension between the two phases, respectively, R is the tube radius, and g is the magnitude of the gravitational acceleration. We first consider the effect of surfactant on the motion of air bubbles. The dependence of the dimensionless terminal velocity on bubble size is shown in FIGURE 3, where the bubble size κ is made dimensionless with the tube radius and the velocity U is scaled with $\Delta \rho g R^2 / \mu$ (where μ denotes the viscosity of the suspending fluid). The solid and dashed curves in this figure represent the best fits to the experimental data in the surfactant-free system and the 0.5% surfactant solution, respectively, taking into account the fact that U must vanish as the bubble size κ tends to zero. The dotted curve represents the corresponding theoretical predictions of Hetsroni et al.[21] for spherical bubbles in a surfactant-free system. For small bubbles, the experimentally measured bubble speeds for the surfactant-free system are in good agreement with the asymptotic predictions of Hetsroni et al. However, as the bubble size becomes larger than about $\kappa = 0.4$, the experimental measurements begin to deviate from the

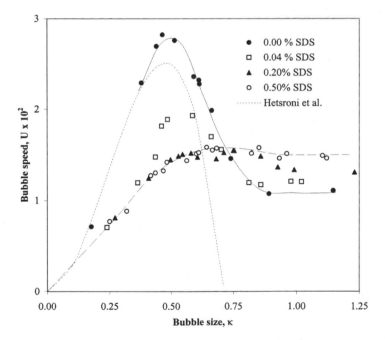

FIGURE 3. Effect of SDS concentration on the rise velocity of air bubbles ($\lambda = 0$) in systems with $Bo = 3.0$–4.4.

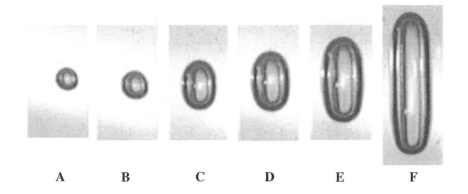

FIGURE 4. Images of rising air bubbles in the 0.2% SDS solution: **(A)** κ = 0.4, **(B)** κ = 0.5, **(C)** κ = 0.7, **(D)** κ = 0.8, **(E)** κ = 1.0, **(F)** κ = 1.2.

theoretical predictions, with the extent of deviation growing substantially for κ > 0.6. This is not surprising, since the asymptotic predictions are valid for small bubbles that remain nearly spherical, whereas significant bubble deformations are observed in the experiments for κ ≥ 0.5, as evidenced by the images of translating bubbles shown in FIGURE 4.

The terminal velocity of a rising bubble initially increases with bubble size until the retarding effect of the tube wall becomes more pronounced and eventually causes the terminal velocity to decrease and approach a limiting value. In the surfactant-free system, the maximum terminal velocity is achieved for a bubble size of about κ = 0.5, and the plateau in the terminal velocity is reached for κ ≥ 0.9. Below CMC, both the location of the maximum terminal velocity and the first appearance of the plateau in the terminal velocity are shifted to slightly larger values of κ as the surfactant concentration increases. In addition, the magnitude of the maximum terminal velocity is reduced with increasing surfactant concentration. The addition of surfactant to the system leads to the development of Marangoni stresses on the surface of the rising bubble as surfactant monomers diffusing to the bubble surface are swept to the trailing end of the bubble by surface flow. These Marangoni stresses oppose the surface flow and retard the motion of the bubble as a whole in the same manner as that observed for the rise of bubbles in an unbounded fluid. Therefore, the reduction in the maximum terminal velocity shown in FIGURE 3 is due to the development of Marangoni stresses that become more pronounced with increasing surfactant concentration up to CMC. Above CMC, the diffusion of surfactant monomers to the bubble surface is augmented by the spontaneous breakdown of micelles caused by the depletion of monomers in their vicinity. Hence, increasing the surfactant concentration beyond the CMC does not affect the mobility of small (κ < 0.7) bubbles.

In contrast to the behavior of small bubbles, increasing the surfactant concentration leads to a slight increase in the terminal velocity of larger (κ > 0.7) bubbles. It appears that the retarding effect of Marangoni stresses is overwhelmed by the influence of surfactants on the shape deformations of large bubbles. In all of the experiments reported here, the shapes of rising bubbles and drops remained axisymmetric

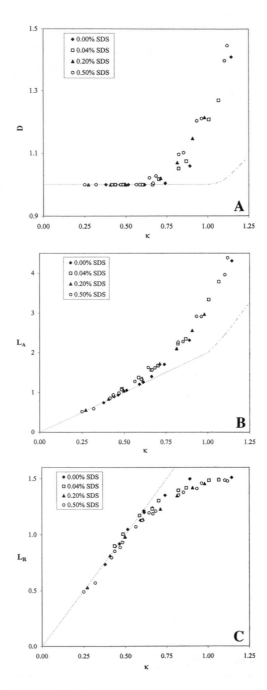

FIGURE 5. Variation of the dimensionless geometric parameters with bubble size for systems with $Bo = 3.0–4.4$: (**A**) deformation parameter, D; (**B**) axial length, L_A; (**C**) radial length, L_R.

as they passed through the tube. Typical profiles (in the meridional plane) of the steady shapes of surfactant-laden air bubbles are shown in FIGURE 4 for various bubble sizes. The steady shape of an air bubble approaches an elongated ellipsoid as the bubble size increases, with a slight loss of fore and aft symmetry. To quantitatively characterize the influence of surfactants on the steady bubble shapes, a deformation parameter $D = P^2/4\pi A$ is defined, where P and A represent the perimeter and area of the deformed drop profile in the meridional plane, respectively. This quantity provides a measure of deviations from a spherical shape (characterized by $D = 1$), and is easily obtained from digitized images of rising bubbles using the Image-Pro image analysis software. FIGURE 5 shows the dependence of the deformation parameter on the dimensionless bubble size for various surfactant concentrations. Also shown in this figure are the geometric parameters L_A and L_R, representing the maximum axial and radial dimensions of the steady bubble profile relative to the tube radius, respectively. The dashed curves in FIGURE 5 denote the values of the geometric parameters for spherical bubbles, whereas the dot–dashed curves represent those corresponding to a stagnant cylindrical bubble with hemispherical end caps. Both the radial and axial dimensions of the bubble initially grow linearly as a function of bubble size, with the first shape transition occurring at a bubble size of about $\kappa = 0.6$, which also represents the point at which the experimental measurements for clean bubbles begin to deviate substantially from the asymptotic predictions (see FIG. 3).

The steady shapes of small ($\kappa < 0.7$) bubbles are insensitive to surfactant concentration. For larger bubbles, increasing the surfactant concentration leads to larger deformations. The radial dimension of a bubble of fixed size decreases, and its axial length grows, with increasing surfactant concentration. The increase in bubble length is accompanied by an increase in the thickness of the liquid film surrounding the bubble. FIGURE 5 also indicates that the transition from a spherical to an elongated shape occurs at smaller bubble sizes as the surfactant concentration increases. Once the transition to a cylindrical shape occurs, an increase in the surfactant concentration leads to a smaller radial dimension (limiting value of L_R) for the cylindrical section of the bubble or, equivalently, a thicker liquid film between the cylindrical bubble and the tube wall. As the surfactant concentration exceeds the CMC, the bubble shape becomes insensitive to the surfactant concentration.

Since the more elongated bubble shapes formed at higher surfactant concentrations are localized near the centerline of the tube, they experience a weaker retarding

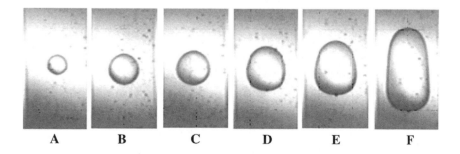

FIGURE 6. Images of rising viscous drops in the 0.2% SDS solution: **(A)** $\kappa = 0.4$, **(B)** $\kappa = 0.5$, **(C)** $\kappa = 0.6$, **(D)** $\kappa = 0.7$, **(E)** $\kappa = 0.8$, **(F)** $\kappa = 1.0$.

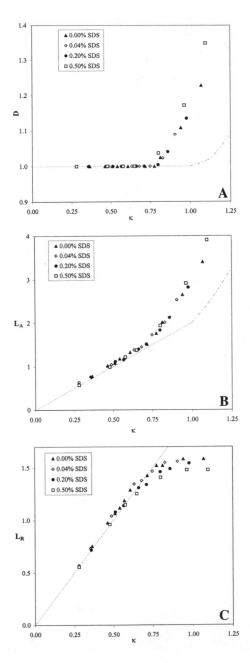

FIGURE 7. Variation of the dimensionless geometric parameters with drop size for systems with $\lambda = 0.23$ and $Bo = 5.3$–6.5: **(A)** deformation parameter, D; **(B)** axial length, L_A; **(C)** radial length, L_R.

effect of the tube wall, and their mobility is enhanced. Hence, the addition of surfactant to the system below CMC can affect the motion of a bubble of fixed size through two distinct mechanisms with opposite effects. Surfactants can retard the motion of the bubble through the action of Marangoni stresses induced by interfacial tension gradients, and the immobilization of the surface of the bubble. Surfactants can also have a favorable effect on the motion of the bubble through the tube by causing a reduction in the surface tension of the bubble, thereby allowing it to more easily deform away from the tube wall. For small bubbles that remain essentially spherical, the latter effect is negligible and the adverse effect of Marangoni stresses dominates. The Marangoni stresses are weakened as the bubble size increases, due to larger shape deformations and the associated surface dilution of surfactant. Hence, the favorable effect of surfactants on bubble deformation eventually becomes dominant when the bubble size becomes comparable to tube size. The value of the bubble size for the onset of surfactant-enhanced bubble motion is dependent on the Bond number of the clean system.

A qualitatively similar behavior is observed for the motion of viscous drops. Typical profiles (in the meridional plane) of the steady shapes of viscous drops rising in a surfactant solution are shown in FIGURE 6 for various drop sizes. A viscous drop evolves into a slightly more tapered shape as its size increases, with its leading end remaining relatively unchanged. Experimental measurements of the geometric parameters D, L_A, and L_R for various drop sizes in systems with various surfactant

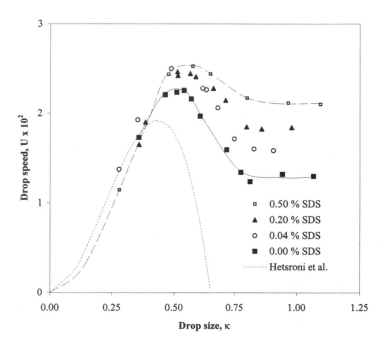

FIGURE 8. Effect of SDS concentration on the rise velocity of viscous drops in systems with $\lambda = 0.23$ and $Bo = 5.3$–6.5.

concentrations are shown in FIGURE 7. Compared to the deformation behavior of an air bubble, the shape of a viscous drop as a whole is less sensitive to the surfactant concentration below CMC. The values of L_R decrease as the surfactant concentration increases to CMC, however, the deformation parameter D and the axial elongation L_A both remain virtually the same, indicating that the drop shape becomes less tapered as the surfactant concentration increases. Because of the viscous nature of the drop phase, even a slight increase in the distance between large drops and the tube wall is sufficient to cause a significant improvement in their mobility, as can be seen from the results in FIGURE 8. The terminal velocity of a large viscous drop in the 0.2% surfactant solution (near CMC) is about 50% larger than that of the corresponding clean drop. In contrast to the behavior of large air bubbles, a notable increase in the deformation of large drops, accompanied by a corresponding increase in their mobility, is observed as the surfactant concentration is increased significantly above CMC in the 0.5% SDS solution. A comparison of FIGURES 3 and 8 shows that the surfactant-enhanced regime of particle motion becomes more pronounced for viscous drops in the sense that the terminal velocity becomes more sensitive to surfactant concentration, whereas the surfactant effect in the surfactant-retarded regime becomes weaker. The larger viscosity of the interior phase weakens the influence of Marangoni stresses and renders the influence of shape deformations on particle mobility more dominant.

ACKNOWLEDGMENTS

Acknowledgment is made to the United Arab Emirates University for providing financial support for one of the authors (EA) during the course of this work.

REFERENCES

1. OLBRICHT, W.L. 1996. Pore-scale prototypes of multiphase flow in porous media. Annu. Rev. Fluid Mech. **28:** 187–214.
2. BRETHERTON, F.P. 1961. The motion of long bubbles in tubes. J. Fluid Mech. **10:** 166–188.
3. TAYLOR, G.I. 1961. Deposition of a viscous fluids on the wall of a tube. J. Fluid Mech. **10:** 161–165.
4. COX, B.G. 1962. On driving a viscous fluid out of a tube. J. Fluid Mech. **14:** 81–98.
5. GOLDSMITH, H.L. & S.G. MASON. 1963. The flow of suspensions through tubes: II single large bubbles. J. Coll. Sci. **18:** 237–261.
6. MARCHESSAULT, R.N. & S.J. MASON. 1960. Flow of entrapped bubbles through a capillary. Ind. Eng. Chem. **52:** 79–81.
7. SCHWARTZ, L.W., H.M. PRINCEN & A.D. KISS. 1986. On the motion of bubbles in capillary tubes. J. Fluid Mech. **172:** 259–275.
8. CHEN, J.D. 1986. Measuring the film thickness surrounding a bubble inside a capillary. J. Coll. Interf. Sci. **109:** 341–349.
9. MARTINEZ, M.J. & K.S. UDELL. 1990. Axisymmetric creeping motion of drops through circular tubes. J. Fluid Mech. **210:** 565–591.
10. RATULOWSKI, J. & H.C. CHANG. 1990. Marangoni effects of trace impurities on the motion of long gas bubbles in capillaries. J. Fluid Mech. **210:** 303–328.
11. PARK, C.-W. 1992. Influence of soluble surfactants on the motion of a finite bubble in a capillary tube. Phys. Fluids A **4**(11): 2335–2346.

12. STEBE, K.J. & D.L. BARTHES-BIESEL. 1995. Marangoni effects of adsorption-desorption controlled surfactants on the leading end of an infinitely long bubble in a capillary. J. Fluid Mech. **286:** 25–48.

13. HO, B.P. & L.G. LEAL. 1975. The creeping motion of liquid drops through a circular tube of comparable diameter. J. Fluid Mech. **71:** 361–384.

14. OLBRICHT, W.L. & L.G. LEAL. 1982. The creeping motion of liquid drops through a circular tube of comparable diameter: the effect of density differences between the fluids. J. Fluid Mech. **115:** 187–216.

15. BORHAN, A. & J. PALLINTI. 1998. Pressure-driven motion of drops and bubbles through cylindrical capillaries: effect of buoyancy. Ind. Eng. Chem. Res. **37:** 3748–3759.

16. BORHAN, A. & C. MAO. 1992. Effect of surfactants on the motion of drops through circular tubes. Phys. Fluids A **4**(12): 2628–2640.

17. JOHNSON, R.A. & A. BORHAN. 2003. Pressure-driven motion of surfactant-laden drops through cylindrical capillaries: effect of surfactant solubility. J. Coll. Interf. Sci. **261:** 529–541.

18. LEVICH, V.G. & V.S. KRYLOV. 1969. Annu. Rev. Fluid Mech. **1:** 293.

19. BORHAN, A. & J. PALLINTI. 1995. Buoyancy-driven motion of viscous drops through cylindrical capillaries at small Reynolds numbers. Ind. Eng. Chem. Res. **34:** 2750.

20. HERNAINZ, F. & A. CARO. 2002. Variation of surface tension in aqueous solutions of sodium dodecyl sulfate in the flotation bath. Coll. Surf. A: Physicochem. Eng. Aspects **196:** 19–24.

21. HETSRONI, G., S. HARBER & E. WACHOLDER. 1970. The flow fields in and around a droplet moving axially within a tube. J. Fluid Mech. **41:** 689–705.

Miscible, Porous Media Displacements with Density Stratification

AMIR RIAZ AND ECKART MEIBURG

Department of Mechanical and Environmental Engineering,
University of California, Santa Barbara, California, USA

ABSTRACT: High accuracy, three-dimensional numerical simulations of miscible displacements with gravity override, in both homogeneous and heterogeneous porous media, are discussed for the quarter five-spot configuration. The
influence of viscous and gravitational effects on the overall displacement
dynamics is described in terms of the vorticity variable. Density differences
influence the flow primarily by establishing a narrow gravity layer, in which
the effective Péclet number is enhanced due to the higher flow rate. Although
this effect plays a dominant role in homogeneous flows, it is suppressed to some
extent in heterogeneous displacements. This is a result of coupling between the
viscous and permeability vorticity fields. When the viscous wavelength is much
larger than the permeability wavelength, gravity override becomes more effective because coupling between the viscous and permeability vorticity fields is
less pronounced. Buoyancy forces of a certain magnitude can lead to a pinch-
off of the gravity layer, thereby slowing it down.

KEYWORDS: miscible, porous media displacements; density stratification

INTRODUCTION

The stability of interfaces separating fluids of different viscosities in porous
media has been the subject of numerous investigations since the pioneering work,[1-3]
established that an adverse mobility ratio, that is, a less viscous fluid displacing a
more viscous one, generates an unstable interface. By means of experiments and,
more recently, numerical simulations, the nonlinear interfacial dynamics has been
studied as well, using a variety of physical models and geometries, compare the
review in Reference 4.

Accurate representation of the diffused interface in miscible displacements
requires a high accuracy numerical method. A combination of spectral methods
and high order, compact finite differences are employed to obtain very high accuracy in simulations of both two-dimensional rectilinear,[5,6] and quarter five-spot
displacements.[7-9] The present investigation extends this approach to three dimensions, in order to analyze the interaction of viscous and gravitational effects in the
quarter five-spot configuration.

Address for correspondence: Eckart Meiburg, Department of Mechanical and Environmental Engineering, University of California, Santa Barbara, CA 93106, USA. Voice/fax: 805-
893-5278.
meiburg@engineering.ucsb.edu

Ann. N.Y. Acad. Sci. 1027: 342–359 (2004). ©2004 New York Academy of Sciences.
doi: 10.1196/annals.1324.029

The paper is organized as follows. The next section presents the governing equations and the boundary and initial conditions. Numerical implementation is discussed next. We then compare neutrally buoyant, two- and three-dimensional homogeneous displacements. The subsequent section presents the results of a parametric study, in which the effects of variation of density difference between the two fluids are investigated from a vorticity-based point of view for homogeneous displacements. The influence of heterogeneous permeability is then described. Finally, the last section highlights the main findings of this investigation and summarizes its most important conclusions.

GOVERNING EQUATIONS

The quarter five-spot arrangement consists of a staggered, doubly periodic array of injection and production wells, as shown in FIGURE 1. The mathematical model, described in detail elsewhere,[10] is based upon the vorticity formulation of Darcy's equation. We assume incompressible flow and use a convection–diffusion equation to advance the concentration field of the injected fluid. The computational domain is shown in FIGURE 2. To derive the dimensionless equations, we choose the side length L of the domain as the characteristic length scale. By denoting the source strength per unit depth as $2\pi Q$, we obtain the time and velocity scales as L^2/Q and Q/L, respectively. The viscosity μ_1 of the injected fluid is taken as the reference value for scaling the viscosities, whereas the difference between the fluid densities $\rho_2 - \rho_1$ provides a characteristic density value. Here the index 1 refers to the injected fluid and 2 denotes the displaced fluid. We define an aspect ratio $A = H/L$, where H is the domain height. The resulting governing equations in nondimensional form are

$$\nabla \cdot \mathbf{u} = 0, \tag{1}$$

$$\omega = \frac{1}{k}\nabla k \times \mathbf{u} + R\nabla c \times \mathbf{u} + \frac{Gk}{\mu}\nabla c \times \nabla z, \tag{2}$$

$$\frac{\partial c}{\partial t} + \mathbf{u} \cdot \nabla c = \frac{1}{Pe}\nabla^2 c. \tag{3}$$

For simplicity, we refer to the terms on the right hand side of the vorticity equation as *permeability vorticity, viscous vorticity,* and *gravitational vorticity.* Three dimensionless parameters appear in these equations

$$Pe = \frac{Q}{D}, \tag{4}$$

$$G = \frac{g(\rho_2 - \rho_1)KL}{Q\mu_1}, \tag{5}$$

$$R = -\frac{1}{\mu}\frac{d\mu}{dc} = \ln\left(\frac{\mu_2}{\mu_1}\right), \tag{6}$$

where c denotes the concentration of the injected fluid, k represents the isotropic permeability, and μ and ρ are the concentration dependent viscosity and density, respectively. We take the scalar diffusion coefficient to be a constant D. The dimensionless

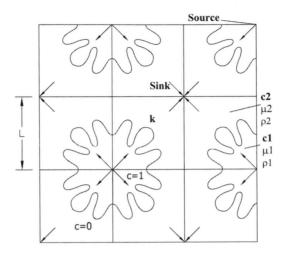

FIGURE 1. The quarter five-spot arrangement of injection and production wells.

parameters are the Péclet number Pe, the gravity parameter G, and the viscosity ratio parameter R.

The velocity is obtained through a three-dimensional vector potential[11] as

$$\mathbf{u} = \nabla \times \psi \tag{7}$$

$$\nabla^2 \psi = -\omega \tag{8}$$

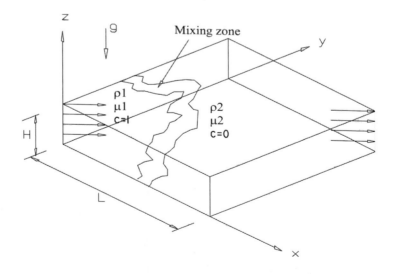

FIGURE 2. The three-dimensional computational domain.

At the vertical boundaries symmetry conditions are assumed. The top and bottom represent no-flux boundaries. By denoting the spatial components of ψ and ω as (ϕ, θ, χ) and (ξ, η, ζ), respectively, we obtain

$$x = 0, 1: \begin{cases} c_x = 0 \ k_x = 0 \\ u = 0 \ v_x = 0 \ w_x = 0 \\ \phi_x = 0 \ \theta = 0 \ \chi = 0 \\ \xi_x = 0 \ \eta = 0 \ \zeta = 0 \end{cases} \tag{9}$$

$$y = 0, 1: \begin{cases} c_y = 0 \ k_y = 0 \\ u_y = 0 \ v = 0 \ w_y = 0 \\ \phi = 0 \ \theta_y = 0 \ \chi = 0 \\ \xi = 0 \ \eta_y = 0 \ \zeta = 0 \end{cases} \tag{10}$$

$$z = 0, A: \begin{cases} c_z = 0 \ k_z = 0 \ w = 0 \\ \phi = 0 \ \theta = 0 \ \chi_z = 0 \end{cases} \tag{11}$$

To avoid an initially singular concentration distribution, we specify as the initial condition, at a small but finite time, the self-similar concentration profile corresponding to the radially symmetric problem.[12] It has the form

$$c_0 = \frac{1}{2}\left[1 + \mathrm{erf}\left(\sqrt{Pe}\left(\frac{r}{r_0} - 1\right)\right)\right], \tag{12}$$

where r_0 represents the initial radial location of the front. It determines the *effective starting time* $t_0 > 0$ of the computation as

$$t_0 = 0.5r_0^2. \tag{13}$$

NUMERICAL IMPLEMENTATION

The numerical solutions are obtained with a combination of sixth-order compact finite difference[13] and spectral methods,[14,15] in conjunction with an explicit third-order time stepping scheme.

Time Stepping Scheme

The concentration field is advanced in time by an explicit third order Runga–Kutta method.[16] Writing the concentration Equation (3) as

$$\frac{\partial c}{\partial t} = F(c), \tag{14}$$

we obtain

$$c_{i,j}^k = c_{i,j}^{k-1} + \Delta t[\alpha_k F(c_{i,j}^{k-1}) + \beta_k F(c_{i,j}^{k-2})], \tag{15}$$

where

$$\alpha_1 = \frac{8}{15}, \; \beta_1 = 0$$

$$\alpha_2 = \frac{5}{12}, \; \beta_2 = -\frac{17}{60}$$

$$\alpha_3 = \frac{3}{4}, \; \beta_3 = -\frac{5}{12}.$$

Solution of the Poisson Equation

The solution procedure for the Poisson equation governing the vector potential is described for the example of the first component of Equation (8)

$$\phi_{xx} + \phi_{yy} + \phi_{zz} = -\xi. \tag{16}$$

Due to the periodicity in the x- and y-directions, this equation can be solved by a Fourier Galerkin method. The Fourier coefficients a_{ij} and b_{ij} of the vector potential and vorticity fields, respectively, are given by

$$\phi = \sum_{j=1}^{n-1} \sum_{i=1}^{n-1} a_{ij}(z)\cos[(i-1)\pi x]\sin[(j-1)\pi y] \tag{17}$$

$$\xi = \sum_{j=1}^{n-1} \sum_{i=1}^{n-1} b_{ij}(z)\cos[(i-1)\pi x]\sin[(j-1)\pi y]. \tag{18}$$

A second order ODE for $a_{ij}(z)$ is obtained by substituting Equations (17) and (18) into (16)

$$[-(i-1)^2\pi^2 - (j-1)^2\pi^2]a_{ij}(z) + a_{ij}''(z) = -b_{ij}(z). \tag{19}$$

By using compact finite differences to approximate d^2a/dz^2, we obtain the following pentadiagonal system

$$\alpha(a_{i,j,k-2} + a_{i,j,k+2}) + \beta(a_{i,j,k-1} + a_{i,j,k+1}) + \gamma a_{i,j,k}$$

$$= -\frac{2}{11}b_{i,j,k-1} - b_{i,j,k} - \frac{2}{11}b_{i,j,k+1},$$

where

$$\alpha = \frac{3}{44\Delta^2}$$

$$\beta = -\frac{2}{11}[(i-1)^2 + (j-1)^2]\pi^2 + \frac{12}{11\Delta^2}$$

$$\gamma = -[(i-1)^2 + (j-1)^2]\pi^2 - \frac{24}{11}\Delta^2 - \frac{3}{22\Delta^2}.$$

A fine spatial and temporal resolution is required in order to resolve accurately all the length scales present in the domain, especially at high Péclet numbers. Time steps as small as $O(10^{-6})$ are used, whereas typical spatial resolutions employ

$256 \times 256 \times 32$ modes; that is, 2×10^6 grid points. For some parameter values a spatial resolution of $513 \times 513 \times 64$ are used. A preliminary analysis was conducted to determine the appropriate grid spacing for various parameter combinations. An important criterion in this regard is the cutoff mode provided by the linear stability analysis.[12,17] The grid spacing is required always to be smaller than the cutoff wavelength. In addition, we also track the energy in the highest Fourier mode and require it to be less than 0.001% of the maximum energy in the spectrum. We also require the maximum and minimum numerical concentration levels to remain between 1.001 and −0.001, respectively, We have found that these particular values of the constraints guarantee convergence and stability of the numerical results. Our numerical algorithm is parallelized for optimal performance.[18]

COMPARISON BETWEEN TWO- AND THREE-DIMENSIONAL, NEUTRALLY BUOYANT DISPLACEMENTS

For a neutrally buoyant displacement without externally imposed initial perturbations in the vertical direction, the present three-dimensional algorithm exactly reproduces the two-dimensional results reported elsewhere.[7] In order to trigger a three-dimensionally evolving flow, we introduce random initial concentration perturbations of the form

$$c(x, y, z, t_0) = c_0(x, y, t_0) + \gamma f(x, y, z) e^{\frac{-(r - r_0)^2}{\sigma^2}}, \qquad (20)$$

where γ denotes the disturbance amplitude, f represents a field of random numbers uniformly distributed in the interval $[-1, 1]$, and σ specifies the width of the initially perturbed layer around the mean interface position r_0. The simulations to be discussed below employ $\gamma = 0.025$ and $\sigma = 0.003$, as well as an initial interface location of $r_0 = 0.2$.

Random initial perturbations of the above form give rise to substantial fingering activity from the very start. FIGURE 3 presents a simulation for $Pe = 800$, $R = 2.5$, and $G = 0$; that is, without density contrast. The number of fingers in both the horizontal and vertical planes at early times are consistent with the linear stability results, as shown in FIGURE 3 A. The number of fingers is greatly reduced at later times, as seen

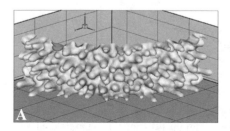 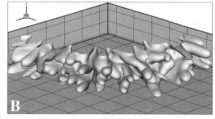

FIGURE 3. Concentration iso-surfaces obtained for random initial perturbations and $G = 0$, $Pe = 800$, $R = 2.5$, and $A = 1/8$, at times **(A)** 0.04 and **(B)** 0.14. Nonlinear interaction mechanisms of merging, shielding, and fading can be observed.

in FIGURE 3 B. This is expected, both on the basis of linear stability results for vertical modes, as well as due to the nonlinear mechanisms of finger interaction, that is, merging and shielding.[5]

The magnitude of disturbances is obtained from the norm of vorticity, which is defined by

$$\|\omega(t)\| = \sqrt{\frac{1}{LNM} \sum_{i=1}^{L} \sum_{j=1}^{N} \sum_{k=i}^{M} \omega(t)_{i,j,k}^2}. \tag{21}$$

A quantitative comparison between the two- and three-dimensional displacements is given in FIGURE 4. The higher vorticity level in the three-dimensional simulation reflects a strong finger growth, which eventually leads to an earlier breakthrough, as shown in FIGURE 4. Here the breakthrough time t_b is defined as the time when the concentration of the injected fluid first reaches 1% somewhere along the height of the production well. Correspondingly, the overall efficiency η of the displacement process is given as the fraction of the total domain volume occupied by the injected fluid at the time of breakthrough, $\eta = \pi t_b/2$.

Comparison of the vorticity magnitudes in FIGURE 4 indicates a significant difference between two- and three-dimensional displacements in the quarter five-spot geometry, even for neutrally buoyant flows. This is somewhat unexpected in light of other findings,[19,20] where only small differences between two- and three-dimensional rectilinear flows were observed. This is attributed to the lack of a vortex stretching term in Darcy's flow,[19] since it is this term that causes quite fundamental differences

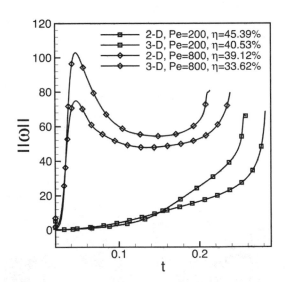

FIGURE 4. Comparison of the vorticity norm for the two- and three-dimensional cases for two different Péclet numbers. Higher vorticity values in the three-dimensional case result in a lower efficiency as compared to the two-dimensional case, for both small and large Pe. The higher vorticity level in the three-dimensional case results from the additional instability toward axial waves.[17]

between two- and three-dimensional flows governed by the Navier–Stokes equations. For quarter-spot displacements on the other hand, the higher vorticity level for three-dimensional displacements is due to an effective redistribution of the concentration gradient associated with the changes in the wavelength ratio of the most amplified vertical and horizontal waves.[17] Consequently, higher overall concentration fluctuations than in the purely two-dimensional case generate faster growing fingers for the three-dimensional case, which leads to an earlier breakthrough.

In order to demonstrate the validity of the simulations for the nonlinear stages, FIGURE 5 compares the present two- and three-dimensional simulation data for the displacement efficiency with experimental results.[21,22] The experiments quoted do not provide sufficient information to calculate a Péclet number. However, numerical simulations performed at $Pe = 800$ agree reasonably well with the experimental data for various viscosity ratios. The three-dimensional simulations are seen to lead to better agreement than their two-dimensional counterparts, especially at high viscosity ratios. Different Pe values result in somewhat different values of η, however, the trend of a continued decrease in η with R at large R values is more accurately represented by the three-dimensional simulations, independent of Pe. Similarly, the inclusion of a dispersion model would be expected to result in somewhat different values of η.

Three-dimensional simulations more accurately represent the displacement process by accounting for interaction between the horizontal and the vertical modes. We quantify this interaction by analyzing the relative magnitude of the norm for the vertical and the horizontal components of viscous vorticity. FIGURE 6 shows that the

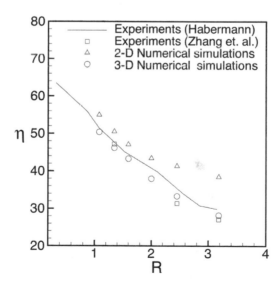

FIGURE 5. Displacement efficiency as a function of the viscosity ratio. Experimental data[21,22] are compared with numerical two-dimensional data for $Pe = 800$ (obtained with the present three-dimensional code using two-dimensional initial perturbations), and with the present three-dimensional data for $Pe = 800$. The three-dimensional results exhibit significantly better agreement with the experimental data.

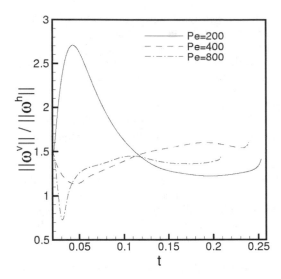

FIGURE 6. Ratio of the vertical to the horizontal vorticity norm for various values of Pe, $R = 2.5$, $G = 0$, and $A = 1/8$. $\|\omega^v\|/\|\omega^h\|$ gives the relative strength of the horizontal and the vertical disturbances, respectively. For initial time $t < 0.05$, horizontal modes dominate for $Pe = 200$ whereas vertical modes gain in strength for $Pe = 400$ and 800. A minimum value of the ratio is observed for $t > 0.12$ as a function of Pe.

horizontal modes, associated with vertical vorticity, dominate early on at time $t < 0.05$ for both $Pe = 200$ and 400, when $\|\omega^v\|/\|\omega^h\| > 1$. However, the negative slope of $\|\omega^v\|/\|\omega^h\|$ for $Pe = 400$ and early times shows that the energy in the vertical modes is growing at a higher rate. For $Pe = 800$, the ratio soon drops below 1, indicating the dominance of the vertical modes. This relative importance of the vertical modes at early times, for $Pe = 400$ and 800, is predicted by linear stability theory.[17] The situation subsequently changes due to the onset of nonlinear behavior. It is important to note that for later times $\|\omega^v\|/\|\omega^h\| > 1$, which indicates that the horizontal modes play a relatively more important part in the displacement dynamics. Also note that at later times $\|\omega^v\|/\|\omega^h\|$ exhibits a minimum. This reflects the fact that for large Pe values the strength of the vertical vorticity is reduced as a result of nonlinear interactions among the horizontal modes. The dominance of the horizontal modes is due to the mean flow in the horizontal planes. Additional simulations show that $\|\omega^v\|/\|\omega^h\|$ is independent of the aspect ratio, but only for neutrally buoyant displacements. Interaction between the horizontal and vertical modes qualitatively changes in displacements with gravity override, as is discussed below.

HOMOGENEOUS DISPLACEMENTS WITH GRAVITY OVERRIDE

In the following, we discuss the influence of the governing dimensionless parameter G on the overall features of the displacement process in a homogeneously permeable domain. How those flows are altered by the effects of density stratification in comparison with the purely viscous instability studied by Chen and Meiburg[7] will be one of the main issues to be analyzed here.

FIGURE 7 presents concentration contours for the case of $Pe = 800$, $R = 2.5$, $G = 0.5$, and $A = 1.8$; that is, all parameters except G have the same values as in FIGURE 3. The simulation employs a grid of size $257 \times 257 \times 32$, as well as a time step of $O(10^{-6})$. The injected fluid, being lighter than the displaced fluid, tends to rise towards the upper boundary of the domain. A *gravity layer* is thus established in which the flow rate is higher than elsewhere in the domain. Both the gravity layer as well as the *underride* region below it give rise to well developed fingers as early as $t = 0.04$. The number of fingers decreases as the interface evolves due to both horizontal and vertical interactions among the fingers (cf., FIG. 7 B). The thickness of the gravity layer is seen to increase, due to multiple mergers with the fingers directly below it. It is important to note the wide range of length scales produced during the displacement.

At $t = 0.14$ the growth of the fingers along the diagonal in the gravity layer is temporarily slowed by fingers approaching from below (FIG. 7C). The resulting buoyancy induced pinch-off effect,[6] cuts off the fluid supply of high velocity fingers in the gravity layer and leads to their gradual fading. This gives a chance to fingers farther away from the diagonal to break through first at $t = 0.21$. The pinch-off mechanism thus delays the time of breakthrough.

Closely related to the structure of the interface is the vorticity field. Equation (2) shows that vorticity is generated due to concentration gradients, which in turn determine the velocity field. Equation (2) also shows that gravitational vorticity compo-

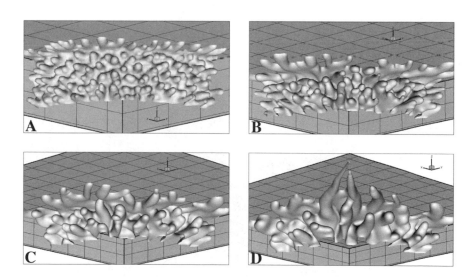

FIGURE 7. Concentration iso-surfaces for $Pe = 800$, $R = 2.5$, $G = 0.5$, and $A = 1/8$ at times 0.04, 0.08, 0.14, and 0.21. A gravity layer with numerous fingers evolves early on and becomes dominant around $t = 0.14$. The coupling of viscous and gravitational instability renders the gravity layer susceptible to a pinch-off by the underride fingers. Breakthrough is achieved by fingers in the gravity layer at $t = 0.21$.

nents can reinforce or cancel directly only the horizontal viscous vorticity components. FIGURE 8 plots isosurfaces of the viscous and gravitational vorticity components for the flow shown in FIGURE 7. Dark (light) shading represents negative (positive) values of vorticity. The horizontal and vertical components of the viscous vorticity are seen to form elongated dipole structures along the edges of the fingers, whereas the gravitational vorticity develops a more complex spatial structure. The buoyancy driven pinch-off shown in FIGURE 7B and C occurs due to the local reinforcement of the horizontal components of viscous and gravitational vorticity. Note that the spatial distribution of the gravitational vorticity shown in the figure is similar, but of opposite sign to that of the vertical viscous vorticity. This reflects the fact that both the vertical viscous vorticity and the gravitational vorticity are associated with horizontal concentration gradients.

We now consider a situation with a larger density contrast. FIGURE 9 depicts concentration isosurfaces from a simulation with the same parameter values as above, except that $G = 2$. The mesh size here is $513\times513\times64$. A strong gravity layer develops, in which the effective local Péclet number is large enough to sustain multiple tip splitting events. Most importantly, the gravity layer becomes increasingly dominant and moves far ahead of the underride region. The fingering activity in the underride region is subdued, as can be seen from a comparison of FIGURES 7C and 9B. Hence we note that increasing G values stabilize the underride region. This form of shear stabilization is studied in detail elsewhere.[23] In the present case, it has a detrimental effect on the overall efficiency of the displacement process, because it

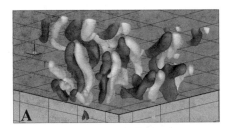 

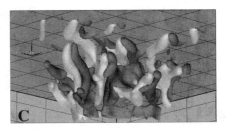

FIGURE 8. Iso-surfaces of **(A)** vertical viscous, **(B)** horizontal viscous, and **(C)** gravitational vorticity components for $Pe = 800$, $R = 2.5$, $G = 0.5$, and $A = 1/8$ at $t = 0.14$. The buoyancy driven interaction occurs due to the local reinforcement of the horizontal components of viscous and gravitational vorticity. The spatial distribution of gravitational vorticity is similar in shape, but of opposite sign to that of the vertical viscous vorticity.

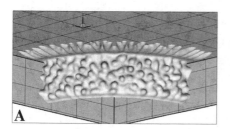 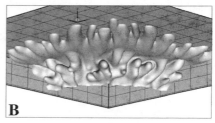

FIGURE 9. Concentration iso-surfaces for $G = 2.0$, $Pe = 800$, $R = 2.5$, and $A = 1/8$ at times **(A)** $t = 0.04$ and **(B)** $t = 0.14$. The gravity layer dominates the displacement. A significant reduction of the fingering activity in the underride region can be observed at this value of G. The breakthrough time is 0.165.

prevents the slowing of the gravity layer by pinch-off events. Consequently, the breakthrough time is reduced by almost 20% as G is increased from 0.5 to 2.

The influence of the gravity override mechanism can be evaluated by considering the relative strengths of waves in the horizontal and vertical directions, as reflected by the ratio of the vertical to the horizontal viscous vorticity in FIGURE 10. It is to be kept in mind that waves in a horizontal plane are associated with vertical vorticity and vice versa. FIGURE 10A shows the cases depicted in FIGURES 7 and 9, as well as a case for $G = 0.25$. The ratio $\|\omega^v\|/\|\omega^h\|$ decreases uniformly with an increase in G, reflecting the emergence of a strong gravity layer. Throughout the displacement process, the interplay of vertical and horizontal vorticity components is strongly affected by G. A comparison of the efficiencies listed in FIGURE 10A shows that the

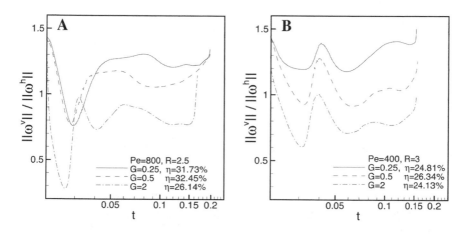

FIGURE 10. Ratio of the norms of vertical to horizontal viscous vorticity components as a function of G. Increase in G uniformly strengthens the vertical component of viscous vorticity. Optimal interaction between the horizontal and the vertical modes is given by $\|\omega^v\|/\|\omega^h\| \approx 1$.

maximum efficiency is achieved at $G = 0.5$, as an optimal interaction between the two components is achieved with the ratio $\|\omega^v\| / \|\omega^h\|$ close to unity throughout the displacement process. A similar optimal behavior is observed in FIGURE 10B for $G = 0.5$. A possible interpretation of the optimal interaction phenomenon is that the initial dominance of vertical viscous vorticity develops the underride fingers, and the later dominance of buoyancy effects allows these fingers to pinch off the gravity layer. If G is too large, the underride fingers do not develop to a point where a meaningful interaction with the gravity layer can be sustained. On the other hand, if G is too small, the gravity effect is not strong enough to allow the underride fingers to curve upward and pinch off the gravity layer. Note that the optimal interaction mechanism is very sensitive to the interfacial structure, which is in turn strongly dependent upon the initial conditions. Given the high cost of numerical simulations, we have not carried out a detailed study of this phenomenon. On the other hand, a gain of a few percent in efficiency can be important for enhanced oil recovery processes, so that a detailed investigation of this phenomenon would appear to be a worthwhile pursuit.

HETEROGENEOUS DISPLACEMENTS WITH GRAVITY OVERRIDE

Spatial permeability variation introduces an additional vorticity component related to permeability, as given by Equation (2). Details about constructing a random permeability field k with horizontal and vertical wavelengths m and n, respectively, and variance s are given elsewhere.[24] From the previous section we see that if the injected fluid is lighter than the displaced fluid gravitational vorticity can give rise to a gravity layer that substantially alters the characteristics of the flow.[6,25] Within the gravity layer the fingers are enhanced, whereas in the underride region they are suppressed.

For homogeneous quarter five-spot displacements, a larger gravity parameter G generally results in earlier breakthrough, although for some parameter combinations intermediate values of G have been observed for which the efficiency is optimized due to specific interactions between the horizontal and vertical modes.[10] Two-dimensional rectilinear displacements[26] demonstrate that the effect of gravity override is considerably reduced by permeability heterogeneities, due to the coupling between the viscous and permeability vorticities. As a result, an optimal efficiency is achieved at an intermediate variance level.

The influence of gravity override is shown in FIGURE 11 for a representative combination of displacement parameters, at various values of the gravity parameter G. Comparison of the $G = 0.5$ case in FIGURE 11B with the $G = 0$ case in FIGURE 11A shows that gravity override strengthens the fingers close to the upper boundary, while weakening those near the lower boundary. Although the gravity layer is not as pronounced for the present, heterogeneous case as it is for the corresponding homogeneous case,[10] a slight diversion of the flow from the underride region to the gravity layer for $G = 0.5$ slows the rapid movement of the dominant fingers in the underride region observed for $G = 0$ in FIGURE 11A. Consequently, for the case $G = 0.5$, FIGURE 11B shows that the gravity layer fingers are slightly stronger as compared to the $G = 0$ case, which results in an improvement of the efficiency. It should be pointed out that the maximum in the recovery curve for an intermediate G in the

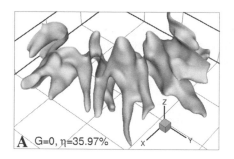
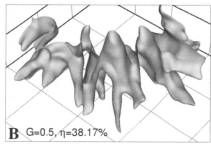

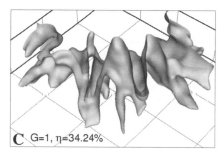

FIGURE 11. Concentration isosurfaces for the gravity override case. $Pe = 400$, $R = 2.5$, $s = 0.1$, $m = 20$, $n = 20$, $t = 0.14$ and various values of G. By encouraging the development of a gravity layer along the top boundary, the dominant fingers close to the lower boundary are weakened resulting in an improvement in efficiency as G goes from 0 to 0.5. Further increase in the gravity override effect at $G = 1$ strengthens the fingers in the gravity layer resulting in a reduction in efficiency. Due to the coupling between viscous and permeability vorticities, the gravity override effect is weaker than in homogeneous displacements.

above case is due to the location of the dominant flow path in the underride region. If the flow path with the lowest resistance were instead located close to the upper boundary, there would be a monotonic decrease in efficiency with increasing G.

Note that the gravity layer is relatively weak for the present, heterogeneous case, as compared to the corresponding homogeneous flow.[10] This fact, which is similar to observations for rectilinear flows, indicates that at the level of $s = 0.1$ the heterogeneity is already too strong for the coupling between the viscous and permeability vorticities to be effectively modified by the gravitational vorticity component. This is confirmed by FIGURE 12, which depicts the same flows as FIGURE 11, except that $s = 0.01$. Here the gravity layer is much stronger, which leads to a lower breakthrough efficiency for $s = 0.01$ than for $s = 0.1$, for both $G = 0.5$ and 1.

The above observation reflects the fact that the dominant path is selected not only on the basis of its permeability. Also important is its potential to support a resonant amplification, which partly depends on its geometric nature, and also on the local flow rate, which in turn is a function of the overall gravitational effect. Hence, the flow can select a relatively low permeability path over one with higher permeability, as long as it supports a strong resonant amplification.

The relatively weak dependence of heterogeneous displacements on the gravity parameter, as compared to their homogeneous counterparts, is also reflected in the norms of the vertical and horizontal viscous vorticity fields. Since gravitational effects primarily result in horizontal vorticity, it is instructive to analyze the ratio of the vertical to the horizontal viscous vorticity norms $\left\|\omega_v^y\right\|/\left\|\omega_v^h\right\|$, shown in FIGURE 13. We vary both the horizontal (m, FIG. 13 A) and the vertical (n, FIG. 13 B) permeability wave number. In each case the homogeneous displacement is more strongly affected by the horizontal vorticity (i.e., by gravitational effects) than any of the heterogeneous displacements. Note that the influence of gravity, $\left\|\omega_v^y\right\|/\left\|\omega_v^h\right\|$ decreases with m and it increases with n.

Individual random realizations of the permeability field can strongly influence the fingering dynamics, and consequently the displacement efficiency.[24] We have not attempted to run sufficiently many simulations in order to obtain statistically significant averages, due to the prohibitive computational expense. Instead, we have limited ourselves to identifying the generic mechanisms that govern heterogeneous displacements. The accurate prediction of the displacement efficiency for a specific permeability field would, of course, require complete knowledge of the permeability distribution.[27]

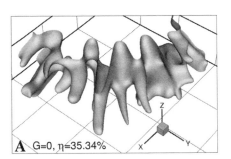
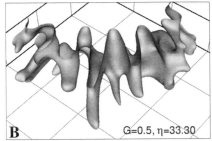
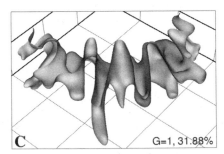

FIGURE 12. Concentration isosurfaces for the same cases as in FIGURE 11, but with a lower variance, $s = 0.01$. The lower level of heterogeneity allows the gravity layer to develop relatively freely, which results in a monotonic decrease in efficiency. The higher efficiency for $s = 0.1$ as compared to $s = 0.01$ for $G = 0$ is due to a weakening of the gravity layer, as well as the strengthening of the off-diagonal fingers, at $s = 0.1$.

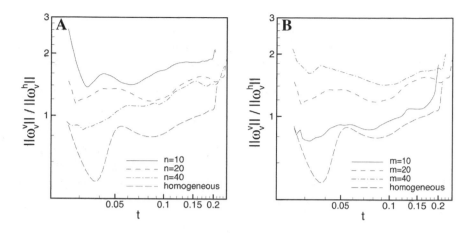

FIGURE 13. Ratio of the vertical and horizontal viscous vorticity norm. $Pe = 400$, $R = 2.5$, $G - 0.5$, $s = 0.1$, $\Lambda = 1.8$, **(A)** $n = 20$, **(B)** $m = 20$. Gravity override increases the relative strength of disturbances in the vertical direction, which are associated with horizontal vorticity. For the homogeneous case $\left\|\omega_v^v\right\| / \left\|\omega_v^h\right\| < 1$. On the other hand, the ratio exceeds unity for most heterogeneous cases, which implies that the perturbations in the horizontal directions are dominant. The ratio decreases with increasing m and it increases with n.

CONCLUSIONS

The present investigation employs high accuracy numerical simulations in order to analyze three-dimensional miscible displacements with gravity override in the quarter five-spot geometry. Even for neutrally buoyant displacements, three-dimensional effects are seen to change the character of the flow in a way that cannot be anticipated from two-dimensional simulations alone. This is in contrast to the case of rectilinear miscible flows, for which the inclusion of the third dimension is found[19,25] to have generally small effects. Part of the difference between two- and three-dimensional quarter five-spot flows can be attributed to the enhanced interaction of disturbances in three dimensions, resulting from the time dependence of the most amplified axial mode.[17] By forcing a large scale redistribution of concentration gradients, the temporal change of the axial wave number introduces an additional source of instability in the system.

Density differences have a significant influence on the displacement, primarily through the production of a narrow gravity layer that bypasses much of the resident fluid to arrive at the production side ahead of the underride region, thereby reducing the displacement efficiency. However, for certain parameter combinations buoyancy forces of the right magnitude can pinch off the gravity layer and hinder its movement towards the production well.

The interaction of viscous and gravitational effects is analyzed on the basis of their respective contributions to the temporal evolution of the vorticity field. The gravitational effect is associated with an increase in the horizontal vorticity component, whereas the viscous effect is related to both horizontal and vertical vorticity components. As expected, the gravitational effect primarily operates on vertical disturbances.

In the presence of density differences, the potential for gravity override becomes important. Although this effect is seen to play a dominant role in homogeneous displacements, it is suppressed to some extent in heterogeneous displacements, even for relatively small values of the heterogeneity variance. This is a result of the coupling between viscous and permeability vorticity fields. For small vertical permeability wavelength relative to the viscous wavelength, gravity override is somewhat more effective because the coupling between viscous and permeability vorticity fields is less pronounced, so that the large scale fingering structures become more responsive to buoyancy effects. This is confirmed by the ratio of the vertical to the horizontal viscous vorticity norm, which decreases with increasing values of the vertical correlation wave number.

ACKNOWLEDGMENT

The authors thank Dr. Hamdi Tchelepi for several helpful discussions. Support for this research by the Petroleum Research Fund, the Department of Energy, Chevron Petroleum Technology Company, as well as through an NSF equipment grant and the San Diego Supercomputer Center is gratefully acknowledged.

REFERENCES

1. HILL, S. 1952. Channeling in packed columns. Chem. Engng. Sci. **1:** 247.
2. SAFFMAN, P.G. & G.I. TAYLOR. 1958. The penetration of a fluid into a porous medium or Hele-Shaw cell containing a more viscous liquid. Proc. R. Soc. London Ser. A **245:** 312.
3. CHOUKE, R.L., P.V. MEURS & C.V.D. POEL. 1959. The instability of slow, immiscible, viscous liquid-liquid displacements in permeable media. Trans. AIME **216:** 188.
4. HOMSY, G.M. 1987. Viscous fingering in porous media. Annu. Rev. Fluid Mech. **19:** 271.
5. TAN, C.T. & G.M. HOMSY. 1988. Simulation of nonlinear viscous fingering in miscible displacement. Phys. Fluids **31**(6): 1330.
6. RUITH, M. & E. MEIBURG. 2000. Miscible rectilinear displacements with gravity override. Part 1. Homogeneous porous medium. J. Fluid Mech. **420:** 225.
7. CHEN, C.-Y. & E. MEIBURG. 1998. Miscible porous media displacements in the quarter five-spot configuration. Part 1. The homogeneous case. J. Fluid Mech. **371:** 233.
8. CHEN, C.-Y. & E. MEIBURG. 1998. Miscible porous media displacements in the quarter five-spot configuration. Part 2. Effect of heterogeneities. J. Fluid Mech. **371:** 269.
9. CHEN, C.-Y. & E. MEIBURG. 2000. High-accuracy implicit finite-difference simulations of homogeneous and heterogeneous miscible porous medium flows. SPE J. **5**(2): 129.
10. RIAZ, A. & E. MEIBURG. 2003. Three-dimensional miscible displacement simulations in homogeneous porous media with gravity override. J. Fluid Mech. **494:** 95.
11. FLETCHER, C.A.J. 1991. Computational techniques for fluid dynamics, second ed., vol. 2 of Springer series in computational physics. Springer.
12. TAN, C.T. & G.M. HOMSY. 1987. Stability of miscible displacements in porous media: radial source flow. Phys. Fluids **30**(5): 1239.
13. LELE, S.K. 1992. Compact finite differences with spectral-like resolution. J. Comput. Phys. **103:** 16.
14. GOTTLIEB, D. & S.A. ORSZAG. 1977. Numerical Analysis of Spectral Methods: Theory and Applications. Society for Industrial and Applied and Mathematics.

15. CANUTO, C. M.Y. HUSSAINI, A. QUARTERONI & T.A. ZANG. 1986. Spectral Methods in Fluid Dynamics. Springer Series in Computational Dynamics. Springer-Verlag.
16. WRAY, A.A. 1991. Minimal storage time-advancement schemes for spectral methods. Preprint.
17. RIAZ, A. & E. MEIBURG. 2003. Radial source flows in porous media: linear stability analysis of axial and helical perturbations in miscible displacements. Phys. Fluids **15**(4): 938.
18. RIAZ, A. 2003. Three-Dimensional Miscible, Porous Media Displacements in the Quarter Five-Spot Geometry. PhD Thesis, Department of Mechanical and Environmental Engineering. University of California, Santa Barbara.
19. ZIMMERMAN, W.B. & G.M. HOMSY. 1992. Three-dimensional viscous fingering: a numerical study. Phys. Fluids **4**(9): 1901.
20. TCHELEPI, H.A. 1994. Viscous Fingering, Gravity Segregation and Permeability Heterogeneity in Two-Dimensional and Three-Dimensional Flows. PhD Thesis, Department of Petroleum Engineering, School of Earth Sciences, Stanford University.
21. HABERMANN, B. 1960. The efficiency of miscible displacement as function of mobility ratio. Trans. AIME **219**: 264.
22. ZHANG, H.R., K.S. SORBIE & N.B. TSIBUKLIS. 1997. Viscous fingering in five-spot experimental porous media: new experimental results and numerical simulations. Chem. Engng. Sci. **52**: 37.
23. ROGERSON, A. & E. MEIBURG. 1993. Shear stabilization of miscible displacements in porous media. Phys. Fluids **5**(6); 1344.
24. RIAZ, A. & E. MEIBURG. 2004. Vorticity interaction mechanisms in variable viscosity, heterogeneous miscible displacements with and without density contrast. J. Fluid Mech. To appear.
25. TCHELEPI, H.A. & F.M.J. ORR. 1994. Interaction of viscous fingering, permeability inhomogeneity and gravity segregation in three dimensions. SPE Res. Eng. **9**(4): 266.
26. CAMHI, E., M. RUITH & E. MEIBURG. 2000. Miscible rectilinear displacements with gravity override. Part 2. Heterogeneous porous media. J. Fluid Mech. **420**: 259.
27. ZHAN, L. & Y.C. YORTSOS. 2001. A direct method for the identification of the permeability field of an anisotropic porous medium. Water Resour. Res. **37**(7): 1929.

Effect of Gravity on the Caloric Stimulation of the Inner Ear

MOHAMMAD KASSEMI,[a] DIMITRI DESERRANNO,[b] AND JOHN G. OAS[c]

[a]National Center for Microgravity Research,
NASA Glenn Research Center, Cleveland, Ohio. USA

[b]Biomedical Engineering Department, Case Western Reserve University

Cleveland, Ohio. USA

[c]Department of Otolaryngology, Cleveland Clinic Foundation,
Cleveland, Ohio. USA

ABSTRACT: Robert Bárány won the 1914 Nobel Prize in medicine for his convection hypothesis for caloric stimulation. Microgravity caloric tests aboard the 1983 SpaceLab 1 mission produced nystagmus results that contradicted the basic premise of Bárány's convection theory. In this paper, we present a fluid structural analysis of the caloric stimulation of the lateral semicircular canal. Direct numerical simulations indicate that on earth, natural convection is the dominant mechanism for endolymphatic flow. However, in the microgravity environment of orbiting spacecraft, where buoyancy effects are mitigated, an expansive convection becomes the sole mechanism for producing endolymph motion and cupular displacement. Transient $1g$ and microgravity case studies are presented to delineate the different dynamic behaviors of the $1g$ and microgravity endolymphatic flows. The associated fluid-structural interactions are also analyzed based on the time evolution of cupular displacements.

KEYWORDS: fluid structural interaction; inner ear; vestibular system; CFD; caloric test; microgravity

INTRODUCTION

The caloric stimulation test is one of the most widely used non-physiological techniques for examining vestibular performance. This procedure uses thermal irrigation of the ear canal with cold and/or hot fluid (water or air) to elicit a vestibular signal. A measure for the intensity of the signal is the associated eye movement or nystagmus. Robert Bárány received the 1914 Nobel Prize in medicine for describing endolymphatic flow and associated cupular deflection during the caloric test and relating it to a buoyancy-driven natural convective mechanism driven by the thermal irrigation.[1,2] Microgravity caloric tests aboard the 1983 SpaceLab 1 mission[3,4] produced nystagmus with an intensity comparable to those elicited during post- and preflight tests in normal gravity of Earth, thus contradicting the basic premise of

Address for correspondence: Mohammad Kassemi, National Center for Microgravity Research, NASA Glenn Research Center, 21000 Brookpark Rd, MS110-3, Cleveland, OH, 44135, USA. Voice: 216-433-5031; fax: 216-433-5033.
    mohammad.kassemi@grc.nasa.gov

Ann. N.Y. Acad. Sci. 1027: 360–370 (2004). ©2004 New York Academy of Sciences.
doi: 10.1196/annals.1324.030

Barany's convection hypothesis and pointing out that the physics of the vestibular apparatus needs to be more precisely deciphered.

More than 100 years ago, Brown and Sequard[5] reported the effects of thermal irrigation on the vestibular system. However, it was not until 1906 that Barany pointed out the clinical significance of this effect when used as a method of assessing vestibular function.[1] He also provided the first detailed description of the physical origins of the caloric test by putting forward the assumption that the small temperature shift brought about by cold or hot irrigation is able to induce a natural convection current in the endolymph.[2] The resulting endolymphatic flow produces a pressure differential across the cupula, deflecting this partition and stimulating the associated ampular sensory hair cells in a manner similar to a rotary test.

Even at its inception Barany's theory was subject of an intense debate. Bartels[6] expressed the opinion that the cause of the caloric reaction was a direct impact of the stimulus on the nerves with heat having stimulating and cold having depressing effects. Kobrak[7] suggested that the caloric response is a product of vascular reactions in form of constrictions or dilations of the vessels that produce an endolymphatic flow deflecting the cupula. Borries[8] stressed the importance of the other labyrinthine structures. In his opinion the caloric response was a reaction of the whole labyrinth, especially the otoliths. Van Canegham[9] put forward the idea that the cause of the caloric reaction is the increase or decrease of the intralabyrinthine pressure due to hot or cold stimulations with its roots in the utricle. Finally, Brunner[10] dismissed altogether the notion that cupular deflections are responsible for the caloric response and proposed that caloric nystagmus must be of a central origin.

Caloric tests performed in the long-duration microgravity environment of orbiting spacecraft have added to the confusion and controversy.[4,11] During the European vestibular experiments on the 1983 SpaceLab 1 mission, crew members were subject to caloric tests. The experiments produced nystagmus results with an intensity comparable, and a direction identical, to those elicited during post- and preflight tests, thus contradicting the basic premise of Barany's convection hypothesis.[3] These results suggested that mechanisms other than thermal convection might be involved in the induction of the caloric response. Scherer and Clarke[3] proposed that the gravity-independent convection brought about by direct volumetric expansion or contraction of the endolymph, upon cooling or heating, is a possible mechanism for provocation of the caloric response in microgravity, whereas Hood[12] and Paige[13] have adhered to the assumption that a thermally-induced neural response brought about by direct thermal stimulation of the hair cells is the source of the vestibular signal.

Ever since Stienheusin[14] presented his torsion pendulum model, fluid and structural dynamics of the vestibular system in general, and the semicircular canals in particular, have been subject to numerous theoretical examinations and mathematical model development. However, theoretical research that directly focuses on the caloric test has been quite scarce, and is basically limited to four doctoral dissertations due to Steer,[15] Young,[16] O'Neill,[17] and Damiano.[18]

The theoretical models presented by Steer,[15] Young,[16] and O'Neil,[17] are based on a lumped parameter approach and, therefore, are limited in their scope due the restrictions inherent to the lumped parameter analyses. Moreover, these models have other limitations from a fluid dynamics point of view. First and foremost is the neglect of viscous effects despite the fact that the endolymphatic flow is character-ized by quite low Reynolds numbers close to or below unity. Young[16] neglected the viscous contributions in determining the pressure differential across the cupula and Steer[15,19] and O'Neil[17] neglected viscous forces in calculating the endolymphatic angular accelerations. Second, in all of the above work, the endolymphatic duct was approximated as a perfect uniform torroid, thus rendering the model to be incapable of anatomical specificity. Oman et al.[20] and Rabbit and Damiano[21] found that the geometry of the canal plays an important role in determining the relative contri-butions of the various terms in the equation of motion. Finally, all the three investigators[15–17] also disregarded the possible contributions of volumetric expan-sion to the endolymphatic motion and cupular deflections.

Damiano[18] developed a continuum model describing the macromechanics of caloric stimulation. He exploited the slender torroidal geometry of the canal and used perturbation analysis to provide a Green's function solution for the dynamic response of the canal system to a singular harmonic thermal line source applied directly to the semicircular duct wall. The endolymph was modeled as a Newtonian fluid subject to both expansive and buoyant convection and the cupula was assumed to be a linearly elastic membrane. He compared the caloric response to the more standard response obtained during rotational tests and was able to characterize it in terms of a single constant that depends on the angle of inclination of the duct and the location of the thermal line source. In this paper, we present a comprehensive fluid-structural-interaction (FSI) analysis of the lateral semicircular canal (LSCC) system outside the restrictive confines of a lumped system approach or perturbation analysis and using a finite element model, we will examine the dynamic coupling that exists between the endolymphatic canal flow and the cupular structural deformation during the caloric stimulation. We consider caloric tests both in 1 g and in microgravity and point out the intricate differences in the endolymph–cupula interaction for these two different gravitational environments.

FSI MODEL AND NUMERICAL METHODOLOGY

A two-dimensional cross section of the LSCC is presented in FIGURE 1. The geometry and all of the associated dimensions are extracted from human data mea-sured by Curthoys and Omen.[22] The horizontal or lateral canal system consists of three main regions: the semicircular duct together with the ampulla and utricle cav-ities. The ampulla cavity is the widened area of the canal at one end just before it communicates with the utricle. A crest-like septum, called crista, transverses the ampulla perpendicular to the longitudinal axis of the canal. The cupula extends from the surface of the crista to the ceiling of the ampulla forming what appears to be a watertight seal. There are five openings on the wall of the utricle as it connects and communicates with both ends of the three semicircular canals. The utricular openings to the two ends of the horizontal canal are explicitly depicted in the two-dimensional cross-section shown in FIGURE 1; the opening to the common crux of the posterior and anterior canals is included implicitly as the flat-shaped area on the lower section of the utricle wall.

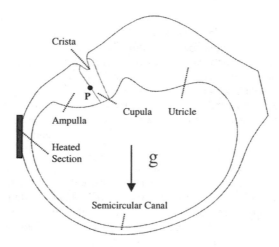

FIGURE 1. The LSCC system.

The caloric test is usually performed in an optimum configuration, with the patient in the supine position and the head tilted 30° upward, thus placing the horizontal canal in a vertical plane aligned with the gravitational field. This is the configuration depicted in FIGURE 1. The ear canal is then irrigated with hot/cold fluid (water/air) that is usually 7° above/below the nominal body temperature of 37o. Due to heat transfer through the temporal bone, the temperature of a segment of the horizontal canal that is closest to the bone (represented by the rectangular area in FIG. 1) changes. The path of heat flow through the porous temporal bone structure is quite complicated and consists of conduction through tissue and bone and convection and radiation in the embedded gas filled cavities. Cawthrone and Cobb,[23] O'Neill,[17] and Young[16] have experimentally shown that an irrigation temperature of about 44°C produces a temperature rise of about 0.5–1°C in the segment of the horizontal canal closest to the temporal bone. Therefore, in all of the simulations presented in this paper it is assumed that all the boundaries of the canal system are at the body temperature, namely 37°C, and at time zero, a one-degree temperature rise or fall is imposed on the section of the canal indicated in FIGURE 1. This thermal impulse is maintained throughout the transient simulations until steady-state conditions are reached. In this fashion, the complexities and time constants associated with the transient three-dimensional heat transfer of the caloric test are isolated from the rest of the problem, so that attention can be focused solely on the dynamics of the endolymph-cupula fluid-structural interaction.

Thermal irrigation instigates an interesting endolymphatic flow in the semicircular canal system comprised of both natural and expansive convections. An arbitrary Euler–Lagrange (AEL) approach is adopted to formulate the governing equations for fluid motion, energy conservation, and structural deformation of the cupula–endolymph system. For a description of the formulation and the detailed equations the reader is referred to Kassemi *et al.*[24] In this formulation, fluid flow in the semicircular canal system is described by the two-dimensional Navier–Stokes equation.

The endolymph is assumed to be a Newtonian weakly compressible fluid with constant properties except for density, which is a linear function of the temperature but not pressure. The values for the endolymph conductivity, volume expansion coefficient, reference density, and viscosity are taken from measured data provided by Steer[19] and its specific heat is assumed to be equal to water. Because of the temperature-dependant density there is a tight coupling between the energy and Navier–Stokes equations through both the mass divergence term in the continuity equation and the buoyancy term in the momentum equations. The energy equation is simply described by a balance between convection and conduction.

Non-slip stationary boundary conditions are applied at all the physical boundaries of the canal except for the wetted surfaces of the cupula where it is specified through a coupling with the structural equations and at the outlet to the common crux where the velocity is left free. All the boundaries of the canal system are assumed to be at the body temperature, namely 37°C, which is also the reference temperature, T_0. At time zero, a one-degree temperature rise or fall is imposed on the segment of the canal shown in FIGURE 1.

Once the endolymph is brought into motion, as instigated by either natural or expansive convections or both, it produces a pressure differential across the cupula causing it to bend. The Navier equation of motion, cast into its weak incremental updated Lagrangian form through the application of the virtual displacement principal and Gauss's theorem, describes the structural deformations in the cupula. For a two-dimensional plane stress case and assuming that the cupula is a linearly elastic material undergoing large deformation involving small strains, the elasticity tensor $\mathbf{D}$ can be simply defined in terms of the Young's modulus and Poisson ratio of the solid. The structural properties of the cupula are taken from Damiano.[18] Its Poisson ratio was set to be 0.49, appropriate for a nearly incompressible membrane and its Young's modulus was varied parametrically between 5–100 dyne/cm^2 with 5 dyne/cm^2 used as the base value.

The movement of cupula is restricted at the top and bottom of its cross-section, where a tight seal is formed with the ampulla wall. Cupula displacements at these two boundaries are set to zero. The strong coupling between the fluid flow and the structural deformations is rigorously preserved through a balance of traction forces and continuity of velocities and displacements at the wetted surfaces of the endolymph–cupula boundaries.

Numerical solutions of the governing system of coupled nonlinear partial differential equations are generated using a customized in-house version of the finite element code Fidap. Transient solutions are generated using the implicit backward Euler time integration scheme for the flow equations and the second order Bossak integration scheme for the structural equation. At each time step (loop) the computational fluid dynamics and computational structural dynamics counterparts of the problem are solved together following the ALE approach based on a moving mesh that deforms according to the pseudoelastic body displacements at the cupula–endolymph surfaces. The strong coupling between fluid flow and structural deformations is rigorously preserved through the balance of the traction forces and the imposition of transfer-compatibility conditions for velocity and displacement at the wetted surfaces through an iterative segregated solution methodology. For the details of the numerical methodology the reader is again referred to Kassemi et al.[24]

The solutions presented in this paper were generated on a nonuniform mesh with 6,374 quadratic elements. At each time step, a convergence tolerance of 0.0001 was used for the velocity and displacement norms and a convergence tolerance of 0.001 was used for the fluid–solid surface norm. All solutions were started with the entire domain at a uniform body temperature and zero velocity, displacement, and stress fields and terminated when steady state conditions were reached. Values for the mesh elasticity and Poisson ratio were chosen to be $1\,dyne/cm^2$ and 0.001, respectively. Comprehensive grid and time-step convergence tests were performed to ensure spatial and temporal resolution of the generated solutions.

RESULTS AND DISCUSSION

The fluid and structural dynamics of the LSCC cupula–endolymph system is examined for typical test conditions encountered during $1\,g$ clinical experiments performed on the ground and microgravity space experiments performed aboard orbiting spacecraft. For both cases, it is assumed that a normal subject is in the supine position with the head tilted 30° upward so that the LSCC is aligned with the gravitational vector in the vertical direction, thus producing optimum conditions for cupula displacement. At time zero, a temperature impulse of 1°C is applied to a section of the LSSC and maintained until steady state conditions are reached. The cupula elasticity for the base case is assumed to be $5\,dyne/cm^2$.

The ground-based caloric test is examined first. The time sequence for the evolution of the flow field and its impact on the cupula displacement and stress fields are shown in FIGURE 2. At $t = 0.1\,sec$, the temperature boundary layer has almost fully penetrated the duct. Consequently, an endolymphatic flow ensues that is driven almost equally by buoyant and expansive convection. At this instance the cupula is still in its unstressed and undeformed state. At $t = 20\,sec$, the recirculating flow due to buoyancy-driven natural convection becomes more pronounced, especially in the vicinity of the heated-section, creating a negative pressure gradient across the cupula. This fluid loading causes slight bending of the membrane, as is evident from the displacement and stress fields at $t = 20\,sec$, and creates stress concentrations in the regions around the crista, along the lower ampulla wall, and near the region of maximum cupular displacement close to the vertical center of the membrane. At $t = 100\,sec$, steady-state conditions are reached whereupon the expansive convection has almost completely dissipated itself and a relatively strong natural convection vortex has emerged as the dominant endolymphatic flow mechanism in the canal. It is interesting to note that the model predicts bulging or bending of the cupula toward the utricle in accordance to experimental observation for a hot caloric stimulation on Earth.

FIGURE 3 contains the time history for the displacement of a point (P) on the cupula surface (see FIG. 1). Our simulations indicate that the highest cupula velocities occur at the initiation of the test and are due to the impact of rapid expansive flow. However, the expansive velocity diminishes quite rapidly and thus contributes very little to the overall displacement of the cupula, which reaches a plateau in about 80 sec, as depicted in FIGURE 3. Thus, in $1\,g$, most of the cupula displacement is due to the natural convective flow that takes over from early on. The cupula deformation

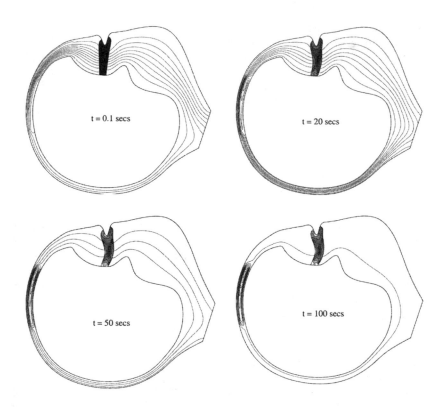

FIGURE 2. Time evolution of endolymph flow and cupula deformation and stress for the ground-based case (an exaggerated scale is used for the axial cupula displacement so that it can be easily discerned).

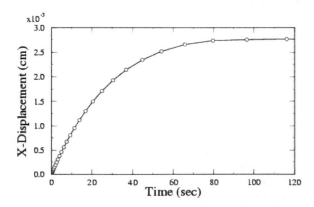

FIGURE 3. Time history of cupula displacement for the ground-based case.

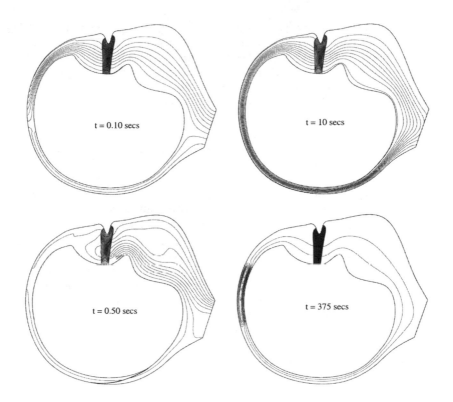

FIGURE 4. Time evolution of endolymph flow and cupula deformation and stress for the microgravity case.

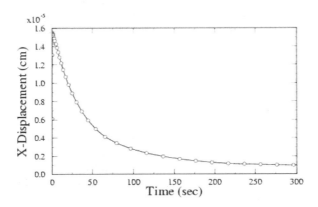

FIGURE 5. Time history of cupula displacement for the microgravity case.

reaches a plateau as a steady state recirculating flow condition is approached and the cupula velocity goes to zero. It should be emphasized that the extent of cupular deflection strongly depends on the elasticity of the membrane. The maximum displacement value that was attained in FIGURE 3 was predicted for a Young's modulus of 5 dyne/cm². Our numerical results show that the maximum cupula displacement decreases logarithmically with an increase in the Young's modulus of the membrane.

The test conditions for the microgravity case are identical to the 1 g case with the subject in the supine position, except for the magnitude of the gravitational field that is reduced to $(10^{-6})g_0$. The time sequence of the transient evolution of the endolymphatic flow and the cupular displacement and stress fields for this case are displayed in FIGURE 4. The most interesting feature of the microgravity caloric test is that, because the buoyancy force is drastically reduced, natural convection effects become negligible and the dynamic behavior of the expansive convection is truly revealed.

Immediately after the application of the thermal impulse, there is pressure increase in the endolymph near the center of the heated section of the canal. The ensuing pressure wave instigates a sudden potential-like flow that emanates from this location and flows towards the cupula partition from both the shorter *canal side* and the longer *utricle side* of the ampulla, as can be easily discerned from the flow field (at time 0.1 sec) in FIGURE 4. The pressure drop along the path to the utricle side of the ampulla is much larger than the pressure drop along the path to the canal side of the ampulla. This, again, creates a negative pressure gradient across the cupula that deforms the partition towards the utricle through a rapid jolt, as is evident from the displacement history displayed in FIGURE 5. The cupula stress fields and displacement history depicted, respectively, in FIGURES 4 and 5 indicate that the peak values occur at about $t = 0.5$ sec. As time evolves, the pressure wave created by the endolymph expansion dissipates itself. Along with this diminishing expansive flow, the cupula partition gradually returns itself to its initial configuration at time zero following a very slow transience, as seen in FIGURES 4 and 5. Finally, at $t = 375$ sec, the expansive convection almost completely disappears and reveals an extremely weak natural convection vortex driven by the residual gravitation field near the heated section. The pressure drop created by this vortex is too small to noticeably affect the cupula as the partition settles back into its original unstressed and undeformed configuration. A comparison between the results shown in FIGURES 3 and 5 indicates that the maximum displacements produced by the expansive convection in microgravity are about two orders of magnitude smaller than their 1 g counterparts produced by buoyancy-driven natural convection.

CONCLUSION

In this paper, we presented a finite element FSI model for the LSCC system with a rigorous treatment of the coupling between the fluid flow and the structural dynamics counterparts. The dynamic behavior of the coupled endolymph–cupula system was captured for both 1 g ground-based caloric tests and microgravity space experiments using nominal endolymph and cupula properties extracted from the literature. Numerical simulations predict the correct trend and behavior for the hot caloric stimulation in supine position on earth. Transient simulations also indicate that there is a significant difference between the dynamic characteristics of the endolymphatic natural convection flow on Earth and those of the expansive convection in microgravity. This leads to entirely different cupular displacement magnitudes and structural dynamics in the two environments. Future microgravity experiments are needed to properly assess the validity of the intricate microgravity fluid structural interactions predicted by the present model.

ACKNOWLEDGMENTS

This work was supported by the Microgravity Research Division at NASA Glenn Research Center (GRC). Computational and system-related support and resources provided by the Computational Microgravity Laboratory at NASA GRC and the Ohio Supercomputer Center (OSC) are also gratefully acknowledged.

REFERENCES

1. BARANY, R. 1906. Untersuchungen uber den vom Vestibularapparat des Obres Reflektorisch Ausgelosten Rhytmischen Nystagmus und seine Begleitertscheinungen. Monatsschr Ohrenheilkd. **40:** 193–297.
2. BARANY, R. 1907. Physiologie und Pathologie (Funktionsprufung) des Bogengangapparatus beim Menschen. Franz Deuticke, Leipzig.
3. SCHERER, H. & A.H. CLARKE. 1985. The caloric vestibular reaction in space. Acta Otolaryngol. (Stockh.) **100:** 328–336.
4. SCHERER, H., U. BRANDT, A.H. CLARKE, *et al.* 1980. European vestibular experiments on Spacelab-1 mission: 3. Caloric nystagmus in microgravity. Exp. Brain Res. **64:** 255–263.
5. BROWN-SEQUARD, C. 1860. Course of Lectures on the Physiology and Pathology of the Central Nervous System. Collins, Philadelphia.
6. BARTELS, M. 1911. Discussion on Barany and Wittmark: Funktionelle Prufung des Vestibular Apparates. Ber d. Deutsch Otol. Gellesch. **20:** 214.
7. KOBRAK, F. 1918. Beitrage zum Experimentellen Nystagmus. Beiter. z. Anat. Physiol. Path. u. Therap. D. Ohres. **10:** 214.
8. BORRIES, G.V.T. 1920. Studier over Vestibular Nystagmus. Thesis, Copenhagen.
9. VAN CANEGHEM, D. 1946. Application du Romberg Amplifie de la Reaction Vestibulaire Sonore sur Diagnostique Differentiel Entre la Tympanosklerose et l'Otoskongiose. Bull. Soc. Belge d'Otol. Rhin. Laryng. **88:** 88.
10. BRUNNER, H. 1925. Handbuch der Neurologie der Kalorischen Nystagmus. Arch Ohr.-Nas. u. KehlkHeilk. **113:** 117.
11. VON BAUMGARTEN, R., A. BENSON & A. BERTHOZ, *et al.* 1984. Effects of rectilinear acceleration and optokinetic and caloric stimulation in space. Science **225:** 208–212.
12. HOOD, J.D. 1989. Evidence of direct thermal action upon the vestibular receptors in the caloric test. Acta Otolaryngol. (Stockh.) **107:** 161–165.
13. PAIGE, G.D. 1985. Caloric responses after horizontal canal inactivation. Acta Otolaryngol. (Stockh.) **100:** 321–327.
14. STEINHAUSEN, W. 1933. Uber die Beobachtung der Cupula in den Bogengangsampullen des Labyrinths des Lebenden Hechts. Pflugers Arch. Ges. Physiol. **23:** 500–512.
15. STEER R.W. 1967. Influence of Angular and Linear Acceleration and Thermal Stimulation on the Human Semicircular Canal. Ph.D. Thesis, MIT, Cambridge, Massachussets.
16. YOUNG, J.D. 1972. Analysis of Vestibular System Responses to Thermal Gradients Induced in the Temporal Bone. Ph.D. Dissertation, University of Michigan, Ann Arbor, Michigan.
17. O'NEILL, G. 1985. Analysis of the Physical and Physiological Events Resulting from Caloric Irrigation of the Human Temporal Bone. Thesis C.N.A.A. London.

18. DAMIANO, E.R. 1993. Continuum Models of Rotational and Caloric Stimulation of The Vestibular Semicircular Canal. Ph.D. Dissertation, Rensselaer Polytechnic Institute, Troy, New York.
19. STEER, R.W., Y.T LI, L.R. YOUNG & J.L. MEIRY. 1967. Physical properties of the labyrinthine fluids and quantification of phenomena of caloric stimulation. Third Symposium on the Role of Vestibular Organs in Space Exploration. NASA SP-152, 409–512.
20. OMAN, C.M., M. MARCUS & I.S. CURTHOYS. 1987. The influence of semicircular canal morphology on endolymphatic flow dynamics. Acta Otolaryngol. (Stoch.) **103:** 1–13.
21. RABBIT, R.D. & E.R. DAMIANO. 1992. A hydroelastic model of the macromechanics in the endolymphatic vestibular canal. J. Fluid Mech. **238:** 337–369.
22. CURTHOYS, I.S. & C.M OMAN 1987. Dimensions of the horizontal semicircular duct, ampulla and utricle in human. Acta Otolaryngol. (Stockh.) **103:** 254–261.
23. CAWTHORNE, T.E. & W.A. COBB. 1954. Temperature changes in the perilymph space in response to caloric stimulation in man. Acta Otolaryngol. **44:** 580–588.
24. KASSEMI, M., D. DESERRANNO, J.G. OAS. 2004. Fluid–structural-interactions in the inner ear. Computers and Structures. In press.

Gravitational Effects on Structure Development in Quenched Complex Fluids

V.E. BADALASSI,[a] H.D. CENICEROS,[b] AND S. BANERJEE[a]

[a]Department of Chemical Engineering,
University of California—Santa Barbara, California, USA

[b]Department of Mathematics,
University of California—Santa Barbara, California, USA

ABSTRACT: When binary liquid mixtures are cooled rapidly from a homogeneous phase into a two-phase system, domains of the two equilibrium phases form and grow (coarsen) with time. In the absence of an external forcing due to gravity or an imposed shear flow, a dynamic scaling regime emerges in which the domain morphology is statistically self-similar at different times with a length-scale that grows with time. In the presence of gravity, however, multiple length scales develop, with the system coarsening more rapidly in the direction of the force. The late-time behavior of such a system is characterized in this study by the calculation of anisotropic growth laws. Gravitation effects significantly affect scaling laws, even with small density mismatch, and the growth mechanism has some similarities to the sedimentation process. However, very few numerical studies have been made of such effects; this is one of the first.

KEYWORDS: Cahn–Hilliard equation; Navier–Stokes equations; Model H; phase separation under gravity

INTRODUCTION

Phase separation during quenching of binary-mixture melts occurs in a variety of practically important systems; for example, during the fabrication of multiphase materials. Physicists have characterized the phase transition of a binary mixture by a universal length scale that has a growth law dependent on the regime of the separation. Using scaling arguments it is possible to find three regimes: diffusive, viscous, and inertial. Siggia[1] added a cutoff length above which gravitational forces have an important effect. However, very little work, either experimental or numerical, has been done to elucidate phase separation phenomena with density mismatch between the domains—a circumstance that often occurs. Here we investigate this problem numerically using a phase field approach for small density mismatch.

Phase field models offer a systematic physical approach for investigating complex multiphase system behavior, such as near-critical interfacial phenomena. However, because interfaces are replaced by thin transition regions (diffuse interfaces), phase field simulations demand robust numerical methods that can efficiently

Address for correspondence: V.E. Badalassi, Department of Chemical Engineering, University of California—Santa Barbara, CA 93106-5080, USA. Voice: 805-893-8614; fax: 805-893-4731.

badalass@engineering.ucsb.edu

Ann. N.Y. Acad. Sci. 1027: 371–382 (2004). ©2004 New York Academy of Sciences.
doi: 10.1196/annals.1324.031

achieve high resolution and accuracy. We use an accurate and efficient numerical method,[2] to solve the coupled Cahn–Hilliard/Navier–Stokes system of equations for binary systems, known as Model H, with small density mismatch between the phases in the Boussinesq approximation. The numerical method is a time-split scheme that combines a novel semi-implicit discretization for the convective Cahn–Hilliard equation with a stiffly stable time-discretization of the projection method for the Navier–Stokes equations. We employ high-resolution spatial discretizations to be able to accurately resolve thin interfaces. The Cahn–Hilliard equation is discretized in space (pseudo) spectrally (via FFT for periodic boundary conditions). We solve the Navier–Stokes modified creeping flow equations with a pseudospectral method.[3] The numerical method is robust and has low computational cost.

We find that gravity effects may significantly affect scaling laws even with a small density mismatch. Furthermore, the effect of gravity during the diffusive and viscous regime is simulated for the first time, and the break down of the usual Siggia-type universal scaling is characterized.

PHASE SEPARATION AND COARSENING

We consider isothermal spinodal decomposition of symmetric binary mixtures under deep quenches. In this case, the fluid motion arise from moving interfaces that create a characteristic *bicontinuous* pattern and thermal fluctuations can be disregarded at long times. The state of the system at any given time can be described by an order parameter ϕ, which is the relative concentration of the two components. A free energy of a bulk system can be defined for times when the system is not in equilibrium[4] and if the effective interactions between the mixture components are short ranged, this free energy can be written as a functional of ϕ

$$F[\phi] = \int_\Omega \left\{ f(\phi(\mathbf{x})) + \frac{1}{2}k|\nabla\phi(\mathbf{x})|^2 \right\}d\mathbf{x}, \tag{1}$$

where Ω is the region of space occupied by the system. The term $f(\phi(\mathbf{x}))$ is the bulk energy density, which depends only on the local concentration and the temperature T. We use the Landau expression that has two minima corresponding to the two stable phases of the fluid

$$f(\phi) = \frac{\alpha}{4}\left(\phi - \sqrt{\frac{\beta}{\alpha}}\right)^2\left(\phi + \left(\phi - \sqrt{\frac{\beta}{\alpha}}\right)^2\right)^2 \tag{2}$$

for the bulk free energy density, where $\alpha > 0$ and $\beta = \beta(T)$; $\beta(T) < 0$ for $T > T_c$ and $\beta(T) > 0$ for $T < T_c$.

Thus, $f(\phi)$ has two nontrivial minima $\phi_{\pm} = \pm\sqrt{\beta/\alpha}$ corresponding to local equilibrium solutions for $T < T_c$. For $T > T_c$ the equilibrium solution is simply $\phi_0 = 0$. The term $k|\nabla\phi(\mathbf{x})|^2/2$ with k a positive constant, quantifies the additional free energy contributions arising from local concentration fluctuations that, in the demixing process, appear in the interfacial regions between the emerging domains of the two stable phases with concentrations $\phi_{\pm}$. The chemical potential μ is defined as

$$\mu(\phi) = \frac{\delta F[\phi]}{\delta\phi(\mathbf{x})} = f'(\phi(\mathbf{x})) - k\nabla^2\phi(\mathbf{x}). \tag{3}$$

The equilibrium interface profile can be found by minimizing the functional $F[\phi]$ with respect to variations of the function ϕ; that is, by solving the equation $\mu(\phi) = \delta F[\phi]/\delta\phi = \alpha\phi^3 - \beta\phi - k\nabla^2\phi = 0$. As well as the two stable uniform solutions $\phi_\pm = \pm\sqrt{\beta/\alpha}$ that represent the coexisting bulk phases, there is a one-dimensional (say along the z-direction) non-uniform solution $\phi_0(z) = \phi_+\tanh(z/\sqrt{2}\,\xi)$ that satisfies the boundary conditions $\phi_0(z \to \pm\infty) = \pm\phi$.[5,6] This solution was first found by Van der Waals[7] to describe the equilibrium profile for a plane interface normal to the z direction, of thickness proportional to $\xi = \sqrt{k/\beta}$, that separates the two bulk phases. Cahn and Hilliard[8,9] generalized the problem to time-dependent situations by approximating interfacial diffusion fluxes as proportional to chemical potential gradients, enforcing conservation of the field. The convective Cahn–Hilliard equation can be written as

$$\frac{\partial\phi}{\partial t} + \mathbf{u}\cdot\nabla\phi = \nabla^2\mu, \tag{4}$$

where $\mathbf{u}$ is the velocity field and $M > 0$ is the mobility or Onsager coefficient. Equation (4) models the creation, evolution, and dissolution of diffusively controlled phase-field interfaces[10] (for a review of the Cahn–Hilliard model see, for example, Ref. 11). We define the interface thickness to be the distance from $0.9\phi_-$ to $0.9\phi_+$ so that the equilibrium interface thickness is $2\sqrt{2}\,\xi\tanh^{-1}(0.9) = 4.164\xi$. This width contains 98.5% of the surface tension stress.[12] In equilibrium, the surface tension σ of an interface is equal to the integral of the free energy density along the interface. For a plane interface, σ is given by[5]

$$\sigma = k\int_{-\infty}^{+\infty}\left(\frac{d\phi_0}{dz}\right)^2 dz = \frac{\sqrt{2}}{3}\frac{k^{1/2}\beta^{3/2}}{\alpha}. \tag{5}$$

It is evident that we can control the surface tension and interface width through the parameters k, α, and β.

We model the fluid dynamics by the Navier–Stokes equations with a phase field-dependent surface force[13] and simplified with the creeping flow approximation:

$$-\nabla p + \eta\nabla^2\mathbf{u} + \mu\nabla\phi + (\rho - \rho_0)\mathbf{a} = 0 \tag{6}$$

$$\nabla\cdot\mathbf{u} = 0, \tag{7}$$

where $\mathbf{u}$ is the velocity field, p is a scalar related to the pressure that enforces the incompressibility constraint (7), η is the viscosity, and $\mathbf{a}$ is the acceleration (gravitational) field (Boussinesq approximation). The coupled Cahn–Hilliard/Navier–Stokes system (4)–(7) is referred to as *Model H* according to the nomenclature of Hohenberg and Halperin.[14]

Nondimensionalization

To nondimensionalize the governing equations (4)–(7) we choose as a convenient characteristic length, L_c, for our simulations the mean field thickness ξ of the interface; that is, $L_c = \xi$. U_c is a characteristic fluid velocity. The characteristic time T_c is the time required for the fluid to be convected a distance of the order of the interface thickness (in the absence of capillarity), $T_c = \xi/U_c$. Local interfacial curvature generates stress that drives fluid motion. It is natural, then, to scale the pseudopressure with surface tension times a term of the same order as the local curvature (i.e., $1/\xi$). The order parameter ϕ is scaled with its mean-field equilibrium value $\phi_+ = \sqrt{\beta/\alpha}$.

With this scaling, $-1 \leq \phi \leq 1$, and the interface separating the two fluids is between $\phi = -0.9$ and $\phi = 0.9$. Summarizing, we have

$$\mathbf{u}' = \frac{\mathbf{u}}{U_c}, \qquad t' = \frac{t}{T_c}, \qquad p' = \frac{p}{\sigma/L_c}.$$

Equations (4)–(7) become

$$\frac{\partial \phi}{\partial t} + \mathbf{u} \cdot \nabla \phi = \frac{1}{Pe} \nabla^2 \mu \tag{8}$$

$$-\nabla p + Ca \nabla^2 \mathbf{u} + \mu \nabla \phi + Bo\, e_z = 0 \tag{9}$$

$$\nabla \cdot \mathbf{u} = 0 \tag{10}$$

where $\mu = \phi^3 - \phi - \nabla^2 \phi$ is the dimensionless chemical potential. The dimensionless groups used above are the Peclet number, Pe, the capillary number, Ca, and the Bond number, Bo, given by

$$Pe = \frac{U_c \xi^2}{M\sigma}, \qquad Ca = \frac{\eta U_c}{\sigma}, \qquad Bo = \frac{(\rho - \rho_0) a \xi^2}{\sigma}, \tag{11}$$

respectively. Physically, the Pe number is the ratio between the diffusive time scale $\xi^3/(M\sigma)$ and the convective time scale ξ/U_c, the capillary number Ca provides a measure of the relative magnitude of viscous and capillary (or interfacial tension) forces at the interface, and the Bond number Bo is the ratio of the acceleration (gravitational) forces and surface tension forces. Note that with this nondimensionalization the length of the fluid domain is interpreted in units of interface thickness ξ.

Dynamic Scaling and Domain Growth

The physics of spinodal decomposition involves diffusion, capillary forces, viscous forces, and fluid inertia. In the case of a deep quench and no external drive we assume that the interface can be characterized by a single length scale that is smooth with a radius of curvature that scales as the domain size itself, which is much larger than the interfacial thickness. This length gives a measure of how fast the domains grow. The domain patterns grow with a time-dependent characteristic length scale

$$L(t) \sim t^n, \tag{12}$$

with a growth exponent n, where n depends on the mechanism controlling growth, for example, diffusion, viscous, or inertial. Model H describes a system order parameter coupled to hydrodynamic flow and in the absence of an external drive gives rise to three different growth regimes. For the diffusive regime $L \ll \sqrt{M\eta}$ and $n = 1/3$, for the viscous regime $\sqrt{M\eta} \ll L \ll \eta^2/(\rho\sigma)$ and $n = 1$, for the inertial regime $L \gg \eta^2/(\rho\sigma)$ and $n = 2/3$.

Finally, we consider the presence of an external drive (i.e., gravity). We can think of a transition to gravity dominated flow arising when heavy domains resting on top of light ones become unstable; that is, when the gravitational force on the more dense phase overcomes the interfacial tension that keeps them suspended. In our model gravity becomes important when[1]

$$L(t) \gg \frac{\sigma}{(\rho - \rho_0)g}. \tag{13}$$

This process can be considered analogous to sedimentation. In fact it is possible to derive similar criteria via the balance of the Stokes friction coefficient and the diffusion time[15] to get (13) (called also the *Laplace length*). Once the instability occurs, we have additional length scales since we have a faster growth in the gravity direction and a method to measure the anisotropic growth laws needs to be defined. From the knowledge of the structure factor one computes the average size of domains in the various directions L_x, L_y, and L_z, for example,

$$L_z(t) = \pi \frac{\int S(\mathbf{k}, t)d\mathbf{k}}{\int |k_x| S(\mathbf{k}, t)d\mathbf{k}}, \tag{14}$$

and analogously for the other directions. In FIGURE 1 we give a qualitative (graphic) interpretation of the various coarsening regimes.

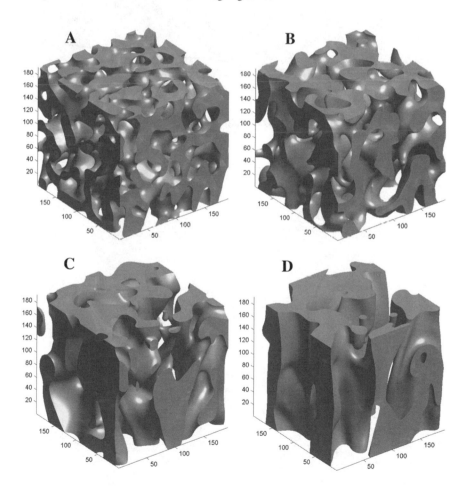

FIGURE 1. Evolution of ϕ represented by isosurfaces of separation of the two fluids at $\phi = 0.0$, at times: (**A**) $t = 800$, (**B**) $t = 2,000$, (**C**) $t = 3,000$, and (**D**) $t = 4,200$; $Pe = 0.5$, $Ca = 70.0$, $Bo = 0.01$, $N = 192$, and $L = 266$.

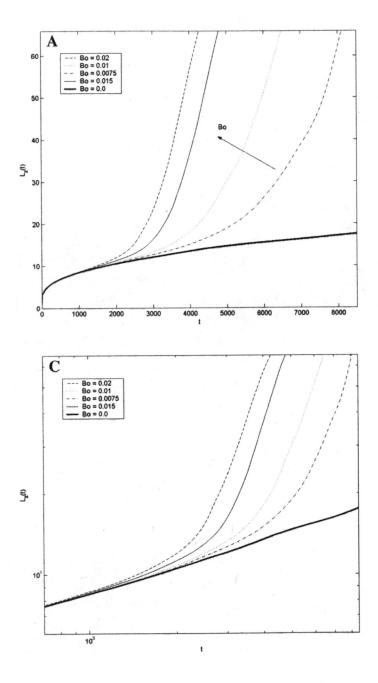

FIGURE 2. Continued opposite.

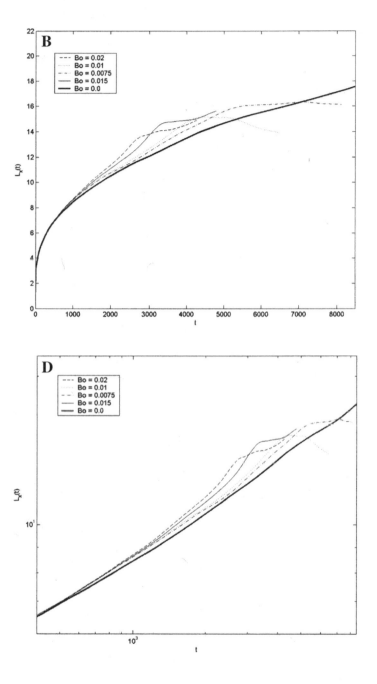

FIGURE 2. Domain size versus time: $Pe = 0.5$, $Ca = 70.0$, and Bo as a parameter; **(A)** $L_z(t)$, **(B)** $L_x(t)$, **(C)** $L_z(t)$ (log–log plot), and **(D)** $L_x(t)$ (log–log plot).

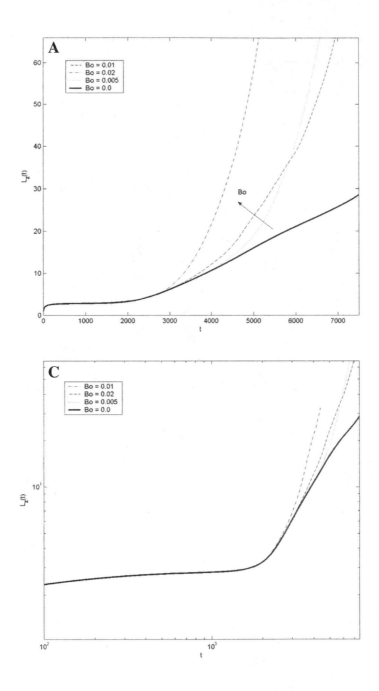

FIGURE 3. Continued opposite.

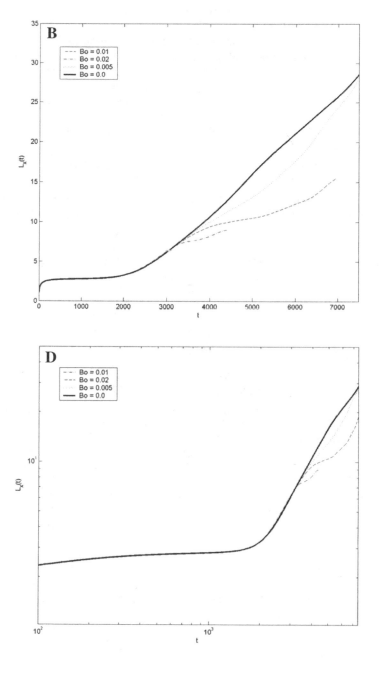

FIGURE 3. Domain size versus time: $Pe = 96.276$, $Ca = 21.26$, and Bo as a parameter; **(A)** $L_z(t)$, **(B)** $L_x(t)$, **(c)** $L_z(t)$ (log–log plot), and **(D)** $L_x(t)$ (log–log plot).

NUMERICAL METHOD

The numerical method we employ is based on that introduced elsewhere[2] with one main modification: because of the periodic boundary conditions and the Stokes flow approximation we can use a fully spectral discretization. The time discretization is based on a semi-implicit scheme combined with a time-split strategy and it effectively decouples Cahn–Hilliard and Navier–Stokes solvers to yield an efficient and robust modular scheme.

The outline of the method is as follows. Given ϕ^n and $\mathbf{u}^n$ the objective is to solve for ϕ^{n+1} and $\mathbf{u}^{n+1}$ with the following steps:

1. Solve the Cahn–Hilliard equation with a semi-implicit method and spectral spatial discretization to obtain ϕ^{n+1}.

2. Using ϕ^{n+1} compute the surface force and solve the Navier–Stokes modified creeping flow equations with a Fourier–Fourier–Fourier (spectral) method.[3]

The C–H step consist of second order semi-implicit semibackward difference formula (SBDF) time advancement.[2] The accuracy and stability of the method are documented in Reference 2. The overall scheme for the convective Cahn–Hilliard equation and the creeping flow equation has only a CFL stability condition,

$$\Delta t_{cfl} \leq \left(\frac{|u|_{max}}{\Delta x} + \frac{|u|_{max}}{\Delta y} + \frac{|u|_{max}}{\Delta} \right)^{-1}, \qquad (15)$$

where (u, v, w) are the components of the velocity field.

NUMERICAL STUDY OF GRAVITATION EFFECTS ON SPINODAL DECOMPOSITION

Our method has allowed us to numerically simulate gravitation effects on spinodal decomposition of a binary mixture with phases of equal viscosity and mobility, in three dimensions. The grid size is $192 \times 192 \times 192$, the domain size is $L = 266$, which corresponds to a three mesh-point thick interface,[2] and the boundary conditions are periodic. The initial conditions consist of uniformly distributed random fluctuations of amplitude 0.05 around a zero background (i.e., critical quench). We implement adaptive time stepping based on the CFL condition (15). FIGURE 1 shows the evolution of the interfaces during the coarsening process. Regions rich in the lighter component ($\phi = 1$) are rendered in dark grey and regions rich in the heavier component ($\phi = -1$) comprise the surrounding (transparent) phase. The two domains become anisotropic in time and elongated in the direction of gravity. We can characterize how the gravitational field interferes with spinodal decomposition by computing the characteristic length scale (growth rate) $L(t)$ as a function of time for diffusive or viscous regimes and for various Bond numbers; that is, various gravitational field strengths. Because of the anisotropy of the system, we calculate growth rates in both the z gravity direction (vertical in FIG. 1) and the x direction (horizontal, normal to gravity in FIG. 1) using definition (14). The growth rate in the y direction is analogous to that in the x direction due to the symmetry of the system.

FIGURE 2 A and C (log–log scale) show the domain growth in the z direction as a function of time starting from a diffusive regime condition ($Pe = 0.5$, $Ca = 70.0$) and

as a function of the *Bo* number. The growth law for *Bo* = 0.0 obeys the classic Lifshitz–Slyozov[17] growth law $L(t) \sim t^{1/3}$, thus validating our computational procedures. At a late stage and *Bo* > 0, the data exhibit an approximately linear growth, typical of the viscous regime and independent of gravity (*Bo*), giving a strong indication that a universal scaling exists in the gravity direction. Furthermore, we find that the growth rate in the gravity (*z*) direction of $L_z(t)$ is considerably faster than that of the no-gravity case. This occurs even with small gravity values, with *Bo* values as low as 0.003. Moreover, $L_z(t) \gg L_x(t)$ (where $L_x(t)$ is the length scale in the *x* direction, see FIG. 2 B and D). This is not surprising since experimental results show similar sensitivity to gravity[15,19] but no measure of the growth law has been reported due to experimental difficulties. The effect of gravity demonstrated by our numerical results is in accord with the analytical result in **(14)**. FIGURE 2 B and D (log–log scale) show the characteristic domain size in the *x* direction $L_x(t)$ as a function of time *t*. In this regime a trend is not evident: the length scale in the *x* direction with gravity (*Bo* > 0) resembles the Lifshitz–Slyozov growth law[17] at the initial stage, but it deviates at later times, becoming slower. To understand more clearly the asymptotic fate of the regime in this direction we need higher resolution simulations to be able to explore a wider range of length scales. In FIGURE 3 we show results corresponding to the viscous regime parameters (*Pe* = 96.0, *Ca* = 20.0). For *Bo* = 0 (no gravity) we find a linear growth $L(t) \sim t$ as predicted.[1] Again the growth of the length scale, $L_z(t)$ is considerably faster than that for the no-gravity case, and in the other direction characterized by lengths scale L_x, even with small gravity values (FIG. 3 A). There is no evidence of collapse to the same universal slope (independent of *Bo*) at later stages (FIG. 3 C). The conditions of deviation from the no-gravity case were characterized using **(14)**. In the *x* direction (FIG. 3 B and 3 D) the slowing down of the growth rate due to gravity is more evident than in the diffusive case (FIG. 2 B and D).

CONCLUDING REMARKS

We used an accurate and efficient numerical method to compute phase ordering kinetics coupled with fluid dynamics to study the effect of gravity on critical spinodal decomposition of a binary mixture. Our results demonstrate that there is reasonable dynamic scaling in the direction of gravity in the diffusive and viscous regime, but a breakdown appears in the dynamic scaling in the transverse directions. Furthermore, growth in the direction of gravity is much faster than that in the transverse directions, and that in the no-gravity case. Our results are in broad agreement and extend numerical results from previous studies of various related models,[18] leading us to believe that all these models share similar asymptotic scaling. Finally, we remark that gravitation effects are very important in the context of the segregation of binary fluids, even when density mismatches are small.

ACKNOWLEDGMENTS

This work was partially supported by the National Aeronautics and Space Administration—Microgravity Research Division, under Contract NAG3-2414. H.D.C.

acknowledges partial support from the National Science Foundation under Grant DMS-0311911 and from an UCSB Faculty Career Development Award Grant.

REFERENCES

1. SIGGIA, E.D. 1979. Late stages of spinodal decomposition in binary mixtures, Phys. Rev. A **20**(2): 595–605.
2. BADALASSI, V.E., H.D. CENICEROS & S. BANERJEE. 2003. Computation of multiphase systems with phase field models, J. Comput. Phys. **190**(2): 371–397.
3. PEYRET, R. 2002. Spectral Methods for Incompressible Viscous Flow. Springer, New York.
4. PENROSE, O., & P. FIFE. 1990. Thermodynamically consistent models of phase-field type for the kinetics of phase transitions, Physica D **43**: 44.
5. BRAY, A.J. 1994. Theory of phase-ordering kinetics, Adv. Phys. **43**(3): 357–459.
6. CHELLA, R. & V. VIÑALS. 1996. Mixing of a two-phase fluid by a cavity flow. Phys. Rev. E **53**: 3832.
7. VAN DER WAALS, J.D. 1879. The thermodynamic theory of capillarity flow under the hypothesis of a continuous variation of density. Verhandel/Konink. Akad. Weten. **1**: 8. (Dutch.)
8. CAHN, J.W. & J.E. HILLIARD. 1958. Free energy of a nonuniform system I. J. Chem. Phys. **28**: 258.
9. CAHN, J.W. & J.E. HILLIARD. 1959. Free energy of a nonuniform system III. J. Chem. Phys. **31**: 688.
10. BATES, P.W. & P.C. FIFE. 1993. The dynamics of nucleation for the Cahn–Hilliard equation, SIAM J. Appl. Math. **53**: 990.
11. ELLIOT, C.M. 1989. The Cahn–Hilliard model for the kinetics of phase separation. *In* Mathematical Models for Phase Change Problems. J.F. Rodrigues, Ed.: 35–72. International Series of Numerical Mathematics, Vol. 88. Bikhäuser Verlag, Basel.
12. JACQMIN, D. 1999. Calculation of two phase Navier–Stokes flows using phase-field modeling. J. Comput. Phys. **115**: 96.
13. GURTIN, M.E., D. POLIGNONE & J. VIÑALS. 1996. Two-phase binary fluids and immiscible fluids described by an order parameter. Math. Models Meth. Appl. Sci. **6**(6): 815.
14. HOHENBERG, P.C. & B.I. HALPERIN. 1977. Theory of dynamic critical phenomena. Rev. Mod. Phys. **49**(3): 435.
15. CHAN, C.K. & W.I. GOLDBURG. 1987. Late-stage phase separation and hydrodynamic flow in a binary liquid mixture. Phys. Rev. Lett. **58**(7): 674–678.
16. KENDON, V.M., M.E. CATES, I. PAGONABARRAGA, *et al.* 2001. Inertial effects in three-dimensional spinodal decomposition of a symmetric binary fluid mixture: a lattice Boltzmann study. J. Fluid Mech. **440** 147.
17. LIFSHITZ, I.M. & V.V. SLYOZOV. 1961. The kinetics of precipitation from supersaturated solid solutions, J. Phys. Chem. Solids **19**(1–2): 35–50.
18. PURI, S., N. PAREKH & S. DATTAGUPTA. 1994. Phase ordering dynamics in a gravitational-field, J. Stat. Physics **75**(5–6): 839–857.
19. BEYSENS, D., P. GUENON & F. PERROT. 1988. Phase separation of critical binary fluids under microgravity: comparison with matched density conditions. Phys. Rev. A **38**(8): 4173–4185.

Density-Driven Instabilities of Variable-Viscosity Miscible Fluids in a Capillary Tube

ECKART MEIBURG, SURYA H. VANAPARTHY,
MATTHIAS D. PAYR, AND DIRK WILHELM

Department of Mechanical and Environmental Engineering,
University of California, Santa Barbara, California, USA

ABSTRACT: A linear stability analysis is presented for variable-viscosity miscible fluids in an unstable configuration; that is, a heavier fluid placed above a
lighter one in a vertically oriented capillary tube. The initial interface thickness is treated as a parameter to the problem. The analysis is based on the
three-dimensional Stokes equations, coupled to a convection-diffusion equation for the concentration field, in cylindrical coordinates. When both fluids
have identical viscosities, the dispersion relations show that for all values of the
governing parameters the three-dimensional mode with an azimuthal wave
number of one represents the most unstable disturbance. The stability results
also indicate the existence of a critical Rayleigh number of about 920, below
which all perturbations are stable. For the variable viscosity case, the growth
rate does not depend on which of the two fluids is more viscous. For every
parameter combination the maximum of the eigenfunctions tends to shift
toward the less viscous fluid. With increasing mobility ratio, the instability is
damped uniformly. We observe a crossover of the most unstable mode from
azimuthal to axisymmetric perturbations for Rayleigh numbers greater than
10^5 and high mobility ratios. Hence, the damping influence is much stronger
on the three-dimensional mode than the corresponding axisymmetric mode for
large Rayleigh numbers. For a fixed mobility ratio, similar to the constant viscosity case, the growth rates are seen to reach a plateau for Rayleigh numbers
in excess of 10^6. At higher mobility ratios, interestingly, the largest growth
rates and unstable wave numbers are obtained for intermediate interface
thicknesses. This demonstrates that, for variable viscosities, thicker interfaces
can be more unstable than their thinner counterparts, which is in contrast to
the constant viscosity result where growth rate was seen to decline monotonically with increasing interface thickness.

KEYWORDS: density-driven instabilities; variable viscosity; miscible fluids;
capillary tube

INTRODUCTION

The capillary tube represents one of the fundamental configurations historically
employed in investigations of interfacial phenomena and diffusive effects in the
region of contact between two fluids. Both hydrodynamic stability problems and

Address for correspondence: Eckart Meiburg, Department of Mechanical and Environmental Engineering, University of California, Santa Barbara, CA 93106, USA. Voice/fax: 805-
893-5278.
  meiburg@engineering.ucsb.edu

Ann. N.Y. Acad. Sci. 1027: 383–402 (2004). ©2004 New York Academy of Sciences.
doi: 10.1196/annals.1324.032

displacement processes have been studied extensively in the above geometry. Hales[1] was among the first to address the stability of an unstably stratified, variable density fluid mixture with a constant density gradient in a vertically oriented capillary tube, in the absence of any net flow through the tube. He found that a stable equilibrium is possible, as long as the density gradient does not exceed a certain critical value. Taylor[2] devised a simple experiment to obtain this critical gradient of density that he argued could be used in order to determine the diffusion coefficient of a fluid pair, see also recent related experimental work.[3] Wooding[4] took an analytical approach to the stability problem for a constant density gradient in a capillary tube. He observed that the three-dimensional mode $\beta = 1$, where β denotes the azimuthal wave number, represents the most unstable disturbance. The above analysis was extended[5] to base states involving density profiles that vary sinusoidally in the vertical direction or deviate from a constant value only in a central layer of small vertical extent. This work confirmed that $\beta = 1$ represents the most unstable mode for the uniform density gradient. Without proof, it was assumed that this also holds for the case of sinusoidally varying density. All of the above investigations were limited to cases in which viscosity variations are absent. To our knowledge, the situation that can be realized most easily in an experiment, namely that of a relatively thin, miscible interface formed by placing a heavier fluid above a lighter one in a capillary tube, has not yet been addressed from a stability theoretical point of view. Variable viscosity influence on density driven instability is an interesting extension. This is the configuration analyzed in the investigation we report here.

The presence of a net flow through the tube complicates the situation considerably. Other investigations[6,7] discussed the fractional amount of viscous fluid left behind on the wall of a tube when it is expelled by an inviscid fluid with which it is immiscible as a function of a suitably defined capillary number Ca. Density effects were deemed unimportant in these studies. Numerical calculations for this case,[8] showed very good agreement with the experimental observations. This classical work has recently been extended to finite viscosity ratios.[9] Petitjeans and Maworthy[10] and Chen and Meiburg[11] analyzed the corresponding miscible problem both experimentally and computationally, based on the Stokes equations (see also the related experiments in Ref. 12). In these flows, a cutoff length is set by diffusive effects rather than surface tension, so that in some sense, a Péclet number Pe takes the place of Ca. The above authors also address the role of density differences by conducting experiments and simulations in vertical tubes. Substantial differences between the experiments and the numerical data are observed at small values of Pe, in that a quasisteady finger emerges for significantly smaller values of Pe in the simulations, as compared to the experiments. This raises the question as to whether nonconventional, so-called Korteweg stresses[13,14] or divergence effects, can be important, an issue that has been addressed elsewhere.[15] A particularly striking finding was reported in the follow-up experiments of Kuang et al.[16] In a vertical capillary tube without net flow, these authors observe that the sharp interface formed by placing a heavier, more viscous silicone oil above a lighter and less viscous one leads to an interfacial instability with an azimuthal wave number $\beta = 1$. However, when a small upward net flow was applied to this gravitationally unstable base state, the interface evolved in an axisymmetric fashion, rather than exhibiting an azimuthal instability mode.

As a first step, in the present work, we perform the linear stability analysis of the density-driven instability for variable-viscosity miscible fluids in a capillary tube. The analysis is based on the three-dimensional Stokes equations, and it proceeds along similar lines as our recent investigation for the corresponding situation in a Hele–Shaw cell,[17,18] as well as the related experiments and nonlinear simulations.[19] Introduction of a net flow will be the next step. It is to be seen if the aforementioned experimental observation reflects an effect of the net flow within the linear framework of the stability problem, or if it represents a nonlinear effect.

The paper is organized as follows: initially the physical problem, the governing equations, and the relevant dimensionless parameters are described in more detail. Next, the linearization is described for both axisymmetric and azimuthal perturbations, and the numerical procedure for solving the resulting eigenvalue problem is outlined. Subsequently, the results of the stability analysis are presented in the form of dispersion relations and associated information for the two cases of identical and different viscosity fluids. Finally, we summarize our main conclusions.

PHYSICAL PROBLEM

Governing Equations

We consider the situation of variable-viscosity miscible fluids in an unstable configuration; that is, a heavier fluid placed above a lighter one in a vertically oriented capillary tube, as shown in FIGURE 1. We assume a suitably defined Reynolds number to be small, so that the motion is governed by the three-dimensional Stokes equations

$$\nabla \cdot \mathbf{u} = 0 \tag{1}$$

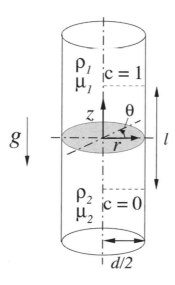

FIGURE 1. Sketch of the vertical capillary tube and the cylindrical coordinate system.

$$\nabla p = \nabla \cdot \underline{\tau} + \rho g \mathbf{e_g} \tag{2}$$

$$\frac{\partial c}{\partial t} + \mathbf{u} \cdot \nabla c = D \nabla^2 c. \tag{3}$$

These equations describe the conservation of mass, momentum, and species, respectively. Here $\mathbf{u}$ represents the flow velocity, the gravitational acceleration g points in the $-z$-direction given by unit vector $\mathbf{e_g}$, c indicates the concentration of the heavier fluid, $\underline{\tau}$ denotes the viscous stress tensor, and D represents the diffusion coefficient, which is assumed constant. Note that implicitly contained in the above set of equations is the Boussinesq approximation, which assumes that density variations are significant in the gravitational term only.

Following other authors, the density ρ and the viscosity μ are assumed to be linear and exponential functions of the concentration c, respectively,

$$\rho = \rho_2 + c(\rho_1 - \rho_2) \tag{4}$$

$$\mu = \mu_2 e^{Rc}, \tag{5}$$

where ρ_1 and μ_1 indicate the density and viscosity of the heavier fluid, and ρ_2 and μ_2 represent the counterparts for the lighter fluid. The mobility ratio R is given by

$$R = \ln \frac{\mu_1}{\mu_2}. \tag{6}$$

The governing equations are rendered dimensionless by introducing a characteristic length L^*, velocity U^*, time T^*, pressure P^*, and density difference G^* in the form

$$L^* = d \tag{7}$$

$$U^* = \frac{\Delta \rho g d^2}{\mu_{min}} \tag{8}$$

$$T^* = \frac{\mu_{min}}{\Delta \rho g d} \tag{9}$$

$$P^* = \Delta \rho g d \tag{10}$$

$$R^* = \Delta \rho = \rho_1 - \rho_2. \tag{11}$$

Note that the nondimensionalization is always carried out with the smaller viscosity, so that a meaningful comparison can be made between cases in which either the lighter or the heavier fluid is the more viscous.

$$\nabla \cdot \mathbf{u} = 0 \tag{12}$$

$$\nabla p = \nabla \cdot \underline{\tau} - c \mathbf{e_g} \tag{13}$$

$$\frac{\partial c}{\partial t} + \mathbf{u} \cdot \nabla c = \frac{1}{Ra} \nabla^2 c, \tag{14}$$

where the Rayleigh number Ra is defined by

$$Ra = \frac{\Delta \rho g d^3}{D \mu_{min}}. \tag{15}$$

The Stokes equations, when formulated in cylindrical coordinates (r, θ, z), exhibit terms of the form r^{-1}, r^{-2}, r^{-3}, which lead to geometric singularities at the axis, $r = 0$. Verzicco and Orlandi[20] proposed rewriting the governing equations by replacing the velocity components v_r, v_θ, and v_z by $q_r = v_r \cdot r$, $q_\theta = v_\theta$, and $q_z = v_z$, respectively. Thus, by definition $q_r = 0$ on the axis, which on a staggered grid avoids the problem of singularities. The authors also demonstrate that alternative formulations (e.g., $q_\theta = v_q \cdot r$) neither enhance the accuracy nor simplify the discretization. Hence, we used this formulation for our numerical analysis.

LINEAR STABILITY ANALYSIS

Linearization and Formulation of the Eigenvalue Problem

We linearize the above set of equations around a quiescent base state

$$
\begin{bmatrix} q_r \\ q_\theta \\ q_z \\ p \\ c \end{bmatrix}(r, \theta, z, t) = \begin{bmatrix} 0 \\ 0 \\ 0 \\ \bar{p} \\ \bar{c} \end{bmatrix}(z) + \begin{bmatrix} q_r' \\ q_\theta' \\ q_z' \\ p' \\ c' \end{bmatrix}(r, \theta, z, t), \tag{16}
$$

where the base concentration profile is given by

$$
\bar{c} = 0.5 + 0.5 \cdot \mathrm{erf}\left(\frac{z}{\delta}\right). \tag{17}
$$

The parameter δ denotes the thickness of the interfacial region. We assume that the diffusive time scale of the base state is much larger than the characteristic time scale of the instability growth, so that the base state can be held constant for the purpose of evaluating the instability growth rate. The perturbations, denoted by a prime, are assumed to have the form

$$
\begin{bmatrix} q_r' \\ q_\theta' \\ q_z' \\ p' \\ c' \end{bmatrix}(r, \theta, z, t) = \begin{bmatrix} \hat{q}_r(r, z) \cdot \cos(\beta\theta) \\ \hat{q}_\theta(r, z) \cdot \sin(\beta\theta) \\ \hat{q}_z(r, z) \cdot \cos(\beta\theta) \\ \hat{p}(r, z) \cdot \cos(\beta\theta) \\ \hat{c}(r, z) \cdot \cos(\beta\theta) \end{bmatrix} e^{\sigma t}, \tag{18}
$$

where the "hatted" quantities represent axisymmetric eigenfunctions and β denotes the azimuthal wave number. It should be noted that, due to the underlying geometry of the problem, only integral values of β have physical significance. By substituting the above relations into the dimensionless conservation equations, subtracting out the base state and linearizing, we obtain the system of linear equations in terms of q_r, q_θ and q_z as

$$
0 = \frac{\partial \hat{q}_r}{\partial r} + \beta \hat{q}_\theta + r \frac{\partial \hat{q}_z}{\partial z} \tag{19}
$$

$$\frac{\partial \hat{p}}{\partial r} = e^{R\bar{c}} \left[\frac{\partial}{\partial r} \left(\frac{1}{r} \frac{\partial \hat{q}_r}{\partial r} \right) - \frac{\beta^2}{r^3} \hat{q}_r + \frac{1}{r} \frac{\partial^2 \hat{q}_r}{\partial z^2} - \frac{2\beta}{r^2} \hat{q}_\theta + R \frac{\partial \bar{c}}{\partial z} \frac{\partial \hat{q}_z}{\partial r} + R \frac{1}{r} \frac{\partial \bar{c}}{\partial z} \frac{\partial \hat{q}_r}{\partial z} \right] \quad (20)$$

$$-\frac{1}{r} \beta \hat{p} = e^{R\bar{c}} \left[\frac{\partial}{\partial r} \left(\frac{1}{r} \frac{\partial}{\partial r} (r \hat{q}_\theta) \right) - \frac{\beta^2}{r^2} \hat{q}_\theta + \frac{\partial^2 \hat{q}_\theta}{\partial z^2} - \frac{2\beta}{r^3} \hat{q}_r + R \frac{\partial \bar{c}}{\partial z} \frac{\partial \hat{q}_\theta}{\partial z} - R \frac{\beta}{r} \frac{\partial \bar{c}}{\partial z} \hat{q}_z \right] \quad (21)$$

$$\frac{\partial \hat{p}}{\partial z} = e^{R\bar{c}} \left[\frac{1}{r} \frac{\partial}{\partial r} \left(r \frac{\partial \hat{q}_z}{\partial r} \right) - \frac{\beta^2}{r^2} \hat{q}_z + \frac{\partial^2 \hat{q}_z}{\partial z^2} + 2R \frac{1}{r} \frac{\partial \bar{c}}{\partial z} \frac{\partial \hat{q}_z}{\partial z} \right] - \hat{c} \quad (22)$$

$$\sigma \hat{c} + \frac{\partial \bar{c}}{\partial z} \hat{q}_z = \frac{1}{Ra} \left[\frac{\partial^2 \hat{c}}{\partial r^2} + \frac{1}{r} \frac{\partial \hat{c}}{\partial r} - \frac{\beta^2}{r^2} \hat{c} + \frac{\partial^2 \hat{c}}{\partial z^2} \right]. \quad (23)$$

This represents an eigenvalue problem with $\hat{p}$, $\hat{q}_r$, $\hat{q}_\theta$, $\hat{q}_z$, and $\hat{c}$ as eigenfunctions and σ as the eigenvalue of the system. There are three externally prescribed parameters in the form of the Rayleigh number Ra, the mobility ratio R, and the initial interfacial thickness δ.

NUMERICAL IMPLEMENTATION AND BOUNDARY CONDITIONS

The computation domain for the solution of the eigenvalue problem extends from the axis to the outer wall in the r-direction, that is, from 0 to 0.5, and from $-l/2$ to $l/2$ in the vertical z-direction, as shown in FIGURE 1. The domain length l has to be chosen sufficiently large for its effect on the numerical eigenvalue and eigenfunction results to be negligible.

Three-Dimensional Perturbations

The linear equations (20)–(23) are discretized by second order finite differences in both the r- and z-directions. To concentrate the numerical resolution in the interfacial region, a non-equidistant grid is taken such that an appropriate concentration of grid points is obtained in the interfacial region. The required numerical resolution is established by means of test calculations. These show that for most cases $N_r = 19$ points in the radial direction is sufficient to keep the error in the eigenvalue to less than 0.1%. The required number of points in the z-direction depends on the domain length and the interface thickness δ. The largest calculations employ up to $N_z = 91$ and $N_r = 23$ points, which results in a matrix of size $5N_zN_r \times 5N_zN_r = 10{,}465 \times 10{,}465$.

At the outer wall of the tube (i.e., at $r = 0.5$), all velocity components are assumed to vanish, as well as the normal derivative of the concentration perturbation. The vertical domain boundaries are sufficiently far away from the interface that we can prescribe homogeneous Dirichlet conditions for all velocity components, as well as for the concentration perturbation. At the axis $r = 0$, $q_r = 0$ since $q_r = v_r \cdot r$. We do not need to specify boundary conditions for the other velocity components or the concentration perturbation, at the axis, since the use of staggered grid implies that only grid points for the radial velocity lie on the axis. For the pressure variable, no boundary conditions are necessary, because we employ a staggered grid.

Axisymmetric Perturbations

To obtain information on the stability of purely axisymmetric perturbations, we consider the case $\beta = 0$ separately. This provides an additional validation of the three-dimensional approach in the limit of small wave numbers. To avoid boundary conditions for pressure and also to save memory by reducing the total number of variables used in the computation, we conveniently rewrite the governing equations in the stream function and vorticity variables.

Vorticity ω and stream function ψ are defined, as usual by

$$v_r = -\frac{1}{r}\frac{\partial \psi}{\partial z}, \quad v_z = \frac{1}{r}\frac{\partial \psi}{\partial r} \tag{24}$$

$$\omega = \frac{\partial v_r}{\partial z} - \frac{\partial v_z}{\partial r}. \tag{25}$$

We assume an axisymmetric disturbance of the form

$$\begin{bmatrix} \psi' \\ \omega' \\ c' \end{bmatrix}(r, z, t) = \begin{bmatrix} \hat{\psi}(r, z) \\ \hat{\omega}(r, z) \\ \hat{c}(r, z) \end{bmatrix} e^{\sigma t}. \tag{26}$$

Using these relations and linearizing, similar to the three-dimensional case, we obtain the system of linear equations

$$0 = \nabla^2 \hat{\psi} + \hat{\omega} \tag{27}$$

$$\begin{aligned} 0 = \nabla^2 \hat{\omega} &+ 2R\frac{\partial \bar{c}}{\partial z}\frac{\partial \hat{\omega}}{\partial z} + \left(\frac{\partial \bar{c}}{\partial z}\right)^2 \hat{\omega} + R^2\frac{\partial^2 \bar{c}}{\partial z^2}\hat{\omega} - 2R^2\frac{1}{r^2}\left(\frac{\partial \bar{c}}{\partial z}\right)^2\frac{\partial \hat{\psi}}{\partial r} \\ &- \frac{2R}{r^2}\frac{\partial^2 \bar{c}}{\partial z^2}\frac{\partial \hat{\psi}}{\partial r} - \frac{2R}{r}\frac{\partial \bar{c}}{\partial z}\frac{\partial^3 \hat{\psi}}{\partial z \partial z} + e^{-R\bar{c}}\frac{\partial \hat{c}}{\partial r} \end{aligned} \tag{28}$$

$$\sigma \hat{c} = \frac{1}{Ra}\nabla^2 \hat{c} - \frac{1}{r}\frac{\partial \bar{c}}{\partial z}\frac{\partial \hat{\psi}}{\partial r}, \tag{29}$$

where $\hat{\psi}$ is set to zero on all domain boundaries and $\hat{\omega}$ vanishes on all boundaries except for the outer wall, where it takes the value $(1/r)(\partial^2\hat{\psi}/\partial r^2)$. At the far-field boundaries the concentration perturbation is assumed to vanish, whereas along the outer wall and the tube axis its normal derivative $\partial \hat{c}/\partial r$ tends to zero. A staggered grid is not required here due to the absence of the pressure variable and of singularities at the axis. Hence, a Chebyshev collocation method is used in the z-direction with two separate subdomains that cover the regions $z \geq 0$ and $z \leq 0$, respectively, in order to concentrate grid points at the interface. In the radial direction a highly accurate, compact finite difference scheme of third order at the wall and up to tenth order in the interior is used.[21] More details about the numerical implementation are published elsewhere.[22]

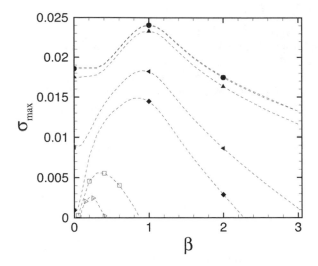

FIGURE 2. Three-dimensional perturbations, dispersion relationships for $\delta = 0.1$ and Ra values: $\triangleright$, 5×10^2; $\square$, 10^3; $\blacklozenge$, 5×10^3; $\blacktriangleleft$, 10^4; $\blacktriangle$, 10^5; $\bullet$, 10^6; $\blacktriangledown$, 10^7. For comparison, the axisymmetric data are plotted as well. The growth rate for $\beta = 1$ is seen to exceed that for $\beta = 0$ for all values of Ra.

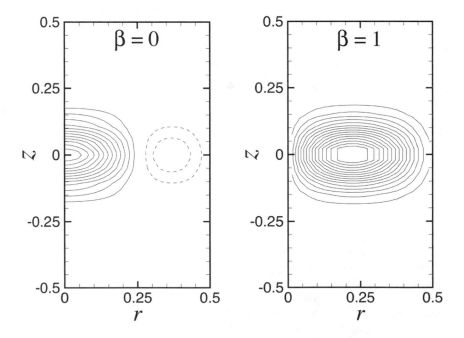

FIGURE 3. Isocontours of the concentration eigenfunction $\hat{c}$ for $\beta = 0$ and 1 for $\delta = 0.1$ and $Ra = 10^5$.

RESULTS

Constant Viscosity

In FIGURE 2 the leading eigenvalue is plotted as a function of the wave number β for several Ra values, ranging from 500 to 10^7, and a constant thickness of the interface $\delta = 0.1$. These dispersion relationships show that for small and intermediate wave numbers, the curves for $Ra > 10^6$ become indistinguishable, implying that an additional increase in Ra affects only the range of unstable wave numbers and the short-wavelength cutoff, but not the most dangerous wave number or its growth rate. It should be noted that the data for noninteger values of β are plotted in FIGURE 2 to guide the eye, because only the integer values are of physical significance. FIGURE 2 demonstrates that the azimuthal perturbation $\beta = 1$ is always more unstable than its axisymmetric counterpart. The concentration eigenfunctions for $\beta = 0$ and 1 are shown in FIGURE 3 for $\delta = 0.1$ and $Ra = 10^5$. The presence of only one sign in the eigenfunction for $\beta = 1$ indicates that the lighter fluid is rising in one half of the capillary tube, with the heavier fluid sinking in the other half. The qualitative form of the fingers produced is illustrated in FIGURE 4.

$$\beta = 0$$

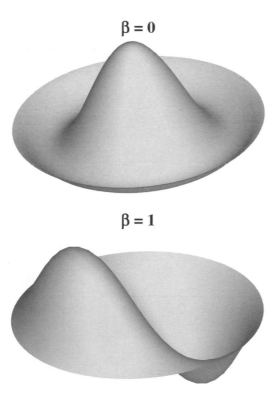

$$\beta = 1$$

FIGURE 4. Qualitative form of finite amplitude fingers for $\delta = 0.1$ and $Ra = 10^7$.

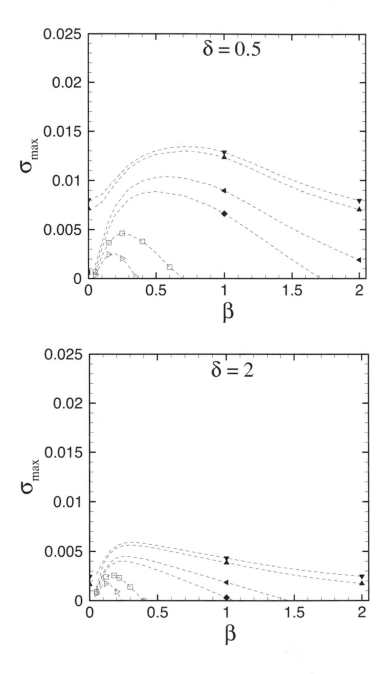

FIGURE 5. Dispersion relationships for various interface thickness values and Rayleigh numbers: $\triangleright$, 5×10^2; $\square$, 10^3; $\blacklozenge$, 5×10^3; $\blacktriangleleft$, 10^4; $\blacktriangle$, 10^5; $\blacktriangledown$, 10^7. $\beta = 1$ represents the most amplified integer mode for all values of δ and Ra.

For larger values of δ, the corresponding dispersion relationships are presented in FIGURE 5. We observe a general trend of decreasing growth rate and smaller cutoff wave number for increasing interface thickness. However, there is no qualitative change in the shape of the concentration eigenfunction $\hat{c}$ with increasing thickness, as is shown in FIGURE 6. From the dispersion relationships shown in FIGURES 2 and 5 we deduce that for all values of the interface thickness, $\beta = 1$ remains the most dangerous mode. For this reason, we consider this wave number in more detail.

By extrapolating the growth rate for $\beta = 1$ and $Ra = 10^7$ to $\delta = 0$, we find that the maximum eigenvalue for a step-like concentration base state is approximately 0.028, as shown in FIGURE 7. The variation of the growth rate σ with Ra, for various δ values and $\beta = 1$, is shown in FIGURE 8. In general, thinner interfaces and larger Ra values are seen to be destabilizing. For $Ra > 10^6$, the growth rate is seen to asymptotically reach a plateau, the value of which depends on δ.

For each δ, a critical value Ra_{crit} can be identified below which the base state is stable to axisymmetric perturbations. The existence of this critical value reflects the stabilizing influence of the outer wall. The Ra_{crit} values found here for capillary tubes are significantly higher than those reported elsewhere[17] for Hele–Shaw cells. This is to be expected, since the stabilizing influence by the walls should be stronger in a capillary tube, where the perturbation is surrounded by the solid wall on all sides. In a Hele–Shaw cell, on the other hand, the perturbations are affected by the walls only in the direction normal to the gap, but not in the spanwise direction. FIGURE 9 depicts the critical Rayleigh number Ra_{crit} as a function of δ, for $\beta = 1$ and $\beta = 0$. For all interface thicknesses, the value of Ra_{crit} for $\beta = 1$ is smaller than the corresponding value for $\beta = 0$. By linear regression through the data points for $\beta = 1$, we obtain the relationship

$$Ra_{crit} = 1800\delta + 920. \tag{30}$$

This relationship indicates that Rayleigh numbers below $O(920)$ are stable for all base concentration profiles, with respect to any axisymmetric or three-dimensional perturbation.

It is interesting to compare the above relationship (30) for the present, error function type base concentration profiles with the classical result[2] dealing with unstable density stratifications with a constant gradient dc/dz in a capillary tube. Taylor demonstrated that such profiles are stable as long as

$$\frac{\frac{dc}{dz}\rho_0 g \alpha d^4}{D\mu} \leq 1087, \tag{31}$$

where

$$\rho = \rho_0(1 + \alpha c). \tag{32}$$

For relatively smooth base concentration profiles of the error function type (i.e., reasonably large values of δ) the present criterion (30) should approach the classical Taylor criterion (31). To check if this is the case, we rewrite (30) for dimensional δ as

$$\left(d\frac{dc}{dz}\right)^{-1} \times \frac{\frac{dc}{dz}\rho_0 g \alpha d^4}{D\mu} \leq 1800\frac{\delta}{d} + 920. \tag{33}$$

In the first term on the left hand side, we approximate dc/dz by the value at $z = 0$

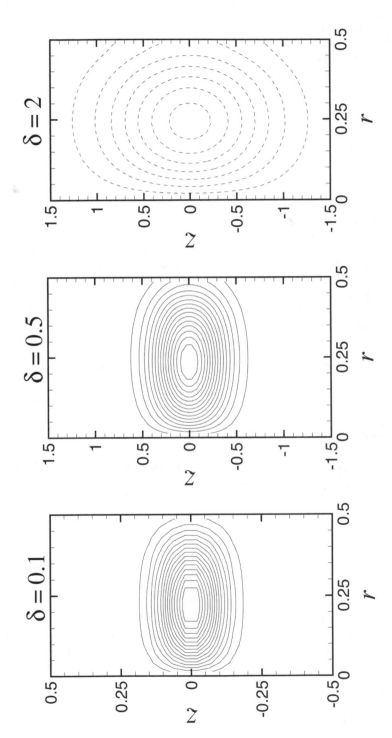

FIGURE 6. Isocontours of the concentration eigenfunction $\hat{c}$ for various values of δ, at $Ra = 10^5$ and $\beta = 1$. Although the isocontours become flatter for decreasing δ, their shapes remain qualitative the same.

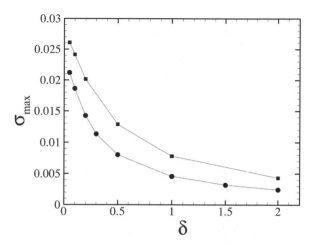

FIGURE 7. The growth rate σ as a function of δ, for $\beta = 0$ (●) and $\beta = 1$ (■), with $Ra = 10^7$. This indicates that for all values of the interface thickness, $\beta = 1$ remains more dangerous than the axisymmetric mode.

$$\left.\frac{dc}{dz}\right|_{z=0} = \frac{1}{\delta\sqrt{\pi}}. \tag{34}$$

For $\delta/d = 2$, we thus obtain

$$\frac{\frac{dc}{dz}\rho_0 g\alpha d^4}{D\mu} \leq 1275, \tag{35}$$

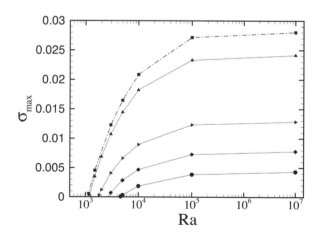

FIGURE 8. The growth rate σ corresponding to the most dangerous wave number $\beta = 1$ as a function of Ra for interface thickness δ: ■, 0; ▲, 0.1; ▶, 0.5; ◆, 1; ●, 2. The *dashed line* corresponds to extrapolated data for the step function profile.

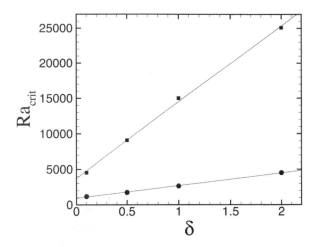

FIGURE 9. The critical Rayleigh number Ra_{crit} as a function of the interfacial thickness parameter δ, for the axisymmetric mode $\beta = 0$ (■) and the most dangerous three-dimensional mode $\beta = 1$ (●). For all values of δ, the axisymmetric mode is seen to have a larger value of Ra_{crit}.

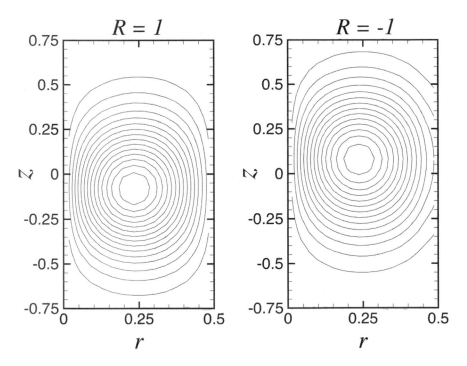

FIGURE 10. Isocontours of the concentration eigenfunction $\hat{c}$ for $R = 1$ and $R = -1$ for $\delta = 0.5$, $\beta = 1$, and $Ra = 10^5$. The eigenvalue corresponding to both cases is $\sigma = 7.97 \times 10^{-3}$. The maxima of the eigenfunctions are always shifted toward the less viscous fluid.

which is indeed close to the relationship, **(31)**, derived by Taylor. The slightly higher value than Taylor's result is expected, since we based the comparison on the largest value of the concentration gradient, rather than its average value.

Variable Viscosity

FIGURE 10 shows the concentration eigenfunction contours for $R = 1$ (less viscous fluid below) and $R = -1$ (less viscous fluid above). Although the growth rates are identical, the maxima of the eigenfunctions are shifted in opposite directions; that is, always toward the less viscous fluid. It has been demonstrated[18] that a simple transformation of the governing equations (Stokes equations) shows the eigenvalues to be equal for both the cases. As a consequence we limit our discussion to positive values of the mobility ratio.

FIGURE 11 displays the dispersion relations for $\delta = 0.1$, $Ra = 10^7$, and various values of the mobility ratio R. An increase in the mobility ratio R dampens the growth of the instability, due to the higher average viscosity of the two-fluid system. However, FIGURE 12 shows a crossover of the most unstable mode from azimuthal to axisymmetric for large values of R and $Ra > 10^5$. Hence the damping influence, with increasing mobility ratio, is much stronger on the three-dimensional mode than the corresponding axisymmetric mode for large Rayleigh numbers. However, no such change of stability occurs for lower Rayleigh numbers, because all the unstable modes in this case are equally damped.

For a constant mobility ratio, similar to the constant viscosity case, the growth rate increases with the Rayleigh number, until it reaches a plateau at about $Ra = 10^5$. Furthermore, the eigenfunctions and finger shapes are similar to the constant viscosity case (cf. FIGS. 3 and 10) except that they are no longer symmetric but shifted toward the less viscous fluid.

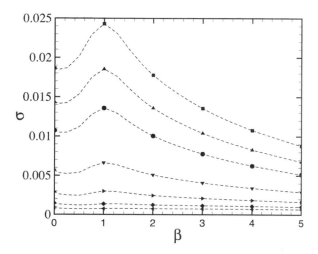

FIGURE 11. Growth rate as a function of β for various mobility ratios R: ■, 0; ▲, 0.5; ●, 1; ▼, 2; ▶, 3; ◆, 4; ◀, 5 with $\delta = 0.1$ and $Ra = 10^7$. Increasing the mobility ratio has a stabilizing effect.

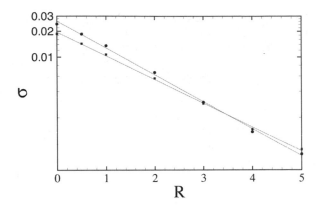

FIGURE 12. Non-uniform damping with increasing mobility ratio results in a crossover of the most unstable mode from $\beta = 1$ (●) to $\beta = 0$ (■). For $\delta = 0.1$ and $Ra = 10^7$, crossover occurs at $R = 3.48$.

FIGURE 13 shows the variation of the maximum growth rate with the interface thickness for $\beta = 0$ and $\beta = 1$, at two different values of the mobility ratio. We find that for larger mobility ratios the highest growth rates occur at an intermediate value of the interfacial thickness. This is in contrast to the constant viscosity case, for which the growth rate is seen to decline monotonically with increasing interface thickness (compare with FIGS. 2 and 5). FIGURE 13 also shows that the most unstable interface thickness increases with the mobility ratio.

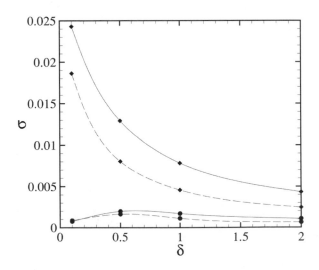

FIGURE 13. The growth rate σ as a function of δ for – – ◆ – –, $R = 0$, $\beta = 0$; ——◆——, $R = 0$, $\beta = 0$; – –●– –, $R = 0$, $\beta = 0$; ——●——, $R = 0$, $\beta = 0$. In all cases $Ra = 10^7$. For larger mobility ratios, interfaces with intermediate thicknesses are seen to be most unstable.

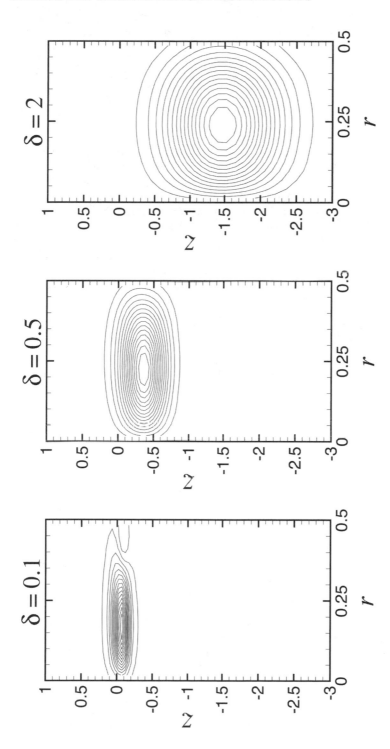

FIGURE 14. Concentration eigenfunctions of the most amplified mode for $Ra = 10^6$, $R = 5$ and different values of δ. For thicker interfaces, the perturbation is located almost entirely in the less viscous fluid.

TABLE 1. Concentration gradient and fluid viscosity at the location of the eigenfunction maximum for $Ra = 10^6$ and $R = 5$

δ	$\bar{c}_z / (\bar{c}_z)_{max}$	μ / μ_{min}
0.1	0.61	2.583
0.5	0.56	2.035

The shapes of the eigenfunctions shown in FIGURE 14 explain the emergence of this most unstable interface thickness. We note that the vertical extent of the eigenfunctions is affected by *both* of the two externally imposed length scales, namely, the tube diameter and the interface thickness. Due to this influence of the tube diameter, for interface thicknesses much smaller than the tube diameter, the vertical scale of the eigenfunction does not decrease at the same rate as δ. Conversely, for interface thicknesses larger than the tube diameter, the size of the eigenfunction does not increase as strongly with δ. As a result, for thin interfaces the eigenfunction extends over a region several times wider than the interface. Since it has to be anchored in the region of the unstable density gradient, it has to extend substantially into the high viscosity fluid, which exerts a stabilizing influence. In contrast, for thick interfaces the eigenfunction can reside almost entirely within the interfacial region, so that its maximum can shift substantially toward the low viscosity region. In other words, the eigenfunction can select a location that represents an optimal combination of unstable density gradient and low viscosity fluid. This is confirmed by TABLE 1, which provides a comparison of the normalized density gradient and the viscosity for two values of δ. The table shows that for the thicker interface the normalized concentration gradient, and hence, the driving force behind the instability, has not decreased much, but the fluid viscosity at the location of the eigenfunction maximum shows that its completely in the lower viscosity fluid for $\delta = 0.5$. Hence, the thicker interface gives rise to a stronger overall instability.

CONCLUSIONS

The current investigation presents linear stability results for the miscible interface formed by placing a heavier fluid above a lighter fluid, with different viscosities, in a vertically oriented capillary tube. The analysis is based on the three-dimensional Stokes equations coupled to a convection–diffusion equation for the concentration field in cylindrical coordinates. By linearizing this set of equations, a generalized eigenvalue problem is formulated, whose numerical solution yields both the growth rate as well as the two-dimensional eigenmodes as functions of the dimensionless parameters characterizing the problem; namely, the Rayleigh number, the mobility ratio, and the interface thickness.

For constant viscosity case, the dispersion relations show that for all Ra values and interface thicknesses the azimuthal mode $\beta = 1$ represents the most unstable disturbance. In particular, its growth rates are consistently higher than those of the axisymmetric mode. The most amplified mode thus corresponds to the formation of one finger of the lighter fluid rising over one half of the tube cross section, with a second finger of the heavier fluid falling in the other half. This is in agreement with other experimental observations.[16] The stability results, furthermore, indicate the existence of a critical Rayleigh number $Ra_{crit} \approx 920$, below which all perturbations are stable. For relatively thick interfaces, the present data for Ra_{crit} are seen to approach the classical value[2] for a uniform density gradient. For a constant interface thickness, the growth rates reach a plateau as $Ra > 10^6$. The detailed numerical analysis for density-driven instability of identical viscosity miscible fluids is given elsewhere.[22]

When the viscosity of either fluid is increased, the instability is damped compared to the constant viscosity case. There is a crossover of the most unstable mode from azimuthal to axisymmetric for large values of R and $Ra > 10^5$. Hence, the damping influence, with variable viscosity, is much stronger on the three-dimensional mode than the corresponding axisymmetric mode for large Rayleigh numbers. It is also seen that the growth rate does not depend on which of the two fluids is the more viscous. For every parameter combination the maximum of the eigenfunctions tends to shift toward the less viscous fluid. For a fixed mobility ratio, similar to the constant viscosity case, the growth rates are seen to reach a plateau for Rayleigh numbers in excess of 10^6. An interesting observation is that at higher mobility ratios the largest growth rates and unstable wave numbers are obtained for intermediate interface thicknesses. Thus, with different viscosity fluids, thicker interfaces can be more unstable than their thinner counterparts.

Other experiments[16] had shown that a small amount of net flow through the capillary tube can stabilize the azimuthal instability mode and maintain an axisymmetric evolution of the flow. To analyze the effect of a net flow onto the linear stability of the interface, we need to include a base flow of Poiseuille type in our analysis. Efforts in this direction are currently underway.

ACKNOWLEDGMENTS

We thank Professors Tony Maxworthy and Bud Homsy for helpful discussions. Support for this research was received from the NASA Microgravity and NSF/ITR programs, as well as from the Department of Energy, the UC Energy Institute, and through an NSF equipment grant.

REFERENCES

1. HALES, A. 1937. Convection currents in geysers. Mon. Not. R. Astron. Soc., Geophys, Suppl. **4:** 122.
2. TAYLOR, G. 1954. Diffusion and mass transport in tubes. Proc. Roy. Soc. B **67:** 857.
3. DEBACQ, M., V. FANGUET, J HULIN & D. SALIN. 2001. Self-similar concentration profiles in buoyant mixing of miscible fluids in a vertical tube. Phys. Fluids **13:** 3097.
4. WOODING, R. 1959. The stability of a viscous liquid in a vertical tube containing porous material. Proc. Roy. Soc. A **252:** 120.
5. BATCHELOR, G. & J. NITSCHE. 1993. Instability of stratified fluid in a vertical cylinder. J. Fluid Mech. **252:** 419.
6. TAYLOR, G. 1960. Deposition of a viscous fluid on the wall of a tube. J. Fluid Mech. **10:** 161.
7. COX, B. 1962. On driving a viscous fluid out of a tube. J. Fluid Mech. **14:** 81.
8. REINELT, D. & P. SAFFMAN. 1985. The penetration of a finger into a viscous fluid in a channel and tube. SIAM J. Sci. Statist. Comput. **6:** 542.

9. SOARES, E., M. CARVALHO & P.S. MENDES. 2004. Immiscible liquid-liquid displacement in capillary tubes. J. Fluid Mech. Submitted.
10. PETITJEANS, P. & T. MAXWORTHY. 1996. Miscible displacements in capillary tubes. Part 1. experiments. J. Fluid Mech. **326:** 37.
11. CHEN, C. & E. MEIBURG. 1996. Miscible displacements in capillary tubes. Part 2. numerical simulations. J. Fluid Mech. **326:** 57.
12. SCOFFONI, J., E. LAJEUNESSE & G. HOMSY. 2000. Interface instabilities during displacements of two miscible fluids in a vertical pipe. Phys. Fluids **13**(3): 553.
13. KORTEWEG, D. 1901. Sur la forme que prennent les équations du mouvement des fluides si l'on tient compte des forces capillaires causées par des variations de densité. Arch. Neel. Sci. Ex. Nat. **II:** 6.
14. JOSEPH, D. & Y. RENARDY. 1992. Fundamentals of Two-Fluid Dynamics. Springer-Verlag, New York.
15. CHEN, C. & E. MEIBURG. 2002. Miscible displacements in capillary tubes: influence of korteweg stresses and divergence effects. Phys. Fluids **14**(7): 2052.
16. KUANG, J., T. MAXWORTHY & P. PETITJEANS. 2003. Miscible displacements between silicone oils in capillary tubes. Eur. J. Mech. B/Fluids **22:** 271–277.
17. GRAF, F., E. MEIBURG & C. HÄRTEL. 2002. Density-driven instabilities of miscible fluids in a hele-shaw cell: linear stability analysis of the three-dimensional stokes equations. J. Fluid Mech. **451:** 261.
18. GOYAL, N. & E. MEIBURG. 2004. Unstable density stratification of miscible fluids in a verical hele-shaw cell: influence of variable viscosity on the linear stability analysis. J. Fluid Mech. **516:** 211.
19. FERNANDEZ, J., P. KUROWSKI, P. PETITJEANS & E. MEIBURG. 2002. Density-driven, unstable flows of miscible fluids in a hele-shaw cell. J. Fluid Mech. **451:** 239.
20. VERZICCO, R. & P. ORLANDI. 1996. A finite-difference scheme for three-dimensional incompressible flows in cylinder coordinates deposition of a viscous fluid on the wall of a tube. J. Comp. Phys. **123:** 402.
21. LELE, S. 1992. Compact finite difference schemes with spectral-like resolution. J. Comp. Phys. **103:** 16.
22. VANAPARTHY, S., E. MEIBURG & D. WILHELM. 2003. Density driven instabilities of miscible fluids in a capillary tube: linear stability analysis. J. Fluid Mech. **497:** 99.

An Experimental Study of the Richtmyer–Meshkov Instability in Microgravity

CHARLES. E. NIEDERHAUS[a] AND JEFFREY W. JACOBS[b]

[a]NASA Glenn Research Center, Cleveland, Ohio, USA

[b]University of Arizona, Tucson, Arizona, USA

ABSTRACT: Richtmyer–Meshkov (RM) instability occurs when a planar interface separating two fluids of different density is impulsively accelerated in the direction of its normal. It is one of the most fundamental fluid instabilities and is of importance to the fields of astrophysics and inertial confinement fusion. Because RM instability experiments are normally carried out in shock tubes, where the generation of a sharp, well-controlled interface between gases is difficult, there is a scarcity of good experimental results. The experiments presented here use a novel technique that circumvents many of the experimental difficulties that have previously limited the study of RM instability in shock tubes. In these experiments, the instability is generated incompressibly, by bouncing a rectangular tank containing two liquids off of a fixed spring. These experiments, which utilize PLIF flow visualization, yield time–motion image sequences of the nonlinear development and transition to turbulence of the instability that are of a quality unattainable in shock tube experiments. Measurements obtained from these images, therefore, provide benchmark data for the evaluation of nonlinear models for the late-time growth of the instability. Because the run time in these experiments is limited, new experiments in the NASA Glenn 2.2 second drop tower, capable of achieving longer run times, are currently under way.

KEYWORDS: Richtmyer–Meshkov instability; microgravity; fluid dynamics; interface stability; Taylor instability

INTRODUCTION

The interface between two fluids of different density can be unstable in the presence of a gravitational field acting normal to the surface. The relatively well-known Rayleigh–Taylor (RT) instability[1,2] results when the gravitational field is constant and the heavier fluid lies over the lighter fluid. An impulsive acceleration applied to the fluids results in an impulsive gravitational field that can produce the Richtmyer–Meshkov (RM) instability.[3,4] Unlike RT instability, RM instability occurs regardless of the relative orientation of the heavy and light fluids. In many systems, the passage of a shock wave through the interface provides the impulsive acceleration. Both RT

Address for correspondence: Charles Niederhaus, M.S. 77-5, NASA Glenn Research Center, 21000 Brookpark Rd., Cleveland, OH 44135, USA. Voice: 216-433-5461; fax: 216-433-8050.

charles.niederhaus@nasa.gov

Ann. N.Y. Acad. Sci. 1027: 403–413 (2004). ©2004 New York Academy of Sciences.

doi: 10.1196/annals.1324.033

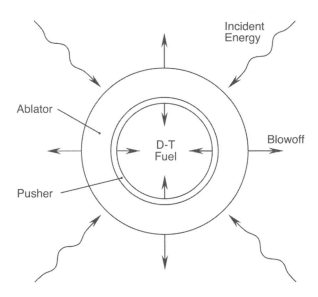

FIGURE 1. Idealized representation of the inertial confinement fusion process.

and RM instabilities result in mixing at the interface. These instabilities arise in a diverse array of circumstances, including supernovæ, stratified lakes and oceans, and supersonic combustion. However, the area of greatest current interest in RT and RM instabilities is inertial confinement fusion (ICF), the goal of which is to produce fusion energy from millimeter size pellets of deuterium and tritium. As is shown in FIGURE 1, ICF uses high-energy laser or X-ray radiation to vaporize an ablative shell encapsulating a higher density pusher sphere. The pusher contains a low-density (less than $1.0 \, mg/cm^3$) deuterium–tritium fuel mixture. The gases expanding from the ablative shell drive the pusher sphere inward to conserve momentum. Thus, the ablator–pusher and pusher–fuel density interfaces undergo a combination of RM and RT instabilities. The turbulent flow that results from these instabilities limits the degree of compression achievable and, thus, the energy output. In the ICF experiments conducted to date, the energy used to power the lasers has exceeded the energy output from the fusion process, a result directly attributable to RM and RT instabilities.[5–8] The National Ignition Facility at Lawrence Livermore National Laboratory is being constructed to study these instabilities and to attempt to achieve net-positive yield in an ICF experiment.

RM instability is most commonly studied in the laboratory using shock tubes in which a shock wave is passed over an interface separating two different gases.[9–14] However, these experiments have been troubled by the difficulty in generating a well-defined interface between the two gases. Other studies have used laser ablation to generate a strong shock that propagates through two solids.[15–21] This technique provides parameter values that are closer to the ICF configuration, but poses measurement difficulties.

The experiments described here use a novel method for generating RM instability in an incompressible liquid–liquid system. Although laser RM experiments are millimeters in size, or smaller, with time scales of nanoseconds, and shock tube RM experiments are centimeters in size with time scales of milliseconds, these liquid–liquid RM experiments occur across a container width of 12 centimeters and have durations of nearly one second. These increases in the dominant length and time scales have permitted much better visualization of the instability, and have provided needed insight into the initial stages of the instability and its transition to turbulence. The experiments were conducted in a three-meter drop tower at the University of Arizona, and the impulsive acceleration was produced by bouncing the fluid-filled container off of a retractable spring. The subsequent free-fall as the experiment travels in the drop tower allows the instability to evolve in effective absence of the Earth gravity.

LINEAR THEORY

The first analysis of RM instability was performed by Richtmyer[3] who focused on modeling the early stages of the phenomena using linear stability theory. His solution, which considers the growth of a small amplitude sinusoidal perturbation, assumes incompressible flow and models the shock interaction as an impulsive acceleration. Although Richtmyer's incompressible assumption has prompted much criticism, his model applies explicitly to the current incompressible experiments.

The analysis of Richtmyer expands upon that developed by Taylor,[2] who studied the stability of an interface between different density fluids in a constant gravitation field. Taylor's analysis considers the two-dimensional inviscid, irrotational flow of incompressible fluids with densities ρ_1 and ρ_2, and velocity potentials ϕ_1 and ϕ_2 ($v = \nabla\phi$), as shown in FIGURE 2. The interface separating the fluids is assumed to have a sinusoidal perturbation, given by

$$\eta = a(t)\cos(kx),\qquad(1)$$

where k is the wave number and $a(t)$ is the amplitude. The amplitude of the disturbance is assumed to be small, such that $|ka| \ll 1$.

The solution is a single equation for the amplitude of the perturbation,

$$\ddot{a}(t) = -kAg(t)a(t),\qquad(2)$$

where

$$A = \frac{\rho_2 - \rho_1}{\rho_2 + \rho_1}\qquad(3)$$

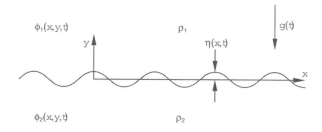

FIGURE 2. Diagram of the configuration used in the linear stability analysis.

is defined as the Atwood number of the system. The solution is valid for an arbitrary wave number and gravitational acceleration.

Richtmyer considered the case where $g(t)$ is impulsive, such that

$$g(t) = \Delta V \delta(t). \tag{4}$$

Solving Equation (2) with the impulsive acceleration (4) yields the solution

$$\dot{a}_0 = -kA\Delta V a_i + \dot{a}_i, \tag{5}$$

where a_i and $\dot{a}_i$ are the amplitude and growth rate (also termed velocity in the literature) of the perturbation prior to the acceleration pulse and $\dot{a}_0$ is the velocity after the acceleration: the result gives a constant postacceleration perturbation growth rate that is a linear function of the wave number, Atwood number, and the initial perturbation amplitude. The experiments described here have an acceleration impulse duration of about 30 msec, too long for a delta function assumption. Therefore, Equation (2) was integrated directly using the preacceleration conditions and the measured acceleration data to obtain the theoretical postacceleration growth rate.

EXPERIMENT DETAILS

The experiments were conducted at the University of Arizona using a purpose-built 3.05-meter vertical drop tower and attached instrumentation, as shown in FIGURE 3. The function of the drop tower is to provide an impulsive acceleration to a two-fluid system, and then allow the system to travel safely in freefall without external disturbance. The tower was designed and built on site and consists of four main parts: a guide rail assembly, a sled assembly, a release mechanism, and a spring mechanism. A shaking mechanism driven by a programmable motion controller introduces internal waves to the two-fluid system that serve as the initial perturbation for the RM instability. Planar laser-induced fluorescence (PLIF) is used for flow visualization. The lower fluid contains a fluorescent dye illuminated by a thin light sheet, and the upper fluid is clear against a black background. The acceleration history of the fluids is measured by two accelerometers mounted to the apparatus.

The most fundamental form of RM instability occurs between two different density, miscible fluids separated by a sharp interface. As already stated, the initial growth rate is proportional to the Atwood number. Therefore, miscible fluids with a large density ratio are desired so as to obtain large growth rates. PLIF introduces the additional requirement of matching indices of refraction. Otherwise, the curved interface between the fluids at large disturbance amplitudes acts as a lens and distorts the laser sheet, impeding the visualization of essential features. The lighter fluid used in these experiments was a water–isopropanol mixture with 70% (by volume) concentration of isopropanol. The heavier fluid was a water–calcium nitrate salt solution with 25% (by weight) calcium nitrate concentration. These two fluids are miscible and, therefore, do not have surface tension. Furthermore, at the concentrations chosen, these fluids have nearly equal indices of refraction and a relatively large density ratio. The actual values varied slightly, but a typical experiment had an index of refraction of 1.3720 and specific gravities of 0.8731 and 1.2025 for the two fluids. The resulting Atwood number was 0.1587. The kinematic viscosity of the lighter fluid was found to be 3.16 centistokes and that of heavier fluid was 1.55 centistokes. The

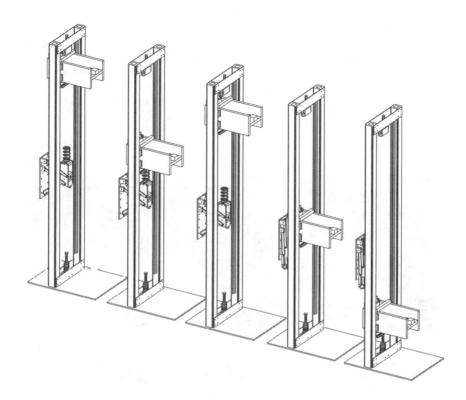

FIGURE 3. A sequence of images showing the sled traveling on the rail system during a typical experiment.

salt solution contained disodium fluorescein dye at a concentration of 0.84 mg/L for use in PLIF.

FIGURE 3 is a sequence of drawings showing the apparatus at various times in during the experiment. The sled is initially raised and the fluid container oscillated to produce the internal waves. At a prescribed time, the sled is released and travels down the rails until it impacts the spring. The fluids thus undergo an impulsive acceleration, followed by a period of freefall as the sled travels up the rails and then down the rails past the retracted spring until it is stopped by a shock absorber at the bottom. A video camera mounted on the sled captures images of the fluids, and accelerometers measure the acceleration history of the fluids. The total time in freefall, during which the instability is allowed to evolve, is 900 msec.

RESULTS AND DISCUSSION

The simplest example of RM instability is that of the small amplitude, single-mode, sinusoidal perturbation with a sharp interface. FIGURE 4 is a sequence of PLIF images showing the evolution of such an instability. The instability is generated from a perturbation with a dimensionless initial amplitude of $ka_i = 0.23$ with $1^1/_2$ waves inside the tank. The first image was taken just before the sled impacts the spring and, thus, shows the initial interface shape. The impulsive acceleration in these experiments is directed from the heavier fluid into the lighter fluid, with the resulting body force on the fluids acting in the opposite direction. This orientation causes the initial perturbation to invert before growing in amplitude. Immediately after inversion, the interface retains a sinusoidal shape, but by image (B) the interface is beginning to become nonsinusodial. Vorticity is deposited along the interface by the baroclinic torque generated during the acceleration pulse. As the instability evolves, the vorticity begins to concentrate midway between the crests and troughs, at the points of

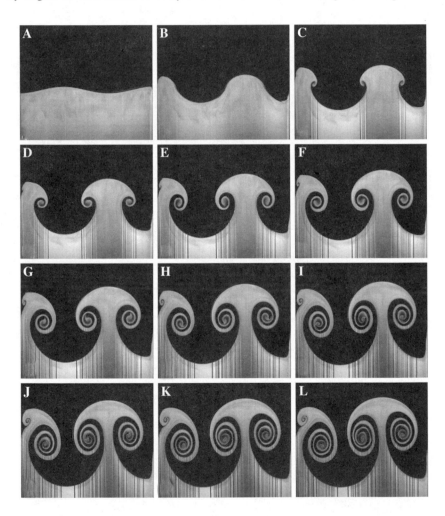

FIGURE 4. A sequence of images from an experiment with $1\frac{1}{2}$ waves and $ka_i = 0.23$. Times relative to the midpoint of spring impact are **(A)** 14 msec, **(B)** 102 msec, **(C)** 186 msec, **(D)** 269 msec, **(E)** 353 msec, **(F)** 436 msec, **(G)** 529 msec, **(H)** 603 msec, **(I)** 686 msec, **(J)** 770 msec, **(K)** 853 msec, and **(L)** 903 msec.

maximum initial interface slope, which are also the locations of the maximum generated vorticity. The resulting vortices produce the symmetric mushroom pattern typical of RT and RM instability.

As time advances, these vortices appear to grow in size as the interface rotates around their centers to form a spiral pattern. The interface is multivalued by image (C), and the vortex has completed several turns by image (L). Note that the interface retains its top-to-bottom symmetry well into the nonlinear regime. This symmetry is characteristic of the RM instability with small density differences. The interface between the two fluids also remains sharp throughout the experiment.

Amplitude Analysis

Richtmyer's linear stability theory[3] shows that the amplitude of the interface satisfies Equation (2). Note that when the gravitational acceleration is zero, as when the sled is in freefall, the theoretical growth rate of the interface is constant. Using the measured amplitude and velocity prior to impact, as well as the time-correlated accelerometer data, this differential equation can be numerically integrated to determine the theoretical postimpulse amplitude and growth rate. Note that in these experiments the acceleration pulse results in the temporary stabilization of the interface. Therefore, the amplitude of the crests and troughs decreases while under acceleration. However, the interfacial velocity achieved is high, and this velocity remains after the acceleration pulse. This results in the inversion of the interface, and the subsequent rapid growth of the perturbation.

The measured overall amplitude (the average of the crest and trough amplitudes found by halving the difference in the crest to trough vertical displacement) is shown in FIGURE 5. The amplitude is made nondimensional in this plot by using the disturbance wave number k and nondimensional time is formed by using the wave number

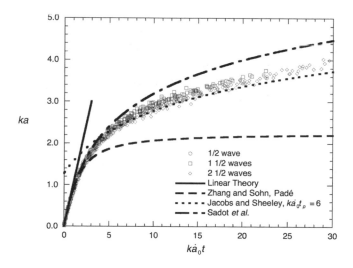

FIGURE 5. Plot of nondimensional amplitude versus nondimensional time data and curves from several models.

k and the theoretical initial growth rate $\dot{a}_0$, derived by integrating (2). Note that $t = 0$ is the theoretical time that the amplitude would be zero, computed using the post-impulse theoretical amplitude and velocity. Linear theory using this nondimensionalization gives a linear stability growth rate of 1, which is shown by the solid straight line. The experiments show excellent agreement with linear theory up to $k\dot{a}_0 t = 0.3$ and are within 10% of the theory at $k\dot{a}_0 t = 0.7$, where nonlinear effects begin to become important. It should be noted that linear theory is derived by assuming $|ka| \ll 1$. Thus, linear theory is surprisingly accurate at moderate values of ka. Also, the ka_i (dimensionless amplitude before impact) value for these experiments ranges from 0.07 to 0.66. Thus, ka_i does not seem to effect the agreement with linear theory.

FIGURE 5 also includes comparisons with three popular nonlinear models. Zhang and Sohn[22] used a Padé approximant to improve the range of validity of their weakly nonlinear perturbation solution (shown as the dashed line). However, this solution has the incorrect asymptotic behavior of the growth rate, decaying as $1/t^2$ instead of $1/t$, as giving by the vortex model of Jacobs and Sheeley,[23] which is an exact solution for $A \to 0$. Sadot et al.[24] developed an empirical model that interpolates between the weakly nonlinear solution and the expected late time solution to obtain much better agreement with the data.

Asymptotic Velocity

To obtain the desired late time behavior in their model, Sadot et al.[24] developed a late time model for the bubble and spike velocities

$$U_{b/s} = \frac{2\pi C(1 + A)}{(1 \pm A)kt}. \tag{6}$$

The values for C given by Sadot et al.[24] are obtained from computations in which they found $C = 1/3\pi$ for $A \geq 0.5$, and $1/2\pi$ for $A \to 0$. There have been several recent attempts to model the late-time asymptotic bubble velocity in fluid systems having an Atwood number that does not equal zero or one. The bubble is defined as portion of the interface in which the light fluid penetrates into the heavy fluid, which in these experiments is the portion where the undyed light fluid extends downward into the heavy fluorescent fluid. For small density differences ($A = 0$), the lighter bubble and the heavier spike have the same growth rate. These experiments have a typical Atwood number of 0.155, and the bubble and spike amplitudes differ by 30% at late time.

Niederhaus and Jacobs[25] used a force-balance model to derive an expression for the bubble velocity,

$$U_b = \frac{1}{(1 + A)kt}. \tag{7}$$

Goncharov[26] extended the approach used by Lazer[27] of expanding the velocity potential near the bubble tip, to obtain

$$U_b = \frac{3 + A}{3(1 + A)kt}. \tag{8}$$

Sohn[28] also used a potential flow model near the bubble tip to obtain

$$U_b = \frac{2}{(2 + A)kt}. \tag{9}$$

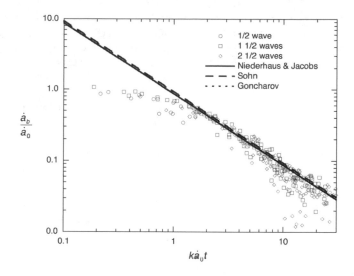

FIGURE 6. Plot of nondimensional bubble velocity versus nondimensional time and three asymptotic models.

FIGURE 6 shows measurements of nondimensional bubble velocity versus time and data from the three models. All the models agree well with the data. However, note that the three models yield values of the bubble velocity that differ by less than 4% at the experimental value of $A = 0.155$. Thus, the experimental uncertainty is too large to ascertain which of the models is more accurate. The data in this plot with lower values of the velocity at late-times are from experiments with longer real time duration (for the same nondimensional time) in which the small amount of bearing drag (about $0.2\,\text{m/sec}^2$) produces its most significant effect. Thus, the presence of bearing drag in these experiments limits our ability to determine the accuracy of these late time models. Future experiments in the GRC 2.2 drop tower in which this effect is removed will yield results better capable of determining this difference.

CONCLUSIONS

The development of single-mode RM instability was investigated using a novel apparatus that allows for quantitative analysis of the two-dimensional instability at very late-times. Miscible liquids with moderate Atwood number were employed, avoiding many of the experimental difficulties encountered by previous studies. The instability was generated by elastically bouncing a fluid-filled container off a vertical spring, imparting an impulsive acceleration. The subsequent freefall as the as the container traveled in the drop tower allowed the instability to develop far into the nonlinear stages. A new apparatus is under construction that will operate in the NASA Glenn 2.2 second drop tower, which will more than double the observation time of the instability in an effort to study the transition to a turbulent flow.

The experimental amplitude measurements were found to be in excellent agreement with linear stability theory for small amplitudes. The measurements were also shown to be in good agreement with the vortex model of Jacobs and Sheeley[23] and the model of Sadot et al.,[24] differing by less than 10% at late times.

Comparisons were made with the asymptotic bubble velocity for three models.[25–27] The models predict a $1/t$ asymptotic velocity, with the magnitude a function of A. All models agree well with the experimental data. However, the models are within $\pm 4\%$ of each other, which is too small for the experimental data to resolve. Experiments at larger values for A (or without bearing drag) are required to more precisely contrast the models. New experiments in the GRC 2.2 drop tower may be able to resolve this.

ACKNOWLEDGMENTS

This work was supported by the NASA Office of Biological and Physical Research, Physical Sciences Research Division and by the Department of Energy Lawrence Livermore National Laboratory.

REFERENCES

1. RAYLEIGH, J.W.S., LORD. 1900. The Scientific Papers of Lord Rayleigh, Vol. II. Cambridge University Press.
2. TAYLOR, G.I. 1950. The instability of liquid surfaces when accelerated in a direction perpendicular to their planes. Proc. Roy. Soc. Lond. A **201:** 192–196.
3. RICHTMYER, R.D. 1960. Taylor instability in shock acceleration of compressible fluids. Commun. Pure Appl. Math. **13:** 297–319.
4. MESHKOV, E.E. 1969. Instability of the interface of two gases accelerated by a shock wave. Izv. Akad. Nauk SSSR Mekh. Zhid. Gaza **4:** 151–157.
5. MCCALL, G.H. 1983. Laser-driven implosion experiments. Plasma Phys. **25:** 237–285.
6. LINDL, J.D., R.L. MCCRORY & E.M. CAMPBELL. 1992. Progress toward ignition and burn propagation in inertial confinement fusion. Phys. Today **45:** 32–40.
7. HOGAN, W.J., R. BANGERTER & G.L. KULCINSKI. 1992. Energy from inertial fusion. Phys. Today **45:** 42–50.
8. LINDL, J.D. 1995. Development of the indirect-drive approach to inertial confinement fusion and the target physics basis for ignition and gain. Phys. Plasmas **2:** 3933–4024.
9. ALESHIN, A.N., et al. 1988. Nonlinear and transitional states in the onset of the Richtmyer–Meshkov instability. Sov. Tech. Phys. Lett. **14:** 466–468. [Russian: Pis'ma Zh. Tekh. Fiz. **14:** 1063–1067].
10. VASSILENKO, A.M., et al. 1992. Experimental research of gravitational instability and turbulization of flow at the noble gases interface. In Advances in Compressible Turbulent Mixing. First International Workshop on the Physics of Compressible Turbulent Mixing. W.P. Dannevik, A.C. Buckingham & C.E. Leith, Eds.: 581–606.
11. BENJAMIN, R.F. 1992. Experimental observations of shock stability and shock-induced turbulence. In Advances in Compressible Turbulent Mixing. Proceedings of the First International Workshop on the Physics of Compressible Turbulent Mixing. W.P. Dannevik, A.C. Buckingham & C.E. Leith, Eds.: 341–348.
12. BROUILLETTE, M. & B. STURTEVANT. 1994. Experiments on the Richtmyer–Meshkov instability: single-scale perturbations on a continuous interface. J. Fluid Mech. **263:** 271–292.
13. BONAZZA, R. & B. STURTEVANT. 1996. X-ray measurements of growth rates at a gas interface accelerated by shock waves. Phys. Fluids **8:** 2496–2512.

14. JONES, M.A. & J.W. JACOBS. 1997. A membraneless experiment for the study of Richtmyer–Meshkov instability of a shock-accelerated gas interface. Phys. Fluids **9:** 3078–3085.
15. DIMONTE, G. & B. REMINGTON. 1993. Richtmyer–Meshkov experiments on the Nova laser at high compression. Phys. Rev. Lett. **70:** 1806–1809.
16. REMINGTON, B.A., et al. 1994. Multimode Rayleigh–Taylor experiments on Nova. Phys. Rev. Lett. **73:** 545–548.
17. DIMONTE, G., C.E. FRERKING & M. SCHNEIDER. 1995. Richtmyer–Meshkov instability in the turbulent regime. Phys. Rev. Lett. **74:** 4855–4858.
18. PEYSER, T.A., et al. 1995. Measurement of radiation-driven shock-induced mixing from nonlinear initial perturbations. Phys. Rev. Lett. **75:** 2332–2335.
19. DIMONTE, G. & M. SCHNEIDER. 1997. Turbulent Richtmyer–Meshkov instability experiments with strong radiatively driven shocks. Phys. Plasmas **4:** 4347–4357.
20. FARLEY, D.R., et al. 1999. High Mach number mix instability experiments of an unstable density interface using a single-mode, nonlinear initial perturbation. Phys. Plasmas **6:** 4304–4317.
21. HOLMES, R.L., et al. 1999. Richtmyer–Meshkov instability growth: experiment, simulation, and theory. J. Fluid Mech. **389:** 55–79.
22. ZHANG, Q. & S. SOHN. 1997. Nonlinear theory of unstable fluid mixing driven by shock wave. Phys. Fluids **9:** 1106 1124.
23. JACOBS, J.W. & J.M. SHEELEY. 1996. Experimental study of incompressible Richtmyer–Meshkov instability. Phys. Fluids **8:** 405–415.
24. SADOT, O., et al. 1998. Study of nonlinear evolution of single-mode and two-bubble interaction under Richtmyer–Meshkov instability. Phys. Rev. Lett. **80:** 1654–1657.
25. NIEDERHAUS, C.E. & J.W. JACOBS. 2003. Experimental study of the Richtmyer–Meshkov instability of incompressible fluids. J. Fluid Mech. **485:** 243–277.
26. GONCHAROV, V.N. 2002. Analytical model of nonlinear, single-mode, classical Rayleigh–Taylor instability of arbitrary Atwood numbers. Phys. Rev. Lett. **88:** 134502.
27. LAYZER, D. 1955. On the instability of superposed fluids in a gravitational field. Astrophys. J. **122:** 1–12.
28. SOHN, S. 2003. Simple potential-flow model of Rayleigh–Taylor and Richtmyer–Meshkov instabilities for all density ratios. Phys. Rev. E **67:** 026301.

Manipulation of Fluid Objects with Acoustic Radiation Pressure

PHILIP L. MARSTON AND DAVID B. THIESSEN

Department of Physics, Washington State University, Pullman, Washington, USA

ABSTRACT: Conditions are summarized for manipulating and stabilizing fluid objects based on the acoustic radiation pressure of standing waves. Examples include (but are not limited to) liquid drops, gas bubbles in liquids, and cylindrical liquid bridges. The emphasis is on situations where the characteristic wavelength of the acoustic field is large in comparison to the relevant dimension of the fluid object. Tables are presented for ease of comparing the signs of qualitatively different radiation force parameters for a variety of fluid objects.

KEYWORDS: acoustic levitation; radiation pressure; drops; bubbles; flames; liquid bridges

Dedicated in memory of Yale University Professor Robert E. Apfel (1943–2002)

INTRODUCTION

Although acoustic radiation pressure appears to be widely used as a technique for manipulating fluid objects in microgravity conditions (as well as in normal gravity), there are several subtle aspects of this method. Some general aspects and recently reported special cases are examined here so as to assist the reader in evaluating potential applications to new areas of research. The emphasis is on the manipulation of nearly spherical objects or on cylindrical objects (such as liquid bridges) in situations where long-wavelength approximations can be used to evaluate the scattering of sound by the object. Some of the applications noted are in areas where the authors have contributed to progress; other applications that are summarized concern recent developments by other groups. Topics and derivations not included in this review have been summarized in other recent reviews.[1–3] The emphasis is on the radiation pressure of standing waves, since it can be shown that ordinarily more acoustical power is needed to provide a given radiation force when traveling waves are used.

RADIATION FORCE ON SMALL COMPRESSIBLE SPHERES IN A STANDING WAVE

For many situations of interest, the direction of the radiation force on a compressible object in an acoustic standing wave can be predicted by applying the results of

Address for correspondence: Philip L. Marston, Department of Physics, Washington State University, Pullman, WA 99164-2814, USA. Voice: 509-335-5343; fax: 509-335-7816.
marston@wsu.edu

Ann. N.Y. Acad. Sci. 1027: 414–434 (2004). ©2004 New York Academy of Sciences.
doi: 10.1196/annals.1324.034

an analysis originally given by Yosioka and Kawasima.[4] In that analysis both the fluid object and the surroundings are taken to be inviscid and the object is a homogeneous sphere in a plane standing wave, such that the radius of the object a is much smaller than the acoustic wavelength $2\pi/k_z$, where k_z is the acoustic wave number. It is also assumed that $\omega = (2\pi \times$ acoustic frequency$)$ is much less than the lowest natural radian frequency for monopole and dipole resonances of the sphere. These frequencies are denoted by ω_0 and ω_1, respectively, and the conditions $\omega \ll \omega_0$ and $\omega \ll \omega_1$ are easily met for liquid drops. Let the incident acoustic standing wave amplitude be

$$p_a(z, t) = p_s \cos[k_z(z + h)]\cos(\omega t), \tag{1}$$

where p_s is the amplitude of the acoustic standing wave of frequency $\omega/2\pi$. Let the sphere be centered at $z = 0$, with $z = -h$ denoting the location of the adjacent pressure antinode of the incident wave. Taking the z-axis to be "vertical", when h is positive the pressure antinode is a distance h below the sphere. The acoustic radiation force is the time-averaged force on the sphere. For the conditions noted, this force is along the z-axis. The result for this force (see Eqs. 61–63 in Ref. 4) to leading order in k_z may be written as follows, using a dimensionless force function Y_{st} for standing-waves:[5]

$$F_z(h) = \frac{\pi}{8}a^2 p_s^2 \beta_o Y_{st} \sin 2k_z h, \tag{2}$$

$$Y_{st} = \frac{8}{3}k_z a \left(f_1 + \frac{3}{2}K^2 f_2\right), \tag{3}$$

$$f_1 = 1 - \frac{\beta_i}{\beta_o}, \tag{4}$$

$$f_2 = 2\frac{d_1}{d_2}, \tag{5}$$

$$d_1 = q - 1, \tag{6}$$

$$d_2 = 2q + 1, \tag{7}$$

$$q = \frac{\rho_i}{\rho_o}, \tag{8}$$

where a is the radius of the sphere, the sphere is taken to be the "inner" fluid having a density ρ_i and adiabatic compressibility β_i, and the corresponding properties of the outer fluid are denoted by ρ_o and β_o. The compressibilities are related to the respective sound speeds c_i and c_o by $\beta_i = 1/\rho_i c_i^2$ and $\beta_o = 1/\rho_o c_o^2$. For the conditions of a plane standing wave considered here, the factor $K = 1$; however, this factor is shown in (3) to facilitate subsequent discussion. The system of equations (2)–(8) was also derived by Gorkov[6] who introduced the notation used here for f_1 and f_2. Inspection of (2)–(8) shows that the radiation force vanishes if the sphere is centered on a pressure antinode (e.g., when $k_z h = 0$ or π) or on a pressure node ($k_z h = \pi/2$).

One reason for writing the system (2)–(8) in the form shown is to display the similarity with certain terms that arise in Rayleigh's analysis[7] of the long wavelength far-field scattering of a plane traveling wave by an inviscid compressible sphere. This may be expressed in terms of the following dimensionless form function:

$$f(ka, \theta) = -\frac{2}{3}(ka)^2\left(f_1 - \frac{3}{2}f_2\cos\theta\right),$$ (9)

where $k = \omega/c_o$ and the normalization is such that the complex scattered pressure p_{sca} at a radius r is related to the incident pressure p_{inc} of the traveling wave by the expression

$$p_{sca} = p_{inc}f(ka, \theta)\frac{a}{2r}e^{i(kr - \omega t)}.$$ (10)

The angle θ in Equation (9) denotes the scattering angle, which vanishes in the case of forward scattering. The physical significance of this normalization is such that, in the case of reflection by a large rigid sphere of radius a calculated from ray optics, $|f(ka, \theta)| = 1$.[8] For a recent discussion of the Rayleigh result, see Roy and Apfel.[9] The terms shown are the leading terms in the low frequency (small ka) expansion of the exact partial wave series (pws) for the scattering by a compressible sphere.[9,10] The other partial waves vanish more rapidly than $(ka)^2$ for $ka \ll 1$. The terms proportional to f_1 and f_2, are respectively, the low ka approximations of the monopole and dipole terms of the pws. These are related to the radial and translational responses (or the lack thereof) of the sphere.[11] The result that the low ka approximation of Y_{st} is associated with the monopole and dipole scattering terms is discussed elsewhere.[6,12,13] An expression for the radiation force using the partial waves for the scattering follows by considering the momentum transport of the scattered wave.[12–14] Another approach for deriving (2)–(8) is to evaluate the appropriate projection of the local radiation stress at the surface of the spherical scatterer. This yields the same result.[2,4,15,16]

GORKOV'S FORMULATION FOR THE RADIATION FORCE

Gorkov[6] derived the result in (2)–(8) for plane standing waves with $K = 1$ by introducing an effective potential energy

$$U = V\left(f_1\langle PE\rangle - \frac{3}{2}f_2\langle KE\rangle\right),$$ (11)

where f_1 and f_2 are defined by Equations (4) and (5), $V = (4/3)\pi a^3$ is the volume of the sphere, $\langle PE\rangle$ is the time average of the acoustic potential energy density evaluated at the location of the center of the sphere (without the sphere present) and $\langle KE\rangle$ is the corresponding time average of the acoustic kinetic energy density. These are related to the first-order velocity u and pressure p_a of the acoustic standing wave by the following time averages:

$$\langle KE\rangle = \frac{\rho_o}{2}\langle u^2\rangle$$ (12)

$$\langle PE\rangle = \frac{\beta_o}{2}\langle p_a^2\rangle.$$ (13)

The radiation force is given by

$$\mathbf{F} = -\nabla U, \tag{14}$$

where the gradient operates on the spatial dependence of $\langle PE \rangle$ and $\langle KE \rangle$. Since $\langle PE \rangle$ and $\langle KE \rangle$ are both positive-definite functions of position, they are maximized at pressure antinodes and velocity antinodes, respectively. Several authors[17,18] have generalized this method to the case of incident waves described by the modes of a closed chamber by replacing the spatial dependence of $\langle PE \rangle$ and $\langle KE \rangle$ with those of the acoustic mode of interest. For the case of compressible spheres situated on the symmetry axis (taken to be the z-axis) of an axisymmetric mode the force reduces to (2)–(8) with $K = k_z/k$, where $k = \omega/c_o$, and for the non-plane wave mode K is less than unity but greater than zero. Ordinarily the standing wave is such that K is close to unity. Although (2)–(8) show only the axial force, Equations (11)–(14) may also be used to describe the lateral restoring force associated with displacing the sphere from the axis of the mode.[17,18] Inspection of Equation (14) shows that the sphere is attracted to minima in U. When f_1 and f_2 are both positive, the sphere is attracted to regions where $\langle KE \rangle$ is large and $\langle PE \rangle$ is small; that is, to velocity antinodes. When f_1 and f_2 are both negative, the sphere is attracted to regions where $\langle PE \rangle$ is large and $\langle KE \rangle$ is small; that is, to pressure antinodes.

EQUILIBRIUM POSITION IN NORMAL GRAVITY

The equilibrium position of the drop is found by balancing the buoyancy with F_z,

$$F_z(h) - \frac{4}{3}\pi a^3 (\rho_i - \rho_o)g = 0, \tag{15}$$

where the acceleration of gravity g is directed along the negative z-axis. Comparing Equations (2)–(8) and (15) shows that equilibrium requires a sufficiently large p_s and the equilibrium value of h does not depend on the sphere radius a, in this the $k_z a \ll 1$ limit. The force is directed away from the closest pressure antinode if Y_{st} is positive and toward it if Y_{st} is negative.

VISCOUS CORRECTIONS FOR STANDING WAVES

Doinikov[19] and others[18,20] have examined the corrections to (2)–(8) resulting from viscosity. Ordinarily, these corrections are small provided the sphere radius a is much larger than the thickness scale δ of the viscous boundary layer that oscillates at the frequency of the acoustic wave. That thickness may be estimated from $\delta = (2\nu/\omega)^{1/2}$, where ν is the kinematic viscosity of the inner or outer fluid. At ultrasonic frequencies, typical values of δ are $1\,\mu m$. Significant deviations from (2)–(8) have been demonstrated for small polystyrene spheres in water having $a < 5\,\mu m$.[21] Viscous corrections to the radiation force are estimated to ordinarily be more significant for traveling waves.[19]

EXAMPLES AND SPECIAL CASES

It is instructive to compare several examples of acoustic radiation force in a standing wave. For convenience, the cases are listed in TABLE 1 showing whether various quantities of interest are positive or negative. The sign of F_z shown is based on the assumption that the equilibrium condition in Equation (15) is achieved in normal gravity.

1. Liquid Drops in Air[2,3,22–27]

In this case, $\beta_i \ll \beta_o$ and $\rho_i \gg \rho_o$ so that $Y_{st} = 20k_z a/3$ when $k = k_z$, in agreement with the result by King for a rigid sphere.[28] The drop is attracted to a velocity antinode (VAN) of the standing wave. In normal gravity, the drop is below the VAN and an upward radiation force is to be expected, since the Bernoulli pressure on the upper surface of the drop is less than on the lower surface, the upper surface being closer to the VAN.

2. Hydrocarbon Drops in Water[29–32]

Typical hydrocarbon liquids such as benzene are less dense and more compressible than water. In normal gravity, the drop is trapped slightly above the pressure antinode (PAN) of the standing wave.

3. Silicone Oil Drops in Water[33–35]

Although this situation is generally similar to Example 2, properties of PDMS silicone oils have made this system advantageous for research. Silicone oils are available for a range of viscosities and they may be diluted with dense liquids to raise the density and lower the standing wave amplitude needed to trap the drop.[33,34] Dyes are

TABLE 1. Signs of radiation force parameters

	Example	Y_{st}	F_z	f_1	f_2
1.	Liquid drops in air	+	+	+	+
2.	Hydrocarbon drops in H_2O	−	−	−	−
3.	Silicone oil drops in H_2O	−	−	−	−
4.	CCl_4 drops in H_2O	−	+	−	+
5.	H_2O drops in decalin	−	+	−	+
6.	Small bubbles in liquids	−	−	−	−
7.	Large bubbles in liquids	{−}	−	−	−
8.	Solid spheres in air	+	+	+	+
9.	Plastic spheres in liquids	+	±	+	±
10.	Small flames in air	−	−	−	−
11.	Dense gas in air	+	+	±	+

NOTE: A minus sign indicates the quantity is negative from Equations (2)–(8) and (15). {} indicates that the sign shown for Y_{st} may be incorrect.

available that are soluble in the oil and virtually insoluble in the water.[35] Silicone oil drops in water were manipulated with radiation pressure in reduced gravity.[36,37]

4. Carbon Tetrachloride Drops in Water[29]

This example is similar to Example 2 except that now ρ_i exceeds ρ_o so that the drop is trapped below the adjacent PAN.

5. Water Drops in Liquid Decalin[38]

This example is analogous to Example 4 and was developed for the purpose of inferring the velocity of sound in supper-cooled drops of water.[38]

6. Small Gas Bubbles in Liquids[39] and SBSL[40–44]

For Equation (2)–(8) to apply to gas bubbles in liquids, it is necessary for ω to greatly exceed the natural radian frequency for radial pulsation of the bubble, that is, approximately

$$\omega_0 \approx \left(\frac{3\gamma P_0}{\rho_o a^2}\right)^{1/2}, \tag{16}$$

where γ is the polytropic exponent (about 1.4 for sufficiently large bubbles of air) and P_0 is the hydrostatic pressure. A correction (usually small) due to surface tension has been neglected. For bubbles of air in water with $P_0 = 1$ atm, $\omega_0/2\pi \approx 3.2$ kHz/a with a in mm and $ka \approx 0.013$. When the bubble is sufficiently small, so that $\omega \ll \omega_0$, the radial response of the bubble is limited by the compressibility of the gas and the approximations leading to Equations (2)–(9) are applicable. Since gas bubbles in water are characterized by $\beta_i \gg \beta_o$, it follows that the radiation force is directed toward the pressure antinode, which may be understood by the following argument. If $V(t)$ denotes the instantaneous bubble volume, the radiation force on small bubbles is approximately the time average, $\langle V(t)\nabla p_a(z, t)\rangle$, where the pressure gradient is evaluated at $z = 0$ (the location of the center of the bubble) and p_a is the local pressure amplitude of the incident acoustic wave. During the part of the acoustic cycle when $p_a(0, t)$ is negative, the bubble volume increases by an amount $-p_a(4\pi a^3/3)\beta_i$ and at that time $-\nabla p_a$ is directed toward the pressure antinode. The direction of $-\langle V(t)\nabla p_a(z,t)\rangle$ is determined by the large-volume phase of the sinusoidal radial oscillations. This explains the sign and magnitude of the contribution to the radiation force in (2)–(8) proportional to β_i. The bubble is attracted to the PAN but comes to equilibrium above the PAN in normal gravity. The aforementioned argument also explains the importance of the contribution to F_z proportional to β_i in Examples 2 and 3. Gaitan et al.[40] discovered that small bubbles acoustically trapped close to a large amplitude pressure antinode can emit a regular light pulse during each acoustic cycle. This stable single bubble sonoluminescence (SBSL) has been widely investigated,[41] including light emission and video records of the bubble pulsation in reduced gravity conditions.[42,43] Modeling the acoustic radiation force on the bubble is complicated by the nonlinear response of the bubble to the sound.[44]

7. Large Gas Bubbles in Liquids[45–47]

When the bubble is sufficiently large that ω is not much smaller than ω_0, the response of the bubble is altered by inertia, and the approximation for f_1 in Equation

(4) is no longer applicable. Various approximations for the radiation force for sizes close-to and above resonance have been discussed.[14,45,48] For bubbles larger than resonant size, $Y_{st} > 0$ in Equation (2). When the bubble is much larger than resonant size, the radial response is limited by inertia and F_z is proportional to radius a, not a^3. Consequently, the equilibrium location h from (15) depends on a. This prediction was confirmed and used by Asaki and Marston[45,47] to verify predictions for F_z not based on (2)–(8). In normal gravity, large bubbles are trapped above the adjacent pressure node. A levitator was developed, having enhanced lateral stability to facilitate trapping of 11.5 mm diameter bubbles in normal gravity.[46] A sealed version of that levitator was developed that could be excited at two widely spaced frequencies. With that chamber, large bubbles were trapped and manipulated in microgravity on USML-1.[36,37] A levitator using a bar resonator has trapped large bubbles in a reduced-gravity drop-tower experiment.[49] The transducer was similar to one designed for liquid bridges.[50]

8. Solid Spheres in Air[51]

This situation is acoustically analogous to Example 1, since $\beta_i \ll \beta_o$. Recent advances have facilitated the levitation of spheres having densities ρ_i of 22 g/cc.[51] On the other extreme, resonators have been fabricated for the acoustic trapping of small spheres with a power requirement of less than 200 mW.[52]

9. Plastic Spheres in Liquids[53,54]

The effective compressibility β_i of a small solid sphere is related to the speeds c_L and c_T of longitudinal and transverse waves, respectively, in the solid by[9,53]

$$\beta_i = \rho_i^{-1}\left(c_L^2 - \frac{4}{3}c_T^2\right)^{-1}. \tag{17}$$

Typically for solid spheres in liquids, where $\beta_i < \beta_o$, the contribution from f_1 in (2)–(8) is not negligible. Plastic spheres in aqueous solutions are attracted to velocity antinodes.

10. Small Diffusion Flames in Air[55]

Thiessen, Wei, and Marston[55] recently demonstrated that small diffusion flames of natural gas in an ultrasonic standing wave in air experience a radiation force that attracts the flame to the adjacent PAN. This observation agrees with the direction of force predicted by (2)–(8). The hydrostatic pressure within the flame is nearly the same as that of the surrounding air although the density of the flame is much less than that of the surrounding air. It may be shown that the flame gas is typically only slightly more compressible than the surrounding air. To estimate the direction and magnitude of the low-frequency radiation pressure, the scattering of sound by the flame is approximated by modeling the flame as a homogeneous sphere of hot gas with $\rho_i/\rho_o \approx 0.2$ and $\beta_i/\beta_o \approx 1.07$. It follows, from Equation (9) that the scattering is primarily associated with the dipole term $f_2 < 0$, and from (2)–(8), that Y_{st} is negative. In this approximation, the effect of the first-order acoustic pressure on the burning rate of the flame is neglected.

11. Dense and Cold Gases in Air[56,57]

A situation that contrasts with the flame case (Example 10) is the radiation force on parcels of dense gas surrounded by air.[56] The sample parcels of gas are made denser than the surroundings, either by cooling and/or selection of the molecular weight of the gas. The dense gas is attracted to VANs as predicted by (2)–(8) with $\rho_i \gg \rho_o$. Typical values of f_2 are as large as 0.3. To maintain a small gas cloud trapped near an antinode, it is necessary to continually supply a flux of gas. The density of the trapped gas is enhanced by using high-molecular mass gases, such as octafluoro-cyclobutane (C_4F_8). The method has also been used to trap clouds of ice aerosols.[57]

BOUNDARIES, BJERKNES FORCES, AND ADIABATIC INVARIANCE

The original derivations of Equations (2)–(8) assumed a plane standing wave for the incident wave and that the acoustic wave scattered by the sphere is never reflected back to the sphere (that is, the boundaries are taken to be perfectly absorbing). Actual acoustic levitators have reflectors for the purposes of establishing the acoustic standing wave. In some cases, the reflectors are enclosed by hard boundaries so as to minimize radiation losses and the power required to excite the standing wave. The wave scattered by the sphere can be reflected from a boundary back to the sphere and that reflected wave alters the spatial dependence of the radiation force. For example, while testing rectangular water-filled resonators during the development of the USML-1 bubble dynamics experiment,[36,37,46] it was discovered that, when a flat solid surface was placed in the resonator close to a levitated bubble, the bubble would be attracted to the surface. This interaction is attributable to the secondary radiation force associated with the reflection off of the inserted surface. In that case, to meet the boundary conditions near a flat surface, it is convenient to introduce a fictitious image bubble having the same size as the original bubble. The interaction of two bubbles having the same radius is known to be attractive as a result of Bjerknes forces.[58] Ordinarily acoustic levitators are configured in such a way that the secondary radiation forces associated with wall reflections are negligible. For situations where boundary interactions need to be considered, in addition to the aforementioned image approach, some of the available methods include generalizations of Gorkov's approach[17] and the analysis of radiation forces based on the principle of adiabatic invariance.[59]

EQUILIBRIUM SHAPE OF LEVITATED DROPS AND BUBBLES

Acoustically levitated drops in air in normal gravity can be significantly flattened, especially in the case of large drops. The equilibrium shape is determined by a balance between capillary forces (which favor a spherical shape) and the detailed distribution of the surface acoustic radiation stress, which in most cases tends to deform a drop or bubble into an oblate shape. The balance between capillary and radiation stresses is such that the aspect ratio tends to increase for larger drops or bubbles. In reduced gravity, the radiation pressure may be primarily needed for positioning the drop, and the radiation pressure may be greatly reduced. Consequently, large drops and bubbles that are nearly spherical may be acoustically trapped in reduced gravity and then deformed upon demand using radiation pressure.[36,37,60,61] Marston *et al.*[15,62,63] developed approximations for the stress distribution and associated deformation. The equilibrium shape is assumed to be only slightly nonspherical so, for the purpose of approximating the stress distribution, the result for a *sphere* having the same volume as the drop or bubble of interest is assumed. The relevant component of the surface stress is given by[15,16,62,63]

$$S_{rr} = \frac{\beta_i - \beta_o}{2} \langle p^2 \rangle + \frac{d_1 \rho_o}{2} (q \langle u_\theta^{(i)2} \rangle + \langle u_r^{(i)2} \rangle). \qquad (18)$$

where the sign convention is such, that when S_{rr} is positive, the force-per-area on the interface is radially outward, p is the first-order acoustic pressure at the interface (which is taken to be continuous across the interface), and u denotes the first order acoustic velocity that has the indicated radial (r) and polar (θ) components. Note that the local radiation pressure p_r is $-S_{rr}$. The other quantities in Equation (18) are defined as in (2)–(8) and it is noteworthy that the stress has been reduced to a form where the only velocity components needed are the inner radial and tangential components at the interface. All of the first-order quantities are found by analyzing the interaction of the acoustic standing wave with an incompressible inviscid sphere. The deformation is approximated by evaluating projections of S_{rr} with spherical harmonic functions $Y_{nm}(\theta, \varphi)$. For the purposes of this review, we restrict attention to the case of a drop or bubble levitated along the axis of an axisymmetric standing wave, so that the only nonvanishing projections have $m = 0$. It is convenient to introduce the following projection notation:

$$p_{n0} = \int_0^\pi S_{rr}(\theta) Y_{n0}(\theta, 0) \sin\theta d\theta, \qquad (19)$$

where, when the corresponding quantity was introduced by Marston,[62] the notation $\hat{p}_{n0}$ was used. Since $Y_{10}(\theta, 0) = (3/4\pi)^{1/2}\cos\theta$, the dipole projection p_{10} is proportional to the radiation force on the sphere, which is given by

$$F_z = 2\pi a^2 \int_0^\pi S_{rr} \cos\theta \sin\theta d\theta. \qquad (20)$$

As previously noted, evaluation of Equation (20) for a sphere in the plane standing wave given by Equation (1) gives (2)–(8) with $K = 1$.[15] The most important stress projection that couples with the shape of a drop or a bubble is the quadrupole stress projection, given by (19) with $n = 2$. Low frequency expansions in ka for p_{20} have been given for drop-like objects[15,63] and for bubbles of gas in liquids.[47] Except when the drop or bubble is on a velocity node, the dominate term is

$$p_{20} \approx -p_s^2 \beta_o (5\pi)^{1/2} \left(\frac{3}{5}\right) d_{1,2}^2 \sin^2 k_z h, \qquad (21)$$

where $d_{1,2} = d_1/d_2$, with d_1, d_2, and the other symbols as defined in (2)–(8), and it has been assumed that $k_z \approx k = \omega/c_o$. The projection approximated in (21) has a maximum magnitude on a VAN, where $k_z h = \pi/2$. For a liquid drop in air, $d_{1,2} \approx 1/2$ and the leading order acoustic correction[15,63] introduces a factor of $[1 + (7/5)(k_z a)^2]$ on the right side of (21) for drops near a VAN.

To approximate the deformation of the fluid object, the equilibrium shape is expanded using spherical harmonics Y_{n0} which for the axisymmetric deformations expresses the surface profile as

$$r(\theta) = a + \sum_{n=0}^{\infty} x_{n0} Y_{n0}(\theta, 0), \tag{22}$$

where the small shape changes are described by the terms with index $n > 1$. A balancing of the purely radial radiation stress with capillary stresses resulting from deformation gives

$$x_{n0} = \frac{a^2 p_{n0}}{(n+2)(n-1)\sigma}, \tag{23}$$

where σ denotes the interfacial tension. An oblate drop or bubble corresponds to one having a negative value of x_{n0} for $n = 2$. Since $Y_{20} = (5/4\pi)^{1/2} P_2(\cos\theta)$, where P_2 is a Legendre polynomial of the indicated argument, the aspect ratio becomes

$$\frac{D}{H} = \frac{r(\pi/2)}{r(0)} = (1+B)(1-2B) \tag{24}$$

for a drop or bubble, where D is the equatorial diameter and H is the height span. From (21)–(23) the normalized acoustic Bond number is

$$B = \left(1 + \frac{7}{5}(k_z a)^2\right) p_s^2 \frac{3\beta_o a}{16\sigma} d_{1,2}^2 \sin^2 k_z h. \tag{25}$$

The aforementioned $(k_z a)^2$ correction is shown for the case of a drop in air. Trinh and Hsu[22] found this prediction to agree approximately with measurements for small drops of volume $(4/3)\pi a^3$ in a calibrated acoustic levitator. Asaki and Marston confirmed the analogous result for bubbles levitated in water.[47] The deformation can be explained in the drop and bubble cases in terms of the pressure reduction near the equator associated with flow relative to a sphere.[3,15,63] The $n = 3$ projection has been calculated[15] for a compressible sphere in a plane standing wave for $ka \ll 1$,

$$p_{30} \simeq -p_s^2 \beta_o ka \pi \left(\frac{\pi}{7}\right)^{1/2} \left(\frac{3qd_{1,2}}{2+3q}\right) \sin 2kh, \tag{26}$$

which vanishes from symmetry at pressure nodes and antinodes. For both the $n = 2$ and 3 cases, the $ka \ll 1$ limits were confirmed by the computations of Jackson et al.,[16] who also considered the projections on spheres when ka is not small. It is not difficult for drops in air to be sufficiently oblate that the results need to be modified to allow for shape change in the evaluation of the radiation stress distribution. Approaches to that problem are discussed by Anilkumar et al.,[64] Shi and Apfel,[65] Yarin et al.,[66] and Lierke.[67] Deformation changes the radiation force because the dipole scattering contribution, f_2 in Equation (11), is altered.[11] For sufficiently large pressure amplitudes, the drop becomes unstable and emits satellite drops.

QUADRUPOLE STRESS PROJECTION ON A FLAME

Thiessen et al.[55] observed that, when a small diffusion flame is positioned at the PAN of an ultrasonic standing wave, the circular profile becomes elliptical. (As previously mentioned in Example 10 of TABLE 1, the flame is attracted to a PAN.) The minor axis of the ellipse lies along the propagation direction of the standing wave, the z-axis in (1), so this deformation is described as "oblate" in analogy to the deformation of drops. On the PAN, $h = 0$ so the contribution to p_{20} shown in Equation (21) vanishes and it is necessary to evaluate a term proportional to $\cos^2 k_z h$, given by Marston et al.[15,63] (see Eq. (2) of Ref. 63). Using the aforementioned, hot-low-density gas approximation for a small diffusion flame gives the prediction that p_{20} is negative, so that the stress distribution is predicted to favor an oblate deformation.[55]

LIGHT SCATTERING BY ACOUSTICALLY DEFORMED DROPS

The ability to continuously vary the aspect ratio of acoustically levitated oblate drops of water enabled the discovery of complicated light scattering patterns that are produced when the drop is sufficiently oblate.[24] The brightest of the patterns are a generalization of the primary rainbow to include cases where a far-field caustic is associated with the merging of three or more rays.[68,69] Related higher-order caustics have been observed for rainbows from rays having as many as six internal chords[70,71] and the resulting caustic structure is extremely sensitive to the shape of the drop.

SHAPE OSCILLATIONS DRIVEN BY
MODULATED RADIATION PRESSURE

As a consequence of the relative magnitudes of surface and volume elastic strain energies, the frequencies of the capillary oscillations of liquid drops ordinarily occur at frequencies much less than the frequencies of the lowest lying acoustic modes. Neglecting viscous corrections gives Lamb's expression[72] for the frequency of the mode associated with the nth spherical harmonic in Equation (22). For mm-radius drops or bubbles, the lowest capillary mode, which is the $n = 2$ quadrupole mode, typically has a frequency less than 200 Hz. Because that frequency is so low, an acoustic wave tuned to that frequency couples weakly to the shape of the drop or bubble. Wang et al.[73] reported that it was possible to excite the $n = 2$ mode of drops in an acoustic standing wave by modulating the amplitude of the acoustic wave. Marston and Apfel[31,32] introduced this type of coupling as a temporal modulation of the radiation stress projections in (19). They confirmed that for immiscible hydrocarbon drops surrounded by water, the modulation frequency for optimum coupling varied with drop radius in a way generally consistent with Lamb's theory. They also demonstrated that drop fission resulted by driving large amplitude oscillations of the $n = 2$ mode. Double-sideband, suppressed-carrier modulation was used to drive the oscillations. A different acoustic field was used to levitate the drop. For the case of drops surrounded by a liquid, the dissipation in the boundary layers within and on the outside of the drop (which oscillate at the frequency of the radiation pressure modulation) can be important in limiting the response of the drop, as discussed by Marston.[62] Other studies of radiation pressure induced shape oscillations of drops include References 15, 25, 33–35, and 74. Although the original analysis[62] included the case of bubbles, it was more than a decade later that observations for bubbles were published.[46] The method was used in microgravity on USML-1 to drive $n = 2$ and $n = 3$ mode oscillations of bubbles having diameters as large as 14.9 mm. One important result obtained by comparing the mode frequencies in microgravity and in normal gravity was that in normal gravity the frequencies were depressed.[36,37,46] This was evidently associated with the oblate shape of large bubbles levitated in normal gravity.[47] To achieve conditions in normal gravity close to those assumed in the analysis by Lamb,[72] it became clear that smaller bubbles would need to be studied. In response to this observation methods were developed[75] for measuring the natural frequency and damping of smaller bubbles. For smaller bubbles, however, the damping is increased, so to calculate the damping it was necessary to improve upon the prior results.[62,74,76] The damping expressions derived in Reference 75 are also useful for viscous drops in a viscous liquid.

INTERFACIAL RHEOLOGY WITH SURFACTANTS

The presence of surfactants alters the interfacial rheology of drops and bubbles. The consequences include enhanced damping and a shifted natural frequency, as well as major modifications to the dynamics when the deformation is large. The drop physics module (DPM) flown in microgravity on USML-1 and USML-2 facilitated studies of the oscillations of large drops in air.[60,61] Experiments with small drops levitated in normal gravity were helpful in identifying parameter ranges of interest.[26] The DPM also enabled the response of large drops to acoustic radiation torques to be investigated.[77] Acoustic levitation has also enabled the effects of surfactants on the capillary modes of bubbles to be investigated. The $n = 2$ and 3 capillary frequencies were measured for an acoustically trapped bubble in reduced gravity on USML-1 in water following the injection of the soluble surfactant SDS. The frequencies were lowered in a manner consistent with theory.[36] A method of recording shape oscillations based on the extinction of light enabled measurements of frequency and damping to be carried out with small bubbles in normal gravity.[75,78,79] After spreading an insoluble surfactant (stearic acid) on a levitated bubble, the surface concentration increases in a known way as the bubble dissolves.[78] In agreement with predictions, a damping maximum was observed associated with a certain stearic acid surface concentration.

RESPONSE OF LIQUID BRIDGES TO RADIATION PRESSURE

A liquid bridge is a column of liquid suspended between two solid supports. The dynamics of liquid bridges have been widely investigated in real and simulated reduced gravity because of potential applications to materials processing and because of fundamental instabilities exhibited by long bridges, such as the Rayleigh–Plateau capillary instability that causes long liquid columns to break up into drops.[80–89] A project was established at Washington State University to use acoustic radiation pressure to stabilize liquid bridges. Some of the underlying theory and some of the results of this project are now summarized.[50,85–87] Related research is being carried out using electrostatic stresses instead of acoustic radiation stresses.[87–89] To summarize the experimental methods and results it is convenient to introduce certain parameters related to the acoustic radiation pressure on a circular column of compressible fluid in a standing wave. For the purposes of discussing the radiation pressure, it is sufficient to limit attention to the two-dimensional situation, where an infinitely long circular cylinder of radius R resides in a plane standing wave. The axis of the column is parallel to the nodal planes of the incident standing wave. To describe the transverse force on the cylinder, it is convenient to introduce a dimensionless force function for cylinders in a standing wave, Y_{stc}. Let the incident wave be given by Equation (1) with the cylinder centered on $z = 0$. In analogy with Equations (2) and (20), the radiation force on a segment of cylinder of length L is

$$F_z = \frac{R}{4} L p_s^2 \beta_o Y_{stc} \sin 2k_z h, \qquad (27)$$

$$F_z = 2RL \int_0^\pi S_{rr} \cos \theta d\theta, \qquad (28)$$

where S_{rr} is given by Equation (18) and θ denotes the polar angle of a patch of surface relative to the z-axis. In the limit $k_z R \ll 1$, the long wavelength limit, we find that

$$Y_{stc} \approx \pi k_z R (f_1 + f_{2c}), \qquad (29)$$

$$f_{2c} \approx \frac{2d_1}{q + 1}, \qquad (30)$$

where f_1, d_1 and q are defined as in (2)–(8) and attention is limited to the plane wave case, where $k_z = \omega/c_o$. In the case of an incompressible cylinder $f_1 = 1$ and the result in Equations (27), (29), and (30) is equivalent to a result given by Wu et al. (the first part of Eq. (24) in Ref. 90). The modification for finite compressibility, given by including the β_i/β_o term in Equation (4) follows by comparing the monopole and dipole terms for low frequency scattering by cylinders. The result in Equations (29) and (30) is important because it may be used to predict the direction of the acoustic radiation force. When Y_{stc} is positive, as in the incompressible case, the cylinder is attracted to a VAN. When Y_{stc} is negative, as in the case of a highly compressible cylinder, the cylinder is attracted to a PAN.

It is convenient to introduce another parameter relevant to the stabilization of liquid bridges. The local radiation pressure $p_r = -S_{rr}$, where S_{rr} is given by Equation (18), depends on the radius R of the cylinder. For the purpose of stabilizing the bridge, it is sufficient to consider the surface average of that dependence, written here as an angular average,

$$\langle p_r \rangle_\theta = \frac{1}{2\pi} \int_0^{2\pi} p_r d\theta. \qquad (31)$$

The new dimensionless acoustic parameter is

$$G_a = \frac{R^2}{\sigma} \frac{d \langle p_r \rangle_\theta}{dR}, \qquad (32)$$

where σ is the surface tension. In the notation used elsewhere,[86] the symbol q was used for G_a. If the radiation pressure tends to squeeze in more when R is large (or equivalently pull radially outward more when R is small) then G_a is positive. The central idea of the suppression of the Rayleigh–Plateau capillary instability of liquid columns is to select the sound field such that G_a is sufficiently positive and the radiation stress pulls outward on the narrow parts of the liquid column while pushing inward on the more rotund parts. Acoustic stabilization has been achieved *passively* through the intrinsic dependence of $\langle p_r \rangle$ on the local radius[86,87] and *actively* by sensing optically where the liquid column is narrow and where it is rotund. The active method uses the shape information to adjust the sound field to bring about the desired change in the local $\langle p_r \rangle$.[85]

To describe the capillary modes of a liquid column of length L suspended between solid circular disks of radius R, it is convenient to restrict attention to the case where the volume of liquid in the column is $V = \pi R^2 L$, which is that of a circular cylinder. Furthermore, it is assumed that the effects of gravitational acceleration are negligible, either because of reduced gravity or because a Plateau tank is used. Let x denote the axial coordinate with $x = 0$ and L denoting the ends of the bridge. Let the bridge modes be described by an axial index N and an azimuthal index m. When the deformation is small, the column surface may be approximated as follows for the purposes of this discussion:

$$r(\theta, x, t) \approx R + \sum_{N, m} X_{Nm}(t) \sin\left(\frac{\pi N x}{L}\right) \cos(m\theta), \qquad (33)$$

where $X_{Nm}(t)$ is an instantaneous mode amplitude. The precise description of the capillary modes of a liquid bridge is complicated because of the fixed-volume condition,[50,83] however, the form used in (33) captures the sense of the deformation. Modes having $m = 0$ are axisymmetric. The dimensionless slenderness of the bridge is

$$S = \frac{L}{2R}. \qquad (34)$$

The natural frequency ω_{20} of the (2,0) capillary mode decreases with increasing S until $S = \pi$, where ω_{20} vanishes. The mode is naturally unstable for $S > \pi$ as a consequence of the Rayleigh–Plateau instability. The modes having $m \neq 0$ are non-axisymmetric and are not normally unstable in the parameter range of interest. To facilitate the discussion of our prior experiments and other feasible effects of radiation pressure on liquid bridges, TABLE 2 shows the signs of relevant parameters for various examples.

TABLE 2. Signs of bridge acoustic parameters

	Example	Y_{stc}	G_a	f_1	f_{2c}
1.	Silicone oil + TBE in H_2O, selective mode excitation	−	negligible	−	0
2.	Silicone oil + TBE in H_2O, active acoustic stabilization	−	+	−	0
3.	Liquid bridge in air, passive acoustic stabilization	+	+	+	+
4.	Liquid bridge in air, (1,1) mode stiffening	+	+	+	+

1. Silicone Oil Bridges in Water: Excitation with Modulated Radiation Pressure

To verify that it is feasible to selectively excite capillary modes of liquid bridges using modulated ultrasonic radiation pressure, Morse *et al.*[50] deployed bridges of a mixture of silicone oil (5cS PDMS) and tetrabromoethane (TBE), density matched to the surrounding water. A 120-kHz ultrasonic standing wave was established that could be modulated at low frequencies. As a consequence of the matched density, $f_{2c} = 0$. Because the compressibility of the bridge mixture exceeded that of water, Y_{stc} is negative and the bridge was deployed at a pressure antinode of the standing wave for lateral stability. The predicted capillary mode frequency as a function of the slenderness S was verified over a range of values for S for the (1,1), (1,2), (2,0), and (3,0) modes. This appears to be the only experiment in which the (1,2) mode of a bridge has been observed, since that mode is not easily excited by vibration of the supports. The experimental configuration was similar to an ordinary Plateau tank except for the presence of an ultrasonic transducer in contact with the outer water bath. The intrinsic value of G_a had a negligible effect on the dynamics of the (2,0) mode.

2. Silicone Oil Bridges in Water: Active Acoustic Stabilization[85]

The bridge mixture and bath were the same as in Example 1 but the ultrasonic transducer was modified to facilitate stronger coupling with the (2,0) mode and also active control of the (2,0) mode stress distribution. The bridge was horizontal and located at (or close to) a pressure antinode of the standing wave. Inspection of Equation **(33)** shows that the presence of the (2,0) mode, associated with the term X_{20}, causes a left–right (LR) asymmetry in the profile of the bridge when the bridge is viewed from the side. In this experiment, the LR asymmetry was detected using a light extinction method and was used to control a LR asymmetry in the ultrasonic field. The asymmetry in the ultrasonic field was associated with a LR asymmetry in the angular average of the radiation pressure on the bridge. As a consequence of the active control, the stress imbalance is proportional to (2,0) mode amplitude X_{20}. The effect on the dynamics of the (2,0) mode is equivalent to a finite value of the acoustic parameter G_a in Equation **(32)**. When the phase of the feedback is selected so that G_a is positive, the Rayleigh–Plateau instability is suppressed and the limiting slenderness for a bridge having a volume of $\pi R^2 L$ is predicted to increase to[86,87]

$$S_{limit} = \frac{\pi}{(1 - G_a)^{1/2}}. \tag{35}$$

Marr-Lyon *et al.*[85] confirmed that bridges having volumes close to $\pi R^2 L$ could be stabilized in this way out to an S value of 4.3. This bridge is significantly longer than the Rayleigh–Plateau limit since $4.3/\pi \approx 1.37$, and from **(35)** requires $G_a > 0.466$. Since only the (2,0) mode was controlled, unstable growth of the (3,0) mode should cause bridges to break if S reaches 4.5.[82,88] Bridges having $S \approx 3.8$ remained stable for up to 45 minutes. Increasing the viscosity of the bridge liquid would increase the damping of the (2,0) mode and decrease the sensitivity of the control system to fluctuations. In addition, the required bandwidth of the feedback control system would be reduced. Active control was also used to shift the frequency of the (2,0) mode for bridges in the naturally stable region, $S < \pi$.[87] When the feedback was phased to give $G_a > 0$, the (2,0) mode was stiffened and the natural frequency of the (2,0) mode could be significantly increased. When the feedback was phased to give $G_a < 0$, the natural frequency of the (2,0) mode could be significantly decreased.

3. Liquid Bridges in Air: Passive Acoustic Stabilization on the KC-135[86]

For the stabilization experiments considered in Example 2, it was necessary to optically sense the (2,0) mode amplitude and to use that information to actively adjust the acoustic field. Another way to achieve significantly positive values for the parameter G_a in (32) is to select the acoustic properties of the surrounding fluid and the bridge such that G_a is positive with a constant acoustic excitation.[86] Evaluation of the integral, Equation (31), for a circular liquid bridge placed on a velocity anti-node in air shows that $\langle p_r \rangle_\theta$ increases with increasing bridge radius R over the range kR from 0.4 to 1.2, where k is the acoustic wave number. The optimum condition is $kR = 0.86$ and the stabilizing influence of the acoustic field is not significantly altered by elliptical deformation of the bridge (see below) from the circular shape used in the original calculations. (The stress distribution on circular cylinders involves partial waves described by Bessel and Hankel functions, whereas the corresponding expansion for an elliptical cylinder involves Mathieu functions.[86,87]) Since $\rho_i \gg \rho_o$, f_{2c} in (29) and (30) is positive and the bridge is attracted to a VAN. This is analogous to the trapping of liquid drops in air (Example 1 of TABLE 1). The liquid column has transverse stability because it has been deployed on a VAN. Attraction of liquid parcels in air to the VAN, implied by (11), is consistent with the suppression of (2,0) capillary mode instability. In order to deploy liquid bridges having an acceptably small value for the gravitational Bond number $B_g = (\rho_i - \rho_o)g_e R^2/\sigma$, where g_e is the effective acceleration of gravity, observations were carried out on NASA KC-135 aircraft. The acoustic frequencies $\omega_a/2\pi$ were 22.1 kHz and 29.4 kHz with R of 2.16 mm and 1.62 mm (giving kR of 0.84 and 0.87). Liquid bridges were stabilized out to $S = 4.5$, which, from Equation (35), corresponds to $G_a > 0.513$. Video records of numerous examples of stabilized bridges have been archived.[87] It appears to be possible, at least in principle, to suppress the unstable growth of all axisymmetric capillary modes $(N,0)$ with $N = 2, 3, \ldots$ by selecting acoustic conditions with $G_a > 1$. For the typical bridge[86] this would require $p_a = 3.2$ kPa (161.2 dB relative to 20 μPa sound-pressure level) in (1). Tests on the KC-135 were limited to a few seconds of stabilization because of the time required to deploy the bridge in reduced gravity conditions. Consequently, it is reasonable to ask whether an analogous passive acoustic stabilization (PAS) may be achieved in a Plateau tank. Unfortunately, the specific stabilization mechanism is suppressed for liquid bridges in a Plateau tank for the following reason. Evaluation of Equation (31) shows that in the kR range of interest, the dipole term of the acoustic partial wave series dominates the contribution to $\langle p_r \rangle_\theta$.[86,87] For small kR, however, the dipole contribution is proportional to f_{2c} in (30), which vanishes for bridges in a Plateau tank. Extended-duration reduced gravity is needed to explore the limitations of the PAS method of liquid bridge stabilization. A secondary response to the standing wave is the elliptical deformation of the bridge, which is analogous to the oblate drop deformation described by Equation (24). This deformation was used to estimate the standing wave amplitude.[86,87] The elliptical deformation and the type of liquid used only weakly affect the surface-averaged radiation stress.

4. Liquid Bridges in Air: Acoustic Stiffening of the (1,1) Mode

Although measurements are not presently available to confirm the following effect, it is discussed to illustrate another application of radiation pressure. The (1,1) mode of a liquid bridge in air involves transverse oscillations of the center-of-mass of the bridge. In the absence of an acoustic field, the restoring force of (1,1) mode oscillations is surface tension. Even for an inviscid bridge, calculation of the (1,1) mode frequency is complicated.[50,83] For long bridges, however, the radian frequency is approximately that of a string of length L with a tension $(2\pi R\sigma)/2 = \pi R\sigma$ and mass-per-length of $\pi R^2 \rho_i$

$$\omega_{1,1} \approx \frac{2\pi}{4S}\left(\frac{\sigma}{\rho_i R^3}\right)^{1/2}, \tag{36}$$

where $S = L/2R$ is the slenderness. In the effective tension, πR is one-half of the perimeter of the bridge. Placing the bridge on a VAN of an acoustic standing wave provides an additional transverse restoring force that raises the (1,1)-mode frequency. The restoring force-per-length is found to be linear in the displacement from the VAN by expanding $\sin(k_z h)$ in (27) about $k_z h = \pi/2$. The resulting spring constant per-unit-length for the radiation force is

$$K_{spring} = \frac{k_z R}{2} p_s^2 \beta_o Y_{stc}. \tag{37}$$

The equation of motion for transverse motion of the bridge is approximately that of a stretched string having a uniform elastic support, which is known to be a Klein–Gordon form of partial differential equation. The lowest mode having vanishing displacements at the supports, at $x = 0$ and $x = L$, has the radian frequency

$$\omega'_{1,1} = \omega_{1,1}\left(1 + \left(\frac{L}{\pi}\right)^2 \frac{K_{spring}}{\pi R\sigma}\right)^{1/2}, \tag{38}$$

where $\omega_{1,1}$ is the unperturbed frequency from Equation (37). For bridges in air, $f_1 + f_{2c} \approx 3$, and when $k_z R \ll 1$, (29) and (30) give $Y_{stc} \approx 3\pi k_z R$. For an acoustic amplitude p_s sufficient to give significant stabilization of the (2,0) mode, (38) implies that there will be a significant increase in the (1,1) mode frequency. For the situation where radiation pressure dominates the restoring force, (38) with (36) reduces to $\omega'_{1,1} \approx (K_{spring}/\pi \rho_i R^2)^{1/2}$. This limit is the cylinder analog of the natural frequency of vertical oscillations of small acoustically levitated drops discussed elsewhere.[92]

ACKNOWLEDGMENTS

We are grateful to NASA for supporting the research on liquid bridges. The bubble research was supported in part by NASA and by the Office of Naval Research. Wei Wei, a student at Washington State University, contributed to the derivation of Equations (29) and (30). A detailed derivation of Equations (29) and (30) with a comparison with the full partial-wave series was recently published.[92]

REFERENCES

1. WANG, T.G. & C.P. LEE. 1998. Radiation pressure and acoustic levitation. *In* Nonlinear Acoustics. M.F. Hamilton & D.T. Blackstock, Eds. Academic Press, San Diego.
2. LIERKE, E.G. 1996. Akustiche Positionierung. Eing umfassender Uberblick uber Grundlagen und Anwendungen. Acustica **82**: 220–237.
3. THIESSEN, D.B. & P.L. MARSTON. 1998. Principles of some acoustical, electrical, and optical manipulation methods with applications to drops, bubbles, and capillary bridges. ASME paper FEDS98-5298.
4. YOSIOKA, K. & Y. KAWASIMA. 1955. Acoustic radiation pressure on a compressible sphere. Acustica **5**: 167–173.
5. HASEGAWA, T. 1979. Acoustic radiation force on a sphere in a quasistationary wave field-theory. J. Acoust. Soc. Am. **65**: 32–40.
6. GOR'KOV, L.P. 1962. On the forces acting on a small particle in an acoustical field in an ideal fluid. Sov. Phys. Dokl. **6**: 773–775.
7. RAYLEIGH, J.W.S. 1945. The Theory of Sound. Dover, New York.
8. MARSTON, P.L. 1997. Quantitative ray methods for scattering. *In* Encyclopedia of Acoustics. M.J. Crocker Ed.: 483–492. John Wiley Press, New York.
9. ROY, R.A. & R.E. APFEL. 1990. Mechanical characterization of microparticles by scattered ultrasound. J. Acoust. Soc. Am. **87**: 2332–2341.
10. ANDERSON, V.C. 1950. Sound scattering from a fluid sphere. J. Acoust. Soc. Am. **22**: 426–431.
11. PIERCE, A.D. 1989. Acoustics. Acoust. Soc. Am., New York.
12. LOFSTEDT, R. & S. PUTTERMAN. 1991. Theory of long wavelength acoustic radiation pressure. J. Acoust. Soc. Am. **90**: 2027–2033.
13. CHEN, X.C. & R.E. APFEL. 1996. Radiation force on a spherical object in an axisymmetric wave field and its application to the calibration of high-frequency transducers. J. Acoust. Soc. Am. **99**: 713–724.
14. LEE, C.P. & T.G. WANG. 1993. Acoustic radiation force on a bubble. J. Acoust. Soc. Am. **93**: 1637–1640.
15. MARSTON, P.L., *et al.* 1982. Resonances, radiation pressure, and optical scattering phenomena of drops and bubbles. *In* Proceedings of the Second International Colloquium on Drops and Bubbles. Jet Prop. Lab. Pub. 82-7, Pasadena, CA. 166–174.
16. JACKSON, H.W., M. BARMATZ & C. SHIPLEY. 1988. Equilibrium shape and location of a liquid drop acoustically positioned in a resonant rectangular chamber. J. Acoust. Soc. Am. **84**: 1845–1862.
17. BARMATZ, M. & P. COLLAS. 1985. Acoustic radiation potential on a sphere in plane, cylindrical, and spherical standing wave fields. J. Acoust. Soc. Am. **77**: 928–945.
18. WHITWORTH, G. & W.T. COAKLEY. 1992. Particle column formation in a stationary ultrasonic-field. J. Acoust. Soc. Am. **91**: 79–85.
19. DOINIKOV, A.A. 1997. Acoustic radiation force on a spherical particle in a viscous heat-conducting fluid. III. Force on a liquid drop. J. Acoust. Soc. Am. **101**: 731–740.
20. WEISER, M.H. & R.E. APFEL. 1982. Extension of acoustic levitation to include the study of micron-size particles in a more compressible host liquid. J. Acoust. Soc. Am. **71**: 1261–1268.
21. YASUDA, K. & T. KAMAKURA. 1997. Acoustic radiation force on micrometer-size particles. Appl. Phys. Lett. **71**: 1771–1773.
22. TRINH, E.H. 1985. Compact acoustic levitation device for studies in fluid dynamics and material science in the laboratory and microgravity. Rev. Sci. Instrum **56**: 2059–2065.
23. TRINH, E.H. & C.J. HSU. 1986. Equilibrium shapes of acoustically levitated drops. J. Acoust. Soc. Am. **79**: 1335–1338.
24. MARSTON, P.L. & E.H. TRINH. 1984. Hyperbolic umbilic diffraction catastrophe and rainbow scattering from spheroidal drops. Nature **312**: 529–531.
25. TRINH, E.H., P.L. MARSTON & J.L. ROBEY. 1988. Acoustic measurement of the surface tension of levitated drops. J. Coll. Interface Sci. **124**: 95–103.

26. TIAN, Y., R.G. HOLT & R.E. APFEL. 1997. Investigation of liquid surface rheology of surfactant solutions by droplet shape oscillations: experiment. J. Coll. Interface Sci. **187:** 1–10.
27. YARIN, A.L., D.A. WEISS, G. BRENN & D. RENSINK. 2002. Acoustically levitated drops: drop oscillation and break-up driven by ultrasound modulation. Int. J. Multiphase Flow **28:** 887–910.
28. KING, L.V. 1934. On the acoustic radiation pressure on spheres. Proc. Roy. Soc. Lond. **147A:** 212–240.
29. CRUM, L.A. 1971. Acoustic force on a liquid droplet in an acoustic stationary wave. J. Acoust. Soc. Am. **50:** 157–163.
30. APFEL, R.E. 1976. Technique for measuring the adiabatic compressibility, density, and sound speed of submicroliter liquid samples. J. Acoust. Soc. Am. **59:** 339–343.
31. MARSTON, P.L. & R.E. APFEL. 1979. Acoustically forced shape oscillations of hydrocarbon drops levitated in water. J. Coll. Interface Sci. **68:** 280–286.
32. MARSTON, P.L. & R.E. APFEL. 1980. Quadrupole resonance of drops driven by modulated acoustic radiation pressure-Experimental properties. J. Acoust. Soc. Am. **67:** 27–37.
33. TRINH, E.H., A. ZWERN & T.G. WANG. 1982. An experimental study of small amplitude drop oscillations in immiscible liquid systems. J. Fluid Mech. **115:** 453–474.
34. TRINH, E. & T.G. WANG. 1982. Large-amplitude free and driven drop-shape oscillations: experimental observations. J. Fluid Mech. **122:** 315–338.
35. MARSTON, P.L. & S.G. GOOSBY. 1985. Ultrasonically stimulated low-frequency oscillation and breakup of immiscible liquid drops: photographs. Phys. Fluids **28:** 1233–1242.
36. MARSTON, P.L., E.H. TRINH, J. DEPEW & T.J. ASAKI. 1994. Response of bubbles to ultrasonic radiation pressure: dynamics in low gravity and shape oscillations. *In* Bubble Dynamics and Interface Phenomena. J.R. Blake, *et al.*, Eds.: 343–353. Kluwer.
37. MARSTON, P.L., E.H. TRINH, J. DEPEW & T.J. ASAKI. 1994. Oscillatory dynamics of single bubbles and agglomeration in a sound field in microgravity. *In* Joint Launch + One Year Science Review of USML-1 and USMP-1 with the Microgravity Measurement Group, Vol. 2. N. Ramachandran, *et al.*, Eds.: 673–690. NASA C-P 3272, Marshall Space Flight Center, AL.
38. TRINH, E. & R.E. APFEL. 1980. Sound velocity of supercooled water down to −33°C using acoustic levitation. J. Chem. Phys. **72:** 6731–6735.
39. ELLER, A. 1968. Force on a bubble in a standing acoustic wave. J. Acoust. Soc. Am. **43:** 170–171.
40. GAITAN, D.F., L.A. CRUM, C.C. CHURCH & R.A. ROY. 1992. Sonoluminescence and bubble dynamics for a single, stable, cavitation bubble. J. Acoust. Soc. Am. **91:** 3166–3183.
41. BRENNER, M.P., S. HILGENFELDT & D. LOHSE. 2002. Single-bubble sonoluminescence. Rev. Mod. Phys. **74:** 425–484.
42. MATULA, T.J. 2000. Single-bubble sonoluminescence in microgravity. Ultrasonics **38:** 559–565.
43. THIESSEN, D.B., B. DZIKOWICZ & P.L. MARSTON. 2001. Simultaneous video and light emission records of single bubble sonoluminescence subjected to variable acceleration. Proceedings of the 17th International Congress on Acoustics-CD. Rome.
44. MATULA, T.J., S.M. CORDRY, R.A. ROY & L.A. CRUM. 1997. Bjerknes force and bubble levitation under single-bubble sonoluminescence conditions. J. Acoust. Soc. Am. **102:** 1522–1527.
45. ASAKI, T.J. & P.L. MARSTON. 1994. Acoustic radiation force on a bubble driven above resonance. J. Acoust. Soc. Am. **96:** 3096–3099.
46. ASAKI, T.J., P.L. MARSTON & E.H. TRINH. 1993. Shape oscillations of bubbles in water driven by modulated ultrasonic radiation pressure: observations and detection with scattered laser light. J. Acoust. Soc. Am. **93:** 706–713.
47. ASAKI, T.J. & P.L. MARSTON. 1995. Equilibrium shape of an acoustically levitated bubble driven above resonance. J. Acoust. Soc. Am. **97:** 2138–2143.
48. CRUM, L.A. & A. PROSPERETTI. 1983. Nonlinear oscillations of gas bubbles in liquids: an interpretation of some experimental results. J. Acoust. Soc. Am. **73:** 121–127.

49. SITTER, J.S., T.J. SNYDER, J.N. CHUNG & P.L. MARSTON. 1998. Acoustic field interaction with a boiling system under terrestrial gravity and microgravity. J. Acoust. Soc. Am. **104**: 2561–2569.

50. MORSE, S.F., D.B. THIESSEN & P.L. MARSTON. 1996. Capillary bridge modes driven with modulated ultrasonic radiation pressure. Phys. Fluids **8**: 3–5.

51. XIE, W.J. & B. WEI. 2002. Dependence of acoustic levitation capabilities on geometric parameters. Phys. Rev. E **66**: 026605.

52. KADUCHAK, G., D.N. SINHA & D.C. LIZON. 2002. Novel cylindrical, air-coupled acoustic levitation/concentration devices. Rev. Sci. Inst. **73**: 1332–1336.

53. SIMPSON, H.J. & P.L. MARSTON. 1995. Ultrasonic four-wave mixing mediated by an aqueous suspension of microspheres: theoretical steady-state properties. J. Acoust. Soc. Am. **98**: 1731–1741.

54. KWIATKOWSKI, C.S. & P.L. MARSTON. 1998. Resonator frequency shift due to ultrasonically induced microparticle migration in an aqueous suspension: observations and model for the maximum frequency shift. J. Acoust. Soc. Am. **103**: 3290–3300.

55. THIESSEN, D.B., W. WEI & P.L. MARSTON. 2003. Some responses of small diffusion flames to ultrasonic radiation. Seventh International Workshop on Microgravity Combustion and Chemically Reacting Systems. 321-324. NASA/CP-2003-212376.

56. TUCKERMANN, R., B. NEIDHART, E.G. LIERKE & S. BAUERECKER. 2002. Trapping of heavy gases in stationary ultrasonic fields. Chem. Phys. Lett. **363**: 349–354.

57. BAUERECKER, S. & B. NEIDHART. 1998. Formation and growth of ice particles in stationary ultrasonic fields. J. Chem. Phys. **109**: 3709–3712.

58. CRUM, L.A. 1975. Bjerknes forces on bubbles in a stationary sound field. J. Acoust. Soc. Am. **57**: 1363–1370.

59. PUTTERMAN, S., J. RUDNICK & M. BARMATZ. 1989. Acoustic levitation and the Boltzmann–Ehrenfest principle. J. Acoust. Soc. Am. **85**: 68–71.

60. WANG, T.G., A.V. ANILKUMAR & C.P. LEE. 1996. Oscillations of liquid drops: results from USML-1 experiments in space. J. Fluid Mech. **308**: 1–14.

61. APFEL, R.E., et al. 1997. Free oscillations and surfactant studies of superdeformed drops in microgravity. Phys. Rev. Lett. **78**: 1912–1915.

62. MARSTON, P.L. 1980. Shape oscillation and static deformation of drops and bubbles driven by modulated radiation stresses—theory. J. Acoust. Soc. Am. **67**: 15–26; erratum **71**: 511-512, 1982.

63. MARSTON, P.L., S.E. LOPORTO & G.L. PULLEN. 1981. Quadrupole projection of the radiation pressure on a compressible sphere. J. Acoust. Soc. Am. **69**: 1499–1501.

64. ANILKUMAR, A.V., C.P. LEE & T.G. WANG. 1993. Stability of an acoustically levitated and flattened drop: an experimental study. Phys. Fluids A **5**: 2763–2774.

65. SHI, W.T. & R.E. APFEL. 1996. Deformation and position of acoustically levitated liquid drops. J. Acoust. Soc. Am. **99**: 1977–1984.

66. YARIN, A.L., M. PFAFFENLEHNER & C. TROPEA. 1998. On the acoustic levitation of droplets. J. Fluid Mech. **356**: 65–91.

67. LIERKE, E.G. 2002. Deformation and displacement of liquid drops in an optimized acoustic standing wave levitator. Acta Acustica. **88**: 206–217.

68. MARSTON, P.L. 1985. Cusp diffraction catastrophe from spheroids: generalized rainbows and inverse scattering. Opt. Lett. **10**: 588–590.

69. NYE, J.F. 1992. Rainbows from ellipsoidal water drops. Proc. Roy. Soc. A **438**: 397–417.

70. MARSTON, P.L. & G. KADUCHAK. 1994. Generalized rainbows and unfolded glories of oblate drops: organization for multiple internal reflections and extension of cusps into Alexander's dark band. Appl. Opt. **33**: 4702–4713.

71. LANGLEY, D.S. & P.L. MARSTON. 1998. Generalized tertiary rainbow of slightly oblate drops: observations with laser illumination. Appl. Opt. **37**: 1520–1526.

72. LAMB, H. 1945. Hydrodynamics. Dover, N.Y.

73. WANG, T.G., M.M. SAFFREN & D.D. ELLEMAN. 1977. Drop dynamics in space. Progr. Astronaut. Aeronaut. **52**: 151–172.

74. HSU, C.J. & R.E. APFEL. 1987. Model for the quadrupole oscillations of drops for determining interfacial tension. J. Acoust. Soc. Am. **82**: 2135–2144.

75. ASAKI, T.J. & P.L. MARSTON. 1995. Free decay of shape oscillations of bubbles acoustically trapped in water and sea water. J. Fluid Mech. **300**: 149–167.

76. MILLER, C.A. & L.E. SCRIVEN. 1968. The oscillations of a fluid droplet immersed in another fluid. J. Fluid Mech. **32:** 417–435.
77. LEE, C.P., A.V. ANILKUMAR, A.B. HMELO & T.G. WANG. 1998. Equilibrium of liquid drops under the effects of rotation and acoustic flattening: results from USML-2 experiments in Space. J. Fluid Mech. **354:** 43–67.
78. ASAKI, T.J., D.B. THIESSEN & P.L. MARSTON. 1995. Effect of an insoluble surfactant on capillary oscillations of bubbles in water: observations of a maximum in the damping. Phys. Rev. Lett. **75:** 2686–2689.
79. ASAKI, T.J. & P.L. MARSTON. 1997. The effects of a soluble surfactant on quadrupole shape oscillations and dissolution of air bubbles in water. J. Acoust. Soc. Am. **102:** 3372–3377.
80. MARTÍNEZ, I. & J.M. PERALES. 2001. Mechanical behaviour of liquid bridges in microgravity. *In* Physics of Fluids in Microgravity, ch. 2. R. Monti, Ed.: Taylor & Francis.
81. LANGBEIN, D. 2002. Capillary Surfaces. Springer-Verlag, Berlin.
82. LOWRY, B.J. & P.H. STEEN. 1995. Capillary surfaces: Stability from families of equilibrium with application to the liquid bridge. Proc. Roy. Soc. London. A **449:** 411–439.
83. SANZ, A. & J.L. DIEZ. 1989. Non-axisymmetric oscillations of liquid bridges. J. Fluid Mech. **205:** 503–520.
84. MAHAJAN, M.P., M. TSIGE, S. ZHANG, *et al.* 2002. Resonance behavior of liquid bridges under axial and lateral oscillating total body forces. Exp. Fluids. **33:** 503–507.
85. MARR-LYON, M.J., D.B. THIESSEN & P.L. MARSTON. 1997. Stabilization of a cylindrical capillary bridge far beyond the Rayleigh-Plateau limit using acoustic radiation pressure and active feedback. J. Fluid Mech. **351:** 345–357.
86. MARR-LYON, M.J., D.B. THIESSEN & P.L. MARSTON. 2001. Passive stabilization of capillary bridges in air with acoustic radiation pressure. Phys. Rev. Lett. **86:** 2293–2296; erratum: **87:** 2099–2101.
87. MARR-LYON, M.J. 2000. Stabilization of Capillary Bridges Far Beyond the Rayleigh-Plateau Limit with Acoustic Radiation Pressure or Electrostatic Stresses. Ph.D. Thesis, Washington State University, Washington.
88. MARR-LYON, M.J., D.B. THIESSEN, F.J. BLONIGEN & P.L. MARSTON. 2000. Stabilization of electrically conducting capillary bridges using feedback control of radial electrostatic stresses and the shapes of extended bridges. Phys. Fluids **12:** 986–995.
89. THIESSEN, D.B., M.J. MARR-LYON & P.L. MARSTON. 2002. Active electrostatic stabilization of liquid bridges in low gravity. J. Fluid Mech. **457:** 285–294.
90. WU, J., G. DU, S. WORK & S. WARSHAW. 1990. Acoustic radiation pressure on a rigid cylinder: an analytical theory and experiments. J. Acoust. Soc. Am. **87:** 581–586.
91. TRINH, E.H. & C.J. HSU. 1986. Acoustic levitation methods for density measurements. J. Acoust. Soc. Am. **80:** 1757–1761.
92. WEI, W., THIESSEN, D.B. & P.L. MARSTON. 2004. Acoustic radiation force on a compressible cylinder in a standing wave. J. Acoust. Soc. Am. **116:** 201–208.

Contrasting Electrostatic and Electromagnetic Levitation Experimental Results for Transformation Kinetics of Steel Alloys

DOUGLAS M. MATSON,[a] DAVID J. FAIR,[a]
ROBERT W. HYERS,[b] AND JAN R. ROGERS[c]

[a]*Tufts University, Medford, Massachusetts, USA*

[b]*University of Massachusetts, Amherst, Massachusetts, USA*

[c]*Marshall Space Flight Center, Huntsville, Alabama, USA*

ABSTRACT: The delay between conversion of metastable ferrite to stable austenite during ternary Fe–Cr–Ni alloy double recalescence is seen to differ by over an order of magnitude for tests conducted using electrostatic and electromagnetic levitation. Several possible reasons for this deviation are proposed. Thermodynamic calculations on evaporation rates indicate that potential composition shifts during testing are minimized by limiting test time and thermal history. Simulation indicates that deviation would be limited to a factor of 1.5 under worst-case conditions. Possible effects due to differences in sample size are also eliminated since the metastable array, where stable phase nucleation must occur, is significantly smaller than the sample. Differences in internal convection are seen to be the most probable reason for the observed deviation.

KEYWORDS: containerless processing; steel transformation; rapid solidification

NOMENCLATURE:

C	molar flux coefficient (experimentally derived), equal to 0.4
D	diffusivity, m/sec^2
I_s	steady-state nucleation rate, nucleii/cm^3·sec
I_t	time-dependent nucleation rate, nucleii/cm^3·sec
J_n	molar flux for species n, mol/m^{2}sec
M	molecular mass, kg/mol
n_c	number of atoms in the critical nucleus, atoms
p_n	vapor pressure, Pa
R	gas constant, 8.4145 J/molK
T	temperature, °C
t	measured delay time, sec
U	fluid recirculation velocity, cm/sec
V	dendrite tip growth velocity, m/sec
ΔT	undercooling relative to stable phase, °C
δ	ferritic phase (bcc) in Fe–Cr–Ni alloys
γ	austenitic phase (fcc) in Fe–Cr–Ni alloys
ϕ	declination angle (from sample equator), degrees
μ	kinetic parameter in BCT growth theory
τ	transformation characteristic incubation time, sec

Address for correspondence: Douglas M. Matson, 200 College Ave., 025 Anderson Hall, Department of Mechanical Engineering, Tufts University, Medford MA 02155, USA. Voice: 617-627-5742; fax: 617-627-3058.

douglas.matson@tufts.edu

Ann. N.Y. Acad. Sci. 1027: 435–446 (2004). ©2004 New York Academy of Sciences.

doi: 10.1196/annals.1324.035

INTRODUCTION AND PROGRAM OVERVIEW

When comparing ground-based results to those obtained in microgravity, a significant difference is seen in the delay between the formation of the metastable ferritic-phase and the stable austenitic-phase in Fe–Cr–Ni alloys. As part of our NASA sponsored program on levitation observation of dendrite evolution in steel ternary alloy rapid solidification (LODESTARS) microgravity testing, using MSL-EML on the International Space Station,[1] experiments directly address the key issues of identifying the solidification path and the role of convection on microstructural evolution. This involves differentiating between the mechanisms governing the formation of the heterogeneous nucleation of the second phase and understanding the significance of the delay between primary and secondary recalescence. Since identification of the solidification path during ground based tests may be biased due to the acknowledged fluid flow environment, space quenching is required to allow an assessment to be made between samples processed under laminar and turbulent conditions. Ground based electrostatic and electromagnetic containerless processing allows for investigation at the extremes in flow condition. Only microgravity

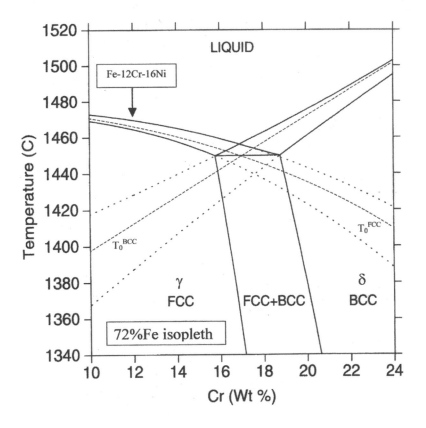

FIGURE 1. Pseudobinary equilibrium phase diagram for the ternary steel alloy system.

electromagnetic testing allows the application of controlled convection over a wide range of flow conditions to investigate the effect of convection, including laminar-turbulent transition, on transformation kinetics.

BACKGROUND

In ternary hypoeutectic steel alloys near the composition of 316 stainless steel, as shown in FIGURE 1, undercooled samples can solidify in a two-step process known as double recalescence. For this process to occur, the undercooling must be sufficient to access the metastable ferrite phase.

Three possible transformations can occur in this system: (1) metastable ferrite can grow into undercooled liquid, (2) stable austenite can grow into the undercooled liquid, and (3) stable austenite can grow into a mixture of metastable ferrite and liquid at a new undercooled condition corresponding to the recalescence temperature following primary metastable solidification.[2]

FIGURE 2 shows an example of double recalescence in an electromagnetically levitated one-gram sample (7 mm diameter) imaged at 40,500 fps with metastable ferrite (grey) growing into undercooled liquid (dark grey) and stable austenite (light grey) growing behind; in both cases the transformation is proceeding from left to right.

In modeling the transformation to final microstructure under conditions of competitive growth of the two phases, two key variables control which phase will form

FIGURE 2. Double recalescence in steel alloys.

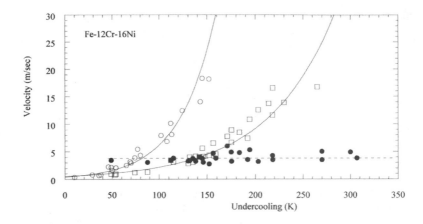

FIGURE 3. Dendrite growth velocity for each phase in ternary alloy: □, δ bcc; ○, γ fcc; ●, semisolid.

as a function of position relative to the primary nucleation site. First, the relative growth velocities must be known, and second, the delay between nucleation of the metastable and stable phases must be determined; both are strong functions of primary undercooling.[2]

Dendrite tip growth velocities, V, can be measured by using a high speed imaging technique that tracks the progress of the thermal front across the exposed surface of the sample.[3] Contrast is developed by differences in radiance due mainly to temperature; brighter is hotter. As mentioned, the undercooled liquid appears dark whereas the metastable solid appears brighter and the stable solid is brightest. Growth of the solid phases appears as expanding regions representing an array of multiple structurally-related dendrites. Velocities for each type of transformation[2] are shown in FIGURE 3 for the Fe–12Cr–16Ni alloy. Growth into the undercooled liquid can be modeled successfully using the Boettinger–Coriell–Trivedi (BCT)[4] model with a kinetic parameter $\mu = 0.30$ for the ferrite phase and $\mu = 0.15$ for the austenite phase,[5] as shown by the solid curves in FIGURE 3. Growth of austenite into the ferrite-liquid semisolid is nearly independent of initial undercooling, as shown by the dashed line in FIGURE 3, since the temperature following primary metastable recalescence is nearly independent of primary undercooling.

In NASA sponsored microgravity research, electromagnetic levitation (EML) experiments were run aboard the shuttle Columbia during the STS-94 mission. The purpose of the work was to investigate how convection changed phase selection in ternary Fe–Cr–Ni alloys. On the ground, the levitation field induces significant convection in the sample, whereas in space the positioning forces to levitate the sample are significantly reduced and less convection is induced. Thus, $U_{\text{space}} \ll U_{\text{ground}}$. Results indicated that growth rates are not significantly changed but that the delay between recalescence events is significantly longer under the reduced flow conditions attainable in space.[6]

Classical nucleation theory[7] predicts that the time-dependent nucleation rate, I_t, is a function of the steady-state nucleation rate, I_s, such that

$$I_t = I_s \exp\left(-\frac{\tau}{t}\right), \tag{1}$$

where τ is the characteristic incubation time and t is the observed delay time. This theory includes the effects of sample size in that the units for I are nucleii/cm^3. Larger samples contain more critical nuclei and the observed delay time will be shorter.

Kantrowitz proposed an expression valid for small times, such that the time dependent nucleation rate is not appreciable until

$$\tau = \frac{n_c^2}{D}, \tag{2}$$

where the characteristic incubation time is related to the diffusivity and the number of atoms in the critical nucleus.[8] Assuming that the theoretical radius of a critical nucleus is inversely proportional to the liquid undercooling, ΔT, and the number of atoms is proportional to the nucleus volume, we can evaluate the shape of the nucleus based on experimental data. For these alloys, a flat plate geometry is indicated[6] and

$$\tau \propto (\Delta T)^{-4}. \tag{3}$$

Note that this relationship requires that the undercooling driving the transformation be known. This undercooling corresponds to the relative temperature difference between primary and secondary recalescence and can be loosely approximated as the difference between T_0 temperatures for each phase, assuming fully partitionless rapid solidification. Current theories do not include the influence of fluid flow on time-dependent nucleation phenomena.

CONVECTIVE ENVIRONMENT

The LODESTARS program aims to investigate how convection influences the transformation from metastable to stable phase in the ternary steel alloys using a single experimental platform, the MSL-EML facility, in space. Since levitation and heating are accomplished separately using a dual induction coil configuration, a wide range of induced flow conditions may be selected. Internal recirculation is an independent experimental variable.

In ground-based electromagnetic levitation (EML) testing, gravity pulls the sample down and the weight must be balanced by application of a strong magnetic field. Levitation and heating are accomplished by using a single induction coil, and the resulting internal flows are significant—on the order of $U = 32$ cm/sec. Turnover occurs at a rate of about 30 Hz in the fluid recirculation loops.[9] Unlike microgravity testing, where radiative and conductive cooling is sufficient, additional cooling by external gas flow is required because of the greater heating from the stronger levitation field required in $1\,g$.

At the other extreme, electrostatic levitation (ESL) provides for conditions where no induced flows exist within a sample, such that U approaches zero for free cooling. Levitation is accomplished by electrostatic attraction while the sample is heated by a laser. Cooling is purely radiative. A comprehensive discussion of the results of magnetohydrodynamic modeling work (MHD) as it pertains to ESL and EML testing is contained in a paper by Hyers.[9] Both methods induce significant surface charges during levitation.

EXPERIMENTAL RESULTS

One-gram samples were prepared by combining elemental stock of 99.995% pure Fe-rod, 99.999% pure Cr-flake, and 99.995% pure Ni-wire to the desired composition and with a target precision limit of ±0.00003 grams (30 ppm by weight). The samples were homogenized by melting in the ground-based EML facility and subsequently remelted twice (nominally) during the EML test program. A typical test involved positioning the sample in the coil, turning on the power to levitate and melt the sample, superheating for approximately two minutes at 100 degrees above the melt plateau, and cooling by increasing the flow of cooling gas (helium with 4% hydrogen) at ambient pressure. Evaporative weight loss over the course of sample fabrication, homogenization, and EML melt processing was held below 0.00100 grams (0.1% by weight). A typical sample had a characteristic diameter of approximately 7 mm with the long axis oriented by gravity along the coil axis.

Following EML testing, the samples were sectioned into eight pieces and forged into a shape reminiscent of Saturn. The rings were removed by cutting and filing to yield a rough blank with a final mass of between 40 and 80 mg for insertion into the ESL. A typical sample had a characteristic diameter of 2 mm and was spherical in shape following melting.

ESL testing involved a series of processing steps, all conducted under a vacuum of 10^{-6} torr. First, a prebake phase is initiated by heating the sample on the transfer pedestal to expel residual volatiles, such as cleaning solvent or adsorbed atmospheric gases. Second, the sample is thermally conditioned. If rapidly heated during initial levitation, the sample will become unstable and fall out at several critical temperatures, affectionately know as "zones-of-death" or ZOD. During subsequent cycles the heating rate can be increased significantly if properly processed. Thus, the sample is slowly heated to just below the melting point to finalize thermal conditioning. Finally, thermal cycling may be accomplished in a rapid manner with the time above the melt plateau minimized to reduce evaporation.

Note that ESL evaporation is particularly significant due to the increase in surface-to-volume ratio as compared to EML samples. Weight loss was controlled to below 2%wt. by limiting testing to three thermal cycles on a sample.

In both ESL and EML experiments, process monitoring consists of two sensors. Temperature measurement is accomplished using one-color pyrometry at 10 Hz and growth velocities and delay times are evaluated using high speed cinematography taken with a Kodak HS 4540 imaging system at 40,500 fps. Even though evaporation has the potential for being significant in terms of sample composition, coating of viewing ports was never observed to be significant in terms of changes to optical path transmissivity.

Ground based ESL and EML delay test results are displayed in FIGURE 4. For the data shown, nucleation was allowed to proceed in a spontaneous manner to

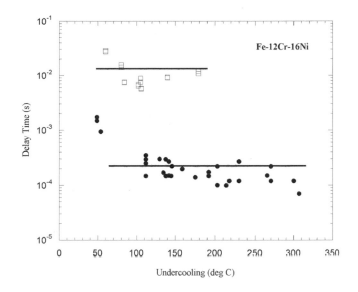

FIGURE 4. ESL and EML ground-based testing: ■, EML; ▽, ESL.

maximize undercooling achieved and no nucleation stimulation (triggered heterogeneous nucleation) was attempted. Lines on the graph indicate a consistent trend in behavior at high undercoolings.

DISCUSSION

As can be seen in FIGURES 3 and 4, no data pertaining to double recalescence behavior is available below an undercooling (with respect to the stable phase) of about 45 degrees, corresponding to the temperature required to access the metastable extension of the equilibrium phase diagram. Double recalescence is not observed unless the metastable-to-stable transformation occurs.

Although growth velocity measurements show no significant difference in dendrite propagation behavior, a significant difference is seen in the delay behavior for samples processed in the ESL as compared to those tested in the EML. Three potential contributions to this deviation have been identified: (1) changes in composition due to preferential evaporative loss of the most volatile species; (2) differences in sample size; and, (3) convection—either external or internal. These will be addressed in order.

Evaporation

Since the sample temperature is known from the output from the one-color pyrometer, an iterative procedure is used to evaluate how much of each chemical species evaporates over each discrete time interval. To accomplish this task, the vapor pressure of each species must be determined as a function of composition and temperature.

The thermodynamic activity coefficient for each species can be calculated as a function of temperature and composition using the computational program ThermoCalc (Royal Institute of Technology, Stockholm, Sweden). From this, the vapor pressure over the alloy is evaluated[10] and the evaporation rate predicted using the Langmuir molar flux equation.

$$J_n = \left(C \cdot \left(\frac{P_n}{\sqrt{2\pi MRT}} \right) \right). \tag{4}$$

Once the evaporation rate is known as a function of chemical species and sample temperature, and given the surface area of the droplet, composition changes can be predicted as a function of actual test thermal history. Model predictions are anchored by conducting a chemical assay on calibration samples.

An example of a nominal thermal history for ESL testing is shown in FIGURE 5 together with the predicted changes in the most volatile component, chromium, from an initial concentration of 11.25%wt Cr. No appreciable loss is seen during the pedestal prebake, the ZOD thermal conditioning, or the 1,000°C thermal hold phases of the testing. Significant loss begins only during actual melt cycles, where the sample is laser heated at a nominal rate of 100°C/sec and radiatively cooled in vacuum at a rate of 60°C/sec. Four melt cycles are shown in FIGURE 5.

This plot depicts Cr-loss as a function of time, but in practice it is easier to track total weight loss. Experience shows that the limit of a maximum loss of 2% of the total mass of the sample is a good conservative estimation technique for maintaining the change in chromium concentration to below 0.5% by weight. Note that either total weight loss or chromium concentration change can readily be predicted using this iterative technique. The concentration limit is predicted to be reached during the fourth cycle and thus testing was curtailed after three cycles.

The 0.5%wt concentration limit is based on an estimate of how composition changes affect the thermodynamic driving force for the transformation. The thermo-

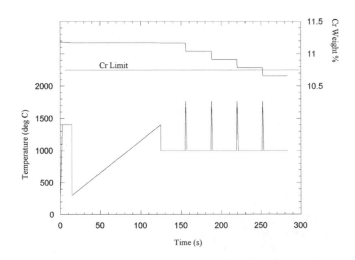

FIGURE 5. Predicted ESL evaporative losses.

dynamic driving force is obtained analytically using ThermoCalc by evaluation of the metastable extension of the equilibrium phase diagram with calculation of the T_0-curve for each phase. A plot of the thermodynamic driving force as a function of composition for this family of alloys yields a value of $dT/dc = -10.4°C/wt\%$.[11] Thus, this negative composition shift results in a positive shift in thermodynamic driving force of 5°C.

For a transformation thermodynamic driving force of 45°C for this alloy, and applying Equation (3), we expect a corresponding decrease in τ by a factor of 1.5. This deviation is predicted to be opposite that observed—preferential loss of Cr should make delay times shorter in ESL testing whereas we observe longer delays. Thus, composition shifts due to evaporation of the most volatile species are not sufficient to explain the observed deviation.

Sample Size

A 1-gram EML sample has a characteristic diameter of approximately 7mm whereas a typical 40-milligram ESL sample has a diameter of 2mm. This large difference in volume may have profound effect on perceived transient nucleation behavior. Nucleation rates are calculated on a unit volume basis and, thus, a smaller sample will have fewer active sites and a longer measured delay time even with the same characteristic incubation time.

The activation of one nucleus per sample equates to the product of the time-dependent nucleation rate, the sample volume, and the delay time. Assuming that the entire volume of a sample were to instantly transform to an array of metastable ferrite and residual liquid during primary recalescence, we can readily solve for the characteristic incubation time given an estimate of the steady-state nucleation rate from Equation (1). With a low activation energy for nucleation, this asymptotically approaches a value of 1,033 nucleii/cm³sec.[12] The correction factor for a 2-mm sample and a 7-mm sample would be 8.5% and 2.5% of the observed delay times, respectively. Note that a 12.4mm diameter sample has a volume of 1cm³ and would not need correction. Our confidence in these predictions is low based on the number of assumptions that are required to predict transient nucleation behavior.

However, for shallow undercoolings, the growth of the metastable array is slow enough compared to the delay time that the sample is not completely transformed. In this case, it is not the sample size that is important in these analyses but rather the size of the metastable dendritic array (which is the region that will transform to the stable phase) at the time when the second recalescence begins. A simple model is proposed to predict the size of the metastable array prior to secondary stable phase nucleation. This is accomplished by combining the observed growth velocity of the metastable phase from FIGURE 3 with the observed delay time from FIGURE 4. These predictions are displayed in FIGURE 6.

For an ESL sample with a diameter of 2mm, when the undercooling is less than 180 degrees, the array size is smaller than the sample. Above this undercooling, the sample is completely transformed before secondary nucleation. This transition is seen to occur at an undercooling of about 300 degrees for an EML sample with a diameter of 7mm.

Although we are ignoring the transient nature of the primary growth as the array evolves from the ferrite nucleation site, no correction to the delay time is required

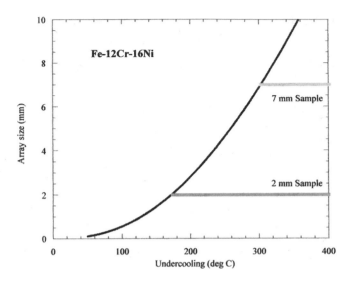

FIGURE 6. The size of the metastable array at the time of secondary nucleation is smaller than the sample diameter for moderate undercooling.

until the critical undercooling for ESL constraint is reached because the ESL and EML samples have the same array size up to this point. From FIGURE 4 we see that difference between EML and ESL delay times is significant at an undercooling smaller than that required to completely transform the smaller sample before secondary nucleation. Thus, sample size differences do not adequately explain the observed deviation.

Convection

The remaining difference between the test conditions in ESL and EML pertains to convective conditions both outside and inside the molten droplet. A statistical analysis of both ESL and EML data[13] shows that a significant portion of tests are characterized by a common nucleation site. Where the first phase nucleates, the second also does. This makes intuitive sense in that this is the position on the droplet where the first phase has existed longest. This observation does not help us to understand the effect of convection on transformation behavior.

First, we consider the potential effect of external gas flow on secondary nucleation. ESL testing is accomplished in a vacuum whereas EML testing involves blowing gas at ambient pressure down onto the top surface of the sample to convectively cool the droplet. An evaluation of the frequency of nucleation events shows that primary nucleation occurs preferentially on the top hemisphere of EML samples, where the gas flow impinges on the surface, whereas no preference is shown in ESL samples. This is to be expected because the EML top surface is maintained at a slightly lower temperature than the bottom hemisphere due to enhanced convective cooling conditions. Unfortunately, this is irrelevant to our discussions on secondary nucleation phenomena, since following primary recalescence the liquid–solid mixture remains at the same temperature across the entire droplet surface (although the fraction of solid on the top may increase at a slightly higher rate than the bottom if the heat extraction rate is locally higher).

If gas flow were to influence nucleation of the second phase, we would expect to observe shorter delay times on the top surfaces of the EML samples. This is not observed; there is no statistical difference in delay times between top (declination angle positive, $\phi > 0$) and bottom (declination angle negative, $\phi < 0$) hemispheres. In a like manner, ESL samples showed a similar trend, which is to be expected in the absence of cooling gas flow.

Additionally, we would expect to observe longer delays on the bottom surface of EML droplets, in the stagnant sample wind-shadow, with delays of similar magnitude to those observed in ESL testing. Clearly, this is not the trend observed in FIGURE 4 and gas flow can not explain the observed differences between ESL and EML delay behavior.

Next, we consider the potential effect of internal fluid flow on secondary nucleation. When the primary and secondary nucleation sites are not collocated, a statistical analysis of the position on the droplet surface again does not show a significant preference for the angle of declination on delay times for either ESL or EML data. This is to be expected for ESL data where no induced flows exist. In EML testing, this is also to be expected because the region on the sample where internal flows are zero are at the stagnation point along the equator, where fluid dives into the droplet center, and at each pole, where fluid flows out. These stagnant regions are small in surface area as compared to the recirculation zones across the top and bottom hemispheres. If delays are longer in the stagnant regions, the lack of data will be masked by the adjacent active recirculation zone.

As expected, analysis of the frequency of both primary and secondary nucleation events shows no correlation with declination angle for ESL samples. For EML tests, the frequency of secondary nucleation appears to disfavor positions along the equator and both poles, where internal flows are minimized, but the apparent trend[13] requires further analysis before confirming that this difference is statistically significant. Internal flows remain as the only viable explanation for the deviation between ESL and EML delay behavior.

CONCLUSIONS

The significant difference between ESL and EML test results for transformation delay times cannot be explained by composition changes, sample size or cooling method. The only unresolved difference between these test conditions is the convection level within the samples. Developing a time-dependent nucleation model that includes fluid flow is the next challenge in explaining transformation behavior in ternary steel alloys.

ACKNOWLEDGMENTS

The authors gratefully acknowledge NASA support through Grant NAG8-1685 as part of the LODESTARS flight program. We also thank the NASA MSFC/ESL team of Trudy Allen, Glenn Fountain, and Tom Rathz for significant contributions to the experimental portion of this work. Finally, we thank Kenny William, who performed thermodynamic modeling as part of his Tufts BSME degree.

REFERENCES

1. FLEMINGS, M.C., et al. 2003. Flight planning for the international space station. In NASA/CP-2003-212339. D. Gilles, et al., Eds.: 221–230. NASA-MSFC, AL.
2. MATSON, D.M. 1999. Growth competition during double recalescence in Fe–Cr–Ni alloys. In Materials in Space—Science, Technology, and Exploration, Vol. 551. A.F. Hepp, et al., Eds.: 227–234. TMS, Warrendale.
3. MATSON, D.M. 1998. The measurement of dendrite tip propagation velocity during growth into undercooled metallic melts. In Solidification 1998. S.P. Marsh, et al., Eds.: 233–244. TMS, Warrendale.
4. BOETTINGER, W.J., et al. 1988. Application of dendrite growth theory to the interpretation of rapid solidification microstructures. In Rapid Solidification Processing: Principles and Technologies IV. R. Mehrabian & P.A. Parrish, Eds.: 13–25. Claitor's Pub. Div., Baton Rouge.
5. KERTZ, J.E. 2001. SM Thesis, Massachusetts Institute of Technology, Cambridge, MA.
6. MATSON, D.M., et al. 1999. Phase selection and rapid solidification of undercooled Fe–Cr–Ni steel alloys in microgravity. In Solidification 1999. W.H. Hofmeister, et al., Eds.: 99–106. TMS, Warrendale.
7. TURNBULL, D. 1956. Solid State Phys. 3: 225.
8. KANTROWITZ, A. 1951. J. Chem. Phys. 19: 1097.
9. HYERS, R.W., et al. 2004. Convection in containerless processing. In Transport Phenomena in Microgravity. S.S. Sadhal, Ed. Ann. N.Y. Acad. Sci. 1027: this volume.
10. HONIG, R.E. & D.A. KRAMER. 1970. Vapor pressure data for the solid and liquid elements. In Techniques for Metals Research, Vol. 4. R.A. Rapp, Ed.: 505–524. Interscience Publishers, New York.
11. KERTZ, J.E. & D.M. MATSON. 2001. Measurement of steel growth kinetics using TEMPUS aboard the NASA KC-135 parabolic aircraft. In Proceedings of the First International Symposium on Microgravity Research and Applications in Physical Sciences and Biotechnology, ESA/SP-454. 639–646. Noordwijk ND.
12. KURZ, W. & D.J. FISHER. 1998. Fundamentals of Solidification, 4th edit. 32–33. Trans Tech Publications, Zurich.
13. VENKATESH, R. & D.M. MATSON. 2004. The influence of internal and external convection in the transformation behavior of Fe–Cr–Ni alloys. In Proceedings of EPD Congress 2004. M.E. Schlesinger, Ed. TMS, Warrendale.

Shape Relaxation of Liquid Drops in a Microgravity Environment

S.S. SADHAL, A. REDNIKOV, AND K. OHSAKA

University of Southern California, Los Angeles, California, USA

ABSTRACT: We investigated shape relaxation of liquid drops in a microgravity environment that was created by letting the drops fall freely. The drops were initially levitated in air by an acoustic/electrostatic hybrid levitator. The levitated drops were deformed due to the force balance among the levitating force, surface tension, and gravity. During the free fall, the deformed drops underwent shape relaxation driven by the surface tension to restore a spherical shape. The progress of the shape relaxation was characterized by measuring the aspect ratio as a function of time, and was compared to a simple linear relaxation model (in which only the fundamental mode was considered) for perfectly conductive drops. The results show that the model quite adequately describes the shape relaxation of uncharged/charged drops released from an acoustically levitated state. However, the model is less successful in describing the relaxation of drops that were levitated electrostatically before the free fall. This may be due to finite electrical conductivities of liquids, which somehow affects the initial stage of the shape relaxation process.

KEYWORDS: shape relaxation; liquid drops; microgravity; levitation; viscosity measurement

INTRODUCTION

Shape relaxation of a deformed drop driven by surface tension toward a spherical shape is of long-standing interest because it plays an important role in natural phenomena as well as in industrial processes. Theoretically, the relaxation of drops with small initial deformation has been studied thoroughly and comprehensive analyses have appeared in the literature.[1–7] Experimentally, however, the subject has been less studied due to complications caused by the gravity of the Earth, which influences the dynamics of relaxation. There are two approaches that can minimize the gravitational influence. One approach is to acoustically levitate a drop in an immiscible liquid whose density is similar to that of the drop.[8–12] This approach is only applicable to selected immiscible liquid systems. The other approach is to employ a microgravity environment. Short-duration microgravity environments up to 30 seconds can be created by drop tube/tower facilities and parabolic flight of aircraft. Long-duration microgravity environments can be created onboard spacecraft orbiting around Earth.[12] Although long-duration microgravity environments have been generally preferable, during our studies on the shape relaxation of acoustically levitated drops, we found that some drops completed the relaxation in short times, of the order of

Address for correspondence: S.S. Sadhal, University of Southern California, Los Angeles, CA 90089-1453, USA.
sadhal@usc.edu

milliseconds.[13] This observation led us to set up an exploratory experiment in which an initially levitated drop was suddenly released and allowed to fall freely for a short distance of several millimeters. The progress in the shape relaxation during the free fall was characterized in terms of the change in the aspect ratio. The results were then compared with a simple linear relaxation model. The study described here is a unique experiment employing a compact and inexpensive tabletop apparatus, in contrast to experiments using conventional microgravity facilities that are structurally extensive and require large resources for operation.

EXPERIMENTAL

We employ an acoustic/electrostatic hybrid levitator to initially suspend a drop in air. The levitated drop is deformed at equilibrium due to the force balance among the levitating force, surface tension, and gravity. As a result, the levitated drop is approximately a prolate spheroid in the electrostatic levitation mode and an oblate spheroid in the acoustic levitation mode. During free fall, the deformed drop relaxes and eventually assumes a spherical shape. The shape relaxation proceeds in either an oscillatory or non-oscillatory manner, depending on the physical properties of the drop, such as volume, surface tension, and viscosity.

FIGURE 1 schematically depicts the present apparatus. Since the operation principle of the apparatus appears in the literature,[14] we only present specific features of the apparatus. It is small enough to be assembled on an optical table. The distance between the top and bottom electrodes is set at 23 mm. A drop is levitated at approximately 11 mm above the bottom electrode (also the transducer head). The apparatus can be operated in the acoustic levitation mode, electrostatic levitation mode, or mixed levitation mode. In the electrostatic levitation mode, the position-tracking unit (laser and detector) monitors the drop position and adjusts the applied voltage 250 times per second to keep the drop at a preset position. The free fall of the drop is initiated by suddenly shutting off the levitation field. The fall lasts roughly 50 milliseconds, until the drop hits the bottom electrode. A high-speed digital camera (operated at 1,000 frame/sec) is used to capture the images of the drop in the levitated state and during approximately the first 20 milliseconds of the free fall. The drop is back-illuminated by white light, which creates a sharply contrasted dark image on bright background.

We prepared distilled water ($\sigma = 71\,\mathrm{mNm^{-1}}$, $\mu = 1\,\mathrm{mPa\cdot sec}$), a water–glycerin solution ($\sigma = 68\,\mathrm{mNm^{-1}}$, $\mu = 31\,\mathrm{mPa\cdot sec}$), and silicone oils ($\sigma = 21\,\mathrm{mNm^{-1}}$, $\mu = 10$ and $100\,\mathrm{mPa\cdot sec}$), where σ and μ are the surface tension and dynamic viscosity, respectively. The experimental procedure was as follows. A small amount of the solution (about 1 μl) was initially suspended in the levitation field. The drop was initially either (1) acoustically levitated and uncharged, (2) acoustically levitated and charged, or (3) electrostatically levitated and (necessarily) charged. After the drop stabilized, the levitation field was shut off so that the drop underwent a free fall. The high-speed camera captured the images of both the levitated and falling drop. These captured images were transferred to a video recorder for analysis. Finally, the images of a solid sphere with a known radius were taken for calibration. All measurements were performed at room temperature.

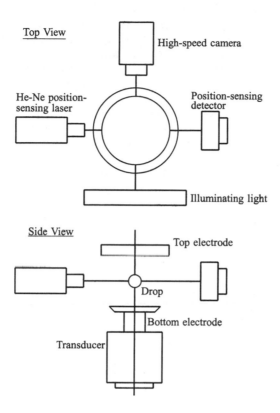

Top View

High-speed camera

He-Ne position-sensing laser

Position-sensing detector

Illuminating light

Side View

Top electrode

Drop

Bottom electrode

Transducer

FIGURE 1. Schematic description of the experimental apparatus

The drop images recorded on videotape were transferred to a computer for size and shape determination by commercially available software. FIGURE 2 shows typical images of levitated drops in the electrostatic levitation mode (left), and in the acoustic levitation mode (right). In general, initial deformation of acoustically levitated drops is larger than that of electrostatically levitated drops. We measured the horizontal axis, W, of the drop at equator and the vertical axis, H, passing the center of the drop. In measuring the axes, we took the following into consideration. The edge of the drop image on screen was not quite sharp but somewhat blurred for several

A   B

FIGURE 2. Drop images: (**A**) an electrostatically levitated drop and (**B**) an acoustically levitated drop.

reasons, including the limited resolution of the camera. The drop images generally covered about 100×100 pixels. The sharpness of the edge also varied among the images. For this reason, it was not possible to locate the exact edge position on the image. Therefore, we decided to obtain two readings for each measurement. The first reading was obtained by locating the edge at the inner side of the blurred region, which yielded the minimum value, W^{min}. The second reading was obtained by locating the edge at the outer side of the blurred region, which yielded the maximum value, W^{max}. The degree of deformation of the drop from spherical form was characterized by the aspect ratio given by W/H. The possible minimum and maximum aspect ratios were calculated as W^{min}/H^{max} and W^{max}/H^{min}, respectively. In addition, we determined the position of the center of mass of a falling drop with respect to that of the initially levitated drop. This measurement allowed us to calculate the acceleration of a falling drop.

THEORETICAL

Work on damped oscillations were previously conducted by Prosperetti.[15] However, the current experimental results are compared with a simple linear model, in line with classical work.[2–6] Only the fundamental mode ($l = 2$), corresponding to the axisymmetric prolate–oblate shape, dynamics is taken into consideration, that is,

$$r = a[1 + \xi_2(t)P_2(\cos\theta)],$$

where r and θ are the spherical coordinates, a is the radius of the drop when spherical, t is the time,

$$P_2(x) = \left(\frac{3}{2}x^2 - \frac{1}{2}\right)$$

is the Legendre Polynomial of order $l = 2$, and $\xi_2(t)$ is the dimensionless amplitude.

Let W and H be the equatorial and the polar diameters of a deformed axisymmetric drop. The aspect ratio W/H is the quantity that is followed in this study. If $W/H > 1$, the drop shape is oblate, if $W/H < 1$, prolate. Within our linear approach, we have

$$\frac{W}{H} = 1 - \frac{3}{2}\xi_2(t).$$

A drop electrostatically levitated in the middle of two parallel electrodes experiences the following force balance:

$$mg = \frac{qV}{L}, \tag{1}$$

where m is the mass of the drop, g is the gravitational acceleration, q is the charge on the drop, V is the applied voltage between the two electrodes, and L is the distance between the two electrodes. Equation (1) is used to calculate the charge of the drop, all the other parameters being known.

After the external electric field is turned off and the drop is released into a free fall, the charge remains an important factor since it influences electrostatically the shape relaxation process. This effect, within the linear mode approach, can be accounted for in the following manner. If the drop is assumed to be a perfect conductor and the surrounding air to be a perfect insulator, it follows from Rayleigh's study[16] that the effect of the electrostatic force is equivalent to the surface tension σ being reduced to

$$\sigma_{ap(l)} = \sigma - \frac{q^2}{(4\pi)^2(l+2)\varepsilon_0 a^3}, \tag{2}$$

where (in SI units) ε_0 is the permittivity of the vacuum and l is the order of the spherical harmonic representing the deviation of the drop shape from the equilibrium sphere. In other words, when studying shape relaxation of a charged drop for each particular mode l, the correct description is attained by formally substituting the apparent surface tension, $\sigma_{ap(l)}$ for σ in the corresponding results obtained for an uncharged drop.

We assume that the shape relaxation proceeds in the fundamental mode ($l = 2$) and define the capillary number by

$$Ca = \frac{\mu^2}{\rho\sigma_{ap(2)}a}, \tag{3}$$

where ρ is the density and $\sigma_{ap(2)}$ is given by Equation (2) with $l = 2$ (for uncharged drops, $\sigma_{ap(2)}$ is replaced by σ). Note that the inviscid limit corresponds to $Ca \to 0$, whereas the highly viscous cases correspond to $Ca \to \infty$. There is a critical combination of parameters that in terms of Ca is $Ca_{cr} = 0.587$, such that for $Ca < Ca_{cr}$ the drop relaxation proceeds in an oscillatory manner, whereas for $Ca > Ca_{cr}$ it occurs in a non-oscillatory manner.

We assume the following expressions to describe the shape relaxation characterized by the aspect ratio, W/H,

$$\frac{W}{H} = 1 + e^{-\delta t}(A_1\cos\omega t + A_2\sin\omega t) \tag{4}$$

for $Ca < Ca_{cr}$, and

$$\frac{W}{H} = 1 + A_1 e^{-\delta_1 t} + A_2 e^{-\delta_2 t} \tag{5}$$

for $Ca > Ca_{cr}$, where A_1 and A_2 are the amplitudes, expected to be small for the linear approach to work. The dimensionless angular frequency, ω^*, is defined by

$$\omega^* = \left(\frac{\rho a^3}{\sigma_{ap(2)}}\right)^{1/2}\omega, \tag{6}$$

where ω is the angular frequency. FIGURE 3 shows ω^* as a function of Ca. As can be seen in the figure, ω^* decreases as Ca increases and vanishes at Ca_{cr}. The dimensionless damping rate, δ^*, is defined by

$$\delta^* = \frac{\rho a^2}{\mu}\delta, \tag{7}$$

where δ is the damping rate. FIGURE 4 shows δ^* as a function of Ca. At $Ca = 0$, $\omega^* = \sqrt{8}$ and $\delta^* = 5$ are established.[1] For $Ca > Ca_{cr}$, there are two non-oscillatory modes (the bottom branch, δ_1^*, and the top branch, δ_2^*). As $Ca \to \infty$, $\delta_1^* \approx (20/19)Ca^{-1}$ controls the damping rate and the second term in Equation (5) can be neglected. When Ca is close to Ca_{cr}, however, both terms contribute to the overall relaxation.

In calculating the theoretical relaxation curves, we first calculate the charge using Equation (1). Second, we determine the apparent surface tension, using Equation (2)

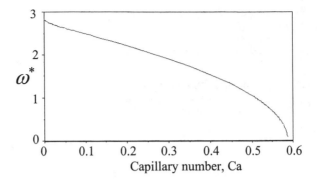

FIGURE 3. Dimensionless frequency of the fundamental mode ($l = 2$) versus the capillary number.

for the fundamental mode, and then the capillary number using Equation (**3**). Third, we read ω^* and δ^* from FIGURES 3 and 4 for the given Ca value to calculate ω and δ using (**6**) and (**7**). Finally, A_1 and A_2 are determined by the value of W/H at $t = 0$. In addition, we assume that $d(W/H)/dt = 0$ at $t = 0$, which corresponds to the relaxation process taking off from the rest state.

RESULTS

FIGURE 5 shows the falling distance of two typical drops as a function of the elapsed time after the release. As can be seen in the figure, they closely follow the curve that represents free fall. This measurement verified that the drops experienced a microgravity environment during the fall.

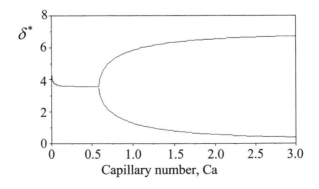

FIGURE 4. Dimensionless damping rate of the fundamental mode ($l = 2$) versus the capillary number.

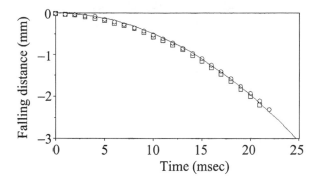

FIGURE 5. Falling distance versus the elapsed time for two typical drops.

We repeated the measurement for each levitation condition to verify the reproducibility and selected representative drops for comparison with the linear model. In the following, we present some of the results. In each figure, the experimentally determined aspect ratios are shown with short vertical bars, whose ends correspond to the possible minimum and maximum aspect ratios. Note that some bars are shorter than the size of the symbol and do not show up in the figure. The curve represents the model prediction and is calculated with experimentally determined parameters. The horizontal line represents the aspect ratio of the completely relaxed drop.

First, we present five figures that show results for acoustically levitated drops. FIGURE 6 shows a result for an uncharged water–glycerin solution drop ($a = 1.20$ mm, $Ca = 0.01$). As can be seen in the figure, the agreement is good in this case. FIGURE 7 shows a result for an uncharged silicone oil drop ($\mu = 10$ mPa·sec, $a = 0.96$ mm, $Ca = 0.0045$). The linear model seems to adequately describe the amplitude and the frequency of the relaxation, excluding the early stage. Discrepancy in the apparent damping rate may be attributed to non-linearity. FIGURE 8 shows a result for an

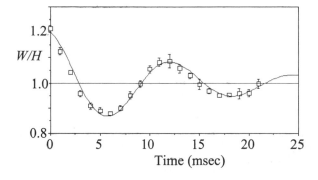

FIGURE 6. Shape relaxation of an acoustically levitated uncharged W/G drop ($a = 1.20$ mm, $Ca = 0.01$).

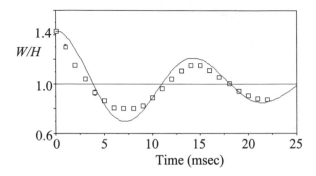

FIGURE 7. Shape relaxation of an acoustically levitated silicone oil (10 mPa·sec) drop (a = 0.96 mm, Ca = 0.0045).

uncharged silicone oil drop (μ = 100 mPa·sec, a = 1.1 mm, Ca = 0.446). The agreement happens to be good despite the relatively large amplitude involved. FIGURE 9 shows a result for an uncharged water drop (a = 1.23 mm, Ca = 0.0000113). Disagreement with the linear model is more pronounced in this case. Discrepancy in the frequency can be attributed to nonlinear effects that tend to decrease the frequency as the amplitude is increased. This effect is expected to be relatively more important for the liquids of low viscosity, such as water. Also for liquids of low viscosity, the dynamic effect of the air (neglected in the present study) can add quite a few percent to the damping rate, as can be estimated by using equation (8) of Reference 17 (the ratio $\alpha\omega^{1/2}/\gamma$ for $n = 2$ in the notation of Ref. 17). Due to the presence of an oscillatory boundary layer at the interface, this effect is somewhat stronger than what could be expected merely on the basis of the ratios of the dynamic properties of air and water.

We can create an acoustically levitated charged drop in the following manner. First, a drop is charged and levitated in the electrostatic levitation mode. After the

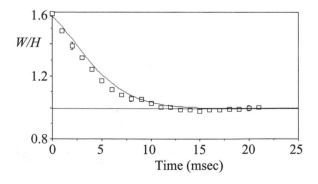

FIGURE 8. Shape relaxation of an acoustically levitated silicone oil (100 mPa·sec) drop (a = 1.1 mm, Ca = 0.446).

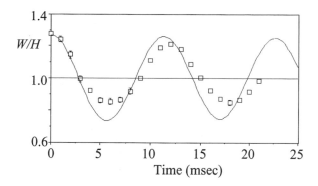

FIGURE 9. Shape relaxation of an acoustically levitated water drop ($a = 1.23\,\text{mm}$, $Ca = 0.0000113$).

applied voltage is determined, the acoustic field is turned on. As its intensity is gradually increased, the applied voltage is accordingly reduced until the drop is purely levitated acoustically. FIGURE 10 shows the result for a charged W/G solution drop ($a = 1.13\,\text{mm}$, $Ca = 0.013$).

The following four figures show results for electrostatically levitated drops. Note that electrostatically levitated drops are necessarily charged. Note also that the initial charge distribution on electrostatically levitated drops is different from that of the acoustically levitated charged drop in FIGURE 10 due to the electrical field necessary for the electrostatic levitation. FIGURE 11 shows a result for a W/G solution drop ($a = 1.3\,\text{mm}$, $Ca = 0.014$). Although there is a discrepancy at the early part of the relaxation, the model seems to adequately describe the subsequent parts of the relaxation. FIGURE 12 shows a result for a silicone oil drop ($\mu = 10\,\text{mPa·sec}$, $a = 1.1\,\text{mm}$, $Ca = 0.008$). FIGURE 13 shows a result for a silicone oil drop ($\mu = 100\,\text{mPa·sec}$,

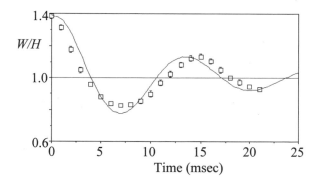

FIGURE 10. Shape relaxation of an acoustically levitated charged W/G drop ($a = 1.13\,\text{mm}$, $Ca = 0.013$).

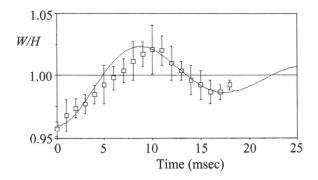

FIGURE 11. Shape relaxation of an electrostatically levitated charged 31 mPa·sec W/G drop ($a = 1.3$ mm, $Ca = 0.014$).

$a = 1.2$ mm, $Ca = 0.4$). FIGURE 14 shows a result for a water drop ($a = 1.3$ mm, $Ca = 0.0000126$).

DISCUSSION

We investigated the shape relaxation of initially levitated drops in three different initial configurations: (1) acoustically levitated and uncharged, (2) acoustically levitated and charged, and (3) electrostatically levitated and charged. We found that the linear model based on the fundamental mode is adequate for acoustically levitated drops when the initial deformation is small. The discrepancies that appeared in drops with large initial deformations can be attributed to nonlinear effects. The amplitude tends to decay faster than the linear model prediction. The frequency is sensitive to the experimentally determined radius. We estimate the error involved in determining the radius to be approximately 5%, which may propagate to generate an 8% error in

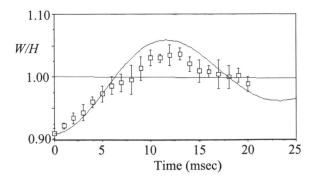

FIGURE 12. Shape relaxation of an electrostatically levitated 10 mPa·sec silicone oil drop ($a = 1.1$ mm, $Ca = 0.008$).

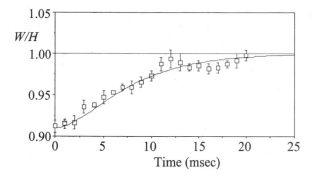

FIGURE 13. Shape relaxation of an electrostatically levitated 100 mPa·sec silicone oil drop ($a = 1.2$ mm, $Ca = 0.4$).

the frequency. The difference in the frequency for the water drop (FIG. 9) is larger than for other drops. This is perhaps because the nonlinear effect of frequency reduction is the most pronounced for drops of low viscosity, and water is indeed the least viscous of the liquids used.

For electrostatically levitated drops, a significant discrepancy is observed at the early stage of relaxation. Since initial deformations are small, the nonlinear effects are unlikely to cause the discrepancy. It might well be that the electrical conductivity plays a role in creating this discrepancy. When a drop is electrostatically levitated, the charge distribution therein is different from that in the absence of the external electric field. When the field is subsequently shut off, the charge undergoes redistribution. If the drop has a finite conductivity, the redistribution is not instantaneous. As a result, it is conceivable that the redistribution couples with the shape relaxation and produces the observed discrepancy. A model that takes into account the finite conductivity is needed to clear up this issue.

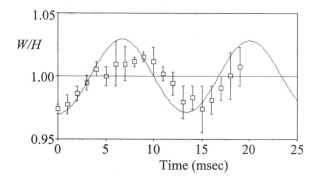

FIGURE 14. Shape relaxation of an electrostatically levitated water drop ($a = 1.3$ mm, $Ca = 0.0000126$).

An intended application of the present technique is to determine the unknown viscosity of a drop by fitting the theoretical model with the experimental result performed on the drop. There is increasing interest in measuring the surface tension and viscosity of undercooled liquids that are in metastable states and exist below their freezing points. Conventional techniques are not applicable to undercooled liquids simply because bulk liquids seldom attain significant undercooling before solidification takes place at one of the numerous solid nucleating sites, including the container walls. It has been demonstrated that levitation of a small liquid sample is an effective way to attain a large degree of undercooling because the small sample volume reduces the number of the potential nucleation sites and a self-contained sample is free from the possibility of nucleation at the container walls. There is an established technique that induces a resonant shape oscillation on a levitated drop and measures the resonant frequency and damping constant after the forced oscillation is terminated.[18] The resonant frequency is related to the surface tension,[19] and the damping constant is related to the viscosity through the Lamb formula.[1] This technique, however, is only applicable to low-viscosity liquids because of the difficulty to induce oscillations on highly viscous liquids. The present technique, on the other hand, is also applicable to highly viscous liquids. We would like to apply this technique to undercooled metallic alloys, specifically glass forming alloys that become highly viscous as the undercooling increases.[20]

ACKNOWLEDGMENTS

This study was supported by the Office of Biological and Physical Research, NASA (Grant No. NAG8-1663).

REFERENCES

1. LAMB, H. 1932. Hydrodynamics. Cambridge University Press.
2. REID, W.H. 1960. The oscillations of a viscous liquid drop. Q. Appl. Math. **18**: 86–89.
3. CHANDRASEKHAR, S. 1959. The oscillations of a viscous liquid globe. Proc. Lond. Math. Soc. **9**(3): 141–149.
4. CHANDRASEKHAR, S. 1961. Hydrodynamic and Hydromagnetic Stability. Dover, New York.
5. MILLER, C.A. & L.E. SCRIVEN. 1968. The oscillations of a fluid droplet immersed in another fluid. J. Fluid Mech. **32**: 417–435.
6. PROSPERETTI, A. 1980. Normal-mode analysis for the oscillations of a viscous liquid drop in an immiscible liquid. J. Mecanique **19**: 149–182.
7. MARSTON, P.L. 1980. Shape oscillation and static deformation of drops and bubbles driven by modulated radiation stresses-theory. J. Acoust. Soc. Am. **67**: 15–26.
8. MARSTON, P.L. & R.E. APFEL. 1980. Quadrupole resonance of drops driven by modulated acoustic radiation pressure—experimental properties. J. Acoust. Soc. Am. **67**: 27–37.
9. TRINH, E., A. ZWERN & T.G. WANG. 1982. An experimental study of small-amplitude drop oscillations in immiscible liquid systems. **115**: 453–474.
10. TRINH, E. & T.G. WANG. 1982. Large-amplitude free and driven drop-shape oscillations: experimental observations. J. Fluid Mech. **122**: 315–338.
11. TRINH, E.H., D.B. THIESSEN & R.G. HOLT. 1998. Driven and freely decaying nonlinear shape oscillations of drops and bubbles immersed in a liquid: experimental results. J. Fluid Mech. **364**: 253–272.

12. APFEL, R.E., Y. TIAN, J. JANKOVSKY, *et al.* 1997. Free oscillations and surfactant studies of superdeformed drops in microgravity. Phys. Rev. Lett. **78:** 1912.
13. OHSAKA, K., A. REDNIKOV, S.S. SADHAL & E.H. TRINH. 2002. Noncontact technique for determining viscosity from the shape relaxation of ultrasonically levitated and initially elongated drops. Rev. Sci. Instrum. **73:** 2091–2096.
14. CHUNG, S.K. & E.H. TRINH. 1998. Containerless protein crystal growth in rotating levitated drops. J. Cryst. Growth **194:** 384–397.
15. PROSPERETTI, A. 1980. Free oscillations of drops and bubbles: the initial-value problem. J. Fluid Mech. **100:** 333–347.
16. RAYLEIGH, J.W.S. 1882. Phil. Mag. **14:** 184.
17. ASAKI, T.J. & P.L. MARSTON. 1995. Free decay of shape oscillations of bubbles acoustically trapped in water and sea water. J. Fluid Mech. **300:** 149–167.
18. RHIM, W.-K., K. OHSAKA, P.-F. PARADIS & R.E. SPJUT. 1999. Noncontact technique of measuring surface tension and viscosity of molten materials using high temperature electrostatic levitation. Rev. Sci. Instrum. **70:** 2796.
19. RAYLEIGH, J.W.S., LORD. 1879. Proc. R. Soc. Lond. **29:** 71.
20. OHSAKA, K., S.K. CHUNG & W.-K. RHIM. 1998. Specific volumes and viscosities of the Ni–Zr alloys and their correlation with the glass formability of the alloys. Acta Mater. **46:** 4535.

APPENDIX

In this appendix, following Prosperetti,[15] we carry out a closer inspection of the initial-value problem for the angular mode $l = 2$, and establish the extent to which the previous simplified approach (leaving only the least-damped, oscillatory, modes) works. Expecting no confusion with the main sections of the paper, here t, ω, and δ refer, respectively, to the dimensionless time, angular frequency, and damping rate, nondimensionalized with the viscous time scale $\rho a^2/\mu$ (or its inverse, accordingly). Let $\xi_l(t)$ be the amplitude of the surface deformation for angular mode l. We are interested in describing the drop shape evolution $\xi_l(t)$ toward a sphere ($\xi_l(\infty) = 0$) when released from a certain initial shape $\xi_l(0)$ and the motionless state of the liquid inside the drop. The latter in particular implies that $d\xi_l/dt = 0$ at $t = 0$. Then, in terms of the Laplace transform,

$$\tilde{\xi}_l(p) = \int_0^{+\infty} \xi_l(t) \exp(-pt)\,dt,$$

the result is (see equation 13 of Ref. 15)

$$\tilde{\xi}_l(p) = \frac{1}{p}\left[1 - \frac{\omega_{l_0}}{\Delta_l(p)}\right]\xi_l(0), \tag{A1}$$

where the denominator is

$$\Delta_l(p) = p^2 + 2p(l-1)\left[2l+1 + \frac{l^2-1}{1 - \frac{1}{2}\sqrt{p}\dfrac{I_{l+1/2}(\sqrt{p})}{I_{l+3/2}(\sqrt{p})}}\right] + \omega_{l_0}^2 \tag{A2}$$

and

$$\omega_{l_0}^2 = \frac{l(l-1)(l+2)}{Ca}$$

corresponds to the frequency of drop oscillations in the ideal-liquid limit. Here, $I_{l+1/2}$ is the modified Bessel function of the first kind of order $l + 1/2$. We note that, in view of the property

$$z\frac{I_{l+1/2}(z)}{I_{l+3/2}(z)} = -z\frac{I_{l+1/2}(-z)}{I_{l+3/2}(-z)}$$

there is no ambiguity associated with the function (A2) along the real negative semi-axis in the complex plane. Accordingly, there is no continuous spectrum (unlike the bubble problem[15]), whereas the inversion of the Laplace transform (A1) and the calculus of residues lead to the summation over the discrete spectrum, which can be written as

$$\xi_l(t) = \xi_l(0)\sum_{k=1}^{\infty} c_k \exp(-\lambda_k t), \tag{A3}$$

where c_k and λ_k are complex constants (dependent on l), $p = -\lambda_k$ are the zeros of the denominator $\Delta_l(p)$, and thus λ_k satisfy

$$\lambda^2 - 2\lambda(l-1)\left[2l+1+\frac{l^2-1}{1-\frac{1}{2}\sqrt{\lambda}\dfrac{I_{l+1/2}(\sqrt{\lambda})}{I_{l+3/2}(\sqrt{\lambda})}}\right]+\omega_{l_0}^2 = 0,$$

which is the characteristic equation as originally obtained elsewhere.[2-4] $J_{l+1/2}$ is the Bessel function of the first kind of order $l+1/2$. For $Ca < Ca_{cr}$ (Ca_{cr} depends on l, of course) the first two modes in (A3) are oscillatory (which can be interpreted as the surface modes) with $\lambda_{1,2} = \delta + i\omega$, where δ is the damping rate, ω is the angular frequency. For $l = 2$ these quantities are shown in FIGURES 3 and 4 as functions of Ca. The remaining eigenvalues are real, and to underscore this we write $\lambda_k = \delta_k$ ($k = 3, 4, \ldots$). These eigenvalues correspond to internal viscous modes. For $Ca > Ca_cr$, all the eigenvalues are real, that is $\lambda_k = \delta_k$ ($k = 1, 2, \ldots$). For the coefficients c_k in (A3), the calculus of residues yields

$$c_k = \frac{\omega_{l_0}^2}{\lambda_k \dfrac{d\Delta_l}{dp}\bigg|_{p=-\lambda_k}}.$$

For $Ca < Ca_{cr}$, when the first two modes are oscillatory, we rewrite (A3) as

$$\xi_l(t) = \xi_l(0)\left[B_1 \exp(-\delta t)\cos\omega t + B_2 \exp(-\delta t)\sin\omega t + \sum_{k=3}^{\infty} c_k \exp(-\delta_k t)\right], \quad \text{(A4)}$$

where all the quantities are real, and

$$B_1 = 2\mathrm{Re}c_1 = 2\mathrm{Re}c_2$$
$$B_2 = 2\mathrm{Im}c_1 = -2\mathrm{Im}c_2$$

We note that all the quantities inside the square brackets in (A4) are functions of l and Ca only (except for t, of course). For the fundamental angular mode $l = 2$, the dependence on Ca has already been shown for ω and δ (FIGS. 3 and 4), and now FIGURES 15 and 16 depict the same for B_1 and B_2. The latter is represented by means of the combination $B_2\omega/\delta$, and it will be clear why from what follows.

The simplified approach we used in this note implies that, instead of (A4), we simply write

$$\xi_l(t) = \xi_l(0)\exp(-\delta t)\left(\cos\omega t + \frac{\delta}{\omega}\sin\omega t\right), \quad \text{(A5)}$$

that is, the internal viscous modes are disregarded and the initial conditions, in particular $d\xi_l/dt = 0$ at $t = 0$, are directly applied to the surface modes. How well the simplified approach works is determined by how far the quantities B_1 and $B_2\omega/\delta$ are from unity. They are not too far, in fact, as we observe from FIGURES 15 and 16. In view of the experimental error involved, bothering with the exact solution of the initial-value problem does not seem to be justified when a simplified approach works that well. Furthermore, even the nonlinear effects, which are neglected here, can potentially have a bigger impact on the result. Note that at one point ($Ca \sim 0.012$), $B_2\omega/\delta$ deviates from unity by about 11%. However, one must keep in

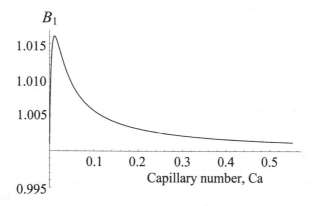

FIGURE 15. Amplitude from **(A4)** as a function of the capillary number.

mind that at this point, B_2 is only about 10% of B_1, which effectively reduces the error to about 1%.

It is interesting to note that both B_1 and $B_2\omega/\delta$ reach their maximum deviation from unity at about the same point ($Ca \sim 0.012$), when the influence of the internal viscous modes on the course of surface evolution is the most pronounced. We can interpret this as follows: this happens at a point where we have a kind of resonance between the surface modes and the internal viscous modes. Indeed, the latter decay e times at $t \sim 1/\delta_3$. Formally the same effect for the surface mode is achieved roughly at 1/8 of the oscillation period, $t \sim 0.25\pi/\omega$. Taking into account that $\omega \approx \sqrt{8}\,Ca^{-1/2}$, $\delta_3 \approx 30$, and $Ca \approx 0.012$, one can see that these times are rather close indeed.

Finally, we mention that some numerical values obtained here are in disagreement with Reference 15. Indeed, figure 3 of Reference 15 shows (in our present notation) that $\xi_2(t)/\xi_2(0)$ for $Ca = 0.36$, the solid line representing the full result and the dashed line referring to the oscillatory modes in **(A4)** taken apart. We observe from

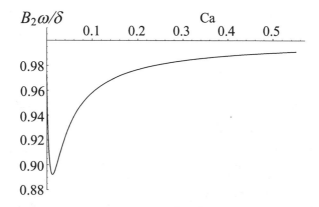

FIGURE 16. As Figure 15.

this figure that B_1 would need to be about 1.08 in order to accommodate an appreciable initial separation between the solid and the dashed curves, whereas here we have obtained $B_1 \sim 1.002$ for the value of Ca in question. In what concerns the simplified approach (**A5**), figure 3 of Reference 15 gives the impression that this approach is about 8% inaccurate, whereas the present evaluation indicates an appreciably smaller error.

Property Measurements and Solidification Studies by Electrostatic Levitation

PAUL-FRANÇOIS PARADIS, JIANDING YU,
TAKEHIKO ISHIKAWA, AND SHINICHI YODA

Japan Aerospace Exploration Agency,
Institute of Space and Astronautical Science, Tsukuba, Japan

ABSTRACT: The National Space Development Agency of Japan has recently developed several electrostatic levitation furnaces and implemented new techniques and procedures for property measurement, solidification studies, and atomic structure research. In addition to the contamination-free environment for undercooled and liquid metals and semiconductors, the newly developed facilities possess the unique capabilities of handling ceramics and high vapor pressure materials, reducing processing time, and imaging high luminosity samples. These are exemplified in this paper with the successful processing of BaTiO₃. This allowed measurement of the density of high temperature solid, liquid, and undercooled phases. Furthermore, the material resulting from containerless solidification consisted of micrometer-size particles and a glass-like phase exhibiting a giant dielectric constant exceeding 100,000.

KEYWORDS: ceramics; containerless; density; levitation; solidification; undercooling

INTRODUCTION

To date, electrostatic levitation has been used in several studies relating to thermophysical property measurements,[1] solidification and nucleation,[2–4] glass formation,[5] neutron scattering investigation,[6] hard sphere model calculation comparative work,[7] and combustion research.[8] However, despite their significant technological importance, very little attention has been devoted to the study of ceramics using this technique, mainly because of problems linked with levitation initiation and lack of charge.

In the study we report here, the use of a hybrid aerodynamic–electrostatic levitation furnace overcame the difficulties associated with high temperature processing of ceramics and freed the sample from any enclosure.[9] The excellent sample position stability and sphericity offered by this levitator make it an attractive platform for the density determination of oxides[5] and for solidification studies.

Barium titanate (BaTiO₃) and its compounds are ferroelectric wide band gap semiconductors that are particularly interesting as nonvolatile ferroelectric memory devices, piezoelectric transducers, thermal switches, and thermistors.[10] When doped

Address for correspondence: Paul-François Paradis, ISS Science Project Office, Tsukuba Space Center, 2-1-1 Sengen, Tsukuba, Ibaraki, Japan 305-8505. Voice: 81-298-68-3813; fax: 81-298-68-3956.
paradis.paulfrancois@jaxa.jp

Ann. N.Y. Acad. Sci. 1027: 464–473 (2004). ©2004 New York Academy of Sciences.
doi: 10.1196/annals.1324.037

with Ce and Rh, $BaTiO_3$ based materials display interesting photorefractive properties, in particular improved phase conjugaison efficiencies.[11]

It is well known that the magnetic properties of $BaTiO_3$[12] arise from the characteristics of the crystal. Knowledge of the single crystal structure is important in the studying the dependence of magnetic and electric properties in the material. Previously, large single crystals of $BaTiO_3$ proved to be very difficult to grow. One of the reasons for this is lack of knowledge of the thermophysical properties (density, viscosity, and surface tension) at high temperature. Therefore, it was decided to establish a thermophysical data base for these materials to aid the ongoing numerical modeling efforts.

This paper briefly describes the unique features of the aerodynamic–electrostatic levitation furnace (levitation initiation, UV imaging) helping to overcome the risk of contamination at elevated temperature ($BaTiO_3$, $T_m = 1,893\,K$),[13] high sample luminosity, and charging problems. We then present the density data for high temperature solid, liquid, and undercooled $BaTiO_3$ and report a gigantic dielectric constant resulting from containerless processing.

EXPERIMENTAL SETUP AND PROCEDURES

For this study, high purity $BaTiO_3$ powder (99.9%, Rare Metallic Co., Tokyo, Japan) was compressed into rods with a about 3 mm diameter at a pressure of 200 MPa. The rods were then sintered for 10 hours at 1,527 K. Cubes with about 2 mm sides were made from these rods and tumbled in a homemade device to obtain spherical polycrystalline samples with diameter of about 2 mm.

The facility[9] used for density measurement (see FIGURE 1) consisted of a stainless steel chamber housing a hybrid aerodynamic–electrostatic levitator and operated at a 450 KPa UHP air pressure. The aerodynamic levitator (see FIGURE 2) permitted a sample to accumulate sufficient electrical charge through thermionic emission by laser heating, allowing subsequent levitation by an electrostatic field. The electrostatic levitator consisted of a pair of parallel disk electrodes between which the positively charged specimen was levitated by means of an active feedback loop (vertical

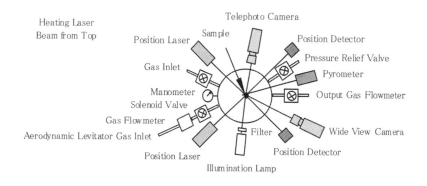

FIGURE 1. Top view of the electrostatic levitation furnace.

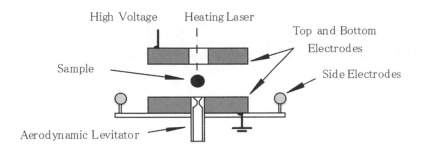

FIGURE 2. Schematic view of the levitator.

control).[14] Four spherical electrodes distributed around the bottom electrode were used for horizontal position control and to provide feedback control.

The sample was heated from the above using unfocused radiation coming from a computer controlled 100W CO_2 laser (Synrad, Evolution 100, 10.6mm). Temperature was recorded over wide ranges using pyrometers (0.90μm, 0.96μm, and 5.14μm). The apparent temperature, obtained from the pyrometer, was calibrated with the release of latent heat of fusion of the material (recalescence peak). Due to lack of data, the emissivity was set at the melting point of the material and assumed constant at all temperatures. This assumption probably holds for the liquid. However, as is evidenced by data listed elsewhere,[15] the transition from the liquid to the solid state in oxides is often accompanied with a marked change in emissivity.

Density measurements were obtained using an imaging technique described in detail elsewhere.[16,17] A high-resolution, black and white CCD video camera equipped with a telephoto objective and a high-pass filter (450nm) was used to obtain a magnified view of the sample illuminated with an intense background UV light from behind. The choice of the UV spectral range allowed perimeter observation with a background practically independent of sample temperature. Hence, excellent imaging was achieved, permitting density measurements for both high temperature solid and liquid materials.[17] In addition to the UV backlighting, the camera was also used with a white light or with no background light to help observe surface features. When a levitated sample was melted, it took a spherical shape due to surface tension and the distribution of surface charge. Once the sample position was stable, images at the rate of 30frames/sec and temperature data were simultaneously recorded with time. The laser beam was then blocked with a mechanical shutter allowing the sample to cool. After the experiment, the video images were digitized and the sample radius r was extracted from each image. The images were then matched with the corresponding cooling curve, obtained from the temperature data. Since the sample was axisymmetric, and because its mass m was known, the density ρ could be found as a function of temperature using the relation

$$\rho(T) = \frac{3m}{4\pi r^3}. \tag{1}$$

The recorded images were calibrated by levitating a brass sphere with a precisely known radius under identical experimental conditions. The samples were recovered and analyzed, as described in the next section.

RESULTS

Density

A BaTiO$_3$ sample (12.16 mg, about 1.7 mm diameter) could be electrostatically levitated with excellent position stability when applying a voltage of -8 KV between the electrodes. From the magnified images, a rotation rate not larger than 1 Hz was observed and the amplitude of oscillation was estimated to be less than 100 μm along both vertical and horizontal directions during heating. Slight instabilities (tumbling and roll over) were observed upon melting BaTiO$_3$. Once the sample was completely melted, it exhibited a stable position. FIGURE 3 A illustrates a temperature profile for a sample experiencing an average radiative and conductive cooling rate of about 500 K/sec. The good signal to noise ratio of this cooling curve is due to the excellent sample position stability during cooling. The recovered samples were spheroids with rough surfaces. The mass of the samples was measured before and after the experiments and, within experimental uncertainties, did not show any

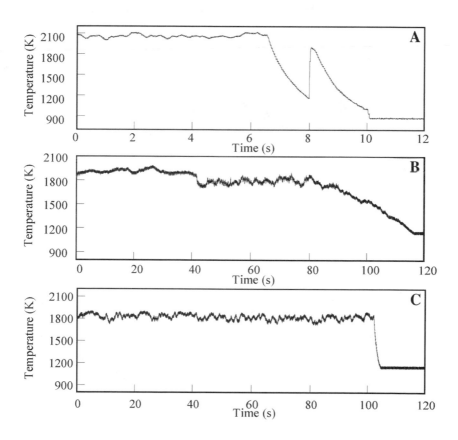

FIGURE 3. Temperature profiles for a BaTiO$_3$ sample: **(A)** BTO-A; **(B)** BTO-B; **(C)** BTO-C.

change. Furthermore, EPMA results did not reveal any evidence of substitution of oxygen by nitrogen in either material.

Density data for solid and liquid BaTiO$_3$ were taken over the temperature intervals 1,220 to 1,893 K and 1,300 to 2,025 K, respectively, (see FIGURE 4). For the liquid, the measurements extended more than 132 K above the melting temperature and 593 K into the undercooled region. The data showed a marked discontinuity, with characteristics of a first-order transition. The density of the solid material exhibited a linear behavior as a function of temperature T that could be fitted by

$$\rho_S(T) = 5.04 \times 10^3 - 0.21(T - T_m) \quad (\text{kg} \cdot \text{m}^{-3}), \tag{2}$$

where $T_m = 1,893$ K. The volume variation $V_S(T)$, normalized with the volume at the melting temperature V_m, was derived from Equation (2), and can be expressed as

$$\frac{V_S(T)}{V_m} = 1 + 4.2 \times 10^{-5}(T - T_m), \tag{3}$$

where $4.2 \times 10^{-5} \text{K}^{-1}$ is the volume expansion coefficient. Similarly, the density of the liquid can be expressed as

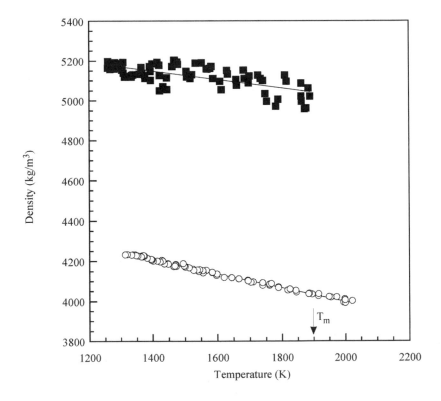

FIGURE 4. Density of high temperature solid (■), liquid (○), and undercooled BaTiO$_3$ versus temperature.

$$\rho_L(T) = 4.04{\times}10^3 - 0.34(T - T_m) \quad (\mathrm{kg \cdot m^{-3}}), \tag{4}$$

with a corresponding volume variation $V_L(T)$ given by

$$\frac{V_L(T)}{V_m} = 1 + 8.4{\times}10^{-5}(T - T_m), \tag{5}$$

where $8.4{\times}10^{-5}\,\mathrm{K}^{-1}$ is the volume expansion coefficient. To the best of our knowledge, these measurements are the first to be reported on such large temperature intervals. However, extrapolating the data for the solid to room temperature yields $5{,}376\,\mathrm{kg \cdot m^{-3}}$, which agrees, within experimental uncertainty, with the value $(5{,}500\,\mathrm{kg \cdot m^{-3}})$ reported elsewhere.[19] In FIGURE 4, it is noticeable that the scatter in the data of the solid is much larger than that of the liquid. This can be explained by the surface and geometry conditions of the solid samples that were not as smooth and spherical as those of the liquid samples. Also, it should be noted that, because the measurement for the solid phase was carried out from a solid sample resulting from a cooling melt, sample porosity could have influenced the data. Furthermore, although X-ray analysis performed on the room temperature sample revealed a tetrahedral structure, the temperature range on which the density of the solid phase was measured was characterized by a cubic crystalline structure, as inferred from the phase diagram.[13]

Solidification

Three types of material processing were performed, as shown in FIGURE 3. The samples were heated either above the melting temperature (FIG. 3A, case 1) or near the melting temperature (FIG. 3B and C, case 2). Following this, morphology, transparency, composition, and microstructure analyses were performed by optical microscopy, EPMA, SEM, and X-ray diffraction. The dielectric constant ε was then measured from 20 K to 300 K with an impedance analyzer, by sputtering the samples with Au electrodes.

For case 1, BTO-A was heated to 2,050 K, 150 K above the melting temperature, and then cooled at a mean rate of 700 K/sec. As the liquid undercooled by about 700 K, recalescence occurred, as expected from the phase diagram. Optical microscopy revealed an opaque and white sample. X-ray diffraction as well as SEM observation revealed that the sample consisted of a columnar crystal microstructure.

Samples BTO-B and BTO-C were heated and kept at the melting temperature for several minutes and were then cooled at respective mean rates of 30 K/sec and 300 K/sec. No recalescence was observed in either profile, although the cooling rate decreased from 300 K/sec to 30 K/sec, indicating that the undercooled liquid could have vitrified. This observation was supported by optical microscopy (yellow, transparent BTO-B and blue, transparent BTO-C). X-ray diffraction as well as SEM observation revealed, however, that both samples were crystalline. BTO-B and BTO-C consisted of a mixed phase microstructure, consisting of 200–500 nm size crystal particles distributed on a glass-like matrix. The particles showed a clear crystalline morphology, but the microstructure of the matrix could not be determined from the resolution of the SEM (20 nm). Hence, this implies that the matrix could be made of glass or nanoparticles with size smaller than 20 nm. TEM analysis are planned to confirm this.

FIGURE 5 shows the temperature dependence of the dielectric constant ε for the three samples in the 20–300 K range for frequencies of 1 kHz, 10 kHz, and 100 kHz. In FIGURE 5 A, the dielectric behavior of BTO-A was characteristic of perovskite $BaTiO_3$, with anomalies (190 K, 273 K) agreeing with the structural phase transitions. The value of ε at room temperature was about 4,000, which is significantly higher compared to p-$BaTiO_3$ ceramics synthesized by conventional techniques. The high value for ε can be explained by an increase in density during solidification. In addition, the columnar crystal growth and possible poling by the facility could also increase ε since the domain growth might be enhanced.

In FIGURE 5 B, the ε of BTO-B was characteristic of h-$BaTiO_3$, exhibiting anomalies at the phase transition temperatures of 220 K and 74 K. However, the ε of BTO-C in FIGURE 5 C illustrated a behavior significantly different from that of BTO-B. A gigantic ε, greater than 100,000 at room temperature, gradually decreased to 60,000 as the temperature decreased to about 70 K and then sharply dropped by more than two orders of magnitude. The temperature drop depended on frequency, increasing from 50 K to 70 K with increasing frequency from 1 kHz to 100 kHz. This relaxation process was not observed in BTO-B. The dielectric behavior for BTO-C was very similar to that of $CaCu_3Ti_4O_{12}$ reported recently.[21]

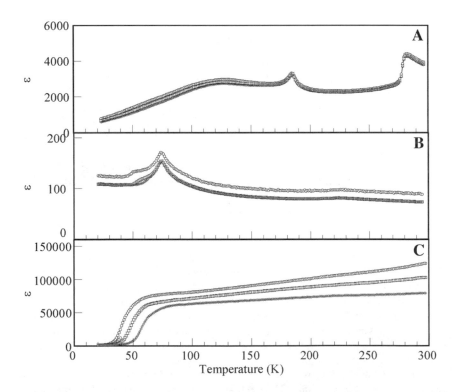

FIGURE 5. Dielectric constant ε as a function of temperature for several frequencies: ○, 1 kHz; □, 10 kHz; and ◇, 100 kHz for samples (**A**) BTO-A, (**B**) BTO-B, and (**C**) BTO-C.

FIGURE 6 shows the temperature dependence of the loss component $\tan\delta$ for the three samples at a frequency of 10 kHz. At room temperature, $\tan\delta$ was less than 0.01 for BTO-A and BTO-B, and was about 0.1 for BTO-C. This makes BTO-C an attractive electronic material compared to BTO-B since $\tan\delta$ only increased by one order whereas ε increased by three orders. FIGURE 6 further reveals that a similar $\tan\delta$ anomaly occurred below 70 K for both BTO-B and BTO-C. It is clear that the $\tan\delta$ anomaly of BTO-B relates to the structural phase transition in h-BaTiO$_3$ at 70 K. Hence, this suggests that the $\tan\delta$ anomaly could also be related to the structural phase transition of BTO-C. Furthermore, this implies that the sharp drop in the value of ε for BTO-C was associated with the structural phase transition in h-BaTiO$_3$.

The giant ε and weak temperature dependence of BTO-C can be attributed to several effects: particle size, oxygen deficiency, and mixed phase microstructure. The effect of particle size on the ferroelectric phase transition and on the dielectric properties of small BaTiO$_3$ particles has been studied for a long time. Experiments with BaTiO$_3$ revealed that particle size plays an important role in the paraelectric–ferroelectric transition and on the dielectric properties.[22] However, since the BTO-B and BTO-C samples had nearly the same particle size and microstructure, the giant ε value for BTO-C could not be simply explained with the effect of particle size. It

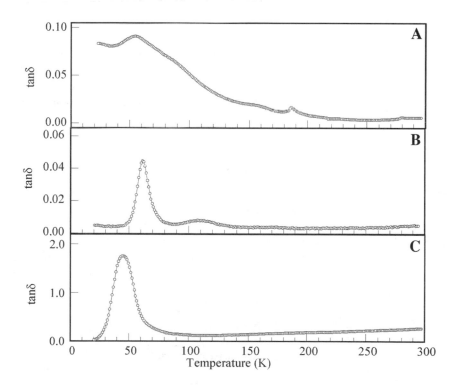

FIGURE 6. Loss component $\tan\delta$ as a function of temperature at a frequency of 10 kHz for samples (**A**) BTO-A, (**B**) BTO-B, and (**C**) BTO-C.

is well known that oxygen deficient $BaTiO_3$ leads to semiconductivity and produces an apparent large ε due to charge transfer. However, the small $\tan\delta$ value of 0.1 in BTO-C indicated that the giant ε value could not arise only from the charge transfer in a semiconducting $BaTiO_3$. The mixed microstructure, consisting of submicrometer particles and a glass-like matrix, suggested that the dielectric behavior of BTO-C can be interpreted in terms of the barrier layer model,[23] based on semiconducting grains with thin insulating grain boundaries. In this model, an effective circuit of parallel capacitors formed from microcrystals gives rise to a weak temperature dependence of ε. The barrier layer model has been applied to fabricate $BaTiO_3$-based ceramic capacitors by multistage processing involving oxygen deficiency and ion doping. For BTO-C, it is thought that the oxygen deficient particles correspond to the semiconducting grains and the glass-like matrix is related to the insulating grain boundaries. According to the barrier layer model, ε mainly depends on the number of capacitors produced by grain boundary and the capacitance of each capacitor. In BTO-C, the submicrometer-size particles are thought to increase the number of capacitors due to an increase in the number of boundaries. The glass-like matrix is also thought to contribute greatly to the high capacitance. Hence, the combined effects of particle size, oxygen deficiency, and mixed phase lead to a high value for ε.

Considering the sharp $\tan\delta$ peak associated with the ferroelectric phase transition occurring in BTO-B and BTO-C, the drastic drop in ε for BTO-C appeared to correspond to the phase transition. However, the relaxation process reflected by the frequency dependence of ε, observed only in BTO-C, indicated that the ε anomaly can also be explained by a relaxation process similar to that reported for $CaCu_3Ti_4O_{12}$.[21,23] This will be further investigated by X-ray diffraction at low temperature.

CONCLUSIONS

Successful processing of $BaTiO_3$ with a pressurized electrostatic levitation furnace allowed measurement of the density of its high temperature solid, liquid, and undercooled phases. In addition, the material resulting from containerless solidification consisted of micrometer-size particles and a glass-like phase exhibiting a giant dielectric constant exceeding 100,000. Future work is currently in progress to devise a model and a theory to explain the results. This will appear in a forthcoming publication.

ACKNOWLEDGMENTS

This work was funded by a Grant-in-Aid for Scientific Research (B) from the Japan Society for the Promotion of Science.

REFERENCES

1. RHIM, W.-K., S.K. CHANG, A.A. RULISON & R.E. SPJUT. 1997. Measurements of thermophysical properties of molten silicon by a high-temperature electrostatic levitator. Int. J. Thermophys. **18:** 459–469.
2. OHSAKA, K., S.K. CHANG, W.-K. RHIM, *et al.* 1997. Specific volumes of the $Zr_{41.2}Ti_{13.8}$ $Cu_{12.5}Ni_{10.0}Be_{22.5}$ alloy in the liquid, glass, and crystalline phases. Appl. Phys. Lett. **70:** 726–728.
3. MORTON, C.W., W.H. HOFMEISTER & R.J. BAYUZICK. 1994. A statistical approach to understanding nucleation phenomena. Mater. Sci. Eng. A **178:** 209–215.
4. RATHZ, T.J., M.B. ROBINSON, R.W. HYERS, *et al.* 2002. Triggered nucleation in $Ni_{60}Nb_{40}$ using an electrostatic levitator. J. Mat. Sci. Lett. **21:** 301–303.
5. PARADIS, P.-F., J. YU, T. ISHIKAWA, *et al.* 2003. Contactless density measurement of superheated and undercooled liquid $Y_3Al_5O_{12}$. J. Cryst. Growth **249:** 523–530.
6. PARADIS, P.-F., T. ISHIKAWA & S. YODA. 2002. Electrostatic levitation furnace for structural studies of high temperature liquid metals by neutron scattering experiments. J. Non-Cryst. Solids **312/314:** 309–313.
7. ISHIKAWA, T., P.-F. PARADIS, T. ITAMI & S. YODA. 2003. Thermophysical properties of liquid refractory metals: comparison between hard sphere model calculation and electrostatic levitation measurements. J. Phys. Chem. **118:** 7912–7920.
8. RHIM, W.-K. 2001. California Institute of Technology, private communication.
9. PARADIS, P.-F., T. ISHIKAWA, J. YU & S. YODA. 2001. An hybrid electrostatic-aerodynamic levitation furnace for the high temperature processing of oxide materials on the ground. Rev. Sci. Instrum. **72:** 2490–2815.
10. DEWITTE, C., J.P. MICHENAUD, & F. DELANNAY. 1989. *In* Euro-Ceramics, Vol. 3. Engineering Ceramics. G. de With, R.A. Terpstra & R. Metselaar, Eds.: 232. Elsevier, London.
11. HUOT, N., J.M.C. JONATHAN, G. PAULIAT, *et al.* 1997. Characterization of a photorefractive rhodium doped barium titanate at 1.06 · m. Opt. Comm. **135:** 133–137.
12. MERZ, W.J. 1949. The electric and optical behavior of $BaTiO_3$ single-domain crystals. Phys. Rev. **76:** 1221–1225.
13. RASE, D.E. & R. ROY. 1955. Phase equilibrium in the system $BaTiO_3$–TiO_2. J. Am. Ceram. Soc. **38:** 110–113.
14. RHIM, W.-K., S.-K. CHUNG, D. BARBER, *et al.* 1993. An electrostatic levitator for high temperature containerless materials processing in 1-*g*. Rev. Sci. Instrum. **64:** 2961–2970.
15. LIDE, D.R. & H.P.R. FREDERIKSE. 1997. CRC Handbook of Chemistry and Physics, 78th edit. CRC Press, Boca Raton.
16. CHUNG, S.K., D.B. THIESSEN & W.-K. RHIM. 1997. A non-contact measurement technique for the density and thermal expansion of molten materials. Rev. Sci. Instrum. **67:** 3175–3181.
17. ISHIKAWA, T., P.-F. PARADIS & S. YODA. 2001. New sample levitation initiation and imaging techniques for the processing of refractory metals with an electrostatic levitator furnace. Rev. Sci. Instrum. **72:** 2490–2495.
18. PARADIS, P.-F., J. YU, T. ISHIKAWA, *et al.* 2003. Contactless density measurement of high temperature $BiFeO_3$ and $BaTiO_3$. Appl. Phys. A **76:** 1–5.
19. CHANNEL INDUSTRIES INC. 839 Ward Drive, Santa Barbara, CA 93111, USA.
20. YU, J., P.-F. PARADIS, T. ISHIKAWA & S. YODA. 2004. Giant dielectric constant of single crystal hexagonal $BaTiO_3$ synthesized by an electrostatic levitation furnace. Chem. Mater. In press.
21. RAMIREZ, A.P., M.A. SUBRAMANIAN, M. GARDEL, *et al.* 2000. Giant dielectric constant response in a copper-titanate. Solid State Commun. **115:** 217–220.
22. SHIH, W.-H. & Q. LU. 1994. Ultrafine titanate powders processed via a precursor-modified sol–gel method. Ferroelectrics **154:** 241–246.
23. SINCLAIR, D.C., T.B. ADAMS, F.D. MORRISON & A.R. WEST. 2002. $CaCu_3Ti_4O_{12}$: one-step internal barrier layer capacitor. Appl. Phys. Lett. **80:** 2153–2155.

Convection in Containerless Processing

ROBERT W. HYERS,[a] DOUGLAS M. MATSON,[b]
KENNETH F. KELTON,[c] AND JAN R. ROGERS[d]

[a]Department of Mechanical and Industrial Engineering,
University of Massachusetts, Amherst, Massachusetts, USA

[b]Tufts University, Medford, Massachusetts, USA

[c]Washington University, St. Louis, Missouri, USA

[d]NASA Marshall Space Flight Center, Huntsville, Alabama, USA

ABSTRACT: Different containerless processing techniques have different strengths and weaknesses. Applying more than one technique allows various parts of a problem to be solved separately. For two research projects, one on phase selection in steels and the other on nucleation and growth of quasicrystals, a combination of experiments using electrostatic levitation (ESL) and electromagnetic levitation (EML) is appropriate. In both experiments, convection is an important variable. The convective conditions achievable with each method are compared for two very different materials: a low-viscosity, high-temperature stainless steel, and a high-viscosity, low-temperature quasicrystal-forming alloy. It is clear that the techniques are complementary when convection is a parameter to be explored in the experiments. For a number of reasons, including the sample size, temperature, and reactivity, direct measurement of the convective velocity is not feasible. Therefore, we must rely on computation techniques to estimate convection in these experiments. These models are an essential part of almost any microgravity investigation. The methods employed and results obtained for the projects levitation observation of dendrite evolution in steel ternary alloy rapid solidification (LODESTARS) and quasicrystalline undercooled alloys for space investigation (QUASI) are explained.

KEYWORDS: electromagnetic levitation; electrostatic levitation; convection; containerless processing; quasicrystals; stainless steel

NOMENCLATURE:

k_B	Boltzmann constant, 1.38×10^{-23} JK^{-1}
e	electron charge, 1.602×10^{-19} C
L_0	(theoretical) Lorentz number, 2.45×10^{-8} WΩK^{-1}
$\vec{J}$	current density
$\vec{H}$	magnetic field
$\vec{B}$	magnetic flux density, $\vec{B} = \mu_0 \vec{H}$
μ_0	permittivity of free space $4\pi \times 10^{-7}$ H/m
$\vec{E}$	electric field
$\vec{u}$	velocity
u_0	reference velocity

Address for correspondence: Robert W. Hyers, Department of Mechanical and Industrial Engineering, University of Massachusetts, Amherst, MA 01003, USA. Voice: 413-545-2253; fax: 413-545-1027.
    hyers@ecs.umass.edu

Ann. N.Y. Acad. Sci. 1027: 474–494 (2004). ©2004 New York Academy of Sciences.
doi: 10.1196/annals.1324.038

l_0	reference length
Re_m	magnetic Reynolds number, $\mu_0 \sigma_{el} u_0 l_0$
Re	Reynolds number, $\rho u_0 R / \mu$
R	radius of the droplet
*	indicates dimensionless quantity
H^*	dimensionless mean curvature
Ca	capillary number, $\gamma / \mu u_0$
Ma	Marangoni number, $|\Delta T \, d\gamma/dT| / \mu u_0$
q	heat flux
σ_{SB}	Stefan–Boltzmann constant, $5.67 \times 10^{-8} \, \text{W/m·K}^4$
θ	angle from sample rotation axis (not the laser axis)
$F(\theta)$	laser flux incident on the droplet at latitude θ
$\tilde{B}o$	dynamic Bond number, $\rho g_0 \beta R^2 |d\gamma/dT|^{-1}$

INTRODUCTION

Experiments on phase selection in steel show a strong dependence of the lifetime of a metastable phase on the convective conditions in the melt.[1] A better understanding of the effect of convection on nucleation of the stable phase depends on testing a wide range of convective conditions. Calculations predict turbulent flow of about 32 cm/sec in a 7 mm sample in ground-based EML, whereas for $1\,g$ ESL with a single laser and a 2 mm diameter sample, the range is 0–4 cm/sec. The ESL range just overlaps the predicted range of microgravity EML with 7 mm diameter samples, which is about 0.5 cm/sec (laminar) to about 17 cm/sec (turbulent).

In the case of quasicrystal-forming samples, nucleation of various phases is being investigated.[2] The larger samples allowed by microgravity EML allow a reduction of almost two orders of magnitude in the shear rate due to internal flow versus single-laser ESL for the conditions of the experiments.

Although some element of computation modeling is often required in the design and/or analysis of microgravity experiments, the models are unusually important to these experiments because they provide the only access to one important variable, the convective velocity.

For the case of electromagnetic levitation of liquid metals, all conventional techniques to measure velocity fail. The samples are opaque and featureless, so laser Doppler velocimetry is not possible. Because the samples are high temperature and highly reactive, and because of the large alternating magnetic field, electromagnetic sensors, such as a Vives probe, do not provide useful data. Finally, seeding the sample with tracer particles for particle imaging velocimetry is not useful because the flow pattern will sweep the particles into the stagnation line(s). However, this method did allow Hyers, Trapaga, and Abedian[3] to determine the transition from laminar to turbulent flow in EML drops by tracking changes in the stagnation line as the Reynolds number of the internal flow increased.

The important phenomena to be captured by the model for EML drops include the electromagnetic force field, the distribution of electromagnetic heating, the recirculation velocity, and the temperature distribution in the drop. The calculation of velocity and temperature is complicated by the transition to turbulence, which occurs within the range of conditions planned for these experiments.[3] Therefore, the convective and thermal models must employ an appropriate turbulence model, as discussed below. The thermal model must also capture the effects of thermal radiation and conduction through the He/H$_2$ process atmosphere.

In modeling the forces and flows in EML, hardware-dependent parameters, including coil geometry, oscillating current frequency, and range of oscillating currents achievable, must be considered. Since the flight hardware for these experiments has not yet been finalized, we have used the hardware-dependent parameters from TEMPUS.

In ESL drops, the positioning forces do not drive convection. Therefore, other physical mechanisms become dominant. The ESL model must capture the effects of the laser heating, including the distribution of the laser heat flux and cooling thermal radiation. Since these ESL experiments are done in vacuum, there is no heat loss by convection or conduction. Flow is still present in ESL, but is driven predominantly by gradients in surface tension due to gradients in temperature. The importance of thermal buoyancy as a driving force for convection must also be evaluated.

THERMOPHYSICAL PROPERTIES

The utility of computation models depends on the accuracy of available property values. The properties used in the simulations for FeCrNi and TiZrNi alloys in the following sections are shown in TABLES 1 and 2, respectively. All properties in the table and their temperature dependences are taken at the melting point, unless otherwise noted.

Except for viscosity and surface tension, the composition dependence was not available. The composition dependence of viscosity over the range of compositions that were employed in LODESTARS is about $\pm 10\%$.[7] The surface tension value and its temperature coefficient were estimated from the Fe–Cr and Fe–Ni binary data collected by Keene.[9] The surface tension changes by about 16 over the range of Cr concentration, and is approximately independent of Ni concentration.

The density, heat capacity, and surface tension are linear with temperature, and for the range of temperatures and compositions in LODESTARS, the curvature of the viscosity–temperature data of Sroka[8] is negligible, consequently viscosity may also be approximated by a linear function.

The temperature dependence of surface tension drives Marangoni convection when there is a thermal gradient along a free surface. This mechanism is the dominant driving force for convection in ESL droplets. Similarly, thermal expansion drives natural convection in $1\,g$ experiments. However, the thermal expansion is modest, and the temperature difference between the hottest and coldest parts of the droplet is small, thus the small density difference may be neglected, except in the term driving natural convection. The temperature dependence of other properties is neglected, resulting in a maximum error of 9% per 100°C from viscosity.

No reliable data on the electrical conductivity of Fe–Cr–Ni ternary alloys were available, hence σ_{el} was estimated from the Wiedemann–Franz–Lorentz law,[10]

$$\frac{k}{\sigma_{el}T} = \frac{\pi^2 k_B^2}{3e^2} \equiv L_0 = 2.45 \times 10^{-8}\,\text{W}\Omega\text{K}^{-1}, \tag{1}$$

where k is the thermal conductivity. Since the measured thermal conductivity is independent of temperature, differentiating (1) with respect to temperature gives

TABLE 1. Thermophysical properties of FeCrNi alloys

Property	Symbol	Value	Reference	Percent per K
Density	ρ	7,040 kg/m^3	4	−0.01%
Thermal expansion	β	1.02×10^{-4} K^{-1}	4	
Thermal conductivity	k	28.4 W/m-K	5	indep. of T
Heat capacity	C_p	830 J/kg-K	6	+0.001%
	$\dfrac{dC_p}{dT}$	7.5×10^{-3} J/kg-K^2	6	
Viscosity	μ	5.59×10^{-3} Pa-sec	7	−0.09%
	$\dfrac{d\mu}{dT}$	-5.15×10^{-6} Pa-sec/K	8	
Surface tension	γ	1.70 N/m	9	−0.02%
	$\dfrac{d\gamma}{dT}$	-3×10^{-3} N/m-K	9	
Electrical conductivity	σ_{el}	$6.63 \times 10^5\,\Omega^{-1}$m^{-1}	see text	−0.06%
	$\dfrac{d\sigma_{el}}{dT}$	$380\,\Omega^{-1}$m^{-1}K^{-1}	see text	
Total hem. emissivity	ε_T	0.33	see text	

TABLE 2. Thermophysical properties of TiZrNi alloys

Property	Symbol	Value	Reference	Percent per K
Density	ρ	5,973 kg/m^3	12	−0.005%
Thermal expansion	β	5.0×10^{-5} K^{-1}	12	
Thermal conductivity	k	18 W/m-K	13	indep. of T
Heat capacity	C_p	630 J/kg-K	2	at 680°C
Viscosity	μ	0.167 Pa-sec	14	at 680°C
Surface tension	γ	1.65 N/m	14	−0.02%
	$\dfrac{d\gamma}{dT}$	-4×10^{-4} N/m-K	14	
Electrical conductivity	σ_{el}	$5.56 \times 10^5\,\Omega^{-1}$m^{-1}	15	
Total hem. emissivity	ε_T	0.25	2	

$$\frac{d\sigma_{el}}{dT} = -\frac{k}{L_0 T^2}. \tag{2}$$

The values for σ_{el} and its temperature dependence obtained by this method are consistent with the experimental data for Fe–Ni.[10] The electrical conductivity values that were calculated for FeCrNi are slightly lower than for Fe–Ni, as would be expected from the inclusion of a third constituent. The total hemispherical emissivity ε_T was estimated from ESL data. The ratio C_p/ε_T was determined from the free cooling rate of liquid samples in vacuum. The known heat capacity C_p was divided by this ratio to obtain ε_T. The weak temperature dependence of ε_T was ignored.

ELECTROMAGNETIC FORCE AND HEATING IN EML DROPLETS

The electromagnetic force and heating of an electromagnetically levitated sample depend on many variables, but these fall into three classes: the geometry of the coils, the size and properties of the sample, and the frequency and magnitude of the applied current. For these calculations, the geometry of the coils from TEMPUS MSL-1 was used (see FIGURE 1) since the final geometry of the coils to be employed in these experiments is not yet determined.

The TEMPUS coils are wound such that the two groups of three coils nearest the equator carry currents in opposite directions to form a quadrupole field. This geometry provides the steepest practical gradient in force versus sample displacement. The positioning coils are operated at about 160kHz, and provide only a small amount of heat to the sample.

The other six coils are all wound in the same direction to create a dipole field. This field geometry provides efficient heating with minimal force on the sample. These heating coils are operated at about 350kHz.

Problem Definition

Given the properties of the sample, the time scale of magnetic diffusion is much greater than the time scale for electromagnetic waves, which in turn is much greater than the time scale for charge relaxation. Under these conditions, it is appropriate to

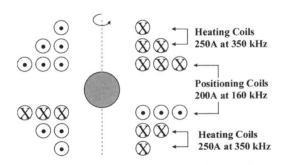

FIGURE 1. TEMPUS MSL-1 coil configuration.[20]

approximate the system as magnetoquasistatic, and the appropriate form of Maxell's equations is

$$\nabla \times \vec{H} = \vec{J}$$

$$\nabla \cdot \vec{B} = 0 \tag{3}$$

$$\nabla \times \vec{E} = -\frac{\partial \vec{B}}{\partial t}.$$

Although the FeCrNi alloys used in this system are ferromagnetic at room temperature, they are non-magnetic above the Curie temperature, about 650°C. The motion of a conductor in a magnetic field influences the current distribution,

$$\vec{J} = \sigma_{el}(\vec{E} + \vec{u} \times \mu_0 \vec{H}). \tag{4}$$

However, the significance of this interaction is determined by the magnetic Reynolds number $re_m = \mu_0 \sigma_{el} u_0 l_0$. For these experiments Re_m varies between 10^{-4} and 10^{-3}. Since this is much less than unity, the flow field does not significantly affect the induced current and a semicoupled model is appropriate, with the magnetic calculations neglecting the effects of fluid flow.

Maxwell's equations are solved by the method of mutual inductances[16] to give the distribution of induced current density in the drop. Then the magnetic field is calculated by means of the Biot–Savart law.

Validation

As a part of the TEMPUS redesign for MSL-1, experimental measurements of the levitation force were made by, and in collaboration with, Dr. G. Lohöfer of the DLR

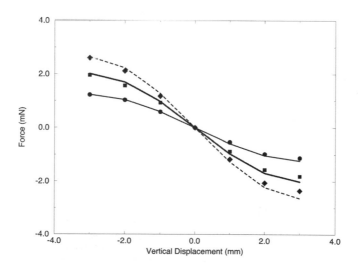

FIGURE 2. Comparison of measurements and predictions for TEMPUS coils; *lines* are predictions and *points* are measurements: ● and ——, 3 V; ■ and ——, 4 V; ◆ and ----, 5 V.[20]

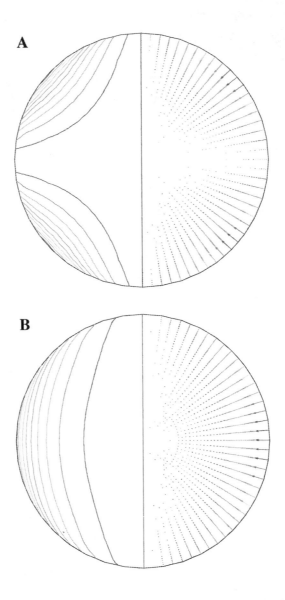

FIGURE 3. Electromagnetic force contours (*left*) and vectors (*right*) for a 7 mm diameter FeCrNi sample in TEMPUS. **(A)** Force from quadrupole positioning field. Positioner current is 150 A, heating current is zero. The maximum force density is 5.7×10^3 N/m³. The force field is mirror symmetric, with the magnitude going to zero at the equator. **(B)** Force from dipole heating field. Positioner current is zero and heater current is 150 A. The force field is mirror symmetric, with the maximum force of 3.6×10^5 N/m³ occurring on the equator of the sphere.[20]

in Cologne, Germany. These measurements were made by measuring the force on a copper sphere suspended in the TEMPUS coils, at various operating currents. The results of the comparison are presented in FIGURE 2. The points represent measurements, and the lines are the predictions generated by the model. The experiments and predictions agree within the error of the experiments, providing some confidence in the predictive capabilities of the model.

Results

The electromagnetic force distribution from the quadrupole positioning field is shown in FIGURE 3 A. The force is directed everywhere inward, and varies from zero at the equator to a maximum at about 45° latitude. The force is zero at the equator because the opposing fields from the upper and lower coils cancel exactly there. The sample is at the center of the coil system, so that the force is balanced—the net translation force on the droplet is zero.

The electromagnetic force distribution from the dipole heating field is shown in FIGURE 3 B. The force is also directed everywhere inward, but is maximum at the equator. For any combination of heating and positioning coil current, the force distribution is the superposition of the two force fields.

MAGNETOHYDRODYNAMIC FLOW IN EML DROPLETS

Problem Definition

The flow in liquid metal droplets is governed by the Navier–Stokes equations. For flow driven by magnetic forces, the nondimensional form of the equations is

$$\nabla^* \cdot \vec{u}^* = 0$$

$$\frac{\partial \vec{u}^*}{\partial t^*} + \vec{u}^* \cdot \nabla^* \vec{u}^* = -\nabla^* P^* + \frac{1}{Re} \nabla^{*2} \vec{u}^* + \vec{F}^*. \tag{5}$$

As shown in the previous section, the magnetic force is independent of the flow. The temperature dependence of the thermophysical properties is weak enough that for the temperature gradients seen in these conditions, the temperature dependence of the properties may be safely ignored in steady-state simulations.

The appropriate boundary conditions are as follows:

1. The flow does not cross the free surface

$$u_r^* \Big|_{r^* = 1} = 0.$$

2. The normal component of stress at the free surface is the sum of ambient pressure and the stress due to surface tension γ, expressed in terms of the dimensionless mean curvature H^* and the Reynolds and Capillary numbers,

$$\tau^* \cdot \hat{i}_n \Big|_{r^* = 1} = -P_0^* + \frac{1}{ReCa} 2H^*.$$

3. The tangential component of stress is the gradient in surface tension due to the temperature gradient along the surface, expressed in terms of the dimensionless temperature gradient, the Reynolds number, and the Marangoni number,

$$\tau^* \cdot \hat{i}_t \Big|_{r^* = 1} = \frac{Ma}{Re} \nabla_S^* T^*.$$

In practice, the equilibrium shape of the droplets in microgravity for moderate settings of the heating current is very nearly spherical, hence the free surface may be reasonably approximated by a fixed spherical surface with no traction.

Solution Method

We use FIDAP, a commercial finite element code for the computational fluid dynamics simulations. The full non-linear Navier–Stokes equations are solved on a two-dimensional, axisymmetric grid. The magnetic force is calculated separately, as described in the previous section, and input via subroutine to drive the flow. The grid used for solution consists of 1,944 elements with 1,741 nodes. Tests show the solution is independent of the grid for this grid density.

Turbulence Modeling

According to the experiments of Hyers et al.,[3] the transition to turbulence in EML droplets occurs at a Reynolds number of about 525. The conditions planned for LODESTARS include a range of Reynolds numbers of 50 to 1,700, so that both laminar and turbulent flows will occur and turbulence must be considered in modeling the flows. The problem of flows near the Reynolds number for transition from laminar to turbulent flow is especially difficult. Berry et al.[17] evaluated a number of

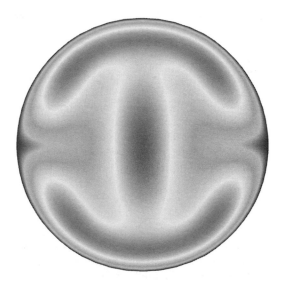

FIGURE 4. Calculated distribution of turbulent viscosity in a FeCrNi droplet, isothermal hold at T_m. The effective viscosity ranges from $1 \times$ laminar to $8.9 \times$ laminar.[20]

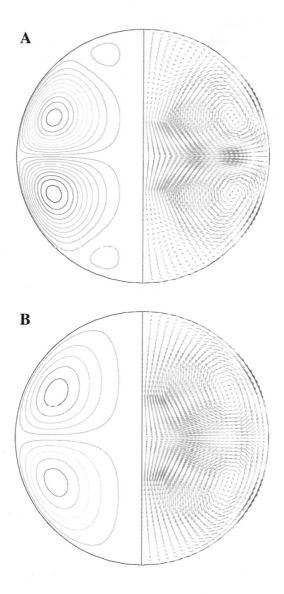

FIGURE 5. Flow in EML droplets. (**A**) Positioner-dominated flow in FeCrNi droplets. The positioner current is 150 A, and the heater current is zero. The droplet is cooling through the melting temperature. The flow is directed outward at the equator because the magnetic force is zero there. The maximum velocity is 1.9 cm/sec. (**B**) Heater-dominated flow in FeCrNi droplets. The positioner current is 150 A and the heater current is 40.0 A. The droplet is cooling through T_m. The flow is directed inward at the equator because the magnetic force is greatest there. The maximum velocity is 5.5 cm/sec.[20]

turbulence models and determined that, for the case of EML drops, the renormalization group (RNG) method[18] is the most appropriate.

The RNG turbulence model shows good agreement with measurements in many systems, including both high and low Reynolds number turbulence. It does not require "wall functions" or parameters that must be empirically adjusted for each flow system. The RNG model also handles non-uniform Reynolds stresses, making it well suited for systems with anisotropic turbulence.[18] The predicted turbulent effective viscosity is plotted in FIGURE 4.

Results

Typical flow patterns for positioner-dominated and heater-dominated flow are shown in FIGURE 5 A and B, respectively. Recall that the electromagnetic force for the positioning field has a maximum at about ±45° latitude and is zero at the equator. Therefore, the positioner-dominated flow at low Reynolds number consists of four loops, directed inward at high latitude, and outward at the equator and poles. At increasing Reynolds number, the inertia of the fluid in the equatorial loops becomes more important, and the polar loops shrink and eventually disappear.

For flow dominated by the heating field, the flow pattern is different. The heating field produces a maximum force directed inward at the equator, so that the flow is also directed inward at the equator. The heater-dominated flow pattern is still mirror-symmetric and axisymmetric, and consists of two loops.

The case shown in FIGURE 5 B is just below the transition from laminar to turbulent flow. At higher velocities, the pattern is qualitatively the same, except that the maximum velocity increases more slowly with increasing field strength.

Another important feature of the flow patterns is that, since the heating flow pattern is opposite the positioning flow pattern, applying a small heating field slows the flow from the positioning field by weakening the equatorial loops. The minimum flow velocity in the droplet occurs for balanced heating and positioning forces, and not for minimum positioning force.

Parametric studies were performed to determine the range of velocities accessible by microgravity EML for various droplet sizes and the functional relation among the different parameters. The maximum flow velocity for various sample sizes and heating field parameters is presented in FIGURE 6.

THERMAL DISTRIBUTION IN EML DROPLETS

In EML testing, heat is added by resistive dissipation of the induced currents and removed by thermal radiation and conduction. There is no appreciable convection in the gas phase in microgravity. However, the conduction of heat from the sample to the water-cooled levitation coils is quite complex. Heat flows from the spherical sample through a gas mixture, the wall of the Si_3N_4 sample holder, another gas layer, and finally to the levitation coils that serve as a heat sink. We represent the resistance to heat flow of all of these components by the single overall heat transfer coefficient U.

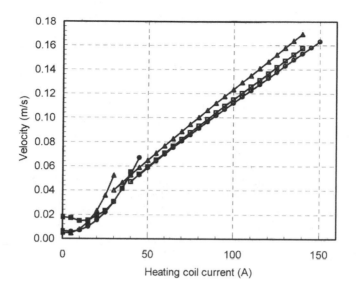

FIGURE 6. Calculated heating-field dominated flow in TEMPUS MSL-1 coil. The *solid symbols* represent laminar flow, the *open symbols* represent turbulent flow: ● and ○, 6 mm; ■ and □, 7 mm; ▲ and △, 8 mm. The positioning coil current is 145 A.[20]

Problem Definition

The heat flow in the drop is governed by the energy equation,

$$\rho C_p \frac{\partial T}{\partial t} + \rho C_p (\vec{u} \cdot \nabla T) = (k\nabla^2 T + Q_{EM}), \tag{6}$$

where all terms have been previously defined except the electromagnetic heat generation term, Q_{EM}. The appropriate boundary condition is

$$-k\frac{dT}{dr}\bigg|_{r=R} = q_{rad} + q_{cond} = \varepsilon_T \sigma_{SB}(T^4 - T_w^4) + U(T - T_c), \tag{7}$$

where the radiative heat flux q_{rad} is the product of the total hemispherical emissivity ε_T, the Stefan–Boltzmann constant σ_{SB}, and the difference between fourth powers of the surface temperature T and the sample holder wall temperature T_w. Since FeCrNi alloys are so high melting, the radiative flux from the sample holder may be reasonably neglected. The conductive heat flux q_{conv} is the product of the overall heat transfer coefficient U and the difference between the sample temperature and the temperature of the water-cooled levitation coils T_c.

Solution Method

The temperature distribution in the EML drops is calculated simultaneously with the fluid flow in the commercial finite element package. The induction heating power distribution is calculated as described above and input via subroutine. For turbulent flow cases, heat transfer within the drop is increased due to turbulent mixing of the fluid. The increased heat transfer is modeled by RNG, as described in the turbulence section.

Results

The temperature distribution is shown in FIGURE 7 an isothermal hold at 250°C below the melting point, where the lowest measurements are made. Isothermal holds represent the worst-case temperature distributions for the planned experiments, since the heat input is greatest under these conditions. Positioning only represents the best-case temperature distribution, and the planned experiments involving transient cooling will cover the range between these cases.

The maximum temperature difference is about 0.1 degrees for the positioner only case, and 3.2 degrees for the isothermal hold condition. In both cases, this temperature difference is too small to affect the nucleation or growth of the primary phase, or the phase selection. The temperature difference across the sample is a weak function of diameter and a strong function of sample average temperature, as shown in FIGURE 8.

MARANGONI FLOW IN ESL DROPLETS

Problem Definition

The governing equations and boundary conditions for electrostatically levitated (ESL) drops are the same as for EML drops except, rather than an electromagnetic force driving the flow, a non-uniform heating flux from a laser causes a temperature gradient that drives Marangoni and natural convection. The heat flux from the laser is approximately Gaussian; the heat flux in a plane perpendicular to the beam direction varies with axial distance r_{cyl} as

$$q_{laser} = \frac{P_0}{\pi b^2} \exp\left(-\left[\frac{r_{cyl}}{b}\right]^2\right),\qquad(8)$$

where P_0 is the total power in the beam and $2b\ln 2$ is the full width at half maximum intensity. The cooling is by thermal radiation. Evaporative cooling was evaluated for this system, but is only a few percent of the radiative cooling at the highest temperatures and much less at typical measurement temperatures.

The heat input into the drop was evaluated for two limits: the droplet not rotating (or the rotation axis parallel to the laser axis), and the droplet rotating quickly on an axis perpendicular to the laser axis. In the case of the droplet not rotating, the laser heat flux absorbed on the surface of the sphere is $aq_{laser}\cos\theta$, where a is the absorption coefficient and θ is the angle from the laser axis.

For the case of a drop that is rotating rapidly about an axis perpendicular to the laser axis, the average laser flux is greatest at the equator, zero at the poles, and a function of latitude only. The rotation of the droplet averages out the laser power around the circumference of the sphere.

For each rotation of the droplet, each point on the surface spends half of a rotation on the dark side of the sphere (opposite the laser), and half of its time traveling through the laser beam. The projection into a plane perpendicular to the laser beam

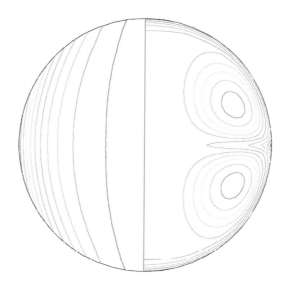

FIGURE 7. Electromagnetic heating distribution (*left*) and temperature distribution (*right*) for isothermal hold at $T_m - 250°C$, FeCrNi droplet. The positioner current is 150 A and the heater current is 135.0 A. The flow has a strong effect on the temperature distribution, sweeping cooler liquid from the poles to cover the equator, even though the heat generation at the equator is greater. The temperatures plotted range from 1,477.9 to 1,480.5 K, with a surface temperature difference of only 1.1 K.[20]

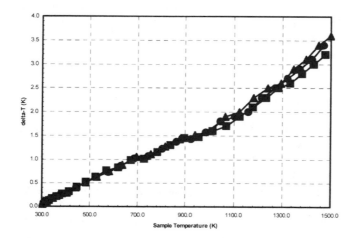

FIGURE 8. Maximum temperature difference versus maximum temperature difference between heating and positioning coils: ●, 6 mm; ■, 7 mm; ▲, 8 mm.[20]

of the path of each point on the sphere is a line through the Gaussian profile, and the average power absorbed is $aF(\theta)\cos\theta/(2\pi)$, where θ is the angle from the rotation axis (not the laser axis) and $F(\theta)$ is the laser flux incident on the droplet at latitude θ, integrated around the circumference of the drop. Substituting the Gaussian beam profile and integrating gives the laser flux incident on the sample,

$$\text{flux}(\theta) = \frac{P_0}{2\pi^{3/2}b}\text{erf}\left(\frac{R}{b}\sin\theta\right)\exp\left(-\left[\frac{R}{b}\cos\theta\right]^2\right). \tag{9}$$

The droplets in real ESL experiments experience conditions between these two limits. The experiments are usually controlled to limit the rotation rate in terrestrial ESL experiments to a few Hz. If the rotation rate is faster than the reciprocal of the thermal equilibration time, then the thermal distribution in the droplet will be close to that predicted by the model for rapidly rotating drops. For 2 mm diameter FeCrNi drops, the time scale for heat transfer across the droplet diameter is about 0.8 sec, so a rotation rate of more than a few Hz will have the averaged temperature profile, whereas drops with a rotation rate much less than 1 Hz will have the temperature profile of a non-rotating drop.

Solution Method

The problems are solved in FIDAP, a commercial finite element code for the computational fluid dynamics simulations. The full non-linear Navier–Stokes equations are solved on a two-dimensional, axisymmetric grid. The absorbed heat flux from the laser is calculated in a boundary condition subroutine.

Results

The quantitative results of the calculations for ESL FeCrNi drops are shown in FIGURE 9. The flow pattern and temperature distribution for the non-rotating FeCrNi droplet at its melting point are qualitatively similar to those shown for the quasicrystal-forming TiZrNi drop, shown in FIGURE 10A.

The heating laser has a full width at half maximum intensity of 0.5 mm, which is a close approximation of the beam of the YAG laser employed at the NASA MSFC ESL during the experiments discussed already. The maximum velocity for FeCrNi at its melting point is 13 cm/sec, and the maximum temperature difference is 29 K. The flow consists of a single recirculation loop, flowing from the hot side to the cold side. The *hot spot* produced by the laser covers a large fraction of the surface area, but is very shallow, in part because of the cold fluid convected from the interior of the drop. The effect of fluid flow on the temperature distribution in the drop is quite large, as shown by the dramatic curvature of the isotherms.

On the other hand, the flow pattern and temperature distribution for the same sample at the same temperature, heated by the same laser, but with the sample rotating is similar to that shown in FIGURE 10B. The flow pattern now shows two loops, due to the mirror symmetry of the thermal distribution. The maximum velocity for FeCrNi is reduced to 5.5 cm/sec, a factor of 2.3, and the temperature difference is only 6.2 K, lower by a factor of 2.4. Both of these calculations are made at T_m, so that both ΔT and v_{max} are slightly lower in undercooled samples.

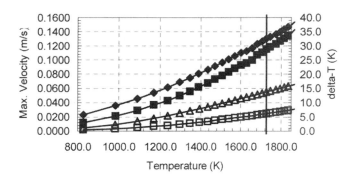

FIGURE 9. Maximum velocity and maximum temperature difference versus average temperature in 2 mm diameter FeCrNi droplets under isothermal holds in ESL, for the limits of fast rotation (*open symbols*) and no rotation (*closed symbols*): ◆ and ◇, maximum velocity; ■ and □, ΔT. Rotation of the drop reduces the temperature difference at the melting point by a factor of 4.6, and reduces the maximum velocity by a factor of 2.3. The vertical line indicates the recalescence temperature, T_0.[20]

For the case of the rotating droplet heated from the side, it was considered that natural convection might also be important in $1\,g$ experiments. However, the calculated flow velocity due to natural convection was 530 times lower than the Marangoni flow for this condition. This result is consistent with expectation, given that the dynamic Bond number

$$\tilde{B}_0 \equiv \frac{\rho g_0 \beta R^2}{\left|\dfrac{d\gamma}{dT}\right|} = 2.5 \times 10^{-3},$$

indicating that the Marangoni force is much greater than the thermal buoyancy force.

Parametric studies exploring the effect of the various process parameters on the temperature and convection in ESL droplets were also performed. These results are presented in FIGURE 9 for droplets heated by a Gaussian laser with a 0.5 mm full width half maximum and a range of laser power suggest that droplet rotation has a large and consistent effect throughout the range of the planned experiments, causing a reduction of 55–80% in ΔT and a reduction of 78–86% in $v_{\max}$.

A third case that must be considered for these experiments is a freely cooling droplet. The droplet is melted and heated above its melting point by the laser, but then the laser is turned off. When the laser is turned off, the temperature distribution on the surface of the droplet quickly becomes uniform, removing the driving force for convection. The viscous drag then quickly reduces the internal flow in the droplet to zero.

The time scale for viscous dissipation of flow is $\tau = \rho l_0^2 / \mu$. For the flow patterns shown in FIGURES 5 and 10, the appropriate length scale is $l_0 = R/3$, hence, for a 2 mm diameter FeCrNi droplets, this time scale is about 0.14 sec. For the ESL experiments reported elsewhere in this paper, the free cooling time is about 4.5 sec, or more than thirty times the viscous dissipation time scale. Consequently, the maximum velocity remaining in the droplet should be lower by orders of magnitude than those predicted for isothermal holds.

Shear Rate in TiZrNi Alloys

For project QUASI, the experiments on nucleation of solid phases drive the limits on convection. Because the critical clusters are very small, they are carried with the flow. Therefore, it is not the absolute flow velocity but the shear rate that is important in these experiments. A large shear rate would cause the diffusion fields of two or more nuclei to overlap, preventing measurements of nucleation kinetics under diffusion-limited conditions.

Taking the effective diffusion length for the calculated critical size of the clusters,[2] the diameter of the diffusion fields around the clusters is approximately 20nm. Using

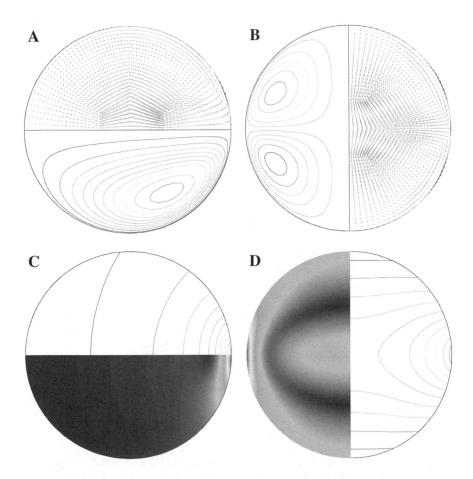

FIGURE 10. Calculated fluid flow and temperature distribution in $Ti_{37}Zr_{42}Ni_{21}$ droplets held at 680°C in ESL with a single heating laser. **(A)** Non-rotating drop, velocity and streamlines, maximum velocity 6.5×10^{-3} m/sec. **(B)** Rotating drop, velocity and streamlines, maximum velocity 8.7×10^{-4} m/sec. **(C)** Non-rotating drop, shear (*bottom*) and temperature (*top*), maximum shear rate 79 sec^{-1}. **(D)** Rotating drop, shear (*left*) and temperature (*right*), maximum shear rate 5.8 sec^{-1}.[20]

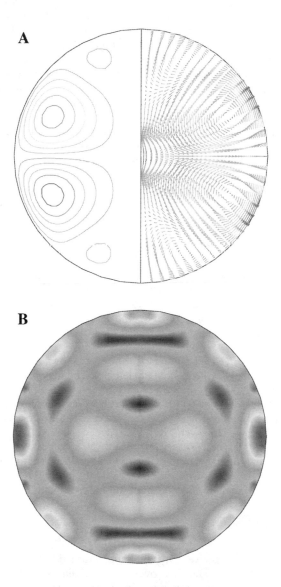

FIGURE 11. Calculated fluid flow and temperature distribution in a $Ti_{37}Zr_{42}Ni_{21}$ droplet cooling through 680°C in EML. TEMPUS coils and parameters are for 50 A positioning and zero for heating currents. **(A)** Velocity and streamlines. The maximum velocity of 1.9×10^{-5} m/sec occurs on the surface. The flow is four loops, outward at the equator and inward at about ±45°. **(B)** Shear rate. The maximum of $0.05\,sec^{-1}$ occurs at the separation point at the equator with secondary maxima at the stagnation points at about ±45°.[20]

an estimate of the critical cluster density for steady-state obtained from the coupled-flux modeling, clusters will be separated by $d = 1\,\mu m$ to $d = 0.1\,\mu m$, if uniformly spaced. For a cooling rate of approximately $10°C/sec$, they should remain apart for at least $t_d = 1\,sec$ to avoid convection contamination in the evolution of the cluster distribution. The maximum allowed shear rate to avoid collisions between the diffusion fields of the critical nuclei is $d/(t_d^* 2L)$, or $5–50\,sec^{-1}$. These are upper limits, given the uncertainties in the assumptions inherent in this estimate. Furthermore, cluster evolution is governed not only by the critical size clusters, but by the entire cluster population, leading to a much smaller estimate of the cluster separation. A rate that is two orders of magnitude less ($0.05–0.5\,sec^{-1}$) than the above estimate is, therefore, deemed necessary to assure a diffusion-controlled experiment.

Marangoni flows in ESL produce sufficient shear to affect the diffusion profiles. Computed fluid dynamic calculations show that the maximum shear rate for a single $0.5\,mm$ diameter Gaussian heating laser is $79\,sec^{-1}$ for a droplet that does not rotate, and $5.8\,sec^{-1}$ for a droplet rotating faster than about $0.5\,Hz$. These results are shown, together with the flow pattern and temperature distribution for each case, in FIGURE 10.

The stationary droplet (FIG. 10 A and C) shows a single recirculation loop from the region heated by the laser toward the cold side of the droplet and a maximum velocity of $6.5\times10^{-3}\,m/sec$. This calculated value is for a $0.5\,mm$ diameter Gaussian laser, and shows reasonable agreement with Bauer's analytical point-source solution.[18,11] The temperature difference across the drop (FIG. 10C, upper half) is large, $15.8\,K$, compared to the analytical value of $15\,K$. The temperature gradient is very steep near the laser spot. The shear rate is also greatest near the laser spot, with a maximum value of $79\,sec^{-1}$.

The rotating droplet (FIG. 10B and D), however, shows a different distribution due to spreading of the laser heat about the equator. The flow pattern consists of two recirculation loops with a maximum velocity of $8.7\times10^{-4}\,m/sec$. The temperature difference across the drop occurs between the equator and poles, and is only $1.7\,K$. The shear rate is still greatest where the flow diverges at the laser spot, but is much smaller, about $5.8\,sec^{-1}$. However, even this reduced value is still too large to meet the requirements of $0.05\,sec^{-1}–0.5\,sec^{-1}$.

On the other hand, microgravity EML produces a shear rate of only $0.05\,sec^{-1}$ at the same cooling rate, as shown in FIGURE 11. Only microgravity EML meets the experimental requirements for QUASI nucleation experiments.

SUMMARY

Direct measurement of the velocity and temperature distribution within EML and ESL is not feasible. Therefore, this investigation requires an extensive modeling component to allow the convection levels and temperature distributions to be quantified.

We have presented a review of thermophysical properties, and techniques and results for:

- electromagnetic force and heating in EML droplets,
- EM-driven flow in EML droplets,
- temperature distribution in microgravity EML droplets accounting for heat loss by various mechanisms,

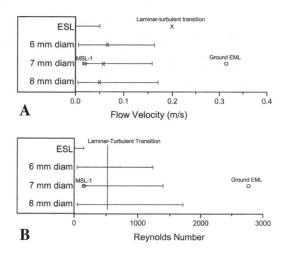

FIGURE 12. Ranges of convection in terms of (**A**) velocity and (**B**) Reynolds number for ESL and microgravity EML. The *circles* indicate the convection levels achieved in MSL-1 microgravity EML and in 1 *g* EML.[20]

- microscopic heat loss to the environment and to the undercooled liquid during recalescence,
- temperature distribution in laser-heated ESL droplets,
- Marangoni convection in ESL droplets, and
- comparison of shear rates in EML and ESL.

The methods described have been employed to calculate the flow ranges achievable for various systems. As shown in FIGURE 12, testing on the ground is limited to very narrow ranges: less than about 4.4 cm/sec ($Re \approx 110$) for ESL and velocity about 32 cm/sec ($Re \approx 2,800$) for 1 *g* EML. In contrast, microgravity EML provides a wide range of accessible convective velocity and Reynolds number permutations.

The limitation of ground-based EML is due to the narrow range of sample size and operating conditions that provide stable levitation. The flow in ESL is dominated by Marangoni convection: the heating laser causes thermal gradients, which cause gradients in surface tension, driving the flow. The maximum Marangoni convection at the required measurement temperature is about 6 cm/sec. When the heating laser is turned off, the flow quickly damps, so the minimum velocity is near zero.

For microgravity EML, the lower limit on convective velocity is determined by the quasistatic acceleration level, which sets the force required to contain the sample. The upper limit on convection is determined by the heat induced in the sample; higher field strengths do not allow sufficient undercooling for the required experiments.

CONCLUSIONS

Convection is an important variable in a number of containerless experiments, even though it often cannot be measured directly. Because the flow characteristics of containerless methods differ, a combination of methods is required to meet the requirements of the experimental program, exploiting the strengths of various techniques.

Computation modeling shows the combination of techniques necessary to meet the LODESTARS requirement of tests over a range of almost two orders of magnitude in velocity. Modeling also shows how to reduce shear rate two orders of magnitude to meet the requirements of QUASI. Experiment-specific modeling of flow conditions is enabling for these two experiments, and is an essential ingredient in almost any program of microgravity experiments.

ACKNOWLEDGMENTS

This work was supported in part by the National Aeronautics and Space Administration under Grants NAG8-1682 and NAG8-1685. The experimental part of this work was performed at the NASA Marshall Space Flight Center Electrostatic Levitation Facility. The authors wish to thank Trudy Allen and Glenn Fountain for their help with the ESL experiments.

REFERENCES

1. MATSON, D.M., D.J. FAIR, R.W. HYERS & J.R. ROGERS. 2004. Contrasting electrostatic and electromagnetic levitation experimental results for transformation kinetics of steel alloys. *In* Transport Phenomena in Microgravity. S.S. Sadhal, Ed. Ann. N.Y. Acad. Sci. **1027:** this volume.
2. KELTON, K.F., *et al.* 2002. Studies of nucleation and growth, specific heat and viscosity of undercooled melts of quasicrystals and polytetrehedral-phase-forming alloys. Science Requirements Document, NASA Document QUASI-RQMT-0001.
3. HYERS, R.W., *et al.* 2003. Met. Trans. **34B:** 29.
4. MIZUKAMI, H., *et al.* 2000. ISIJ Intl. **40:** 987.
5. MIETTINEN, J. 1997. Met. Trans. **28B:** 281.
6. BOGAARD, R.H., *et al.* 1993. Thermochem. Acta **218:** 373.
7. SROKA, M. & J. SKALA. 1982. Freiberger Forschung. B **B229:** 17.
8. SROKA, M. & L. DOBROVSKY. 1989. Kovove Materialy **27:** 731.
9. KEENE, B.J. 1988. Intl. Mater. Rev. **33:** 1.
10. IIDA, T. & R.I.L. GUTHRIE. 1988. The Physical Properties of Liquid Metals. Oxford University Press. 235–241.
11. KITA, Y. & Z. MORITA. 1984. J. Non-Cryst. Solids **61/62:** 1079.
12. BRADSHAW, R.C., *et al.* 2004. Intermetallics. Submitted.
13. JOHNSON, W.L., *et al.* 1999. Thermophysical properties of bulk metallic glass forming liquids: transport properties and atomic diffusion. Science Requirements Document, NASA.
14. HYERS, R.W., R.C. BRADSHAW, J.R. ROGERS, *et al.* 2004. Intl. J. Thermophys. **25**(4): 1155.
15. MOLOKANOV, V.V., V.N. CHEBOTNIKOV & Y.K. KOVERNISTYI. 1989. Inorg. Mater. **25:** 46.
16. ZONG, Z.H., *et al.* 1992. IEEE Trans. Magnet. **28**(3): 1833.
17. BERRY, S., *et al.* 2000. Met. Trans. B **31B:** 171.
18. YAKHOT, V. & S.A. ORSAG. 1986. J. Sci. Comp. **1**(3): 3.
19. BAUER, H.F. 1985. Appl. Math. Mech. **65:** 461.
20. FLEMINGS, M.C., *et al.* 2003. Levitation observation of dendrite evolution in steel ternary alloy rapid solidification (LODESTARS). Science Requirements Document, NASA Document LODESTARS-RQMT-0001.

Active Electrostatic Control of Liquid Bridge Dynamics and Stability

DAVID B. THIESSEN, WEI WEI, AND PHILIP L. MARSTON

Department of Physics, Washington State University, Pullman, Washington, USA

ABSTRACT: Stabilization of cylindrical liquid bridges beyond the Rayleigh–Plateau limit has been demonstrated in both Plateau-tank experiments and in short-duration low gravity on NASA KC-135 aircraft using an active electrostatic control method. The method controls the (2,0) capillary mode using an optical modal-amplitude detector and mode-coupled electrostatic feedback stress. The application of mode-coupled stresses to a liquid bridge is also a very useful way to study mode dynamics. A pure (2,0)-mode oscillation can be excited by periodic forcing and then the forcing can be turned off to allow for a free decay from which the frequency and damping of the mode is measured. This can be done in the presence or absence of feedback control. Mode-coupled feedback stress applied in proportion to modal amplitude with appropriate gain leads to stiffening of the mode allowing for stabilization beyond the Rayleigh–Plateau limit. If the opposite sign of gain is applied the mode frequency is reduced. It has also been demonstrated that, by applying feedback in proportion to the modal velocity, the damping of the mode can be increased or decreased depending on the velocity gain. Thus, both the mode frequency and damping can be independently controlled at the same time and this has been demonstrated in Plateau-tank experiments. The International Space Station (ISS) has its own modes of oscillation, some of which are in a low frequency range comparable to the (2,0)-mode frequency of typical liquid bridges. In the event that a vibration mode of the ISS were close to the frequency of a capillary mode it would be possible, with active electrostatic control, to shift the capillary-mode frequency away from that of the disturbance and simultaneously add artificial damping to further reduce the effect of the g-jitter. In principle, this method could be applied to any fluid configuration with a free surface.

KEYWORDS: liquid bridges; active control; Rayleigh–Plateau instability

NOMENCLATURE:

S	slenderness, ratio of bridge length to diameter of supports
V	bridge volume normalized to that of a cylinder of the same length and support diameter
C	Ohnesorge number
μ	ratio of surrounding fluid viscosity to bridge viscosity
ρ	ratio of surrounding fluid density to bridge density

INTRODUCTION

The goal of the work described here has been to control the stability and dynamics of cylindrical liquid bridges using applied mode-coupled surface stresses. A

Address for correspondence: David Thiessen, Department of Physics, Washington State University, Pullman, WA 99164-2814, USA. Voice: 509-335-4908; fax: 509-335-7816.
thiessen@wsu.edu

Ann. N.Y. Acad. Sci. 1027: 495–510 (2004). ©2004 New York Academy of Sciences.
doi: 10.1196/annals.1324.039

cylindrical volume of liquid bridging two solid circular disks of equal diameter in zero gravity is unstable when the bridge length exceeds the circumference of the disks. This is known as the Rayleigh–Plateau instability. Thus, under quiescent zero gravity conditions the bridge stability depends only on a single parameter, known as the slenderness, which is defined as the ratio of the bridge length to diameter ($S = L/D$). The critical slenderness is, therefore, $S = \pi$. Furthermore, a liquid bridge is very susceptible to vibration at frequencies near one of the capillary modes of the bridge. The capillary modes are characterized in terms of a pair of indices (n,m), where the axial index, n, gives the number of half-wavelengths in the axial direction, and the azimuthal index, m, gives the number of wavelengths in the azimuthal direction. Axisymmetric modes have $m = 0$. The Rayleigh–Plateau instability involves the spontaneous exponential growth of the $(2,0)$ axisymmetric mode leading to breaking of the bridge. This mode shape can be seen in FIGURE 1.

Axial vibrations easily couple into the $(2,0)$ mode and can lead to bridge rupture even when the slenderness is less than critical. The $(2,0)$ capillary mode is, therefore, seen to be very important to bridge stability in that it is linearly unstable for $S > \pi$ and easily excited by finite amplitude disturbances even when $S < \pi$. An active electrostatic control method has been developed that applies a surface stress distribution coupling into the $(2,0)$ mode exclusively. The method uses a pair of ring electrodes concentric with the grounded conducting bridge, as shown in FIGURE 1. The stress is applied by a feedback system in response to the signal from a $(2,0)$-mode amplitude detector. The feedback stress can be applied in such a way as to change the frequency and damping of the $(2,0)$ mode, as well as to stabilize the bridge beyond the Rayleigh–Plateau (RP) slenderness limit. A liquid bridge surrounded by air in low gravity (parabolic flight) is stabilized well beyond the Rayleigh–Plateau limit for several seconds (see FIGURE 2). In this case the bridge slenderness was 4.4, and it

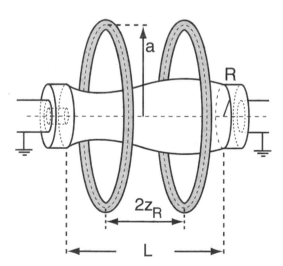

FIGURE 1. Schematic of liquid bridge and concentric ring electrodes. The bridge is deformed in a $(2,0)$-mode shape, which is corrected in this case by applying a potential to the left electrode.

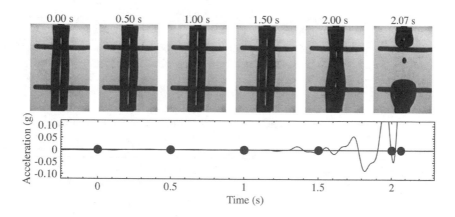

FIGURE 2. Stabilized liquid bridge in air in low gravity ($S = 4.4$, $V = 1.026$).[21] *Dots on accelerometer trace correspond to times at which images were taken.*

only broke after experiencing an axial acceleration, as seen in the accelerometer record. Although the current work concentrates on one capillary mode of a particular fluid configuration, the principle can be applied to any of the modes and to other fluid configurations. The only requirements are that the modal amplitude can be detected and a mode-coupled feedback stress can be applied with minimal delay. It should also be possible to control several modes at once. In general, the low-index modes have the least natural damping and are, therefore, the most desirable to control.

Studies of liquid bridges are largely motivated by the float-zone (FZ) process of crystal growth, involving a molten bridge between a feed rod and growing crystal. Space-based FZ growth has potential advantages for producing larger, more homogeneous crystals. Control of the stability, dynamics, and static shape of the molten surface would be helpful in optimizing the crystal growth process. The slenderness of the molten bridge has been shown to affect the shape of the solid–melt interface.[1] Thermocapillary-flow experiments in model half-zones show that the complexity of the flow depends on the slenderness, with the complexity increasing for decreasing slenderness.[2-4] Active electrostatic control of the molten zone could allow for a larger range of molten zone lengths to be used in the FZ process. In the absence of control, the practical length of the molten zone would be limited, not only by the Rayleigh–Plateau instability, but by increased susceptibility to *g*-jitter for bridges approaching the RP limit.[5] Although the Rayleigh–Plateau stability boundary can be extended by increasing the volume of liquid in the bridge, the stability of steady thermocapillary flow appears to be maximized for volumes near the cylindrical volume for model half-zone experiments.[6-8] Control of the grown crystal diameter requires controlling the so-called growth angle of the liquid meniscus at the solid–melt–gas trijunction.[9,10] The electrostatic control system can be used to control the static shape of a liquid bridge at the same time that it is controlling the dynamics and stability. This could be used to adjust the growth angle in a FZ process. Longer molten zones allowed by a control system have the additional advantage that thermal gradients could be reduced in the vicinity of the growing crystal. The processing of some oxide materials require that temperature gradients in the solid be small in order to avoid fracture.[11]

Static magnetic fields have been applied to counter gravity for small-scale paramagnetic liquid bridges.[12] By effectively reducing the Bond number to zero, bridges in air (as opposed to a Plateau tank) could be extended near to the RP slenderness limit.[12] Modulation of the magnetic field also allows for studies of mode dynamics as a function of Bond number.[13] The resonance frequency was found to decrease as the Bond number increased and the bridge became closer to its stability limit.[13] Axial electric fields have been applied to stabilize dielectric liquid bridges beyond the RP limit in both a Plateau tank[14,15] and for a bridge surrounded by a dielectric gas in zero gravity.[16] Bridges with a slenderness as great as 4.32 were stabilized in zero gravity.[16] Annular bridges in a Plateau tank with small but non-zero Bond number have been stabilized beyond the RP limit by a combination of buoyant force and a controlled shear stress.[17] The controller could be relatively slow in this case, because the upward flow in an outer annulus required to counter the gravitational deformation of the bridge induces a recirculating flow in the annular bridge causing a significant increase in the characteristic time of collapse for an unstable bridge. An active acoustic control method, based on the same principle as the active electrostatic control described here, was demonstrated to stabilize bridges beyond the RP limit in a Plateau tank.[18] A passive acoustic stabilization method was demonstrated for bridges in air under zero gravity conditions.[19]

EXPERIMENTAL METHOD

Results from both ground based (Plateau tank) and short-duration reduced gravity (parabolic flights on NASA KC-135) are discussed here. Aside from the Plateau tank, the hardware is identical for the two types of experiments.[20,21] The major components of the experiment are the bridge deployment apparatus, modal amplitude detector, and feedback electronics. Because of time constraints associated with the KC-135 experiments, the bridge deployment apparatus must be capable of deploying a cylindrical bridge in about two seconds. Fluid-injection and length-extension motors are synchronized to maintain a cylindrical bridge volume during extension. Bridges are formed between circular support disks, which are 0.432cm diameter, and the electrodes have an inner diameter of approximately 1.2cm for both the Plateau-tank and KC-135 experiments. Both wire ring electrodes and thin annular disk electrodes have been used. The (2,0)-mode amplitude is detected by a light-extinction method.[20] The signal from this detector is then sent to a feedback electronics box that, in turn, sends a signal to a high-voltage amplifier. Details of the feedback algorithm and high-voltage electronics are given elsewhere.[20]

Because of the sensitivity of long liquid bridges to g-jitter on parabolic flights, the experiment was built into a free-float rack that could be released from contact with the aircraft during a parabola. For a good parabola, as much as 10sec of free-float time can be achieved. This allows time to deploy a bridge to a length less than critical, activate feedback control, extend the bridge length beyond the Rayleigh–Plateau limit, and maintain the bridge in a stable condition for a few seconds before the end of the parabola.

The electrostatic system used to apply feedback stresses requires an electrically conducting bridge liquid surrounded by vacuum or a dielectric fluid. The electrical conductivity of the bridge fluid need not be very high.[20] The Plateau tank experiments were conducted with a bridge liquid of water–methanol–NaCl density matched to a 20cS silicone oil outer bath. A second Plateau tank system with lower viscosity consisted of a bridge liquid of an aqueous CsCl solution density matched to a fluorinated heat-transfer liquid (3M HFE-7500). For the KC-135 experiments, bridges consisting of glycerol–water–NaCl solutions were deployed in air.

THEORY

The effect of feedback on modal dynamics can be understood by analogy with a driven, damped harmonic oscillator. Consider a mass and spring system with dashpot damping and a delayed feedback force given by

$$F_{fb} = -gx(t - \tau) - g_v x_t(t - \tau_v),$$

where $x(t)$ is the displacement of the mass from its equilibrium position at time t, g is the amplitude gain, and g_v is the velocity gain. The feedback is delayed by τ and τ_v for amplitude and velocity feedback, respectively. The solution to the equation of motion for the displacement of the mass is

$$x(t) = x_0 e^{i\Omega t}.$$

Assuming small feedback delays, the characteristic equation for the complex frequency response of the mass-spring system is

$$(k + g) - (m - \tau_v g_v)\Omega^2 + i\Omega(\gamma + g_v - g\tau) = 0,$$

where k is the spring constant, m the mass of the oscillator, and γ is the damping coefficient. It can be seen that the main effect of amplitude gain is to increase the effective spring constant of the system, whereas the main effect of velocity gain is to increase the damping. The effects of feedback delay are also obvious, with delay in the amplitude feedback causing a reduction in damping and velocity-feedback delay causing a reduction in the effective mass. A very similar characteristic equation is obtained for the (2,0)-mode oscillation of a liquid bridge with mode-coupled feedback stress. The instantaneous modal amplitude corresponds to the displacement of the oscillator. The characteristic equation is

$$k_e - (m_b + m_a)\Omega^2 + \alpha(i\Omega)^{3/2} + i\Omega(\gamma + \gamma_{fb}) = 0,$$

where k_e is the effective modal spring constant for the (2,0) mode including the effect of mode-coupled feedback, m_b is the modal mass for the case of inviscid fluid and the absence of feedback, m_a is an added-mass term that includes the effects of viscosity and feedback, α is a boundary-layer damping coefficient, γ is the damping coefficient associated largely with bulk flow, and γ_{fb} is a damping coefficient associated with feedback. The dynamics of a bridge of cylindrical volume depends only on the following parameters:

$$\text{slenderness} \quad S = \frac{L}{D}$$

$$\text{Ohnesorge number} \quad C = \frac{\mu_i}{(\rho_i \sigma R)^{1/2}}$$

$$\text{viscosity ratio (outer/inner)} \quad \mu = \frac{\mu_o}{\mu_i}$$

$$\text{density ratio (outer/inner)} \quad \rho = \frac{\rho_o}{\rho_i}$$

where σ is the interfacial tension between the inner (bridge) fluid and outer (surrounding) fluid. The bulk damping coefficient, γ, is proportional to C, and the boundary-layer damping coefficient, α, is proportional to $C^{1/2}$. Parameters in the characteristic equation have more complicated dependencies on the viscosity and density ratios. The effective modal spring constant, k_e, for the $(2,0)$ mode has the following dependence on slenderness and dimensionless feedback gain, G,

$$k_e \propto \left(\frac{\pi^2}{S^2} - 1 + G \right).$$

In the absence of feedback, the spring constant of the $(2,0)$ mode is seen to be reduced as slenderness is increased toward π, becoming negative when $S > \pi$. The point at which $k_e = 0$ is the point at which the $(2,0)$ mode becomes unstable. Amplitude feedback can be seen to stabilize the system by turning a negative spring constant into a positive spring constant. The feedback damping coefficient has the form

$$\gamma_{fb} \propto (G_v - G\tau).$$

The added mass term, m_a, has a contribution that reduces the effective modal mass in proportion to $G_v \tau_v$. Thus, the effects of feedback on modal dynamics are seen to be qualitatively the same as for a mass and spring system.

The effective feedback force on the modal mass is generated by applying a mode-coupled distribution of Maxwell stresses to the liquid surface of the bridge. The Maxwell stress at the surface of the grounded conducting bridge liquid arises from an electric field generated in the surrounding fluid (air or dielectric liquid). An electric potential is applied to either the left or right electrode, producing a radially outward stress on the surface beneath that electrode. The Maxwell stress on the conducting surface is

$$P_M = -\frac{1}{2} \varepsilon \varepsilon_0 E^2,$$

where ε_0 is the permittivity of free space, ε the dielectric constant of the outer fluid, and E the magnitude of the electric field at the surface. Note that the Maxwell stress is always an outward force on the surface regardless of the sign of the electric field. Thus, a potential is applied to only one electrode at a time. It should also be pointed out that the Maxwell stress varies as the square of the field. The effective modal force on the bridge, therefore, also varies as the square of the electrode potential. A square-root feedback circuit was developed, such that the feedback force is linear in the modal amplitude or modal velocity.[20,22] The Fourier coefficient for the Maxwell stress coupling to the $(2,0)$ mode is

$$P_2 = -\frac{\varepsilon\varepsilon_0}{L}\int_{-L/2}^{L/2} E^2(z)\sin\left(\frac{2\pi z}{L}\right)dz,$$

where $E(z)$ is the magnitude of the electric field at the surface of the bridge as a function of axial distance from the middle of the bridge. The magnitude of the Fourier coefficient of the applied stress is modified by changing the electrode potentials in response to the modal amplitude detector signal. The modal amplitude and velocity gains are defined by

$$G_2 = \frac{dP_2}{dx}, \qquad G_{v2} = \frac{dP_2}{d(dx/dt)},$$

where x is the instantaneous mode amplitude. A mode-coupling parameter has been defined as

$$\eta = -\frac{R^2}{S}\int_{-S}^{S} E_{1L}^2(\tilde{z};\tilde{a}, \tilde{z}_R)\sin\left(\frac{\pi\tilde{z}}{S}\right)d\tilde{z},$$

in which $\tilde{z} = z/R$ is the dimensionless axial distance from the middle of the bridge, E_{1L} is the electric field at the surface of the bridge when the left electrode has a potential of one volt and the right electrode is at ground. In the simplest case of ring electrodes made of infinitely thin wire and treating the bridge as an infinite-length cylinder, the electric field depends on the geometry of the bridge–electrode system only through the two parameters $\tilde{a} = a/R$ and $\tilde{z}_R = z_R/R$, where a is the radius of the electrode ring, R is the radius of the bridge, and z_R is one-half the electrode spacing (see FIG. 1). The electrode radius was chosen so that the mode-coupling parameter was maximized.[20] Note that this parameter depends only on relative dimensions and is, therefore, independent of the overall length scale of the system. The system gains are then

$$G_2 = \frac{\varepsilon\varepsilon_0\alpha_D\eta}{2R^2}K, \qquad G_{v2} = \frac{\varepsilon\varepsilon_0\alpha_D\eta}{2R^2}K_v,$$

where α_D is the sensitivity of the modal amplitude detector and K and K_v are adjustable electronic gains for the amplitude and velocity error signals, respectively. The dimensionless gains are

$$G = \frac{\varepsilon\varepsilon_0\alpha_D\eta}{2\sigma}K, \qquad G_v = \frac{\varepsilon\varepsilon_0\alpha_D\eta}{2(\rho_i\sigma R^3)^{1/2}}K_v.$$

For the current apparatus, a typical value of the coupling parameter is $\eta = 0.2$ and for the detector sensitivity $\alpha_D = 600\,V/m$. According to the simple model considered above, a non dimensional amplitude gain of 1.0 would be sufficient to stabilize a bridge to infinite length; however, this neglects instabilities for higher mode numbers $((3,0)$ and above). A non dimensional gain of $G = 0.511$ would suffice to stabilize a bridge to $S = 4.4934$, which is the point at which the $(3,0)$-mode becomes unstable (see below). Notice that G is independent of the size of the system apart from whatever dependence the detector sensitivity may have on size. On the other hand, the velocity gain is reduced for larger bridges and would require an increased electronic gain to maintain the same dimensionless velocity gain for larger systems. The gain in the Plateau tank system is enhanced over that of a bridge in air because of the high dielectric constant of the bath liquid ($\varepsilon = 5.8$ for HFE-7500).

The dynamic model of the (2,0) mode on a cylindrical bridge is useful, however, in practice, liquid bridges often have a non cylindrical volume. For zero gravity, the linear stability of a fixed-contact-line liquid bridge between equal-diameter circular supports can be characterized by two parameters, the non dimensional volume, $V = V_{liq}/\pi R^2 L$, and the slenderness, $S = L/2R$, where R is the radius of the supports and L the distance between the supports. Equilibrium bridge shapes can be computed from the Young–Laplace equation. An equilibrium shape may be stable to all infinitesimal disturbances or may be unstable to one or more eigenmodes. A method of determining stability boundaries for liquid bridges based on computing families of equilibria is given by Lowry and Steen.[23] FIGURE 3 shows stability boundaries for zero gravity bridges over a range of volume and slenderness. For the region labeled *stable*, bridges are linearly stable to all disturbances. The region marked *(2,0)-unstable* is unstable to one eigenmode (the (2,0) mode), whereas the region labeled *(3,0)-unstable* is unstable to both the (2,0) and (3,0) eigenmodes. The horizontal dashed line indicates the cylindrical volume. This line crosses the (2,0)-mode stability boundary at $S = \pi$, the Rayleigh–Plateau limit.

The electrode system shown in FIGURE 1 was designed to optimally couple into the (2,0) mode for bridges with cylindrical volume. It would be expected then, that for large enough amplitude gain, cylindrical bridges could be stabilized for slenderness values up to the (3,0)-mode boundary ($S = 4.4934$ for $V = 1.0$). It is reasonable to assume that (2,0)-mode coupling would still be strong for bridges with somewhat greater or lesser volume than the cylinder and that, therefore, a significant region of the space labeled *(2,0)-unstable* could be accessed. Evidence of this is presented subsequently. It might also be possible, by adding a third electrode, a (3,0)-mode amplitude detector, and supporting electronics, to stabilize a bridge against both (2,0)- and (3,0)-mode instabilities and therefore extend bridges into the region marked *(3,0)-unstable*.

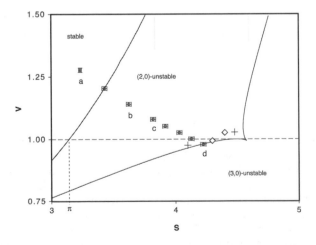

FIGURE 3. Stability boundaries and stabilization results: ○, Plateau tank (low viscosity); ◇, KC-135 data; +, Plateau tank (high viscosity).

RESULTS OF STABILIZATION EXPERIMENTS

Plateau Tank

Bridges of 62.0wt% water, 36.0wt% methanol, and 2.0wt% sodium chloride in a density matched bath of 20cS silicone oil were stabilized to slenderness values close to the theoretical limit. The stabilized bridges could be held for many minutes. FIGURE 4A shows a $S = 4.48$ stabilized bridge that has slightly more volume than the cylinder. The static shape seen is close to that predicted by the Young–Laplace equation for a bridge of that volume. FIGURE 4B shows a bridge with $S = 4.1$, but with less than cylindrical volume. The static shape is a $(3,0)$ mode shape with opposite deformation from the over-volume case. Both cases were near a stability boundary, as can be seen from FIGURE 3 (+ symbols).

The Plateau tank experiments described above used a high viscosity outer bath (20cS) to make stabilization easier by damping any vibrations. A fluid system with a much lower viscosity bath was also used in the Plateau tank primarily for dynamics studies. The bridge consisted of an aqueous solution of 52wt% CsCl and the bath was a fluorinated liquid (3M HFE-7500). The use of velocity feedback to increase the mode damping helped with the stabilization process. FIGURE 5A shows a bridge for the low-viscosity system stabilized to a slenderness of 4.46. FIGURE 5B shows the same system with a stabilized bridge at slenderness 3.78, but with an offset voltage applied to the right electrode to cause a static $(2,0)$-mode shape. This causes the

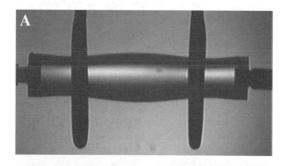

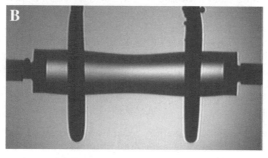

FIGURE 4. Images of stabilized bridges in Plateau tank. **(A)** $S = 4.48$, $V = 1.028$ and **(B)** $S = 4.1$, $V = 0.975$.

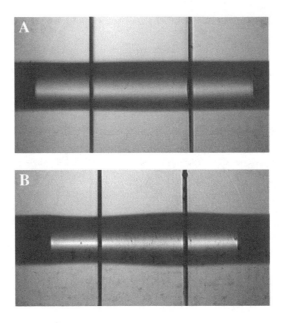

FIGURE 5. Long stabilized bridges with the low viscosity Plateau tank system using velocity feedback in addition to amplitude feedback to make the system more stable. **(A)** $S = 4.46$ and **(B)** $S = 3.78$ and offset voltage purposely applied to right electrode to give the bridge a static $(2,0)$-mode shape.

meniscus angle at the right support to be about three degrees relative to the bridge axis. Control of the meniscus angle is important in floating zone crystal growth.

The electrostatic control system is seen to work, even for bridges of non cylindrical volume. FIGURE 6 shows a sequence of images taken for a fixed volume of liquid, starting with the tips close enough together that the bridge is naturally stable. In FIGURE 6A the non dimensional volume is 1.27 and the slenderness is 3.22. Feedback control is activated and then the tips are moved apart incrementally without changing the absolute volume of liquid in the bridge. Thus, S increases and the non dimensional volume, V, decreases, as can be seen in FIGURE 3. Not all of the images from the sequence are shown in FIGURE 6. The bridges in FIGURE 6B and C are within the region marked *(2,0)-unstable* of FIGURE 3. The bridge in FIGURE 6D is very close to the $(3,0)$-mode stability boundary and a further increment in length led to immediate bridge breakage. No stabilized bridges have been observed in the region labeled *(3,0)-unstable* in FIGURE 3.

KC-135

The active electrostatic stabilization system was developed and tested on 11 different KC-135 flights comprising over 400 parabolas. Short duration low gravity tests were performed for bridges consisting of aqueous solutions of 43.5wt% glycerol and 1.0wt% sodium chloride surrounded by air. The Ohnesorge number for these bridges was approximately 0.01. Only the effects of amplitude gain have

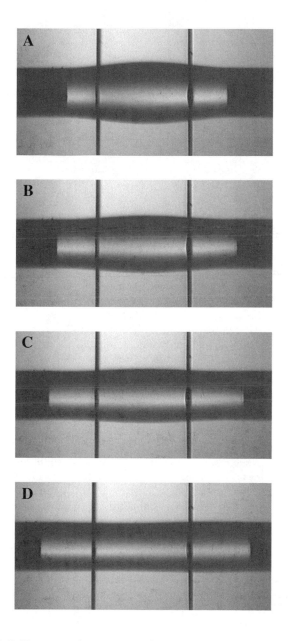

FIGURE 6. Bridge extension sequence in a low viscosity Plateau tank system: (**A**) $S = 3.22$, $V = 1.13$; (**B**) $S = 3.62$, $V = 1.13$; (**C**) $S = 3.92$, $V = 1.04$; (**D**) $S = 4.22$, $V = 0.98$.

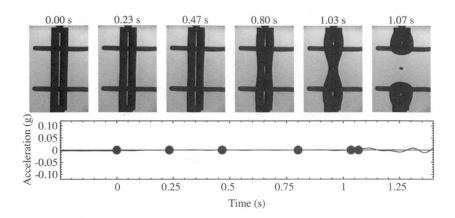

FIGURE 7. Stabilized liquid bridge in air in low gravity ($S = 4.3$, $V = 0.993$).[21] *Dots* on accelerometer trace correspond to times at which images were taken.

been tested on the KC-135. Two of the best results, out of many parabolas flown, are shown in FIGURES 2 and 7. FIGURE 2 shows a bridge of slenderness 4.4 with $V = 1.026$. The bridge was stabilized for several seconds before a large axial acceleration caused it to break by growth of the $(2,0)$ mode. FIGURE 7 shows a slenderness 4.3 bridge with $V = 0.993$ that breaks by growth of the $(3,0)$ mode. Although this bridge is slightly shorter than that in FIGURE 2, it is actually closer to the $(3,0)$-mode stability boundary because it is lower in volume (see FIG. 3).

DYNAMICS EXPERIMENTS

The dynamics of the $(2,0)$-mode oscillations were generally studied by purposely exciting the oscillation with modulation of the electrode potentials. This can be accomplished simultaneously with feedback control by adding a sine wave to the feedback signal prior to the square-root part of the feedback circuit. As soon as the mode oscillation reaches a steady state, the excitation is cut off at a zero crossing and the oscillation is allowed to decay. The signal from the modal amplitude detector can then be analyzed to measure the frequency and damping of the oscillation. Feedback control can be left on during the decay to study the effect of feedback on the frequency and damping of the mode. Dynamics experiments are very difficult to carry out on the KC-135 owing to the lack of time in low gravity. A few results of free decays for purposely excited oscillations in the absence of feedback were obtained to demonstrate the feasibility of these kinds of measurements. In a few cases, on the KC-135 flight experiments, the $(2,0)$ mode would be accidently excited on a stabilized bridge by a bump to the apparatus during free float.

Plateau Tank

To test active damping it was desired to study a bridge system with small natural damping (low Ohnesorge number). The system chosen consists of a 52 wt% CsCl aqueous solution (0.99 cp) for the bridge in a bath of HFE-7500 (1.24 cp). The density of the system is 1.61 g/cm^3. The qualitative effect of velocity feedback is illustrated by the decay records in FIGURE 8. FIGURE 8 A shows a free decay for a bridge of slenderness 2.63 in the absence of feedback. It can be seen that the oscillation decays within a period of about five seconds. Next, the $(2,0)$-mode oscillation was excited on the same bridge with *negative* velocity feedback. The decay in this case, with negative velocity feedback active, is shown in FIGURE 8 B. The initial mode amplitude is seen to be approximately twice as high as in the case without feedback, and the decay takes many times longer. On the other hand, by adding *positive* velocity gain, the damping can be increased, as shown in FIGURE 8 C. In this case, for the same bridge as above, the initial amplitude is reduced by nearly a factor of ten from the case of no feedback and the decay time is greatly reduced. Velocity feedback alone has little effect on mode frequency. The simultaneous application of amplitude and velocity feedback has been shown to increase the mode frequency and damping at the same time.

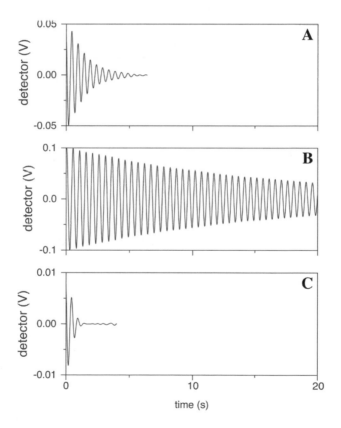

FIGURE 8. Modal amplitude detector signals for decays of $(2,0)$-mode oscillations on a slenderness 2.63 bridge in cases **(A)** without feedback, **(B)** negative velocity feedback, and **(C)** positive velocity feedback. Velocity feedback is seen to strongly influence the mode damping without much effect on the frequency.

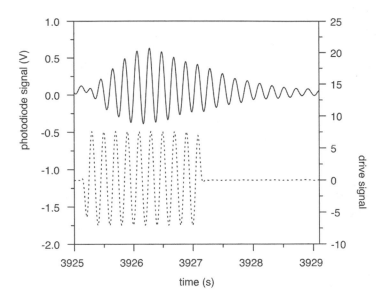

FIGURE 9. Driven oscillation and subsequent free decay of $(2,0)$-mode oscillation for bridge in air on KC-135 ($S = 2.4$):[21] ——, photodiode signal; ·····, drive signal.

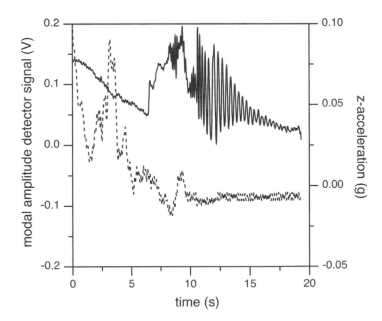

FIGURE 10. A $(2,0)$-mode oscillation and decay for a stabilized bridge of slenderness 4.1 on KC-135 excited by a bump to the apparatus: ——, modal amplitude detector; - - - -, z-acceleration.

KC-135

During a good parabola, it was possible to excite a (2,0)-mode oscillation on a naturally stable bridge of slenderness 2.4 by modulation of the electrode potentials. The mode oscillation was then allowed to decay in the absence of feedback. The excitation signal and corresponding modal amplitude signal are shown in FIGURE 9. The frequency of the decaying oscillation was measured as 4.9 Hz and the decay time (1/e) as 0.92 sec. Based on measurements of the fluid properties and bridge geometry, a semi-analytical, linear theory of bridge oscillations[24] predicts a frequency of 5.2 Hz and decay time of 1.4 sec. Experimental uncertainties may account for the difference with theory. A (2,0)-mode oscillation for a stabilized bridge with slenderness 4.1 was accidently excited when the free-float rack was bumped. The modal amplitude detector signal and corresponding accelerometer signal are shown in FIGURE 10. Amplitude feedback was active to stabilize the bridge, but velocity feedback was not implemented on the flight experiment. The frequency of the decay was measured as 3.1 Hz and the characteristic decay time as 1.6 sec. Active damping has not yet been tested for a bridge in air in low gravity.

SUMMARY

A qualitative understanding of the effect of mode-coupled feedback on the (2,0)-mode dynamics and stability is given by analogy with a mass and spring system. Adding a modal feedback force in proportion to modal amplitude with appropriate gain can stiffen the effective spring and stabilize the bridge against breakup at lengths beyond the Rayleigh–Plateau limit. Stabilization has been demonstrated for both Plateau tank bridges and bridges in air under low gravity conditions, for slenderness values very near to the theoretical stability boundary of the next higher mode. Velocity feedback has been demonstrated to add damping to the mode oscillation and has been found helpful for stabilizing bridges in a low viscosity Plateau tank system to lengths that are very near the (3,0)-mode stability boundary. It has further been demonstrated that the static shape of the bridge can be modified even while feedback is active. This could potentially be used to control the meniscus angle in a float zone crystal growth process.

ACKNOWLEDGMENT

This work was supported by NASA.

REFERENCES

1. LOPEZ, C., J. MILCHAM & R. ABBASCHIAN. 1999. Microgravity growth of GaSb single crystals by the liquid encapsulated melt zone (LEMZ) technique. J. Cryst. Growth **200:** 1–12.
2. MONTI, R., R. SAVINO & M. LAPPA. 2000. Influence of geometrical aspect ratio on the oscillatory Marangoni convection in liquid bridges. Acta Astronaut. **47:** 753–761.

3. LAPPA, M., R. SAVINO & R. MONTI. 2001. Three-dimensional numerical simulation of Marangoni instabilities in liquid bridges: influence of geometrical aspect ratio. Intl. J. Num. Meth. Fluids **36**: 53–90.
4. KAWAMURA, H., I. UENO & T. ISHIKAWA. 2002. Study of thermocapillary flow in a liquid bridge towards an on-orbit experiment aboard the International Space Station. Adv. Space Res. **29**: 611–618.
5. LANGBEIN, D. 1990. Crystal growth from liquid columns. J. Cryst. Growth **104**: 47–59.
6. SHEVTSOVA, V.M., M. MOJAHED & J.C. LEGROS. 1999. The loss of stability in ground based experiments in liquid bridges. Acta Astronaut. **44**: 625–634.
7. SUMNER, L.B.S., G.P. NEITZEL, J.-P. FONTAINE, *et al.* 2001. Oscillatory thermocapillary convection in liquid bridges with highly deformed free surfaces: experiments and energy-stability analysis. Phys. Fluids **13**: 107–120.
8. KUHLMANN, H.C., C. NIENHUSER, H.J. RATH, *et al.* 2002. Influence of the volume of liquid on the onset of three-dimensional flow in thermocapillary liquid bridges. Adv. Space Res. **29**: 639–644.
9. SUREK, T. 1976. Theory of shape stability in crystal growth from the melt. J. Appl. Phys. **47**: 4384–4393.
10. SUREK, T. & S.R. CORIELL. 1977. Shape stability in float zoning of silicon crystals. J. Cryst. Growth **37**: 253–271.
11. COCKAYNE, B. 1968. Developments in melt-grown oxide crystals. J. Cryst. Growth **4**: 60–70.
12. MAHAJAN, M.P., M. TSIGE, P.L. TAYLOR, *et al.* 1998. Paramagnetic liquid bridge in a gravity-compensating magnetic field. Phys. Fluids **10**: 2208–2211.
13. MAHAJAN, M.P., M. TSIGE, S. ZHANG, *et al.* 2002. Resonance behavior of liquid bridges under axial and lateral oscillating total body forces. Expts. Fluids **33**: 503–507.
14. SANKARAN, S. & D.A. SAVILLE. 1993. Experiments on the stability of a liquid bridge in an axial electric field. Phys. Fluids A **5**: 1081–1083.
15. RAMOS, A., H. GONZALEZ & A. CASTELLANOS. 1994. Experiments on dielectric liquid bridges subjected to axial electric fields. Phys. Fluids **6**: 3206–3208.
16. BURCHAM, C.L. & D.A. SAVILLE. 2000. The electrohydrodynamic stability of a liquid bridge: microgravity experiments on a bridge suspended in a dielectric gas. J. Fluid Mech. **405**: 37–56.
17. ROBINSON, N.D. 2001. Experiments in Film and Liquid Bridge Dynamics and Stability. Ph.D. Thesis, Cornell University.
18. MARR-LYON, M., D.B. THIESSEN & P.L. MARSTON. 1997. Stabilization of a cylindrical capillary bridge far beyond the Rayleigh–Plateau limit using acoustic radiation pressure and active feedback. J. Fluid Mech. **351**: 345–357.
19. MARR-LYON, M., D.B. THIESSEN & P.L. MARSTON. Passive stabilization of capillary bridges in air with acoustic radiation pressure. Phys. Rev. Lett. **86**: 2293–2295; erratum **87**(20): 9001.
20. MARR-LYON, M., D.B. THIESSEN, F.J. BLONIGEN, *et al.* 2000. Stabilization of electrically conducting capillary bridges using feedback control of radial electrostatic stresses and the shapes of extended bridges. Phys. Fluids **12**: 986–995.
21. THIESSEN, D.B., M. MARR-LYON & P.L. MARSTON. 2002. Active electrostatic stabilization of liquid bridges in low gravity. J. Fluid Mech. **457**: 285–294.
22. MARR-LYON, M.J. 2000. Stabilization of Capillary Bridges Far Beyond the Rayleigh–Plateau Limit With Acoustic Radiation Pressure or Electrostatic Stresses. Ph.D. Thesis, Washington State University.
23. LOWRY, B.J. & P.H. STEEN. 1995. Capillary surfaces: stability from families of equilibria with application to the liquid bridge. Proc. Roy. Soc. Lond. A **449**: 411–439.
24. NICOLAS, J.A. & J.M. VEGA. 2000. Linear oscillations of axisymmetric viscous liquid bridges. Z. Angew. Math. Phys. **51**: 701–731.

Ventless Pressure Control of Two-Phase Propellant Tanks in Microgravity

MOHAMMAD KASSEMI[a] AND CHARLES H. PANZARELLA[b]

[a]*National Center for Microgravity Research, NASA Glenn Research Center, Cleveland, Ohio, USA*

[b]*Ohio Aerospace Institute, Cleveland, Ohio, USA*

ABSTRACT: This work studies pressurization and pressure control of a large liquid hydrogen storage tank. A finite element model is developed that couples a lumped thermodynamic formulation for the vapor region with a complete solution of the Navier–Stokes and energy equations for the flow and temperature fields in the liquid. Numerical results show that buoyancy effects are strong, even in microgravity, and can reposition a vapor bubble that is initially at the center of the tank to a region near the tank wall in a relatively short time. Long-term tank pressurization with the vapor bubble at the tank wall shows that after an initial transient lasting about a week, the final rate of pressure increase agrees with a purely thermodynamic analysis of the entire tank. However, the final pressure levels are quite different from thermodynamic predictions. Numerical results also show that there is significant thermal stratification in the liquid due to the effects of natural convection. A subcooled jet is used to provide simultaneous cooling and mixing in order to bring the tank pressure back down to its initial value. Three different jet speeds are examined. Although the lowest jet speed is ineffective at controlling the pressure because of insufficient penetration into the liquid region, the highest jet speed is shown to be quite effective at disrupting thermal stratification and reducing the tank pressure in reasonable time.

KEYWORDS: pressure control; two-phase storage; microgravity; propellant tanks; computational fluid dynamics; space cryogenic tanks

NOMENCLATURE:

e	specific heat, erg/K·g
d	diameter, cm
g	gravitational acceleration, cm/sec^2
Gr	Grashof number
J	evaporative flux, $g\beta_l \rho_l^2 q_w R^4/k_l \mu_l^2$
k	conductivity, erg/cm·sec·K
L	latent heat, erg/g
m	molar mass, g/mol
n	normal vector
p	pressure, Pa
q	heat flux vector, mW/cm^2
Q	total heat flow, W
r, z	cylindrical coordinates, cm
R	spherical tank radius, cm

Address for correspondence: Mohammad Kassemi, National Center for Microgravity Research, NASA Glenn Research Center, 21000 Brookpark Rd, MS110-3, Cleveland, OH 44135, USA. Voice: 216-433-5031; fax: 216-433-5033.

mohammad.kassemi@grc.nasa.gov

Ann. N.Y. Acad. Sci. 1027: 511–528 (2004). ©2004 New York Academy of Sciences.

doi: 10.1196/annals.1324.040

R_G	ideal gas constant, erg/K·mol
t	time, sec
T	temperature, K
u, w	velocity components, cm/sec
V	volume, cm^3
We	Weber number, $\rho_l \bar{w}_j^2 r_j^2 / 2\sigma R$
β	expansion coefficient, 1/K
ε	time truncation tolerance
μ	dynamic viscosity, g/cm·sec
ρ	density, g/cm^3

Subscripts

b	normal boiling point
c	conduction
e	on the Earth
i	at the liquid–vapor interface
l	liquid phase
s	saturation point
w	tank wall
v	vapor phase

INTRODUCTION

Integral to all phases of future human space and planetary expeditions is effective, affordable, and reliable cryogenic fluid management for use in the propellant and life-support systems. Without safe and efficient cryogen storage, economically feasible and justified human missions will not be possible. With the exception of extremely short-duration missions, significant cost savings can be achieved if the launch mass can be reduced by improving the cryogenic storage and transfer technologies.[1]

Cryogen vaporization is one of the main causes of mass loss and leads to the self-pressurization of storage tanks.[2] Vaporization can occur during the filling process or may be caused by heat leaks into the tank from the surrounding environment. Ordinarily, direct venting to the environment can be used to relieve the excess pressure. For on-surface applications, such as those on Earth, Moon, or Mars, the spatial configuration of liquid and vapor is dictated by gravity and is well known. In this situation, continuous venting can be easily accomplished, but over a significant length of time it results in considerable cryogen mass loss. For in-space applications, the spatial configuration of liquid and vapor is generally unknown, and direct venting without prepositioning of the two phases is precluded due to the possibility of expelling liquid along with the vapor. Moreover, venting in space is also undesirable because it prohibits manned flight operations around the storage tanks. Therefore, from both safety and cryogen conservation viewpoints, a ventless pressure control strategy is highly desirable for both on-surface and in-space applications.

The zero boil-off (ZBO) pressure control strategy has been proposed as an effective means of achieving ventless storage through the synergetic application of active cooling and mixing.[2] Cooling can be achieved by using cryocoolers, and mixing is normally provided by impellers or forced liquid jets. The transport mechanisms in this situation can be extremely complex and require hand-in-hand experimental and theoretical elucidation before being applied in practice.

There are a number of experiments and thermodynamic studies of cryogen storage tanks in normal and reduced gravity. The most notable[3] shows that the initial rate of pressure rise is lower in reduced-gravity conditions, because of an increase in the liquid-wetted wall areas and increased boiling. The experimental results were shown to lie somewhere between those obtained from two simplified thermodynamic analyses. Further experimental work[4,5] investigated the effect of a mixing jet. Dimensionless parameters are developed that characterize the four different liquid flow patterns observed and their effect on the bulk mixing behavior. Additional experiments investigated the application of jet mixing to control the tank pressure.[6] The results show that the effects of natural-convection boundary layers forming at the wall on the vapor pressure rise can be countered by a subcooled jet flow emerging from the center of the tank. The results also suggest that a thermal equilibrium state is hard to achieve and that the existing correlations for mixing time and vapor-condensation rates based on small-scale tanks may not be applicable to large-scale liquid hydrogen systems. The correct extrapolation can only be determined when the interaction between the forced and natural-convective flows is properly understood. Other experimental investigations[7] also showed that there are significant departures from thermodynamic equilibrium and that a mixing jet could be used to minimize thermal stratification and reduce the tank pressure.

The theoretical and numerical treatments of cryogenic storage tanks can be separated into three main categories. The first category consists of tank pressurization studies that compute the pressure rise in the vapor mainly in terms of thermodynamic considerations. This method was used to determine the self-pressurization of a partially filled liquid-hydrogen storage tank under microgravity conditions.[8] The effects of tank size, liquid fill level, and wall heat flux on the tank pressure rise are theoretically studied. They show that liquid thermal expansion tends to cause vapor condensation and wall heat flux leads to liquid evaporation at the interface. However, this approach is limited because the problem is assumed to be one-dimensional and that there is no convection in the liquid. Another model uses a lumped analysis of a no-vent fill process in a ground-based environment.[9] It accounts for several major effects, such as fluid inlet temperature, interfacial mass transfer, and inlet jet characteristics, that influence the fill process. It includes a semi-empirical condensation model based on universal submerged jet theory that considers condensation to be a function of bulk fluid properties and the liquid turbulence to be induced by the jet geometry and orientation. This work[9] shows very good agreement between the model used and experimental results using Freon-114. Finally, a thermodynamic analysis of cryogen boil-off in a dewar is used to correlate the pressure variations in the container to the latent heat of vaporization during an experiment with liquid helium.[10]

The second category is composed of investigations that examine just the fluid flow and thermal stratification in the liquid cryogen without any thermodynamic consideration of the vapor phase. These investigations focus on the fluid flow in the liquid in terms of either mainly natural convection[11–13] or in terms of forced flows caused by jets or external thrusts.[14–16] In these representations, the transport processes in the vapor phase are ignored and the temperature of the liquid–vapor interface is assumed uniform and equal to its initial saturation value. Consequently, these investigations divulge no information with respect to the pressure rise in the vapor as a function of the various flow parameters in the liquid.

Finally, the third category investigations mainly examine the behavior and evolution of the liquid–vapor interface, excluding any thermal or pressurization effects. The analyses are performed for both ground-based and microgravity applications and are mainly based on the volume-of-fluid (VOF) approach as embodied by the Los Alamos code RIPPLE and its derivatives.[17] Investigations in this category have focused on: evolution of the free surface as influenced by the microgravity environment,[18] reorientation of the vapor subject to spacecraft thrust,[19] free surface deformation as affected by the jet flow and geysers[17,20] or by external forces, such as magnetic fields[21,22] and fluid slosh coupled to gravity–gradient accelerations or spacecraft dynamics.[23,24] The studies in this category are all limited to isothermal models, and, again, divulge no information with respect to tank pressurization.

Panzarella and Kassemi[25] have developed a more comprehensive numerical model by coupling a lumped thermodynamic model of the vapor region to a direct numerical simulation of the Navier–Stokes and energy conservation equations in the liquid region. The lumped model assumes the temperature and pressure throughout the vapor region are spatially uniform but may increase over time due to any net heat and mass transfer coming into the vapor region. The coupling between the liquid and vapor regions is due to the fact that as the vapor pressure rises, so does the corresponding saturation temperature, and this is the temperature boundary condition prescribed at the free surface separating the liquid and vapor regions. Thus, as the interfacial temperature rises, the temperature distribution in the liquid is altered, and this, in turn, may change the net rate of vaporization.

This model was used to investigate the pressurization of a small ground-based tank. It was shown that the initial rate of pressure rise depends on the particular heat flux distribution prescribed on the tank wall even though the final rate of pressure rise agrees with a purely thermodynamic description of both the liquid and vapor regions. It was also shown that a subcooled liquid jet that simultaneously cools and mixes the liquid could be used effectively to prevent the pressure rise in the tank.

In this paper, the same approach is used to investigate the pressure rise and pressure control of larger cryogenic tanks under microgravity conditions. The primary difference is that in microgravity the vapor region is free to move around inside the tank, and the corresponding free surface displacements are larger and more complicated. Therefore, the finite-element model is extended to handle these larger free-surface displacements by using an interface tracking method that moves the computation nodes along with the free surface as it evolves. Interface tracking is ideal for this class of problems, since there are no significant topology changes. In addition, it avoids additional errors that are likely to be introduced if one of the diffusive interface capturing methods, such as VOF, were used.

The primary objectives of the present work are twofold: first, to delineate the role of the various transport mechanisms that affect tank pressurization in microgravity; second, to examine the feasibility of using forced mixing provided by a subcooled liquid jet as a means of controlling the tank pressure in space.

MATHEMATICAL FORMULATION

Consider a spherical cryogenic tank partially filled with liquid, as shown in FIGURE 1. In the microgravity configuration, the vapor is assumed to initially occupy a single spherical region completely surrounded by liquid, although it is free to move and deform after that. A cylindrical coordinate system is used with the origin at the center of the tank and with the z-axis parallel to the direction of gravity. The solutions are assumed to be axisymmetric around the z-axis. The details of this model are described elsewhere,[25] and only a brief summary of the essential elements is presented here. TABLE 1 lists the material properties of liquid hydrogen and the values of other parameters used in this paper.

The liquid is treated as an incompressible fluid, described by the Navier–Stokes equations together with the standard mass and energy conservation equations. The Boussinesq approximation is used to account for buoyancy effects. The vapor is treated as an inviscid compressible ideal gas with spatially uniform temperature, pressure, and density. The vapor density is related to its temperature and pressure through the ideal gas law, and the temperature and pressure are further related by the saturation condition given by the Clausius–Clapeyron equation for an ideal gas,

$$T_v = T_s = \left(\frac{1}{T_b} - \frac{R_G}{Lm} \ln \frac{p_v}{p_b} \right)^{-1}. \tag{1}$$

Obviously, the immediate consequence of this and the uniformity constraint is that the temperature in the vapor is everywhere equal to its saturation value. By making these assumptions, the state of the vapor is solely determined by its pressure.

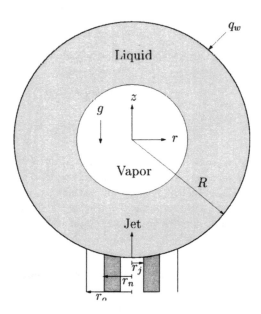

FIGURE 1. The single vapor bubble configuration for the space-based cryogenic tank. A cylindrical coordinate system is used with the symmetry axis along the direction of gravity.

TABLE 1. Material properties of hydrogen at its normal boiling point and other relevant parameter values

Parameter	Value
β_l	$0.0175\,\mathrm{K}^{-1}$
c_v	$1.012\times10^8\,\mathrm{erg/K\cdot g}$
c_l	$9.7\times10^7\,\mathrm{erg/K\cdot g}$
k_l	$12{,}440\,\mathrm{erg/cm\cdot sec\cdot K}$
μ_l	$1.327\times10^{-4}\,\mathrm{g/cm\cdot sec}$
ρ_v	$0.00133\,\mathrm{g/cm}^3$
ρ_l	$0.07047\,\mathrm{g/cm}^3$
g_e	$981\,\mathrm{cm/sec}^6$
L	$4.456\times10^9\,\mathrm{erg/g}$
m	$2.0\,\mathrm{g/mol}$
p_b	$1.014\times10^5\,\mathrm{Pa}$
T_b	$20.27\,\mathrm{K}$
R_G	$8.31\times10^7\,\mathrm{erg/K\cdot mol}$
σ	$1.93\,\mathrm{dyn/cm}$
R	$150\,\mathrm{cm}$
r_j	$15\,\mathrm{cm}$
r_n	$21\,\mathrm{cm}$
r_o	$33\,\mathrm{cm}$
q_w	$0.01\,\mathrm{mW/cm}^2$
Q_w	$2.827\,\mathrm{W}$
$\overline{w}_j$	$0.005, 0.05, 0.5\,\mathrm{cm/sec}$
T_j	$20\,\mathrm{K}$

At the free surface, there is the usual jump in normal stress due to surface tension forces, and the vapor shear stress is assumed to be negligible. The complete mass balance boundary condition accounts for the interfacial evaporative mass flux, J, but since J is so small here, it is appropriate to just use the standard kinematic boundary condition for a nonvolatile fluid. The energy boundary condition is modified to account for evaporation and is given by

$$LJ = (\mathbf{q}_l - \mathbf{q}_v) \cdot \mathbf{n}_i, \tag{2}$$

and the temperature at the free surface is set equal to the saturation temperature,

$$T_i = T_s. \tag{3}$$

The boundary conditions at the tank wall, with the exception of the jet inlet and outlet surface, are no-slip for the velocity, and a prescribed heat flux of q_w applied uniformly over the entire surface. This results in a total heat input of $Q_w = 4\pi R^2 q_w$.

When the jet is turned off, the no-slip condition applies everywhere on the tank wall, but when it is turned on, a parabolic profile is prescribed across the inlet area ($r < r_j$) with an average vertical speed $\bar{w}_j$ and an inlet fluid temperature T_j. A similar profile is prescribed over the outlet ($r_n < r < r_o$) with the same mass flow rate to ensure that the amount of liquid in the tank remains constant. The temperature gradient is assumed to be zero at the outlet surface. Because of a change in the heat transfer surface area due to the jet orifice, the heat flux imposed on the remainder of the tank wall is adjusted so that the total heat input is the same for both the jet and no-jet cases.

A lumped vapor model is used to determine the vapor pressure rise due to any net heat and mass transfer into the vapor region. This is done by setting the rate of change of total vapor energy equal to the net energy coming in through its boundaries. Likewise, the rate of change of total vapor mass is set equal to the net mass transfer across the liquid–vapor interface due to evaporation or condensation. Then, by assuming that the vapor temperature and pressure are spatially uniform across the entire region, a single nonlinear equation can be derived for the rate of vapor pressure rise. The original derivation of this lumped model can be found elsewhere,[25] where it was used to study the pressurization of a smaller ground-based cryogenic tank.

This ultimately leads to a single evolution equation for the vapor pressure rise,

$$\frac{dp_v}{dt} = F(p_v)Q, \tag{4}$$

where Q is the total heat entering the vapor region and F is a complicated function given by

$$F(p_v) = \frac{L}{V_v}\left\{c_v T_s + \left(\frac{L_m}{R_G T_s} - 1\right)\frac{\rho_l}{\rho_l - \rho_s}\left[L - \rho_v\left(\frac{1}{\rho_s} - \frac{1}{\rho_l}\right)\right]\right\}^{-1}. \tag{5}$$

Once entering the vapor region Q is split between the energy required for evaporation and that required to raise the vapor temperature at its current rate. This is done automatically in the term F. It can be shown that V_v, ρ_s, and T_s are all functions of p_v alone, and, therefore, F is a function that depends only on p_v and other material properties.

Comparisons are also made to a lumped thermodynamic model of the entire tank, which assumes that the temperature and pressure are uniform throughout the liquid region as well. The pressure rise for this model is obtained from the following global energy, mass, and volume balances:

$$\frac{d}{dt}(\rho_l V_l c_l T + \rho_v V_v c_v T) = Q_w \tag{6}$$

$$\frac{d}{dt}(\rho_l V_l + \rho_v V_v) = 0 \tag{7}$$

$$\frac{d}{dt}(V_l + V_v) = 0. \tag{8}$$

These equations are solved, together with the ideal gas law and the saturation temperature relationship, to determine the pressure rise.

NUMERICAL METHODOLOGY

The liquid equations are solved by using an in-house modified version of the Galerkin finite element code FIDAP[26] that has been extended by providing coupling with the lumped-vapor model. The temperature and velocity fields are discretized using nine-node quadratic elements, and the pressure is discretized using a linear discontinuous approximation (the three pressure unknowns per element are the coefficients of the linear polynomial that approximates the pressure). The position of each node on the free surface is adjusted by using a front-tracking approach. The interior nodes are moved using the method of straight spines. They are shifted proportionally along straight, nearly radial lines passing through the free surface.

The spatial mesh is successively refined until grid convergence is achieved as was demonstrated for a ground-based tank.[25] This same approach is used here and the final computation mesh contains between 2,500 and 5,000 elements, depending on the particular case. The mesh is generally graded near the free surface, the tank wall, and toward the top of the tank when the vapor region is close to the wall.

The implicit backward-Euler method (with a forward-Euler predictor) is used for the time discretization, and the time steps are chosen adaptively by using a method[27] that keeps the relative time truncation error less than a prescribed tolerance ε. It was also shown[25] that a value $\varepsilon = 10^{-4}$ is sufficient to resolve the temporal behavior of the ground-based solutions. This same value is used here except for certain cases for which the solution is more rapidly varying. Then, the more conservative value, $\varepsilon = 10^{-5}$, is used instead. At each time step, the resulting nonlinear system of equations is solved using an iterative quasi-Newton method,[28] and convergence is assumed when the relative change in the solution norm is less than 10^{-5}. The Jacobian is updated using Broyden's method after every five iterations.

The approximate computation time for the following cases varied from about 15 CPU hours for the simpler no-jet cases to as long as 4 CPU days for some of the more complicated cases, such as a moving vapor region with a jet. These results were computed on a 1.6 GHz Intel Xeon system with 16 GB of memory running Red Hat Linux 8.0.

RESULTS AND DISCUSSION

On the ground, the liquid normally settles to the bottom of the tank, but in microgravity, it may be anywhere. Since the contact angle of liquid hydrogen is nearly zero, it is likely that it will completely wet the tank wall. Thus, the initial configuration considered in this paper corresponds to the situation where the vapor has accumulated into a single spherical vapor region completely surrounded by liquid. The vapor region is free to evolve over time and it will be shown that under normal microgravity conditions it will reach the tank wall before there is any appreciable pressure rise.

In space, the acceleration due to gravity or other sources may be identically zero or negligibly small. If a nearly zero-gravity environment is assumed, there would be no preferred equilibrium location for the vapor region and it would remain stationary at its initial location. However, under microgravity conditions, the vapor region will not be stationary and it will move toward the tank wall in a direction dictated by the average residual acceleration vector. As the vapor region approaches the tank wall it deforms while the liquid trapped between the bubble and the wall drains. If the liquid layer is completely drained or evaporated, then a configuration similar to the ground-based situation is approached. This was studied in detail in a previous paper.[25]

Three cases are considered so as to study the problem over all the relevant times-cales. In the first case, the spherical vapor bubble is assumed to be initially at the center of the tank and the history of the tank pressure, as well as the evolution of the liquid flow and thermal fields, are examined by following the deforming bubble as it approaches the tank wall. Its motion is driven solely by the buoyancy force resulting from the density jump across the liquid–vapor interface.

Since the time scale for the vapor bubble to reach the tank wall is much shorter than the conduction or convection time scales, the second case study focuses on the temperature, pressure, and flow fields that develop over a longer time span while the vapor bubble remains in a fixed position near the tank wall. Fluid is still allowed to slip over the interface by assuming the vapor shear stress is negligible. This permits us to study the average long-term pressurization of the tank.

Finally, the third case examines the possibility of controlling the tank pressure in microgravity, with the mixing provided by a subcooled liquid jet. Again, the bubble is positioned near the wall, and the tank is pressurized for an additional 75 days. Three different jet speeds are considered, spanning three orders of magnitude. For each jet speed, the time required to bring the pressure back down to its initial value is determined.

In this section, the results of the numerical simulations are presented in terms of the history of the vapor pressure, saturation temperature, and the net heat flow into the vapor region. Representative flow and temperature fields in the liquid region are also presented at given times. These fields are depicted by ten equally spaced temperature and/or streamline contour lines between the minimum and maximum values. To save space and because the solution is axisymmetric, the isotherm and streamlines contour plots are combined into a single image with the isotherms on the left and streamlines on the right. The geometric and thermophysical properties of the system are given in TABLE 1. The fill ratio is 75% for all cases and the Bond number is about 6.

Case 1: Moving Vapor Region

The vapor region is located initially at the center of the tank and, as a consequence of the interfacial buoyancy force, it will float upward in the direction opposite to the residual acceleration vector that, in this case, is pointing in the negative z-direction. The bubble will also deform along the way due to surface-tension forces.

The temperature and streamline distributions for three successive times are shown in FIGURE 2 and the resulting pressure and temperature change is plotted in FIGURE 3. After only a few minutes, the bubble is well on its way toward the wall, as shown in FIGURE 2A. At the intermediate time, shown in FIGURE 2B, the vapor region is clearly being pushed up against the wall with only a very thin liquid film in between. Eventually, surface tension pulls the vapor region back into a nearly spherical shape, as shown in FIGURE 2C, at which point the simulation is terminated because the liquid layer between the interface and the tank wall becomes too thin to be meaningfully resolved.

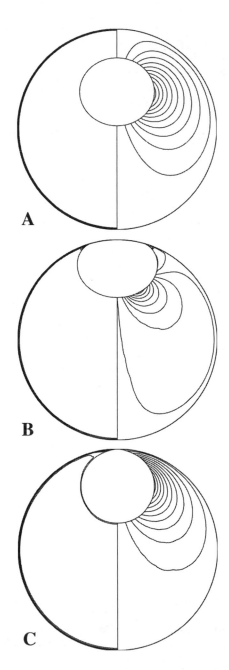

FIGURE 2. Isotherms and streamlines for **(A)** $t = 259$ sec, **(B)** $t = 463$ sec, and **(C)** $t = 567$ sec, as the initially centered spherical vapor bubble rises only due to buoyancy. The minimum temperature is 20 K and the maximum temperatures are **(A)** 20.00012 K, **(B)** 20.0058 K, and **(C)** 20.0072 K.

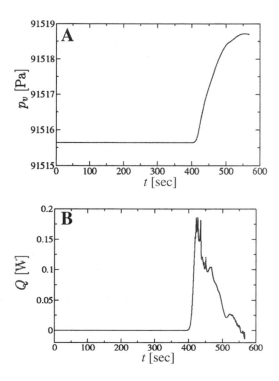

FIGURE 3. The pressure rise (**A**) and net heat flow (**B**) into the vapor region for the no-jet case depicted in FIGURE 2.

One interesting feature of this case is that while the vapor region is being pushed up against the wall it is receiving an extra amount of heat, as evidenced by the sharp upward spike in Q seen in FIGURE 3. However, once it begins to retract from the wall it no longer receives as much heat and Q decreases. Near the end of the simulation the bubble has moved so far away from the wall that it receives hardly any additional heat, and, in fact, Q becomes slightly negative as the excess heat stored in the vapor region gets released into the subcooled bulk liquid.

The high frequency oscillations observed in the interfacial heat flow of FIGURE 3 are caused by fluctuations of the free surface and the associated shedding of minor vortices. It is interesting to note that, because the pressure rise is related to the temporal and spatial integral of heat flow (see Equation (4)), the pressure curve in FIGURE 3 does not exhibit similar sensitivity to the fluctuating free surface.

Case 2: Tank Pressurization with the Vapor Near the Wall

The previous case study implied that after a relatively short period of time in microgravity the vapor region will migrate close to the tank wall provided there is no major change in the average direction of the residual acceleration vector during that time. To investigate the long-term pressurization without having to take the

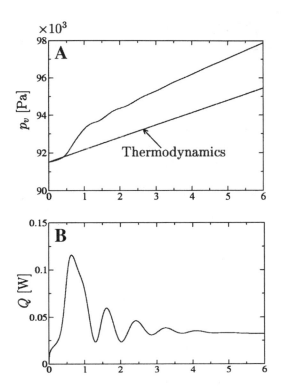

FIGURE 4. The initial (**A**) pressure rise and (**B**) net heat flow into the vapor with the vapor region at the wall.

FIGURE 5. Final isotherms (*right*) and streamlines (*left*) at $t = 75$ days (immediately before the jet is turned on) with the spherical vapor bubble at the wall. The minimum and maximum temperatures are 21.58 K and 22.27 K, respectively.

small time steps required to resolve the high frequency free-surface oscillations, the spherical vapor bubble is positioned at the wall with only a 1-mm liquid gap between the interface and the wall. The vapor region is not allowed to move from this position or deform from its spherical shape, but liquid is allowed to slip over its surface because of the zero shear-stress boundary condition at the interface.

At first, there is an oscillatory transient response as the fluid flow and temperature fields adjust to this sudden change in heat input. This is exhibited by oscillations in the pressure and net heat flow depicted in FIGURE 4 and is due primarily to a competition between the two convection cells in the upper tank region shown in FIGURE 5. The vortex near the wall is driven by the hotspot, rotates in a counterclockwise direction, and takes the warm fluid from a region close to the hotspot on the wall to the interface. The second cell is near the liquid–vapor interface, it rotates in a clockwise direction and is driven by the heat flow from the vapor to the liquid near the lower region of the bubble. Eventually, the temperature and flow fields settle down into the nearly stationary spatial configuration seen in FIGURE 5, and consequently the heat flux approaches a nearly constant value. The rate of pressurization agrees well with that predicted by a purely thermodynamic model of the entire tank although the final pressure levels are quite different, as shown in FIGURE 6.

The temperature fields in FIGURE 5 indicate that there is significant thermal stratification in the liquid due to natural convection generated by the residual gravity. In a previous paper[29] it was shown that there is no such stratification in a zero-gravity situation. This shows that buoyancy still plays a significant role in microgravity even though it acts over a time scale that is much larger than on the ground. The maximum flow speed due to natural convection is only 0.00313 cm/sec for the case shown in FIGURE 5, but it has an accumulative effect over the long periods of time considered here. This can be understood by considering that the relative strength of natural convection, as indicated by the Grashof number, is a very rapidly increasing function of the tank radius whereas it is only linearly proportional to the gravitational acceleration. Thus, the product gR^4 for a three meter diameter tank in microgravity is comparable to a 9.5 cm diameter tank in normal gravity. For the same wall heat flux they would have identical Grashof numbers and the extent of thermal stratification would be similar. This is corroborated by comparison of present microgravity results for a large tank with those predicted for a small tank in a previous 1 g study.[25] Of course, slight differences exist because of the different liquid–vapor configurations in the two environments.

Case 3: ZBO Pressure Control With a Subcooled Mixing Jet

After 75 days, the subcooled jet is turned on and its effect on the pressure and temperature rise is tracked for an additional 75 days, as shown in FIGURE 6. This is done for jet speeds spanning three orders of magnitude. The jet is subcooled, since its temperature is less than the corresponding saturation temperature at that time. Note that the thermodynamic curve does not continue after the first 75 days since there is no easy way to account for the transient cooling effect of the jet using a simple thermodynamic analysis. It is also assumed that the vapor region remains nearly spherical despite the influence of the jet, since the ratio of the inertia of the jet to surface tension forces, as given by the Weber number, is only 0.00685 and thus much less than unity even for the fastest jet speed considered here.

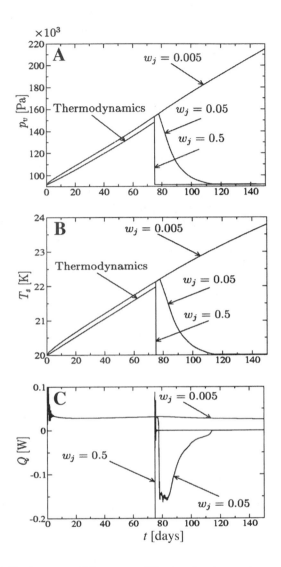

FIGURE 6. The long-term (**A**) pressure, (**B**) saturation temperature, and (**C**) total heat before and after the subcooled jet is turned on at $t = 75$ days.

For the lowest jet speed of 0.005 cm/sec, the pressure and temperature continue to rise at nearly the same rate as before since the jet is unable to sufficiently penetrate into the liquid region to reach the vapor. FIGURE 7 A shows that the temperature and flow field surrounding the vapor region are nearly the same as they are before the jet is turned on and the thermal stratification is hardly disrupted, since the cooling effect of the jet is limited to the bottom of the tank. As a result, the net heat flow into the vapor only slightly decreases, as shown in FIGURE 6C, and this is insufficient to

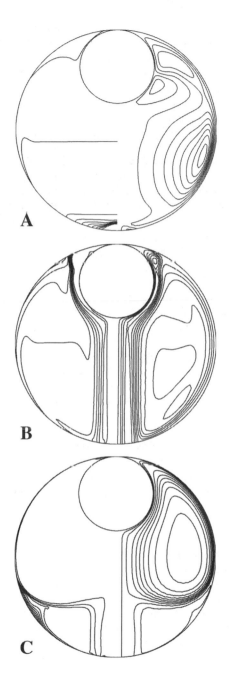

FIGURE 7. Final isotherm and streamline at $t = 150$ days with jet speeds of **(A)** $\bar{w}_j = 0.005$ cm/sec, **(B)** $\bar{w}_j = 0.05$ cm/sec, and **(C)** $\bar{w}_j = 0.5$ cm/sec. The minimum temperature is 20 K and the maximum temperatures are **(A)** 23.95 K, **(B)** 20.4 K, and **(C)** 20.13 K.

cause any noticeable change in the pressure rise. At its lowest speed, the jet cannot effectively control the pressure over the time span considered here.

When the jet speed is increased by an order of magnitude to 0.05cm/sec, it becomes much more effective. It still takes about 2.8 days before there is any significant cooling effect, since it takes that long for the jet to reach the interface. Obviously, this is much slower than the value of about one hour that is predicted by a simple calculation based on the jet inlet velocity and the distance from the bottom of the tank to the interface. This discrepancy is due primarily to the counterflow generated by natural convection that tends to resist jet penetration. Once cooling begins, it takes about 45 days for the jet to bring the saturation temperature and vapor pressure back down to their initial values. This time, the cooler jet fluid penetrates into the liquid region and encapsulates the entire vapor region, thus isolating it from the hot spot near the upper tank wall, as indicated by the final isotherms in FIGURE 7B. Even so, there is still noticeable thermal stratification in the rest of the liquid.

The cooling effect is further enhanced when the jet speed is increased by another order of magnitude, to 0.5cm/sec. In this case, the net heat flow into the vapor drops after only six minutes, which is in better agreement with the value calculated simply on the basis of the distance and the jet speed, because the jet flow is now considerably stronger than the counterclockwise vortex created by natural convection. For this case, once the cooling/mixing begins, it only takes about five hours for the jet to bring the saturation temperature and vapor pressure back down to their initial values. The final temperature profile also shows that the recirculation of cooler fluid, due to the fact that jet flow now encompasses over half of the liquid volume. The circulation cell due to natural convection associated with the hot spot is noticeably weaker and only limited to a small region near the bottom of the tank. As a result, thermal stratification in the liquid has been almost entirely disrupted by the jet.

CONCLUSIONS

In this paper, we have carefully examined the pressurization of a large liquid hydrogen storage tank in microgravity, and we have assessed the feasibility of zero boil-off (ZBO) pressure control using a subcooled liquid jet through a series of parametric numerical case studies.

The results show, somewhat surprisingly, that buoyancy and natural convective effects are still important in microgravity and cannot be ignored when predicting the pressurization of large cryogenic tanks in space. First and foremost, it is clear that for such large tanks in space typical average residual accelerations are sufficient to rapidly move the vapor region up to the tank wall as compared to the conduction and natural convection time scales. Thus, when considering this class of problems, the overall location of the vapor region can be predicted with some degree of certainty if the temporally averaged direction and magnitude of the acceleration vector are known.

It was also shown that for relatively large tanks, such as that considered here, natural convection plays a significant role because it is sufficiently strong to change the thermal stratification in the liquid, even in microgravity. This is a consequence of the fact that the Grashof number is a rapidly increasing function of the tank radius. After a lengthy initial transient response, the rate of tank pressurization in microgravity eventually agrees with purely thermodynamic predictions even though the final pressure levels are quite different, mainly due to the initial transients.

Finally, It is also shown that a subcooled jet can be used to effectively control tank pressurization in microgravity as long as it is sufficiently strong to penetrate through the liquid and encapsulate the vapor region. However, to accomplish this the jet must overcome the resistance created by the counter-flowing natural convective vortex. Largely due to the long time scales associated with heat conduction and convection for large tanks in microgravity, the relatively mild jet velocities considered in this paper are still effective even though they are in the laminar regime.

ACKNOWLEDGMENTS

This work was supported by NASA OBPR through the Microgravity Division at NASA Glenn Research Center. Additional resources were provided by the Computational Microgravity Laboratory and the Microgravity Fluids Physics Branch at NASA Glenn Research Center.

REFERENCES

1. SALERNO, L.J. & P. KITTEL. 1999. Cryogenics and the human exploration of Mars. Cryogenics **39:** 381–388.
2. KITTEL, P. & D.W. PLACHTA. 2000, Propellant preservation for mars missions, Adv. Cryogen. Eng. **45:** 443.
3. AYDELOTT, J.C. 1967. Effect of gravity on self-pressurization of spherical liquid-hydrogen tankage. NASA TN-D-4286.
4. AYDELOTT, J.C. 1979. Axial jet mixing of ethanol in cylindrical containers during weightlessness. NASA TP-1487.
5. AYDELOTT, J.C. 1983. Modeling of space vehicle propellant mixing. NASA TP-2107.
6. LIN, J.C., M.M. HASAN & N.T. VAN DRESAR. 1994, Experimental investigation of jet-induced mixing of a large liquid hydrogen storage tank. AIAA Paper 94-2079.
7. POTH, L.J. & J.R. VAN HOOK. 1972. Control of the thermodynamic state of space-stored cryogens by jet mixing. J. Spacecraft **9:** 332–336.
8. LIN, C.S. & M.M. HASAN. 1992. Self pressurization of a spherical liquid hydrogen storage tank in a microgravity environment. AIAA-92-0363.
9. VAUGHAN, D.A. & G.R. SCHMIDT. 1991. Analytical modeling of no-vent fill process. J. Spacecraft **28:** 574–579.
10. CHA, D.A., R.C. NEIMAN & J.R. HULL. 1993, Thermodynamic analysis of helium boil-off experiments with pressure variations. Cryogenics **33:** 675–679.
11. GRAYSON, G.D., D.A. WATTS & J.M. JURNS. 1997. Thermo-fluid-dynamic modeling of a contained liquid in variable heating and acceleration environments. ASME Publication FEDSM97-3567.
12. LIN, C.S. & M.M. HASAN. 1990. Numerical investigation of the thermal stratification in cryogenic tanks subjected to wall heat flux. AIAA-90-2375.
13. NAVICKAS, J. 1988. Prediction of a liquid tank thermal stratification by a finite difference computing method. AIAA-88-2917.
14. GRAYSON, G.D. & J. NAVICKAS. 1993. Interaction between fluid dynamic and thermodynamic phenomena in a cryogenic upper stage. AIAA-93-2753.
15. LIN, C.S. & M.M. HASAN. 1990. Vapor condensation on liquid surface due to laminar jet-induced mixing: the effects of system parameters. AIAA-90-0354.
16. HOCHSTEIN, J.I., P.M. GERHART & J.C. AYDELOTT. 1984. Computational modeling of jet induced mixing of cryogenic propellants in low-g. AIAA 84-1344.

17. KOTHE, D.B., C.R. MJOLSNESS & M.D. TORREY. 1991. Ripple: a computer program for incompressible flows with free surfaces. Los Alamos National Laboratory, LA-12007-MS.
18. LIU, C.H. 1994. A numerical calculation of time dependant dynamical behavior of liquid propellants in a microgravity environment. Micrograv. Sci. Technol. VII/2: 169-172.
19. HUNG, R.J. & K.L. SHYU. 1992. Constant reverse thrust activated reorientation of liquid hydrogen with geyser initiation. J. Spacecraft Rockets 29: 279–285.
20. THORNTON, R.J. & J.I. HOCHSTEIN. 2001. Microgravity propellant tank geyser analysis and prediction. AIAA-2001-1132.
21. MARCHETTA, J.G. & J.I. HOCHSTEIN. 2000. Simulation and dimensionless modeling of magnetically induced reorientation. AIAA 2000-0700.
22. MARCHETTA, J.G., J.I. HOCHSTEIN & D.R. SAUTER. 2001. Simulation and prediction of magnetic cryogenic propellant positioning in reduced gravity. AIAA-2001-0930.
23. PETERSON, L.D., E.F. CRAWLEY & R.J. HANSMAN. 1989. Nonlinear fluid slosh coupled to the dynamics of a spacecraft. AIAA J. 88-2470: 1230–1240.
24. HUNG, R.J. & C.C. LEE. 1994. Effect of a baffle on slosh waves excited by gravity-gradient acceleration in microgravity. J. Spacecraft Rockets 31: 1107–1114.
25. PANZARELLA, C.H. & M. KASSEMI. 2003. On the validity of purely thermodynamic descriptions of two-phase cryogenic fluid storage. J. Fluid Mechan. 484: 136–148.
26. ENGELMAN, M.S. & R.L. SANI. 1984. Finite element simulation of incompressible flows with free/moving surface. In Numerical Methods in Laminar and Turbulent Flows. Pineridge Press, Swansea.
27. GRESHO, P.M., R.L. LEE & R.C. SANI. 1979. On the time-dependent solution of the incompressible Navier–Stokes equations in two and three dimensions. In Recent Advances in Numerical Methods in Fluids, Vol. 1. 27–79. Pineridge Press, Swansea.
28. ENGELMAN, M.S., G. STRANG & K.J. BATHE. 1981. The application of quasi-Newton methods in fluid mechanics. Intl. J. Numerical Meth. Engin. 17: 707–718.
29. PANZARELLA, C.H. & M. KASSEMI. 2004. Pressurization of spherical cryogenic tanks in space. J. Spacecraft Rockets. In press.

Gravitational Effects on the Weld Pool Shape and Microstructural Evolution During Gas Tungsten Arc and Laser Beam Welding of 304 Stainless Steel and Al–4wt% Cu Alloy

NAMHYUN KANG,[a] JOGENDER SINGH,[a] AND ANIL K. KULKARNI[b]

[a]*Department of Materials Science and Engineering,*
The Pennsylvania State University, University Park, Pennsylvania, USA

[b]*Department of Mechanical and Nuclear Engineering,*
The Pennsylvania State University, University Park, Pennsylvania, USA

ABSTRACT: Effects of gravitational acceleration were investigated on the weld pool shape and microstructural evolution for 304 stainless steel and Al–4wt% Cu alloy. Effects of welding heat source were investigated by using laser beam welding (LBW) and gas tungsten arc welding (GTAW). As the gravitational level was increased from low gravity (LG ~ 1.2g) to high gravity (HG ~ 1.8g) using a NASA KC-135 aircraft, the weld pool shape for 304 stainless steel was influenced considerably during GTAW. However, insignificant change in the microstructure and solute distribution was observed at gravitational levels between LG and HG. The GTAW on Al–4wt% Cu alloy was used to investigate the effect of gravitational orientation on the weld solidification behavior. Gravitational orientation was manipulated by varying the welding direction with respect to gravity vector; that is, by welding upward opposing gravity (‖-U) and downward with gravity (‖-D) on a vertical weld piece and welding perpendicular to gravity (⊥) on a horizontal weld piece. Under the same welding conditions, a larger primary dendrite spacing in the ‖-U weld was observed near the weld pool surface and the fusion boundary than in the case of ⊥ or ‖-D welds. The ‖-D weld exhibited different solidification morphology and abnormal S shape of solidification rate curve during its growth. For 304 stainless steel GTAW, significant effects of gravitational orientation were observed on the weld pool shape that was associated with weld surface morphology and convection flow. However, the weld pool shape for LBW was mostly constant with respect to the gravitational orientation.

KEYWORDS: gravity; laser beam welding; gas tungsten arc welding; microstructure; weld pool shape; 304 stainless steel; Al–Cu alloy

INTRODUCTION

The effects of gravitational (acceleration) level on a welding process have been studied for the past two decades because it is expected to play a significant role in

Address for correspondence: Namhyun Kang, Korea Institute of Industrial Technology, DongChun, Yeon Su, Incheon, 406-130 Korea. Voice: 82-32-850-0218; fax: 82-32-850-0210.
nhkang@kitech.re.kr

Ann. N.Y. Acad. Sci. 1027: 529–549 (2004). ©2004 New York Academy of Sciences.
doi: 10.1196/annals.1324.041

space construction. In a terrestrial environment, gravitational (acceleration) orientation influences the performance on circumferential pipe welding. For the performance of these processes, it is necessary to understand gravitational effects on welding phenomena, such as convection flow and weld pool dimension, microstructural evolution, and segregation. Although several micro- and multigravity studies have been conducted experimentally[1–8] and numerically,[9,10] there is a lack of studies that combine various results for weld pool shape with respect to gravitational level.

Considerable research has been conducted to understand convection in a liquid weld pool. Surface tension driven flow was determined to be dominant and the gravity driven buoyancy flow to be minimal in driving convection under Earth gravity.[11,12] However, our experiments showed that gravity plays an influential role in weld surface deformation under certain welding conditions, and this, in turn, affects weld pool shape.[13] Previously, we found that CO_2 laser welding on polypropylene and low carbon steel exhibits greater penetration but narrower bead width for the parallel-up (∥-U) orientation than those for the parallel-down (∥-D) orientation. To better understand gravitational effects, it is necessary to consider the orientation and the level of gravity independently. Weld pool shape can then be studied more systematically as a function of gravity and is associated with weld surface deformation.

There are a number of resources reporting the gravitational effects on weld pool shape and convection flows.[1,2,14,15] During unidirectional solidification, the absence of gravity-driven convection flow increased the primary dendrite spacing (λ_1) in Pb–Sn alloys[16] and Al–Cu alloys,[17] but it showed an opposite result, decreasing λ_1, in $Pd_{40}Ni_{40}P_{20}$ alloys.[18] No gravitational effects on the macrosegregation were reported in the Al–Cu alloy,[19] however, Liu et al. reported lower P solute content in the primary phase as gravitational level decreased.[18] For Nd–YAG laser welding on 304 stainless steel, high gravity produced an equiaxial dendritic microstructure transformed from a radial columnar structure in low gravity welds.[20] Fine equiaxial grains located along the solid–liquid boundary disappeared in high gravity due to enhanced convection in GTA welded Al alloys.[21] However, systematic studies of the gravitational effect, either orientation or level, on the weld microstructural evolution have not been conducted, to the best of our knowledge, so as to resolve the above controversy.

The objective of this paper is to summarize the gravitation effects on welding phenomena from our previous results.[8,13,22,23] The effects of gravitational level and orientation on weld pool shape and microstructural evolution are discussed during GTAW and LBW for 304 stainless steel and Al–4wt% Cu alloy. The study on weld pool shape is associated with free surface deformation and convection flow as a function of gravity. The microstructural study is focused on the behavior of solidification morphology, orientation, and λ_1 in Al–4wt% Cu alloy. The Al–Cu system was chosen for this study because it could help amplify gravitation effects on solute segregation by virtue of the high density difference between solute atoms. Studies of gravity on the welding process are expected to play a significant role in space station construction and circumferential pipe welding on Earth.

EXPERIMENTAL DETAILS AND METHODS

Stainless steel (304) welds were produced in NASA KC-135 aircraft two decades ago using parabolic trajectories to simulate acceleration levels from about $0.1g$ to about $1.8g$. An onboard accelerometer recording during a flight is shown in FIGURE 1. More specific details about the KC-135 flight experiment are published elsewhere.[24] The base metal had a diameter of 75 mm and a thickness of 3 mm. Bead-on-plate welds were produced by using the GTAW with a rotating tungsten–2% thorium electrode under the conditions: 75 amps, 10–12 volts, and 5 mm/sec translational velocity. To insure that the gravitational level was the only variable, the welds were conducted so that the transition between μg and $1.8g$ occurred within the weld. For experimental details see Reference 8.

To examine effects of gravitational orientation on the weld pool shape and microstructural evolution, experiments were designed so that the relation between the gravity vector and the arc translation direction could be varied. FIGURE 2 depicts the weld orientation used to simulate various gravitational conditions: welding upward in a direction opposing gravity (parallel-up or ‖-U weld), welding downward in the direction of gravity (parallel-down or ‖-D weld), and welding perpendicular to the direction of gravity (perpendicular or $\perp$ weld).

During weld solidification, columnar-dendritic structures are the frequently observed morphology. The microstructural size of dendrites characterizes the solute segregation pattern that largely determines the properties of the material.[25,26] One of the most important quantities used to describe dendritic structures in columnar growth is the primary dendrite spacing (λ_1). Most experimental studies have shown that λ_1 decreases as the solidification parameter increases. The following equation shown λ_1 as a function of solidification parameters, thermal gradient (G_L) and solidification rate (V_S),

$$\lambda_1 = C_1 G_L^{-a} V_S^{b}, \tag{1}$$

where C_1 is a kinetic constant that is characteristic of the alloy system under consideration. The constants a and b are defined, respectively, as 0.5 and 0.25 from theoretical models.[27–29]

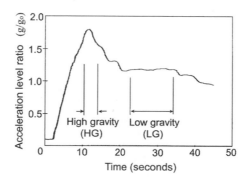

FIGURE 1. Accelerometer data during a KC-135 flight.

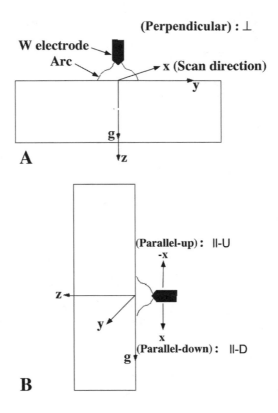

FIGURE 2. Welding direction: **(A)** perpendicular orientation to gravity and **(B)** parallel-up and parallel-down orientations to gravity.

The solidification behavior of the weld has also been studied by applying rapid solidification theories.[30] Kurz, Giovanola, and Trivedi (KGT) developed a theoretical model for dendritic growth under rapid solidification conditions.[31] This model shows that the solidification rate (V_S) is the key factor in microstructural size. According to the KGT model, the experimentally measured λ_1, as a function of V_S, obeys the relationship,

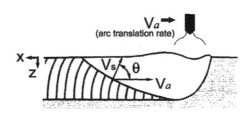

FIGURE 3. Schematic diagram of the molten pool at the center plane of the welded track.

$$\lambda_1^2 V_S = \text{constant.} \tag{2}$$

V_S is obtained by measuring the grain orientation in a longitudinal cross-section through the center-plane of the welded track, as shown in FIGURE 3. The microstructural orientation tends to be perpendicular to the local solid–liquid interface because the grains are known to grow parallel to the direction of heat dissipation. Therefore, the orientation of the grain boundary or that of the interdendritic phase is measured with respect to the arc translation direction. The relationship between the arc translation rate (V_a) and the local solidification rate (V_S) is

$$V_S = V_a \cos\theta, \tag{3}$$

where θ is the angle between the vectors representing V_S (parallel to the microstructure) and V_a (parallel to the arc translation). The details of this method are described elsewhere.[30] The above observations have been applied successfully to the welding microstructures of Al–Cu alloys.[32] However, Equation (3) has not been exploited to investigate the influence of gravitational orientation on the weld solidification and microstructural behavior.

EFFECTS OF GRAVITATIONAL LEVEL ON 304 STAINLESS STEEL DURING GTAW

Gravitational effects on the GTAW of 304 stainless steels were studied to examine the behavior of weld pool shape and its impact on microstructure and solute dis-

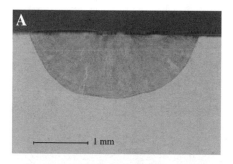

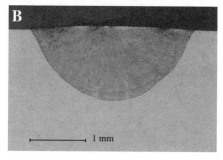

FIGURE 4. Cross-section of weld pool developed in (**A**) high gravity and (**B**) low gravity.

tribution. This was accomplished during both NASA KC-135 flights and in ground-based experiments.

Effects of Gravitational Level on Weld Pool Shape

The fraction of the weld data for which the influence of gravity was found to be significant was separated into two classes: high gravity (HG ~ 1.8 g) and low gravity (LG ~ 1.2 g). Comparing the weld pool shape of HG and LG, the average width increased by 10% and the average depth decreased by 10% at the HG. However, the area of the weld pool remained constant between the HG and the LG. FIGURE 4 shows the cross-section transverse to the arc translation axis, of the weld pool developed in HG and LG. The shape of the weld pool at LG is more hemispherical than that at HG. The hemispherical shape of the weld pool at LG is caused by weak buoyancy convection, but at HG buoyancy convection is probably increased by influencing the weld pool shape. However, this behavior of the weld pool shape as a function of gravitational level is probably more significantly associated with other factors, such as the degree of free-surface deformation and convection flows influenced by arc current distribution. This is because surface-tension driven convection flow dominates over electromagnetic flow and gravity-driven buoyancy flow in the absence of a surface-active element.[11,12]

Effects of Gravitational Level on Solute Distribution

Gravitational effects on the solute distribution in the duplex microstructure (ferrite and austenite) were further investigated by using scanning transmission electron microscopy (STEM). FIGURE 5A shows a typical micrograph of the HG weld (1.8 g) containing a region of δ-ferrite between two γ-austenites. Note the line of contamination spots marking the analysis of energy-dispersive spectrometry. FIGURE 5B shows the measured Cr and Ni composition profiles corresponding to the points shown in FIGURE 5A. The average Cr compositions in the ferrite and the austenite of the HG were about 25.5 wt% and 18.9 wt%, respectively. A micrograph and composition profiles of the LG weld (1.2 g) are indicated in FIGURE 6. The average values of Cr in the ferrite and austenite were about 25.8 wt% and 18.9 wt%, respectively. A measurement of Ni was also made. The compositions of Cr and Ni are indicated in TABLE 1 both for HG and LG. Thus, the average composition of Cr and Ni at the core of the ferrite and austenite dendrites showed unnoticeable effect resulting from gravity within the measurement scatter. The reason for minimal gravitational effects on the solute distribution is partly associated with the solidification path taken by 304 stainless steel and the similar densities among the major alloying components (Fe, Cr, and Ni). The measured compositions of Cr and Ni would contain the effects of solid-state diffusion as well as gravity-induced effects in the liquid melt.

Effects of Gravitational Level on Ferrite Content

A microstructural analysis was conducted to observe the amount of retained ferrite as a function of gravitational level by using X-ray diffraction. HG and LG showed volume percentages of ferrite equal to 13.7 and 13.5, respectively. The measured range of the ferrite content matched reasonably with a prediction from a Schaeffler constitution diagram and WRC-1992.[33,34] The alloy composition rather than the heat input or the cooling rate was known to control the ferrite content.[35] However, the solute distribution showed minimal changes in alloy composition, as mentioned in the previous section. As a result, it was concluded that ferrite content was not significantly affected as the gravitational level varied from HG to LG.

To better understand gravitational effects on weld microstructure and solute distribution, more critical preparation must precede experiments in the selection of weld materials and process parameters. To study solute distribution under the gravitational variation, a binary alloy experiencing no solid-state transformation will make the investigation easier than multicomponent alloy. Large density difference of solute atoms may be better, so as to amplify the gravitation effects on solute distribution; for example, Al–Cu binary alloy. Based on our evidence.[13,23] high heat input

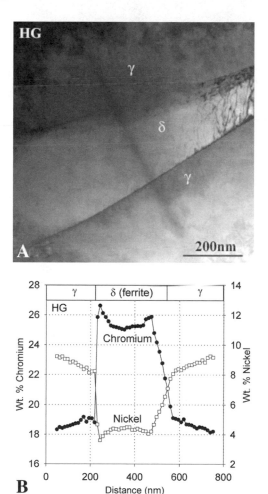

FIGURE 5. STEM microanalysis between two austenite regions in HG 304 stainless steel (300 kV). (**A**) micrograph and (**B**) composition versus distance profiles for Cr and Ni.

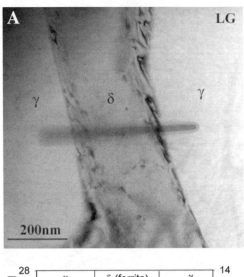

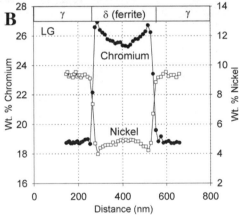

FIGURE 6. STEM microanalysis between two austenite regions in LG 304 stainless steel (300 kV). **(A)** micrograph and **(B)** composition versus distance profiles for Cr and Ni.

and larger weld pool also increased the impact of gravity, because gravity-driven force in convection flow and on the free surface depends upon the volume of liquid weld metal. The effects of gravitational orientation on weld pool shape are discussed further in the following section in connection with free-surface deformation.

TABLE 1. **Mean composition in 304 stainless steel welds**

	δ (HG)	γ (HG)	δ (LG)	γ (LG)
Cr (wt%)	25.5 ± 0.5	18.9 ± 0.5	25.8 ± 0.4	18.9 ± 0.2
Ni (wt.%)	4.3 ± 0.3	8.9 ± 0.3	4.8 ± 0.4	9.6 ± 0.4

EFFECTS OF GRAVITATIONAL ORIENTATION
ON WELD POOL SHAPE

Effects of gravitational orientation were investigated on the behavior of the weld pool shape associated with free surface deformation for 304 stainless steel. For various welding orientations, the behavior of the weld pool shape is associated with weld surface deformation and correspondingly the values of the Peclet (*Pe*) and the Marangoni (*Ma*) numbers. Two welding processes (GTAW and LBW) were employed to study the effects of heat source on the behavior of weld pool shape.

Gas Tungsten Arc Welding (GTAW) on 304 Stainless Steel

The weld pool shape on 304 stainless steel was investigated to determine the effects of welding orientation. FIGURE 7 shows the weld pool shape for various translation velocities (V_a) and two extreme welding orientations (∥-U and ∥-D). In region (III), in which V_a is between 8 and 10 mm/sec, less convexity in the ∥-U weld produced significantly larger depth (30–40%) in the weld pool center than that in the ∥-D weld. Thus, the ∥-U weld shows almost flat pool bottom, whereas the ∥-D weld clearly indicates dual penetrations away from the center (convex shape). The perpendicular weld lies between two extreme welding orientations (∥-U and ∥-D), although this is not included in FIGURE 7. Comparing the weld pool shape between the ∥-U and ∥-D, the ∥-D weld must have stronger outward convection flows (i.e., higher *Pe* and *Ma*) on the weld surface than in the ∥-U case. However, in region (III), there were no apparent effects of gravitational orientation on the weld width, deepest penetration, or cross-sectional area.

In region (II) of FIGURE 7, in which V_a is between 4 and 5 mm/sec, convexity mostly disappeared in the weld pool bottom and transformed to the flat or concave shape. The ∥-U weld showed a narrower (6–13%) width than that in the perpendicular and

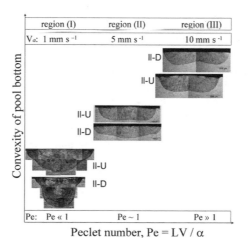

FIGURE 7. Effects of gravitational orientation on weld pool convexity for extreme welding orientation (∥-U and ∥-D) and 165 A GTAW on 304 stainless steel.

||-D welds. However, there were no apparent effects of gravitational orientation on the penetration and weld area. In regions (II) and (III), minor effects of gravitational orientation were observed on the weld pool shape; narrower width in the ||-U or more convexity in the ||-D weld, respectively. Despite this abnormal behavior, the weld pool shape was mostly independent of gravitational orientation within the V_a range 4–10 mm/sec.

In region (I) of FIGURE 7, in contrast, the gravitational orientation changed the weld pool shape significantly at 2 mm/sec and it became more obvious at the smallest value of V_a (1 mm/sec) in the present study. At 1 mm/sec, the ||-D weld showed 28% narrower width but 31% deeper penetration than perpendicular and ||-U welds. It is worth mentioning that the weld pool area and arc power remained constant (±3%) with respect to the welding orientation. In this region (I), the weld pool may have a low $Pe \ll 1$ due to low V_a. Low Pe indicates that the dominating heat transport is by conduction in the weld pool, resulting in concave or hemispherical weld pool shape. As the beam radius (d_a) decreased, the weld pool shape approached a more hemispherical shape.[36] Based on the weld pool shape during the present study, the ||-D weld demonstrated more hemispherical shape than the ||-U weld pool. Therefore, it is speculated that the ||-D weld must have smaller d_a than that in the ||-U weld. For the present study, d_a denotes the diameter of the electric arc, because GTAW is employed instead of LBW. Despite providing the same welding heat input, it is now important to understand the cause of the weld pool shape variation as a function of gravitation. This is believed to be associated with weld surface deformation and correspondingly the value of Pe.

FIGURE 8 shows surface deformation plotted in the axial directions for various V_a. The weld pool surface was measured at the end of welded track after the welding arc was extinguished. Regardless of the translational velocity (V_a), the ||-U weld showed deepest surface depression and the ||-D weld displayed maximum mass accumulating ahead of the arc due to gravity. The perpendicular weld surface is between that in the ||-U and ||-D welds, as shown in FIGURE 8. Based on the degree of surface depression, it is fairly certain that the ||-U weld will have greater gap between the electrode tip and the weld piece during welding than in the ||-D case. Correspondingly, the ||-U weld will have larger arc diameter (d_a), compared to the ||-D weld because the electric arc has a divergent current distribution. The larger d_a, however, will contribute to the weld pool shape differently depending on V_a and resultant Pe, as discussed elsewhere.[36] In the case of $Pe \gg 1$ associated with high V_a, convection dominates the weld pool shape. The larger d_a in the ||-U weld produces smaller V_{max} resulting in smaller Pe. Therefore, the ||-U weld will have lower Pe than the ||-D weld in regime (III), as shown in FIGURE 7. The ||-U weld showed less convexity, due to its lower Pe and Ma, compared to the ||-D weld pool shape (FIG. 7). For $Pe \ll 1$, associated with low V_a, in contrast, conduction dominates the weld pool shape. The ||-U weld showed less hemispherical shape, due to its larger d_a than the shape of the ||-D weld in region (I), as indicated in FIGURE 7. Between these regimes, that is, medium V_a with $Pe \approx 1$ in region (II), both conduction and convection contribute to a weld pool shape that is independent of gravitational orientation. To enhance understanding, the relation of Pe and weld pool shape is summarized with respect to gravitational orientation as follows:

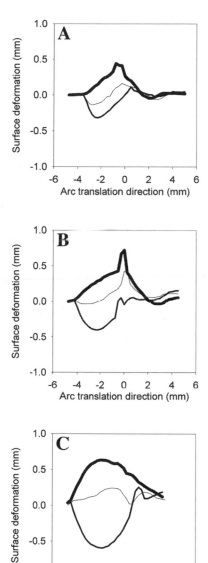

FIGURE 8. Weld pool surface morphology along the weld center line at various translational velocities: (A) 10 mm/sec, (B) 4 mm/sec, and (C) 1 mm/sec; ——— , ⊥; ━━━ , ‖-U; ━━━ , ‖-D.

If $Pe \gg 1$ (at high V_a)

- ‖-U weld: a deeper surface depression $\Rightarrow$ larger d_a $\Rightarrow$ smaller V_{max} $\Rightarrow$ smaller Pe and Ma $\Rightarrow$ less convexity;

- ‖-D weld: maximum mass accumulating ahead of the arc $\Rightarrow$ smaller d_a $\Rightarrow$ larger V_{max} $\Rightarrow$ larger Pe and Ma $\Rightarrow$ more convexity.

If $Pe \ll 1$ (at low V_a)

- ‖-U weld: a deeper surface depression $\Rightarrow$ larger d_a $\Rightarrow$ less hemispherical shape;

- ‖-D weld: maximum mass accumulating ahead of the arc $\Rightarrow$ smaller d_a $\Rightarrow$ more hemispherical shape.

Thus, surface morphology was determined to be a significant factor on the weld pool shape as a function of gravitational orientation. As the translation velocity decreases, the difference in the surface morphology is more evident between the ‖-U and ‖-D welds (FIG. 8C). This result implies that gravitation effects are more significant on the weld pool surface as the weld pool size increases. The value of the Peclet number (Pe) and convexity associated with weld surface morphology successfully explained the behavior of weld pool shape as a function of gravitational orientation.

Laser Beam Welding (LBW) on 304 Stainless Steel

Effects of the heat source on weld pool shape were investigated by using a laser beam instead of a gas tungsten arc. Experimental details are indicated in FIGURE 9. In the previous section, a more significant influence of gravity was observed as weld pool size, thus maximum laser power and a focused beam were used for LBW to magnify gravitational effects on weld pool shape. For fully focused 2.6kW LBW, weld pool shape was studied more specifically by using weld pool width (W_1 and W_2), penetration, and cross-sectional area as a function of gravitational orientation. Fully focused LBW showed a nail-shaped weld composed of weld bead and weld root. The weld root was caused by keyhole formation from the intense laser beam. W_1 indicates the width of the weld bead, which is normally known as the weld pool width for GTAW. W_2 is the width of the weld root. No variation on the weld pool shape (W_1, W_2, penetration, or cross-sectional area) was observed as a function of gravitational orientation. Detailed results on the weld pool shape for LBW are given elsewhere.[23] The weld pool shape remained stable regardless of the gravitational orientation throughout the translational velocity range (8–42mm/sec).

To reduce or avoid the keyhole formation during LBW, the laser power was reduced to 1.6kW and the laser beam was unfocused 7mm below the work piece. A detailed description of a 7mm unfocused beam is depicted in FIGURE 9C as compared with fully focused beam. In the case of *parallel LB*, the laser beam direction for ‖-U is opposite to the gravity vector, whereas a ‖-D weld has the same direction of laser beam as the gravity vector. To study effects of gravitational orientation without the influence of laser beam direction, *side LB* was used in order to have a constant relation between the laser beam direction and gravity vector, as shown in FIGURE 9B. The objective of reducing keyhole formation is to produce a weld pool governed by the convection flows. Convection flows within the weld pool are changed by varying the gravitation condition, as discussed in the previous section. The weld pool for 1.6 kW and 7 mm unfocused LBW showed two classes of shape: (1) a hemispherical shape at 0.5–1 mm/sec and (2) a nail shape at 3–7 mm/sec. Although slight increases (about 10%) in the value of W_1 were observed for the $\perp$ weld between 3 and 7 mm/sec, in general there were insignificant effects of gravitational orientation on the weld pool shape, such as stable regions (II) and (III) shown in FIGURE 7. Regardless of the laser beam direction (*parallel LB* or *side LB*), laser power (2.6 or 1.6 kW), and the degree of laser focusing (fully focused or 7 mm unfocused), the weld pool shape was mostly constant with respect to the gravitational orientation.

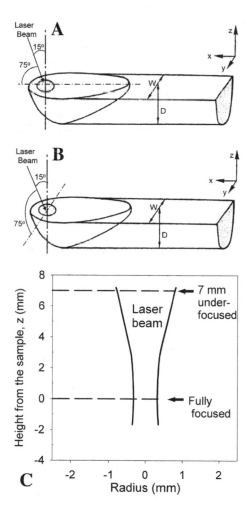

FIGURE 9. Welding setup for the LBW at perpendicular welding orientations: (**A**) 15° beam direction from the welding front (called *parallel LB*); (**B**) 15° beam direction from the side (called *side LB*); and (**C**) focused and 7 mm-unfocused beam position on the sample.

EFFECTS OF GRAVITATIONAL ORIENTATION ON
MICROSTRUCTURAL EVOLUTION DURING GTAW

The present study evolved from a previous Ni GTAW study[13] with the following hypothesis: as gravitational orientation varies, the weld pool shape is changed as a result of variation in convection flows. It is believed that convection flow is associated with solidification rate (V_S) and thermal gradient (G_L) at the solid–liquid interface. Therefore, the weld microstructure and solidification orientation will be influenced as a function of the gravitational orientation. The present study focused on the specific welding conditions (slow V_a and high power) for the purpose of magnifying the gravitation effects on weld pool shape and microstructure. Bead-on-plate welds were produced on Al–4wt% Cu alloy autogenously for a constant direct current 185 A and a variable arc voltage of 16.5 V (±0.2 V). The tungsten–2% thorium electrode was held stationary and the sample plate (50mm × 150mm × 6.5mm) mounted on the working table was translated with a constant velocity 3mm/sec.

Weld Pool Shape

An analysis of the gravitational effects demonstrated that the weld pool shape varied considerably when changing the welding orientation depicted in FIGURE 2. TABLE 2 shows the average depth, width, depth to width ratio, and transverse cross-sectional area of the fusion zone (FZ). The ||-U weld had 15% deeper penetration and 22% larger FZ area than those of the ⊥ and ||-D welds. The width and the depth to width ratio of the welds, in contrast, showed little to no variation with welding orientation. These results were comparable to the previous investigation of GTA welds on nickel.[13] Regardless of the effects due to convection[2,10] or surface deformation,[13,23] it is fairly certain that the G_L and cooling rate (C_R) in the weld pool will vary with respect to the gravitational orientation. This is because the weld pool shape varied under the constant heat input. Furthermore, it is anticipated that the variation of G_L and C_R will have an influence on the solidification morphology and primary dendrite spacing, as is presented in the next two sections.

Weld Solidification Morphology and Orientation

Considering significant gravitational effects on the FZ shape, microstructural studies on the solidification morphology were conducted from the grain structures taken from the longitudinal cross-section at the center plane. The columnar grains from the fusion boundary had a coarse and elongated shape for all welding orientations. As solidification proceeded, the grain became finer continuously toward the weld surface. The ||-D grains on the surface remained elongated in shape. However,

TABLE 2. **Measured weld pool shape as a function of gravitational orientation**

	Depth (mm)	Width (mm)	Depth/Width	Area (mm²)
⊥	2.35 ± 0.1	6.60 ± 0.1	0.36 ± 0.03	11.0 ± 0.4
||-U	2.75 ± 0.1	6.75 ± 0.2	0.41 ± 0.05	13.6 ± 0.5
||-D	2.35 ± 0.1	6.50 ± 0.1	0.36 ± 0.03	10.6 ± 0.3

the ∥-U grains mostly lost the directionality of the columnar structure, that is, approached the equiaxial structure, although it still seems to retain the columnar structure because it maintains directionality. The grain shape of the ⊥ weld is between that of the ∥-U and ∥-D grains.

Additional investigation of the longitudinal cross-section was conducted to determine gravitation effects on weld solidification orientation. The solidification orientation was measured by the angle θ between the columnar growth direction and the arc translation direction, as illustrated in FIGURE 3. For the ⊥ and ∥-U welds, the orientation of the columnar-dendritic grains decreased continuously from the fusion boundary to the weld pool surface. However, the ∥-D weld showed an abnormal behavior of the orientation between the boundary and the surface. FIGURE 10A depicts the grain orientation measured through the thickness at the center-plane of the welded track. The orientations of the columnar grains with respect to the arc translation direction were converted to the solidification rate (V_S) using Equation **(3)**. FIGURE 10B shows the calculated V_S, which is correlated with the grain orientation.

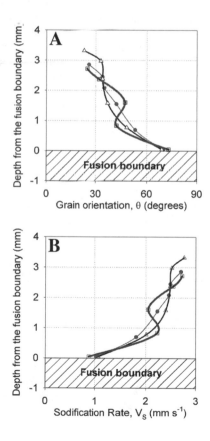

FIGURE 10. Effects of gravitational orientation (●, perpendicular; △, parallel-up; and □, parallel-down) on the solidification behavior. **(A)** Measured angle θ between V_S and V_a. **(B)** Calculated V_S with respect to the depth from the fusion boundary.

The ∥-D weld clearly exhibited an abnormal S shape of the V_S function, whereas the ⊥ and ∥-U welds had continuously increasing V_S from the boundary to the surface. Except for the abnormal V_S curvature of the ∥-D weld, all welding orientations showed almost same values for V_S at both the ends of the fusion boundary and the weld pool surface.

The variation of solidification morphology and orientation is explained by using a concept of convection flows and solidification parameters (G_L and V_S) with respect to the gravitational orientation. The convection flow within the weld pool is illustrated for both the ∥-U and ∥-D welds in FIGURE 11. This includes the shape of the trailing solid–liquid interface and the dimension of the weld pool associated with convection flow. The shape and dimension of the weld pool have been exaggerated by scaling for the purpose of comparing the ∥-D and ∥-U welds. Thermal gradients (G_L) in the x- and z-directions are also indicated as separate graphs to explain the solidification morphology later in this section and the primary dendrite spacing (λ_1) in the next section. For the ∥-U weld, gravity may promote the flow of liquid metal toward the rear of the weld pool. Due to its promotion on the outward convection flow, the trailing solid–liquid boundary will exhibit a circular shape with no abnormal trend (FIG. 11 A). So does V_S, because it is directly related to the shape of the trailing solid–liquid interface. In contrast, outward convection flow in the ∥-D weld may be inhibited because its direction, from the center to the rear of the weld, is opposite that of the gravity vector. The inhibited outward convection flow will produce a receded solid–liquid interface toward to the center of the weld pool, as shown in FIGURE 11 B. The receded interface will also influence the abnormal S shape of the solidification orientation and V_S, as shown in FIGURE 10 A and B, respectively. The behavior of the ⊥ weld lies between the ∥-U and ∥-D cases.

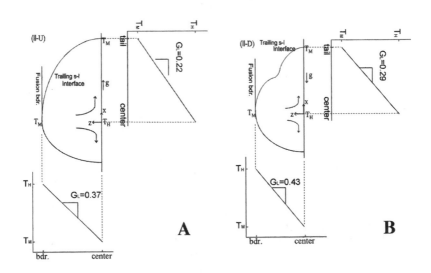

FIGURE 11. Illustration of the trailing solid–liquid interface for **(A)** the ∥-U weld and **(B)** the ∥-D weld. The schematic diagrams to estimate G_L ($K\,\mu m^{-1}$) are located on the right-hand side and on the bottom of the weld pool illustration.

Comparison of G_L between the ‖-U and ‖-D welds can be visualized more clearly from the illustration on thermal behavior (FIG. 11). T_H is the highest temperature in the center of the weld pool surface and T_M is the melting temperature. The highest temperature (T_H) was set to 1,900 K, lying between the melting temperature (870 K) and vaporization temperature (2,700 K) of Al alloys. A linear temperature distribution between T_H and T_M was applied to simplify the calculation. The slope of the lines represents G_L in FIGURE 11. Despite the assumptions made for G_L, the experimental findings can be adequately explained in the present study. The G_L range was determined to be 0.22–0.37 K·µm^{-1} and 0.29–0.43 K·µm^{-1} for the ‖-U and ‖-D welds, respectively. The larger dimension of the ‖-U weld produced a smaller range for G_L. The ⊥ weld had a G_L range of 0.29–0.43, the same as the ‖-D weld, because they had nearly the same penetration and width of the weld pool. As a result, the small weld pool dimension in the ‖-D weld produced a larger G_L than that of the ‖-U weld. This implies that the ‖-D weld has more columnar structure near the surface and the ‖-U weld loses the columnar structure. In summary, it appears that the shape of the solid–liquid interface associated with convection flows significantly influences V_S and G_L for the welding conditions considered in this study. This confirms the hypothesis postulated in the beginning of this section along with experimental findings.

Primary Dendrite Arm Spacing

The effects of gravitational orientation on the grain substructure were investigated because the variation of solidification morphology and convection flows plays a role in modifying the microstructure size.[37] FIGURE 12 shows a logarithmic scale plot of the behavior of λ_1 with V_S in a longitudinal cross-section through the thickness. The ‖-U weld exhibited approximately 18% larger averaged value for λ_1 both near the fusion boundary and at the weld pool surface, compared to that of the ‖-D and the ⊥ welds. It was observed that the ‖-D and ⊥ welds had almost the same λ_1 values. This result is probably because of the slower cooling rate in the ‖-U weld. The ‖-U weld had approximately 22% larger weld pool area although the welding power remained constant for all experiments. This suggests that the slower cooling rate in the ‖-U weld has a longer time for diffusion and coarsening, therefore producing a larger microstructure size. A linear regression analysis on the λ_1 behavior (FIG. 12) is based on previous work by Gremaud *et al.*;[38] the G_L effects on λ_1 could be safely ignored because the solidification rate (V_S) is known to be the key variable in microstructure selection under rapid solidification conditions. Although this analysis conflicts with significant G_L influences on the solidification morphology, mentioned in the previous section, it shows fairly good agreement between the fitted lines and the experimental λ_1 values for the ⊥ and ‖-U welds. The λ_1 values for the ‖-D weld (FIG. 12C) show a large scatter compared with the regression line, which is associated with the abnormal S shape of the V_S curve. Except for λ_1 at $V_S = 2.2$ mm/sec, the fitted line also showed good agreement with the experimental λ_1 values. The exponent b (= 0.54) for the ⊥ weld was in a good agreement with the previous experimental relation ($\lambda_1^2 V_S = $ constant), as shown in Equation (2). The ‖-D weld clearly showed a smaller exponent ($b = 0.38$) than that for the ‖-U weld ($b = 0.61$). The λ_1 in the ‖-U orientation was more significantly affected by the V_S variation than that in the ‖-D orientation.

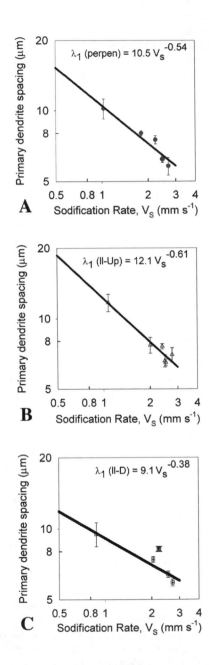

FIGURE 12. Effects of gravitational orientation on the behavior of λ_1 for **(A)** $\perp$, **(B)** $\|$-U, and **(C)** $\|$-D welds as a function of V_S.

The good agreement of the λ_1 fitted lines with respect to V_S shows that the G_L variation along the solid–liquid interface may not be significant on λ_1 and rapid solidification conditions can be applied to the present study. However, the G_L contribution with respect to gravitational orientation should not be ignored in the λ_1 behavior based on the significantly different pre-exponential constants; that is, 10.5 for $\perp$, 12.1 for ||-U, and 9.1 for ||-D. This G_L effect on λ_1 is addressed elsewhere[22] in combination with the V_S effect. The theoretical model with no consideration of convection flows indicates the exponent $b = 0.25$, as shown in Equation (1), which is closer to the b value for the ||-D weld. Therefore, the ||-D weld will have relatively less convection flow when compared to the ||-U weld. This result also conforms with the morphology behavior; that is, the appearance of a more columnar structure near the surface in the ||-D weld. The Hunt model[27] predicted the trend of λ_1 as a function of gravitational orientation within a fairly reasonable range. More details on this quantitative λ_1 analysis are published elsewhere.[22]

In summary, the λ_1 analysis neglecting the G_L variation showed reasonable agreement with experimental λ_1 values except for the abnormal λ_1 behavior in the ||-D weld. Normal welding orientation ($\perp$) showed good agreement with the relation $\lambda_1^2 V_S = $ constant. However, λ_1 in the ||-U weld was more significantly affected by V_S based on the larger exponent b compared to that in the ||-D weld. The abnormal λ_1 behavior in the ||-D weld might have occurred from the abnormal S shape of the V_S variation. The larger λ_1 value in the ||-U weld pool surface might be associated with a morphology change and a variation in both V_S and G_L. This result also confirms our findings and the hypothesis postulated at the beginning of this section.

CONCLUSION

The effects of gravitational level were studied for GTAW of 304 stainless steels. As the gravitational level increased from LG (about $1.2g$) to HG (about $1.8g$), the weld pool shape was affected considerably; specifically, a 10% decrease in weld depth and 10% increase in weld width. However, no significant change in microstructure and solute distribution was observed at gravitation levels between LG and HG. Ferrite content and concentration of the solutes (Cr and Ni) in the duplex weld microstructure remained constant, within experimental scatter, as the gravitational level varied.

For the 165 A GTAW on 304 stainless steel, the weld pool shape was mostly independent of the gravitational orientation within the V_a range 4–10mm/sec. As the weld pool dimensions increased, the gravitational orientation produced significant effects on the weld pool shape; that is, deeper penetration and narrower width in the ||-D weld as compared to the ||-U weld pool shape. In the ||-D weld, the accumulated liquid piled up at the surface, resulting in a smaller gap between the welding electrode and the welding piece. Correspondingly, the arc diameter (d_a) became smaller than that in the ||-U weld, which caused the weld pool in ||-D to become more hemispherical in shape. LBW on 304 stainless steel did not produce significant variations in the weld pool shape as a function of gravitational orientation. This is because keyhole formation dominates in determining the weld pool shape during LBW and laser beam diameter is mostly constant with respect to the z-direction, as compared to the arc diameter during GTAW.

The GTAW microstructure for Al–4wt% Cu alloy was investigated to determine the impact of gravitational orientation on the weld solidification behavior. This was accomplished through GTAW and an analytical study of the weld microstructure. Using a heat input of relatively high power ($185\,A \times 16.5\,V$) and slow arc translation velocity ($3\,mm/sec$), a bead-on-plate welding experiment was performed under a wide range of observed V_S (0.8–$3\,mm/sec$) with a calculated G_L (150–$380\,K/mm$). The ‖-U weld showed 22% larger weld pool area than that of the ⊥ and ‖-D welds. A larger value for λ_1 in the ‖-U weld was observed near the weld pool surface and the fusion boundary than in the case of ⊥ and ‖-D welds. This was because of the morphology change (loss of the columnar structure) and the smaller G_L range induced by larger weld pool dimensions. The ‖-D weld exhibited different solidification morphology—more columnar structure near the weld pool surface and abnormal S shape of the solidification rate curve during its growth. This result might be associated with relatively mild convection flows and receded solid–liquid interface due to the gravity. In summary, the gravitational orientation changed the weld pool shape associated with convection flows. This variation on the convection flow influenced the shape of the trailing solid–liquid interface. Therefore, the solidification morphology and primary dendrite spacing (λ_1) were modified because the solidification rate (V_S) and thermal gradient (G_L) were affected by convection flow.

ACKNOWLEDGMENT

The authors acknowledge financial support from NASA Materials Microgravity Division under Grant NAG8-1272.

REFERENCES

1. PATON, B.E. 1972. Weld. Engin. **57**(1): 25–29.
2. AIDUN, D.K., J.J. DOMEY & G. AHMADI. 2000. Weld. J. **79**(6): 145s–150s.
3. OKHOTIN, A.S., V.F. LAPTCHINSKY & G.S. SHONIN. 1976. COSPAR Symposium on Materials Sciences in Space: 355–362. Philadelphia, PA, USA.
4. WANG, G. & K.N. TANDON. 1995. Micrograv. Sci. Tec. **VIII**(2): 131–133.
5. WORKMAN, G.L. & W.F. KAUKLER. 1990. Laser Materials Processing. Conference ICALEO 1990. **71**: 430–440. Boston, Massachusetts, USA.
6. MCKOWEN, C.R., M.H. MCCAY & C.M. SHARP. 1993. Applying Lasers in Education Symposium. Conference ICALEO 1993. **78**: 766–776. Orlando, FL, USA.
7. AIDUN, D.K. 2001. Acta Astronaut. **48**(2–3): 153–156.
8. KANG, N., J. SINGH, & A.K. KULKARNI. 2003. Mater. Manuf. Process. **18**(4): 549–561.
9. KEANINI, R.G. & B. RUBINSKY. 1990. Weld. J. **69**(6): 41–50.
10. DOMEY, J., D.K. AIDUN, G. AHMADI, et al. 1995. Weld. J. **74**(8): 263s–268s.
11. KOU, S. & D.K. SUN. 1985. Metall. Trans. **16A**: 203–213.
12. CHAN, C., J. MAZUMDER & M.M. CHEN. 1984. Metall. Trans. **15A**: 2175–2184.
13. KANG, N., T.A. MAHANK, A.K. KULKARNI & J. SINGH. 2003. Mater. Manuf. Process. **18**(2): 169–180.
14. NOGI, K., Y. AOKI, H. FUJII & K. NAKATA. 1998. Acta Mater. **46**(12): 4405–4413.
15. FOLEY, J.S. & C.M. BANAS. 1987. ICALEO'87. 47–54.
16. BATTAILE, C.C., R.N. GRUGEL, A.B. HMELO & T.G. WANG. 1994. Metall. Mater. Trans. **25A**: 865–870.

17. YU, H., K.N. TANDON & J.R. CAHOON. 1997. Metall. Trans. **28A:** 1245–1250.
18. LIU, R.P., J.H. ZHAO, X.Y. ZHANG, *et al.* 1998. J. Mater. Sci. **33:** 2679–2682.
19. TANDON, K.N., F. SAADAT, M.C. CHATURVEDI & J.R. CAHOON. 1991. Micrograv. Sci. Tec. **VI**(1): 19–25.
20. WORKMAN, G.L. & W.F. KAUKLER. 1990. ICALEO'90. 430–440.
21. AIDUN, D.K. & J.P. DEAN. 1999. Weld. J. **78**(10): 349s–354s.
22. KANG, N., J. SINGH & A.K. KULKARNI. 2003. J. Mater. Sci. **38**(17): 3579–3589.
23. KANG, N., J. SINGH & A.K. KULKARNI. 2004. J. Mater. Sci. In press.
24. ANTAR, B.N. & V.S. NUOTIO-ANTAR. 1993. Fundamentals of Low Gravity Fluid Dynamics and Heat Transfer. CRC Press Inc.
25. FLEMINGS, M.C. 1974. Solidification Processing. McGraw-Hill, New York.
26. KURZ, W. & D.J. FISHER. 1992. Fundamentals of Solidification. Trans. Tech. Publications. Aedermannsdorf, Switzerland.
27. HUNT, J.D. 1979. Solidification and Casting of Metals. The Metal Society, London.
28. KURZ, W. & D.J. FISHER. 1981. Acta Metal. **29:** 11–20.
29. TRIVEDI, R. 1984. Metal. Trans. **15A:** 977–982.
30. RAPPAZ, M., S.A. DAVID, J.M. VITEK & L.A. BOATNER. 1989. Metal. Trans. **20A**(6): 1125–1138.
31. KURZ, W., B. GIOVANOLA & R. TRIVEDI. 1986. Acta Metal. **34:** 823–830.
32. BROOKS, J.A., M. LI & N.C.Y. YANG. 1997. Mathematical Modelling of Weld Phenomena 4 184–198. Institute of Materials.
33. SCHAEFFLER, A.L. 1949. Met. Prog. **56:** 680–680B.
34. DAVIS, J.R. 1994. Stainless steels. *In* ASM Specialty Handbook. **577:** 340–341. Materials Park, Ohio.
35. BARNHOUSE, E.J. & J.C. LIPPOLD. 1998. Weld. J. **77**(12): 477s–487s.
36. LIMMANEEVICHITR, C. & S. KOU. 2000. Weld. J. **79**(8): 231s–237s.
37. CHAO, L.S. & J.A. DANTZIG. 1986. Proceedings of 10th National Congress on Applied Mechanics. 249–255. Austin, TX.
38. GREMAUD, M., M. CARRARD & W. KURZ. 1990. Acta Metal. Mater. **38**(12): 2587–2599.

Design and Preparation of a Particle Dynamics Space Flight Experiment, SHIVA

JAMES D. TROLINGER,[a] DREW L'ESPERANCE,[a] ROGER H. RANGEL,[b]
CARLOS F.M. COIMBRA,[c] AND WILLIAM K. WITHEROW[d]

[a]*MetroLaser Incorporated, Irvine, California, USA*

[b]*University of California at Irvine, Irvine, California, USA*

[c]*University of Hawaii, Honolulu, Hawaii, USA*

[d]*NASA Marshall Space Flight Center, Huntsville, Alabama, USA*

ABSTRACT: This paper describes the flight experiment, supporting ground science, and the design rationale for a project on spaceflight holography investigation in a virtual apparatus (SHIVA). SHIVA is a fundamental study of particle dynamics in fluids in microgravity. Gravitation effects and steady Stokes drag often dominate the equations of motion of a particle in a fluid and consequently microgravity provides an ideal environment in which to study the other forces, such as the pressure and viscous drag and especially the Basset history force. We have developed diagnostic recording methods using holography to save all of the particle field optical characteristics, essentially allowing the experiment to be transferred from space back to Earth in what we call the "virtual apparatus" for microgravity experiments on Earth. We can quantify precisely the three-dimensional motion of sets of particles, allowing us to test and apply new analytic solutions developed by members of the team. In addition to employing microgravity to augment the fundamental study of these forces, the resulting data will allow us to quantify and understand the ISS environment with great accuracy. This paper shows how we used both experiment and theory to identify and resolve critical issues and to produce an optimal experimental design that exploits microgravity for the study. We examined the response of particles of specific gravity from 0.1 to 20, with radii from 0.2 to 2 mm, to fluid oscillation at frequencies up to 80 Hz with amplitudes up to 200 microns. To observe some of the interesting effects predicted by the new solutions requires the precise location of the position of a particle in three dimensions. To this end we have developed digital holography algorithms that enable particle position location to a small fraction of a pixel in a CCD array. The spaceflight system will record holograms both on film and electronically. The electronic holograms can be downlinked providing real-time data, essentially acting like a remote window into the ISS experimental chamber. Ground experiments have provided input to a flight system design that can meet the requirements for a successful experiment on ISS. Moreover the ground experiments have provided a definitive, quantitative observation of the Basset history force over a wide range of conditions. Results of the ground experiments, the flight experiment design, preliminary flight hardware design, and data analysis procedures are reported.

KEYWORDS: particle dynamics; Stokes flow; microgravity; holography

Address for correspondence: James Trolinger, Director of Research, MetroLaser Inc., 2572 White Road, Irvine, CA 92614, USA. Voice: 949-553-0668; fax: 949-553-0495.
jtrolinger@metrolaserinc.com

Ann. N.Y. Acad. Sci. 1027: 550–566 (2004). ©2004 New York Academy of Sciences.
doi: 10.1196/annals.1324.042

INTRODUCTION: THE SCIENCE PROBLEM

Particles suspended in a fluid react and move under numerous forces imposed upon them. For a sphere that is not spinning in a uniform fluid and in low Reynolds number conditions, these forces can be summarized as follows:

- Gravitational: In a gravitational field this force causes the particle to sink if it is heavier than the fluid. A constant force.

- Buoyant: In a gravitational field this force causes the particle to rise if it is lighter than the fluid; a constant force.

- Pressure: If the fluid accelerates, then it applies a pressure on the particle surface. This is the same as a buoyant force, except in the direction of the fluid acceleration. It is proportional to the first derivative of the absolute fluid velocity with respect to time.

- Stokes drag: This force is due to a type of friction of the fluid on the particle as it moves through the fluid. It is proportional to the relative velocity (velocity difference between the fluid and the particle), or the zeroth derivative of the velocity. It also depends also on the viscosity of the fluid and the particle radius.

- Virtual mass: This force of the fluid on the particle occurs because the particle must accelerate some of the surrounding fluid to move it out of the way. Due to the inertia of the fluid, the force is proportional to the particle acceleration relative to the fluid or the derivative of the relative velocity.

- Basset history drag: When a particle is accelerated in a fluid, a finite amount of time is required for the forces to transfer through the fluid to the particle and vice versa, since the fluid is deformable. The history force accounts for this time difference. It can depend on both relative velocity and acceleration and is proportional to the half derivative of the relative velocity. When the fluid surrounding the particle moves sinusoidally in time, the half derivative is approximately proportional to the sum of the velocity and acceleration.

- Inertial: The inertia of the particle, due to its mass, produces a force that is equal and opposite to the sum of the forces accelerating the particle and it is proportional to the acceleration of the particle, the derivative of the particle absolute velocity with respect to time.

The history force is the most illusive of the forces and is the most difficult to observe. Observing, quantifying, interpreting, understanding, and evaluating the importance of this force are a primary focus of our investigation. More physical insight into the history force results from a few interpretations. Since the term accounts for the finite time required for the transfer of viscous and pressure forces between the particle and the entire volume of surrounding fluid, one would expect the actual forces and related movements of the particle to be less than those predicted without accounting for the history term. This is only significant when the movements associated with velocity and acceleration are in the same time scale as the viscous transport time, the time required for information to move from the surface of a particle to a particle radius away. Unless relative acceleration exists between the particle and fluid the history force quickly decays to zero.

Historical Perspective of Particle Equations of Motion and Their Solutions

A detailed literature review, provided by Coimbra and Rangel,[1] is briefly highlighted here. The earliest significant work on particles in fluids was done in 1845 by Stokes,[2] who proposed an expression for the drag force at low Reynolds numbers. The Stokes equation is not an equation of motion for the particle but rather a simplified momentum equation for the fluid (the Navier–Stokes equation without any inertial term).

A significant improvement on the work of Stokes came more than 40 years later when, independently, Basset[3] and Boussinesq[4] (1887 and 1885) added the local derivative term to the Stokes equation (meaning now that the equation allows for time-dependent adjustment of the flow field). The resulting partial differential equation of motion for the particle moving in a quiescent fluid comprised inertial, viscous, pressure, and virtual mass terms. This was an important step because an integro-differential equation for the motion of the particle was generated, representing the first equation of motion that took into account the relative acceleration between the fluid and the particle.

In 1947 Tchen[5] expanded the Basset and Boussinesq result to include a uniform but non-quiescent flow field (the background flow field is then a function of time but not of position). Tchen produced a stationary solution (independent of initial conditions) to his first equation (this is key to us because this solution is the one that interests us so far regarding SHIVA). Tchen attempted to expand the technique to a nonuniform flow field but made small mistakes in the derivation leading to an inconsistent equation.

Note that the flow field, which is necessarily uniform for Tchen's equation to be valid, is the *background* flow, or the flow that exists without the presence of the particle. Since the background flow is assumed to be uniform, the contribution from the presence of the particles dies out at large distances and thus both the flow field without the particle and the flow field at large distances are equivalent. The contribution from the nonuniform, radially symmetric inner field is included in Tchen's equation, which is consistent with the Stokes limit (vanishing particle Reynolds number, $Re_p \to 0$). In a fluid cell without free surfaces and with the walls at relatively large distances from the particle (or conversely, at very large frequencies given by large Strouhal numbers, Sl), the far-field uniformity condition is well satisfied for small Re_p and large $Re_p Sl$. The local acceleration term in the dimensionless streamfunction equation is of order $Re_p Sl$, and the convective terms are of order Re_p. The local acceleration term scales with the time scale factor defined in the next section.

In 1983 Maxey and Riley[6] corrected Tchen's second derivation for non-uniform flows, generating a consistent equation for non-uniform flows that is valid only when Re_p approaches 0 since there are no convective effects accounted for. Even at very small Re_p, small convective effects can substantially change the behavior of particles in non-uniform flow.[7]

In 1998 Coimbra and Rangel[8] recognized that the history force is proportional to the half derivative of the relative velocity, possibly the first discovery of a force in nature that is proportional to a fractional derivative. This allowed them to solve, analytically, Tchen's first equation (that for uniform flow), including initial transients. This allows one to calculate the motion of particle from a given initial condition, instead of only the long-term (stationary) solution provided by Tchen.

In terms of the relative dimensionless velocities $w = v - u$, where v and u are the particle and the fluid velocity, respectively, Tchen's equation can be written as follows:

$$\frac{dv}{d\hat{t}} = \alpha\frac{du}{d\hat{t}} - \frac{\alpha}{2}\frac{dw}{d\hat{t}} - w - \sqrt{\frac{9\alpha}{2\pi}}\frac{d^{1/2}w}{d\hat{t}^{1/2}} + (1-\alpha)\frac{\tau_p g}{U_0}. \tag{1}$$

The history term is expressed with the half derivative[9] and α is the ratio of the particle to fluid densities. Velocities are made dimensionless by the flow characteristic velocity U_0 and time is made dimensionless by defining a particle characteristic time, $\tau_p = 2\rho_p a^2/9\mu$, g is the gravity acceleration, a is the radius of the spherical particle, ρ_p is the specific mass of the particle, and μ is the dynamic viscosity of the fluid. The terms in Equation (1) are, in order, inertial, forcing (pressure), virtual mass, Stokes drag, history drag, and gravity and buoyancy.

EXPERIMENTAL STUDY: FOCUSING ON THE BASSET HISTORY FORCE

One of our early objectives in the SHIVA project was to assess the importance of each of the force terms and determine if a microgravity environment can be exploited to study those forces that would otherwise be dominated by gravity and steady-state drag. Since the history force is the least understood and least investigated force, we elected to emphasize quantifying and understanding this force and its importance. The theoretical study was tasked with identifying the regions where the term is the most significant, as well as being separable from other forces. This included identifying measurands, parameters, and conditions for meaningful experiments that would confirm the effects of the history term and its importance in the equation of motion.

Ground experiments were designed to establish that the necessary measurements could be made with sufficient accuracy to isolate and quantify the force. Removing gravity would clearly be a major experimental step in relaxing the measurement requirements (for studying any of the other forces) as well as allowing appropriate experimental parameters to be attained. In the presence of gravity all except neutrally buoyant particles sink or rise to the surface, making the measurement more difficult as well as increasing the Reynolds number beyond the limits of validity of the equation.

Coimbra and Rangel[9] were able to show through a scaling analysis that the history drag on a particle in an oscillating fluid can amount to about 60% of the total force, depending on the relative velocity when the oscillation time is equal to the viscous transport time. This result is a primary basis of the experimental design for the SHIVA experiments to date because it identifies the optimal region in which to experimentally verify the magnitude of the history force as predicted by Tchen's equation.

The resulting analysis defines the time scale factor, S (ratio of viscous transport time to oscillation time),

$$S = \frac{\Omega}{v}\left(\frac{a}{3}\right)^2, \tag{2}$$

where Ω is the forcing oscillation angular frequency and v is the kinematic fluid viscosity. The relative size of the history force is maximized when S is unity. When the period of oscillation and the viscous transport time are of the same order of magnitude, the unsteadiness of the near field becomes relevant and, thus, the history drag reaches its maximum. FIGURE 1 summarizes the results of the scaling analysis pointing out that the experiments should be designed to operate with values near $S = 1$, where the history force dominates.

Experiment Design

The selected experimental parameters are, α, Re, S, and the measurands are η, the ratio of particle to fluid amplitude, and θ, the phase difference between the fluid and the particle. In searching for an optimal forcing function it was recognized that sinusoidal forcing functions are relatively easy to provide experimentally, retain all of the forces in the equation of motion, and can be selected for $S = 1$. Since the particle and fluid are continuously undergoing changes in velocity and acceleration, the history term is always present. Moreover, in principle any steady forcing function can be represented as a series of sinusoids, thus the study represents a broad range of forcing functions.

This choice provides additional insight into the history force by reducing it to an expression that is proportional to the sum of velocity and acceleration. By definition,

$$\frac{d^n \sin \omega t}{dt^n} = \omega^n \sin\left(\omega t + \frac{n\pi}{2}\right). \tag{3}$$

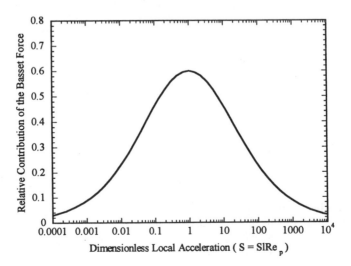

FIGURE 1. Relative contribution of the Basset force for harmonic particle motion: ——, Basset force/(Stokes drag + virtual mass + Basset force.

We know from the stationary solution to the equation of motion that the resulting particle velocity also can be approximated by a sinusoid, and therefore, the difference between the particle and fluid velocities will also be sinusoidal. Consequently,

1. zeroth derivative—forces proportional to velocity or velocity difference, $\sin(\omega t)$.
2. half derivative—the history force, $\omega^{1/2}\sin(\omega t + \pi/4) = 0.707\omega^{1/2}(\sin\omega t + \cos\omega t)$.
3. first derivative—forces proportional to acceleration. $\omega\sin(\omega t + \pi/2)$.

Note that if the relative velocity is sinusoidal, then the history drag is proportional to the sum of the relative velocity and acceleration. This means that the history drag is out of phase with both the velocity and the acceleration. One would expect that it would be most observable in a transition region between conditions where velocity and acceleration dominate, and that is exactly what the Coimbra–Rangel scaling analysis showed. It told us precisely where to look.

In a sinusoidal fluid movement the three forces (history drag, virtual mass, and Stokes drag) vary from zero to a maximum at different times in each cycle. The virtual mass drag peaks when the movement is at the largest amplitude (maximum acceleration), the Stokes drag peaks at the zero position (maximum velocity), and the history drag peaks halfway between (maximum sum of velocity and acceleration). Interestingly, the history and virtual mass forces can be in the opposite direction to the fluid motion.

A typical case selected is for polypropylene particles ($\alpha = 2$) in Krytox oscillating at 60 Hz, which leads to a value of $S = 1$. Maintaining attainable, measurable amplitudes and frequencies for a fluid cell required that we operate in a range below 100 Hz for amplitudes of a few hundred microns. Even this combination introduces a severe vibration problem to deal with, especially when precision measurements are required. To produce the measurements specified from the that are theory necessary to isolate the history force required locating the sphere in the fluids to micron precision.

Ground Experiments

We designed ground experiments to support the flight design. The objectives of the ground experiments were to:

- establish that the measurement accuracy and resolution required to achieve our project goals are achievable;

- prove that the flight experiment is feasible with attainable hardware;

- evaluate the capability of critical hardware components to meet experiment requirements (e.g., laser power and coherence, photographic film properties);

- conduct the experiment and specifically quantify the history force to the greatest extent possible on Earth;

- gain experience and test concepts that will be necessary for conducting a space flight experiment;

- develop data reduction and interpretation methods

- identify and resolve critical issues; and

- determine how much can be gained by conducting the experiment in space.

Critical Issues Resolved in Ground Experiments

The following are a few of the many critical issues that were identified and resolved in ground experiments:

1. Conducting experiments in a gravitational field that would be representative of microgravity results, while also retaining low *Re* conditions.

2. Locating the particles with micron precision, at least some of the time on video cameras in a form that can be downlinked. Since a typical pixel size for a digital camera is 5–10 microns, either premagnification or subpixel measurements would be required. Since particles are free floating, their exact position is unknown so focused imaging would require continuous refocusing.

3. Oscillating an acceptable experimental fluid/particle chamber sinusoidally with frequencies up to 100 Hz at amplitudes up to 200 microns, without vibrating the optical system unacceptably, sending unacceptable forces into the space station, or introducing harmonics. Maintaining optics stable to micron precision could require extremely rigid mounting. The forcing mechanism experiences equal and opposite forces, so it will move opposite to the chamber unless much more massive.

4. Keeping a particle at least 10 diameters away from a wall or nearest neighbor to prevent interaction and interference. This requires experiments with particles sufficiently small to retain a small cell.

Although many other minor issues arose in the course of the experiment design, these were the most challenging.

The Tethered Particle Method

The problems of conducting the experiments on Earth include particles rising or falling too fast, thus violating the low Reynolds number assumption, and the difficulty in locating the particle position with great accuracy when the particle is rising or falling. In a gravitational field, since the fluid to particle density ratio, α, approaches unity, the rate of rising or sinking of the particle is small, a precise measurement of particle position is easy to make, and the Reynolds number is small. However, to study the history drag as α approaches unity requires better measurement accuracy because η also approaches unity. Our ground investigation has addressed this in various ways, but the most versatile to date has been a tethered particle method. Particles are tethered on a fiber, attached with superglue, that holds them in the proper position for recording. Tiny holes were drilled in the particles at the attachment points. This simulates microgravity better, gives much more time, and simplifies recording at the expense of the interference of the tethering fiber.

We modeled the effects of the tethering fiber and also conducted control experiments to identify where they could not be neglected. Analysis suggested that tether effects could be significant. We have observed tether effects experimentally by noting where the results differ from theory (this is not guaranteed to be a tether effect) and by tethering neutrally buoyant particles and comparing with untethered results. In the latter experiments, tether effects were shown to be severe for tethers larger than 50 microns in diameter, especially for particles less than one millimeter in diameter and at frequencies above a few Hertz. Tether effects for 50-micron diameter particles were negligible for particles larger than one millimeter in diameter and frequencies below 60 Hz. We tried a range of materials starting with human hairs (about 100 micron diameter), nylon strings (50 micron), and ultimately copper wire manufactured for use with MEM devices (20 micron diameter). The results with the smallest copper wires appear to be excellent with little or no tether effect for particles larger than 500 microns in diameter, at frequencies up to 80 Hz. and amplitudes up to 100 microns. FIGURE 2 illustrates a four-centimeter cubic cell with both heavy and lightweight particles tethered from the top and bottom of the cell, respectively.

Shaker Systems

We attempted to use various shaker systems to provide the sinusoidal forcing function, including mechanical (cam) and electromagnetic forcing of various types. Each has a characteristic problem and all added significant vibration to the optical table, making precision measurements difficult. The most reliable system, finally

FIGURE 2. Cell showing tethered polypropylene and steel particles (note particles near the wall).

FIGURE 3. SHIVA ground cell showing two spheres (2 mm radii) tethered in the flu-
id- filled cell by fibers to counteract gravity driven buoyancy.

adopted, employed a voice coil driven by an electromagnetic field coil. FIGURE 3
shows the system with two spheres (2 mm radius) tethered in the fluid.

Small spheres attached to the cell window serve as a reference. The voice coil
shown at left drives the cell at frequencies up to 100 Hz and amplitudes up to
200 microns. Digital holograms of the cell are recorded and used to determine the
relative movement of the spheres and the fluid. A piezoelectric sensor (shown upper
left) measures the cell position at all times. A Fourier analysis of this signal determines the frequency content in the system and allows us to avoid resonances, harmonics, and unwanted frequencies.

Minimizing Vibration

The vibration problem was severe and would become worse in space because of
weight restrictions. Vibration added errors to the position measurements by transfer-
ring energy into the optical table. Also, the ISS "Good Neighbor Policy" requires
that the system be isolated from the ISS at all times. We solved the problem in the
ground experiments at first by making the mounts heavy and rigid, but this solution
would not apply in space. We resolve this issue by mounting both the shaker and the
cell on a dual traverse, so that the two could act and react with the transfer of vibrational energy into the optical table due only to friction in the traverses.

FIGURE 4 schematically illustrates the cell and reactionary mass mounted on
traverses, FIGURE 5 shows the actual cell in use. The cell and voice coil are
each mounted on traverses and are driven in opposite directions by the voice coil at

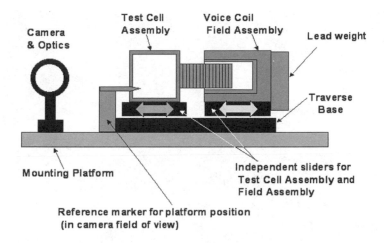

FIGURE 4. Schematic illustration of the counter mass system for minimizing vibration transfer to the optical table.

frequencies up to 100 Hz and amplitudes up to 200 microns. This configuration eliminates potential unwanted vibration that would be transferred into the apparatus as well as to other experiments on the Space Station. The counter mass is about 10

FIGURE 5. SHIVA ground cell showing voice coil and brass reaction mass at *right* and cell at *left*.

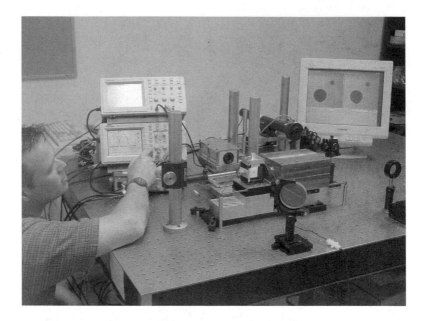

FIGURE 6. SHIVA ground experiment showing cell, voice coil, and reaction mass electronic holograms of spheres tethered in the fluid-filled cell (*upper right*), and instruments to determine precise cell position at all times (*upper left*).

times that of the test cell, so it moves with an amplitude of about 10 microns. The traverses also constrain the movement precisely to a straight line. With this configuration negligible vibration is experienced in the optical table and the Good Neighbor Policy will be easily satisfied.

FIGURE 6 shows the entire apparatus including monitoring instruments and optics. Electronic holograms of spheres in the cell are recorded on the camera, shown upper left, allowing a precise measurement of the sphere and cell positions.

Locating Particles with Micron (Subpixel) Precision

The problem is to locate the two extreme oscillation positions of the particle in the fluid with great precision and compare that with the fluid amplitude of oscillation. We selected holographic recording as a method to achieve microscopic position location without focusing. We have used holography to track the precise position of particles in a fluid in previous applications,[10] including spaceflights.[11,12] Very short exposures are made at the oscillation extrema by shuttering a laser beam used for the recording with the timing signals extracted from the piezo-device that locates the cell position. We can also determine the phase difference of the cell and particle movement by making several measurements and varying the shutter time delay.

Achieving the desired resolution over the entire test chamber is not difficult with recording on photographic materials. The resolution is limited only by the effective hologram diameter and distance from the particle. However, to process data during the spaceflight requires a downlink of information. This can be achieved with CCDs and digital holography, but the resolution limit and size of the detector arrays (typically 5–10 micron pixel sizes) over one square centimeter presents a challenge when micron precision is desired. In digital holography the problem reduces to that of locating the center (or amount of movement) of a diffraction pattern to 0.1 pixel or better.

By using cross-correlation methods we achieved better than 0.1 pixel position location. FIGURE 7 shows the appearance of an electronic hologram of a tethered 2-mm radius sphere in an extreme position during oscillation. The large sphere is tethered in a fluid-filled cell by a fiber (shown at "six o'clock") to counteract gravity, and the small sphere is fixed to the cell window for reference to determine fluid movement. The right hand figure is the superposition of two electronic holograms made in extreme opposite cell positions as the cell oscillates at 60 Hertz. The bright spots at the centers of the diffraction patterns, the "Poisson spot" (enhanced here), show the relative movement of the fluid and the tethered sphere. The amount of movement is determined precisely by cross correlating the diffraction patterns of the holograms made at various times, enabling resolution that is much better than the camera pixel size.

A smaller sphere at "one o'clock" is attached to the cell window for reference (the fluid and cell are assumed to move together). The diffraction pattern inside and outside the particle shadow can be used to produce a correlation mask. The bright spot at the center of the shadow is the Poisson spot. A cross correlation of this mask has a distinct peak at the particle center, with better than 0.1 pixel position sensitivity. The difference in fluid and particle movement is clearly evident.

Results From Ground Experiments

Experiments were completed for a full range of frequency as well as three particle types. Our ground experiments have established and quantified experimentally, with

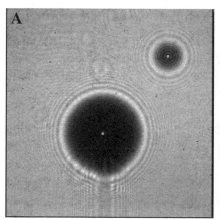 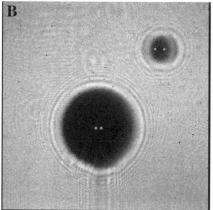

FIGURE 7. Electronic holograms of 1-mm and 2-mm radii spheres (contrast enhanced): **(A)** spheres in extreme left position; **(B)** superposition of two electronic holograms of spheres in extreme left and right positions.

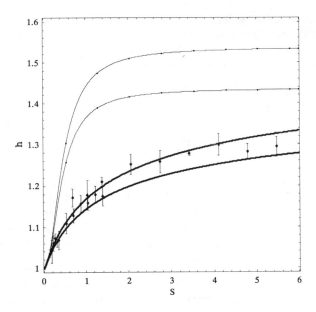

FIGURE 8. Displacement ratio (η) as a function of dimensionless frequency (S) for values of fluid-to-particle density ratio (α) close to 2. All experimental data points are for $1.835 < \alpha < 2.100$.

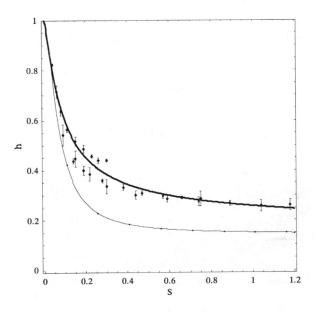

FIGURE 9. Displacement ratio (η) as a function of dimensionless frequency (S) for values of fluid-to-particle density ratio (α) close to 0.1.

a large amount of high quality data and with a clear margin, the existence of the history drag, which has escaped definitive observation since it was first proposed more than 50 years ago. Unlike other forces on particles in fluids, the history force, although anticipated through theory, has never been established or even clearly observed experimentally because it is usually masked by other forces and is extremely illusive.

Our findings are compared with theory in FIGURE 8 for particles lighter than the fluid, and in FIGURE 9 for the case of particles heavier than the fluid. The amplitude ratios of particle to fluid, η, are plotted versus the time scaling factor, S. In each case, the lightly drawn curves are the expected values without the influence of the history force. These figures clearly show the importance of the history force in this region.

In FIGURE 8, the upper thin curves correspond to the solution without the history drag for $\alpha = 2.100$ (the top curve) and for $\alpha = 1.835$ (the second curve from top to bottom). The thick curves correspond to the full solution for the same limiting cases. In FIGURE 9, the lower thin curve corresponds to the solution without the history drag for $\alpha = 0.1$. The thick curve corresponds to the full solution for the same limiting cases.

We have demonstrated in ground experiments that we can locate a particle in a vibrating test cell with a precision sufficient to measure the effect of the history term on the particle amplitude. We tested extensively the cases of one- and two-mm radius polypropylene particles in a Krytox liquid filled cell that is shaken at frequencies up to 100 Hz with amplitude of up to 0.2 mm. The required precision is about five microns depending on the amplitude of vibration and the value of α. We have run many experiments near the region where the scaling function $S = 1$ (S is the ratio of viscous time to vibration period, and at $S = 1$ the motion information transfers about one particle diameter during one cycle of oscillation) and with $\alpha = 2.1$. The effect of the history term peaks at about 60 Hz for this case, making this an optimum frequency for observing the history term effects. A vibrational amplitude of 100 microns results in a 15-micron difference with and without the history term. We have shown in our ground experiments that we are at least a factor of two better than the resolution required to quantify the effect.

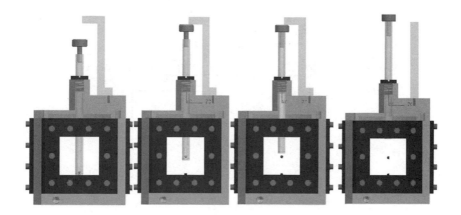

FIGURE 10. Particle injection procedure.

The Flight Experiment

The flight experiment includes wide ranges of the parameters α, Re, S, while measuring η, the ratio of particle to fluid amplitude, and θ, the phase difference. In addition, the particle movement will be tracked over long periods of time to examine the influence of the space station movement. All of the parameters will be varied well outside of the range of validity of existing theory to test the limits of these theories. Precision tracking of particle movement over long periods of time under the residual accelerations of the space station will allow us to make some of the most precise measurements of low frequency residual acceleration of the space station ever performed. Particle number ranges from single particles injected into the cell to many particles of different size in the cell at once. A special particle injection mechanism, shown in FIGURE 10, was developed to allow placing one or two particles near the center of the cell. These will allow us to study the simplest case of a particle with no interactions to more complex cases that include interactions.

The particle is contained at the end of a cylindrical sleeve, shown in the leftmost figure, which is withdrawn by a screw mechanism, as shown in the second figure from left. When the sleeve reaches the cell center, as shown in the third figure, a plunger pushes the particle into the fluid. Then the cylinder is withdrawn completely from the cell leaving it suspended at the cell center. A bellows attached to the cell allows the cell to be completely filled with fluid.

FIGURE 11 diagrams a simple and yet comprehensive flight experiment optical apparatus. We selected hologram recording both on film for the highest resolution and volume coverage and a digital camera to allow us to downlink particle position data. Holograms will be recorded on 70-mm film with two lines of sight through the cell (2) to cover particle movement in all directions with the highest fidelity. A digital camera (4) located above the optical line of sight shown in the figure records the position of particles that lie in the central one square centimeter cross section cov-

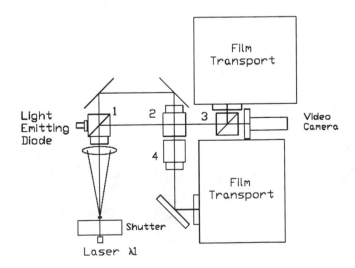

FIGURE 11. Spaceflight experiment configuration.

FIGURE 12. Flight experiment hardware design.

ered by the camera sensor. A second video camera will provide a view of the entire cell to allow a continuous view of particles in the cell at lower resolution. A light emitting diode illuminator is provided for this purpose. Laser light is separated from the LED light by beamsplitters (1), (2), and (3) and filters.

The cells are designed so that they can be easily installed in the system with a quick release mechanism and a specially designed alignment feature that positions the cell accurately. The experiment will be conducted in a microgravity sciences glovebox (MSG). A preliminary design is illustrated in FIGURE 12, which incorporates the optical system shown in FIGURE 11, with the dual axis traverse, electromagnetic shaker, and 16 different cells preloaded with various particle and fluid types that cover the selected parameters for the study.

CONCLUSIONS

We have presented some results of a ground based experimental program that is designed to support the flight experiment SHIVA. Using a tethered particle method to simulate microgravity we conducted a wide range of experiments, identified and resolved what appear as the most challenging obstacles to a successful experiment, and have developed the necessary methods to meet experimental goals. The tether has a measurable influence on smaller particle motion, both as predicted by theory and as observed in experiments. We have proven that we can determine the position of a particle and fluid with sufficient accuracy to meet the objectives of the experiment. We have verified the ability to quantify the elusive history force for a limited range of experimental parameters that can be met on Earth. We then presented a preliminary flight experiment and optical system design that can provide a comprehensive coverage of parameters.

ACKNOWLEDGMENTS

This is a NASA flight definition NRA (NAS8-98091) under topic NRA-96 HEDS-02. The NASA management team includes Melanie Bodiford, Project Manager, Dr. David Smith, Project Scientist, Bill Patterson, Systems Engineer, all from Marshall Space Flight Center. The authors wish to acknowledge Michael Dempsey of MetroLaser for conducting much of the experimental work and Jason Waggoner and Michael Karigan of NASA for flight system design.

REFERENCES

1. COIMBRA C.F.M. & R.H. RANGEL. 2001. Spherical particle motion in harmonic Stokes flows. AIAA J. **39:** 1673–1682.
2. STOKES, G.G. 1845. On the theories of internal friction of fluids in motion. Trans. Camb. Phil. Soc. **8:** 287–305.
3. BASSET, A.B. 1888. On the motion of a sphere in a viscous liquid. Phil. Trans. R. Soc. Lond. A **179:** 43–63. (Also in A Treatise on Hydrodynamics, Chap. 22. Dover, 1961).
4. BOUSSINESQ, J. 1885. Sur la résistance qu'oppose un liquide indéfini en repos, sans pesanteur, au mouvement varié d'une sphère solide qu'il mouille sur toute sa surface, quand les vitesses restent bien continues et assez faibles pour que leurs carrés et produits soient négligeables. C.R. Acad. Sci. Paris, **100:** 935–937.
5. TCHEN, C.M. 1947. Mean Value and Correlation Problems Connected with the Motion of Small Particles Suspended in a Turbulent Fluid. Ph.D. Thesis, Delft University.
6. MAXEY, M.R. & J.J. RILEY. 1983. Equation of motion for a small rigid sphere in a nonuniform flow. Phys. Fluids **26:** 883–888.
7. COIMBRA, C.F.M. & M.M. KOBAYASHI. 2002. On the viscous motion of a small particle in a rotating cylinder. J. Fluid Mech. **469:** 257–286.
8. COIMBRA, C.F.M. & R.H. RANGEL. 1998. General solution of the particle momentum equation in unsteady Stokes flows. J. Fluid Mech. **370:** 53–72.
9. RANGEL, R.H., J.D. TROLINGER, C.F.M. COIMBRA, *et al.* 2001. Studies of fundamental particle dynamics in microgravity. UEF: MTP-01-32 Proceedings of Microgravity Transport Processes in Fluid, Thermal, Biological and Materials Sciences II, Banff, Alberta, Canada, September 30–October 5, 2001.
10. TROLINGER, J.D., W.M. FARMER & R.A. BELTZ. 1968. Multiple exposure holography of time varying, three-dimensional fields. Appl. Optics **7**(8): 1640–1641.
11. TROLINGER, J.D., M. ROTTENKOLBER & F. ELANDALOUSSI. 1997. Development and application of holographic particle image velocimetry techniques for microgravity applications. Meas. Sci. Tech. **8:** 1573–1583.
12. TROLINGER, J.D., R.B. LAL, D. MCINTOSH & W.K. WITHEROW. 1996. Holographic particle image velocimetry in the first international microgravity laboratory aboard the Space Shuttle Discovery. Appl. Optics **35**(4): 681–689.

Index of Contributors